STRUCTURE REPORTS

for 1972

Volume 38B

Part 1

Structure Reports is prepared under the guidance of a Commission of the International Union of Crystallography. The members of the Commission sometime concerned with the preparation of this volume are listed below.

COMMISSION ON STRUCTURE REPORTS
during the preparation of Volume 38B

B. Beagley	W. B. Pearson
L. D. Calvert	S. E. Rasmussen (ex officio)
G. Ferguson	J. M. Robertson
A. J. Frueh	C. B. Shoemaker
E. J. Gabe	J. Trotter (Chairman)
J. Iball	N. G. Vannerberg

STRUCTURE REPORTS

for 1972

Volume 38 B (Part 1)

ORGANIC SECTION

General editor

J. Trotter

Section editor

G. Ferguson

Springer Science+Business Media, B.V.

First published in 1975

ISBN 978-94-017-3117-1 ISBN 978-94-017-3115-7 (eBook)
DOI 10.1007/978-94-017-3115-7

Copyright 1975 by Springer Science+Business Media Dordrecht
Originally published by the International Union of Crystallography in 1975
Softcover reprint of the hardcover 1st edition 1975

Short extracts and single
illustrations may be reproduced without formality,
provided that the source is acknowledged, but
substantial portions may not be reproduced
by any process without written permission
from the International Union of Crystallography

Koninklijke Drukkerij Van de Garde BV., Zaltbommel

TABLE OF CONTENTS

SYMBOLS

The letters a, b, c, α, β, γ are used consistently for the edges and angles of the unit cell. Other letters used consistently are:

D_m	Measured density in g/cm^3
Z	Number of formula units per unit cell
x, y, z	Atomic coordinates as fractions of cell edge
A, B, C	Types of layer in layer structures
M, A, B	Variable metal atom(s) in a sequence of related structures
X	Variable non-metals, usually halogen, in a sequence of related structures
Ln	Lanthanon, rare-earth
R	Discrepancy factor (for diffractometer data unless otherwise indicated); also variable organic radical

LIMITS OF ERROR

Errors are quoted as standard deviations in units of the last place. Thus 1.542(3) Å means 1.542 Å, standard deviation 0.003 Å. Usually only an indication of bond length standard deviation is given.

TRANSLITERATION OF RUSSIAN

Transliteration is in accordance with draft recommendation no. 6 of the International Organization for Standardization, and the scheme is reproduced in previous volumes of Structure Reports (see e.g. Vol. 29, page VI). There are some apparent inconsistencies, since the names of Russian authors which appear in English language Journals are given as in the original; only one form is listed in the Author Index.

INTRODUCTION

In the past the aim of Structure Reports has been to present critical reports on all work of crystallographic structural interest, whether it is derived directly from X-ray, electron, or neutron diffraction, or even indirectly from other experiments. The reports were intended to be critical and not mere abstracts, except in some cases when a brief indication of the content of a paper of related interest was included in the form of an abstract. In selecting topics for reporting, the criterion 'of structural interest' was freely interpreted in terms of what was topically interesting. However, the amount of literature covering matters of structural interest became so large that this policy could no longer be followed, and from Volume 28 onwards, critical reports are given only on actual structure determinations. Only in this way was it possible to keep yearly volumes to a fairly uniform and usable size.

Starting with Volume 30, Structure Reports is produced in a new format by photo-offset printing from typed manuscript with unjustified lines. At the time when the decision for this change was taken, the cost of setting the manuscript in type was becoming so high as to render the cost of individual subscription prohibitive. At that time automatic typing methods giving justified lines, etc. for photo-offset reproduction did not offer any saving over type-setting, but hand typing of the manuscript could give a considerable saving in production costs. In the belief that a publication that is too expensive to buy is of little value, the format was changed, sacrificing elegance to availability. The new format did not lead to increased length of the volumes since the information content of the typed and typeset pages is practically identical. However, the amount of work to be reported demanded the eventual separation of Structure Reports into two volumes, A. Metals and Inorganic, and B. Organic, and it was convenient to introduce this change also at Volume 30.

Ideally, the reports have been prepared in such a way that no further structural information would be gained by consulting the original paper, although from Volume 21 onwards, atomic positional parameters are not generally reproduced for structures containing more than about 30 independent atoms. The two main reasons for this are that such tables occupy a great deal of space, making the volumes very bulky, and that the chance of including typographical errors in reproducing extensive tables of data is such, that anybody wishing to make detailed use of them would in any case consult the primary reference.

The phenomenal increase in the number of structural papers in the late 1960's has made it necessary to pursue these policies to their final conclusion, in order to keep the volumes to a usable size. The beginning of a new decade affords an opportunity to consolidate the various economies in space and format. Starting with Volumes 37, the main economy is omission of tables of atomic positional parameters (usually meaningless numbers in themselves) except for simple structures (where atoms are often in special positions). This means that atomic parameters are usually given in full in the Metals section, infrequently in the Inorganic section, and very seldom in the Organic section. If the actual values of the atomic parameters are required, it is now therefore often necessary to consult the original paper (or in the future computer data banks). To compensate for the omission of this tabular material, efforts have been made to describe the structures more fully, with illustrations if appropriate. In addition, interatomic distances (especially in the Organic section) are usually not given in full, but the significant ones are discussed in the descriptions of the structures. Details of the structure analyses are now given only briefly, since methods have become fairly standardized; unless otherwise stated it may be assumed that the analyses involve single-crystal diffractometer data, Patterson or direct methods of structure solution, and least-squares refinement.

Although the data must be presented as briefly as possible, every effort is made to avoid jargon, so that the information is readily understandable by the non-crystallographer as well as the crystallographer (a copy of the International Tables for X-ray Crystallography, Volume I, Kynoch Press, Birmingham, would, nevertheless, be useful). The arrangement in individual reports is generally: name, formula, paper(s) reported, unit cell and space group data, brief details of analysis, atomic positions (if given in full), interatomic distances and angles (if given in full), description and discussion of the structure (with diagrams if appropriate), and additional references. Editorial comments are enclosed in square brackets, and it may be assumed that material not distinguished in this manner is based directly on the papers reported. The Volumes are divided into three main sections: Metals, Inorganic Compounds (in the A volumes), and Organic Compounds (including organometallic compounds, in the B volumes). The arrangement in the Metals section is roughly alphabetical, and that of the Inorganic and Organic sections is roughly in order of increasing complexity of composition, related substances and related structures being kept together as far as possible. The Subject Indexes are arranged alphabetically by the names printed as the headings of the reports, and also include other common names and information. The Formula Index in the A volumes is arranged in alphabetical order of the chemical symbols; in the B volumes the classification is by the number of carbon atoms and secondary classification by the number of hydrogen atoms; other constituents then follow alphabetically.

University of British Columbia
Vancouver, Canada

J. TROTTER

22 January 1975

STRUCTURE REPORTS

SECTION III

ORGANIC COMPOUNDS

Edited by

G. Ferguson

with the assistance of

M. Bartlett

M. Currie

D. M. Hawley

F. C. March

W. C. Marsh

M. Napier

D. F. Rendle

R. J. Restivo

S. J. Rettig

J. Trotter

ARRANGEMENT

To find a particular organic compound the subject index or the formula index should be used. The general arrangement is: aliphatic or open-chain compounds; open chains with N, S; benzene derivatives; cyclic hydrocarbons; condensed ring systems; heterocyclic compounds; carbohydrates, amino-acids, natural products; molecular complexes; organometallic compounds - B, Si, P, As, Sb, groups IA, IIA, III, IV, VI; transition metal complexes - π-complexes, other ligands; polymers, proteins, electron-diffraction and microwave studies. Only complete structure analyses are described; compounds for which only lattice parameters are determined, and those which have been described only in preliminary communications and for which details will appear at a later date, have not been reported.

TRICHLOROACETIC ACID

$C_2HC\ell_3O_2$ $\cdot CC\ell_3COOH$

P.-G. JÖNSSON and W.C. HAMILTON, 1972. J. Chem. Phys., 56, 4433-4439.

Monoclinic, $P2_1/c$, a = 6.152, b = 9.705, c = 12.020 Å, β = 124.29°, D_m = 1.81, Z = 4. Neutron radiation, R = 0.057 for 615 reflexions.

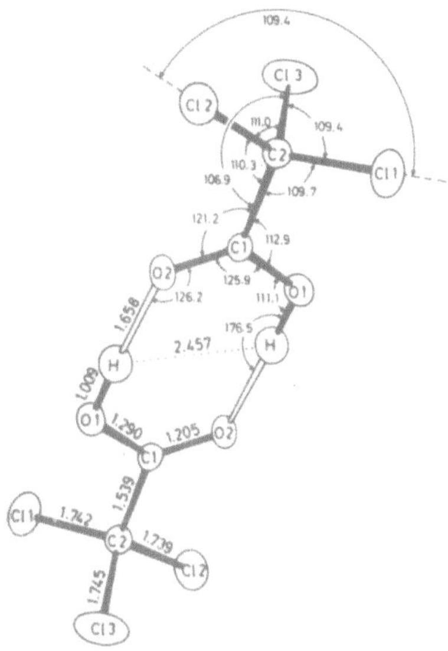

Fig. 1. Dimeric structure of trichloroacetic acid with bond lengths (σ = 0.004 Å for C-O C-Cℓ, 0.01 Å for O-H) and bond angles (σ = 0.3 - 0.4°).

The structure is composed of hydrogen bonded dimers (Fig. 1) with normal van der Waals separations between them. The carboxyl group is planar and the dimer is nearly planar, with the obvious exception of the Cℓ atoms. The torsion angle Cℓ(2)-C(2)-C(1)-O(2) of 11.5(4)° is similar to the corresponding angle 6.3(16)° in acetic acid (1).

1. Structure Reports, 37B, 3.

MONOFLUOROACETIC ACID

$C_2H_3FO_2$ FCH_2CO_2H

J.A. KANTERS and J. KROON, 1972. Acta Cryst., B28, 1946-1949.

Monoclinic, $P2_1/c$, a = 4.30, b = 7.55, c = 9.98 Å, β = 85.2°, Z = 4. Mo radiation, R = 0.106 for 500 reflexions.

Molecules of fluoroacetic acid are hydrogen bonded across centres of symmetry, thus forming planar dimers (with O-H···O = 2.64 Å). The C-C bond length is short (1.46(1) Å); other bond lengths and angles have normal values (C=O = 1.22(1), C-O(H) = 1.31(1), C-F = 1.37(1) Å). Apart from the hydrogen bond linking the molecules the next shortest intermolecular contact is between a carbonyl oxygen and CH (C···O = 3.36 Å).

DIFLUOROACETAMIDE

$C_2H_3F_2NO$ $CF_2H \cdot CO \cdot NH_2$

D.O. HUGHES and R.W.H. SMALL, 1972. Acta Cryst., B28, 2520-2524.

Monoclinic, $P2_1/c$, a = 5.143, b = 12.809, c = 7.037 Å, β = 128.3°, D_m = 1.72, Z = 4. Cu radiation, R = 0.089 for 554 reflexions.

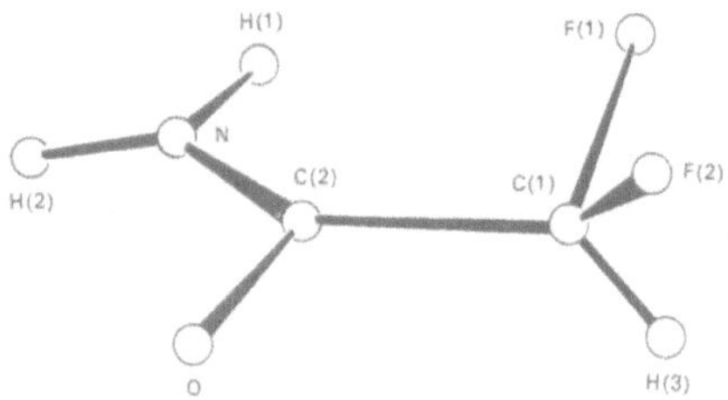

Fig. 1. Conformation of difluoroacetamide.

The molecules form centrosymmetric hydrogen bonded dimers (N-H···O = 3.011(6) Å). These dimers form further hydrogen bonds sideways to similar dimers related by the a translation (N-H···O = 2.924(9) Å). The carboamide group is effectively planar and the torsion angles involving the fluorine atoms O-C-C-F are 156 and 39° (Fig. 1). Principal bond lengths are C-C = 1.543(7), C-F = 1.364(9), C-N = 1.334(8) and C-O = 1.247(8) Å.

ACETIC ACID - PHOSPHORIC ACID

(1:1 ADDITION COMPOUND)

$C_2H_7O_6P$ $CH_3COOH \cdot H_3PO_4$

P.-G. JÖNSSON, 1972. Acta Chem. Scand., 26, 1599-1619.

Triclinic, $P\bar{1}$, a = 9.627, b = 11.505, c = 6.456 Å, α = 97.98°, β = 90.11°, γ = 68.89°, D_m = 1.59, Z = 4. Cu radiation, R = 0.042 for 1730 X-ray reflexions, R = 0.056 for 1845 neutron reflexions.

The structure is composed of uncharged acetic acid and phosphoric acid molecules. The asymmetric unit of the structure contains two acetic acid molecules and two phosphoric acid molecules. The two independent acetic acid molecules are connected by O-H···O hydrogen bonds to form acetic acid dimers. Each acetic acid molecule is joined by an O-H···O bond to a phosphoric acid molecule (Figs. 1 and 2). The remaining two hydrogen atoms in each phosphoric acid molecule are engaged in O-H···O bonds with phosphoric acid neighbours (see Fig. 3). A two-dimensional system of hydrogen-bonded phosphoric acid molecules is thus formed giving infinite puckered layers of phosphoric acid molecules in the structure. The acetic acid dimers are situated between the phosphoric acid layers (Fig. 4).

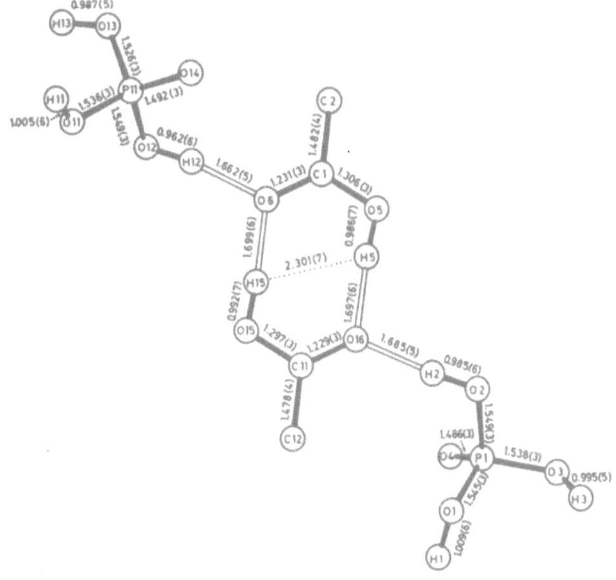

Fig. 1. Bond lengths (neutron results) for the acetic acid - phosphoric acid system.

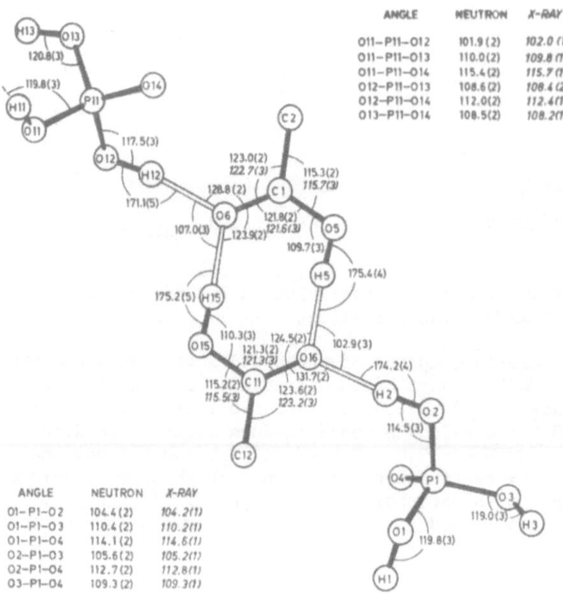

ANGLE NEUTRON X-RAY
O11–P11–O12 101.9(2) 102.0(1)
O11–P11–O13 110.0(2) 109.8(1)
O11–P11–O14 115.4(2) 115.7(1)
O12–P11–O13 108.6(2) 108.4(2)
O12–P11–O14 112.0(2) 112.4(1)
O13–P11–O14 108.5(2) 108.2(1)
ANGLE NEUTRON X-RAY
O1–P1–O2 104.4(2) 104.2(1)
O1–P1–O3 110.4(2) 110.2(1)
O1–P1–O4 114.1(2) 114.6(1)
O2–P1–O3 105.6(2) 105.2(1)
O2–P1–O4 112.7(2) 112.8(1)
O3–P1–O4 109.3(2) 109.3(1)

Fig. 2. Bond angles (neutron results, with X-ray values in italics) for the acetic acid - phosphoric acid system.

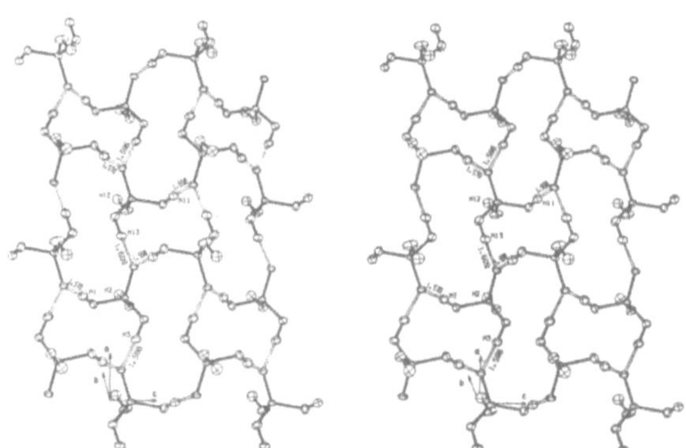

Fig. 3. Illustration showing part of a hydrogen-bonded layer of phosphoric acid molecules. Thermal ellipsoids are scaled to include 20% probability.

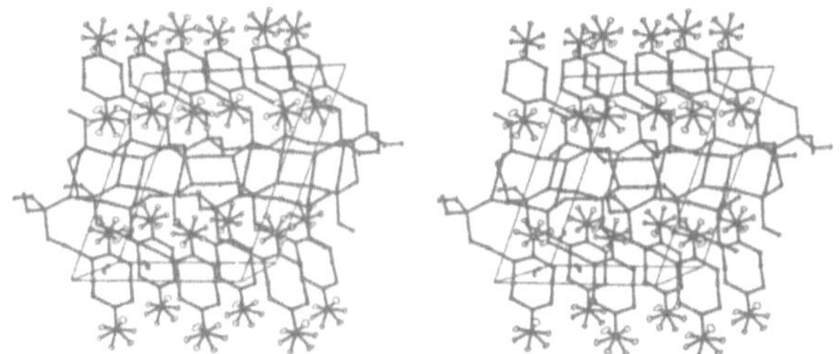

Fig. 4. Illustration of the packing in $CH_3COOH \cdot H_3PO_4$. Thermal ellipsoids are scaled to include 2% probability. The methyl hydrogen atoms are disordered over two sites (mean C-H distances are 1.04(1) and 1.05(1) Å).

DIFLUOROMALONIC ACID

$C_3H_2F_2O_4$ $HOOC \cdot CF_2 \cdot COOH$

J.A. KANTERS and J. KROON, 1972. Acta Cryst., B28, 1345-1349.

Monoclinic, $P2_1/c$, a = 9.59, b = 5.61, c = 9.80 Å, β = 115.1°, D_m not given, Z = 4. Mo radiation, R = 0.078 for 1493 reflexions.

The crystal is a racemate of two conformational antipodes. Molecules are linked in infinite chains along *a* by hydrogen bonds (O(1)-H···O(2') = 2.704 and O(3)-H···O(4') = 2.690 Å) between carboxyl groups across centres of symmetry. One carboxyl group, C(3)O(3)O(4), lies nearly in the carbon-chain plane (deviation 11°) with C=O group *cis* to the C-C bond, the other group being turned nearly perpendicular (84°) to it. Bond lengths are C-O(H) = 1.270, 1.288, C=O = 1.207, 1.213, C-F = 1.340, 1.345, C-C = 1.537, 1.541 Å (σ = 0.004 Å).

ACETYLENEDICARBOXYLIC ACID

$C_4H_2O_4$ $HO_2C\text{-}C{\equiv}C\text{-}CO_2H$

V. BENGHIAT, L. LEISEROWITZ and G.M.J. SCHMIDT, 1972. J. Chem. Soc., Perkin II, 1769-1772.

Monoclinic, $P2_1/n$, a = 14.894, b = 6.420, c = 4.862 Å, β = 90.90°, Z = 4. Cu radiation, R = 0.045 for 1031 reflexions.

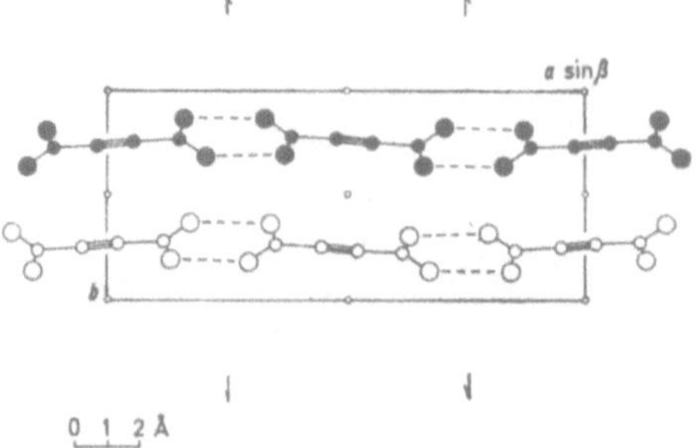

Fig. 1. The packing arrangement of acetylenedicarboxylic acid seen along [001].

In this crystal structure the carboxylic acid groups are orientationally disordered so that all four C-O bond lengths are nearly equal (mean value 1.254(10) Å). The structure (Fig. 1) consists of hydrogen bonded chains (mean O···O = 2.668 Å) lying parallel to the [10$\bar{1}$] *n*-glide direction. The resulting carboxylic acid group pairs which are formed via the *n*-glide operation are nearly coplanar. The molecule itself is not planar; the dihedral angle between the planes of the carboxy groups is 57.8°. The mean C-C bond length is 1.458(1) Å but the C≡C bond distance (1.168(2) Å) is apparently 0.04 Å shorter than the commonly accepted value.

BUT-3-YNOIC ACID

$C_4H_4O_2$ $HC{\equiv}C\text{-}CH_2\text{-}CO_2H$

V. BENGHIAT and L. LEISEROWITZ, 1972. J. Chem. Soc., Perkin II, 1772-1778.

Orthorhombic, $Pca2_1$, a = 8.060, b = 4.195, c = 25.433 Å, D_m = 1.32, Z = 8. Mo radiation, R = 0.067 for 968 reflexions.

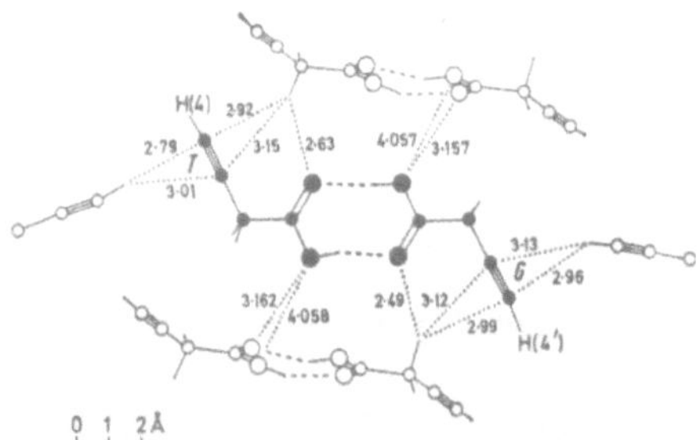

Fig. 1. Packing arrangement in but-3-ynoic acid seen along the normal to the best plane through the carboxylic acid dimer of the shaded asymmetric pair. Some short intermolecular contacts (Å) are shown. In both molecules the C≡C-C-C=O system is *syn*-planar.

The molecules are hydrogen-bonded (O-H···O = 2.69 and 2.65 Å) in pseudo-centrosymmetric pairs. In the crystal there are no intermolecular ethynyl-H···O=C interactions; rather, the acetylenic C-H bond points at the C≡C triple bond (H···C≡C(midpoint) 3.0 Å) (Fig. 1) and a C(sp^3)-H bond is directed towards a carbonyl oxygen (C-H···O = 3.45 Å, H···O = 2.6 Å). The two molecules in the asymmetric unit are not individually planar; the dihedral angles between the carboxy and propynyl groups are 8.3 and 8.8°, and are mostly due to a twist about each C-CO_2H bond. Mean bond lengths are C≡C = 1.177(8), ≡C-C = 1.454(8), C-CO_2H = 1.495(8), C=O = 1.218(8), C-OH = 1.283(8) Å. But-3-ynoic acid also appears in a pseudo-orthorhombic form showing order-disorder.

TETROLIC ACID (α- and β-forms)

$C_4H_4O_2$ CH_3-C≡C-CO_2H

V. BENGHIAT and L. LEISEROWITZ, 1972. J. Chem. Soc., Perkin II, 1763-1768.

α- TETROLIC ACID

Triclinic, P$\bar{1}$, a = 7.320, b = 5.099, c = 7.226 Å, α = 83.97, β = 117.46, γ = 112.00°, Z = 2. Cu radiation, R = 0.07 for 985 reflexions.

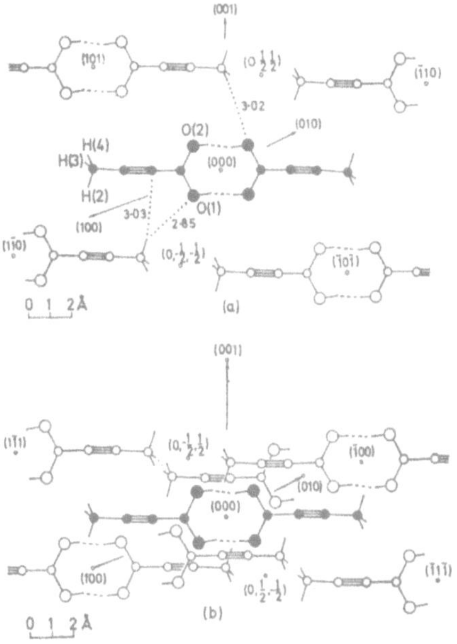

Fig. 1. α-Tetrolic acid: packing arrangement seen perpendicular to the best plane of the shaded dimer which is close to the (11$\bar{1}$) plane. (a) Neighbouring molecules which lie almost within the (11$\bar{1}$) plane and (b) molecules which lie in sheets above and below the shaded dimer.

In the crystal structure of the α-modification (Fig. 1) the molecules form hydrogen-bonded cyclic dimers (O···O = 2.649 Å) in which the C-O bond lengths are almost equal (1.252, 1.265 Å). 'Half hydrogens' were attached to each oxygen, their parameters refined, and their presence substantiated by a difference-Fourier synthesis. The C-C bond lengths of the α-form are C-CH_3 = 1.458(3), C≡C = 1.182(3) and C-CO_2H = 1.441(3) Å.

β-TETROLIC ACID

Monoclinic, $P2_1$, a = 7.887, b = 7.121, c = 3.937 Å, β = 100.18°, Z = 2. Mo radiation, R = 0.040 for 687 reflexions.

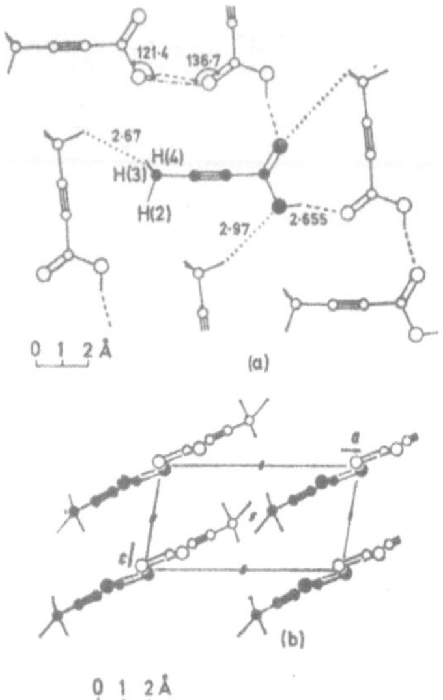

Fig. 2. β-Tetrolic acid: (a) packing arrangement seen perpendicular to the best plane of the shaded molecule showing some intermolecular contacts (Å) and angles (°) and (b) packing arrangement seen along [010].

In the β-modification of the structure the bond lengths within the carboxylic group are distinct: C=O = 1.204, C-O = 1.310 Å; the hydroxy-hydrogen was clearly evident. The molecules form a continuous array of O-H···O= hydrogen bonds (O···O = 2.655 Å) between molecules related by a twofold screw axis (Fig. 2), similar to the arrangement found in e.g. acetic acid (1). The C-C bond lengths in the β-form are: C-CH_3 = 1.455(4), C≡C = 1.178(4), C-CO_2H = 1.441(4) Å.

1. I. NAHRINGBAUER, 1970. Acta Chem. Scand., 24, 453.

CHLOROSUCCINIC ACID

$C_4H_5C\ell O_4$ $HOOC \cdot CH_2 \cdot CHC\ell \cdot COOH$

L. KRYGER, S.E. RASMUSSEN and J. DANIELSEN, 1972. Acta Chem. Scand., 26, 2339-2348.

Orthorhombic, $P2_12_12_1$, a = 9.202, b = 8.699, c = 7.591 Å, Z = 4. Mo radiation, R = 0.027.

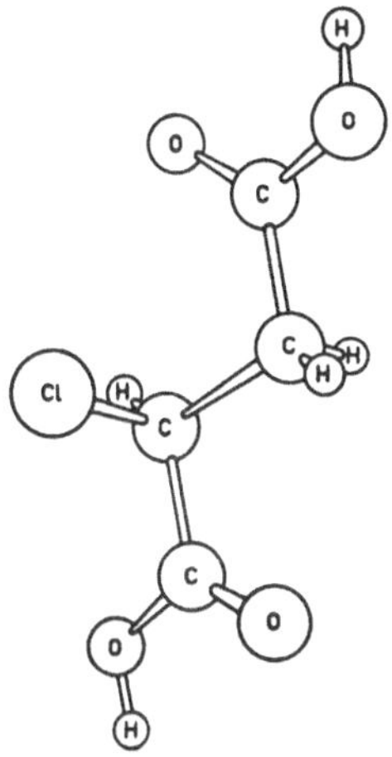

Fig. 1. (—)-Chlorosuccinic acid absolute configuration.

The absolute configuration (Fig. 1) was determined by the anomalous dispersion method. The four carbon atoms in the chain are almost coplanar and nearly coplanar with the two carboxyl groups. Bond lengths are Cℓ-C(2) = 1.815(3), C(1)-C(2) = 1.530(3), C(2)-C(3) = 1.504(3), C(3)-C(4) = 1.527(3), C-O(H) = 1.233(3) (mean value), C=O = 1.282(3) Å (mean value). There may be some slight degree of double bonding in the central C(2)-C(3) bond. The molecules are linked into chains along c by O-H···O hydrogen bonds (2.67 Å).

FUMARAMIC ACID

$C_4H_5NO_3$

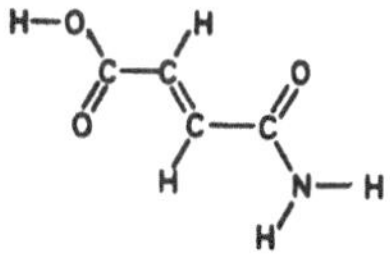

V. BENGHIAT, H.W. KAUFMAN, L. LEISEROWITZ and G.M.J. SCHMIDT, 1972. J. Chem. Soc., Perkin II, 1758-1763.

Monoclinic, $P2_1$, a = 7.367, b = 9.041, c = 3.742 Å, β = 99.39°, D_m = 1.54, Z = 2. Mo radiation, R = 0.038 for 1182 reflexions.

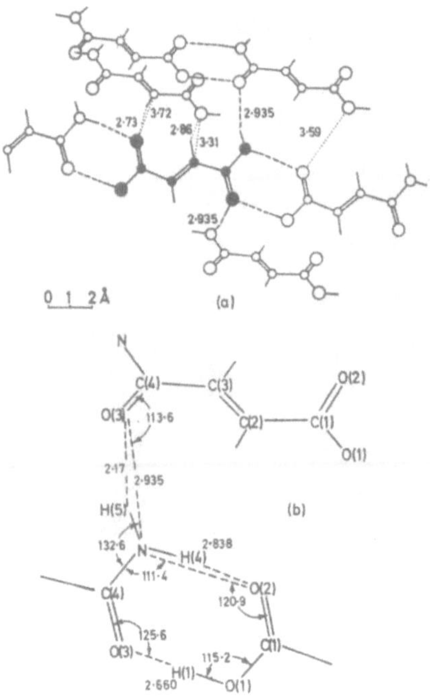

Fig. 1. Fumaramic acid. (a) Packing arrangement seen along the normal to the best plane of the shaded molecule. Some intermolecular contacts (Å) are shown. (b) Interatomic distances (Å) and angles (°) of the hydrogen-bonded system.

The molecular packing and hydrogen bonding in this crystal structure are shown in Fig. 1. The hydrogen-bond network is two-dimensional. The amide and carboxy-groups of molecules related by translation form cyclic hydrogen-bonded (asymmetric) pairs (OH···O = 2.660; NH···O = 2.838 Å) generating chains which are interlinked, along the 2_1 axis, by NH···O (2.935 Å) bonds. The conformations of the groups C:C·C(NH_2):O and C:C·C(OH):O are syn-planar, in contrast to the anti-planar C:C·C:O arrangements in the α- and β-forms of fumaric acid.

The bond lengths are: C=C = 1.320, C(OH)-C = 1.497, C(NH_2)-C = 1.486, C(OH)=O = 1.209, C(NH_2)=O = 1.247, C-OH = 1.310, and C-NH_2 = 1.323 Å; the mean σ for bonds not involving hydrogen is 0.002 Å.

MESACONIC ACID

$C_5H_6O_4$

M.P. GUPTA and S.R.P. YADAU, 1972. Acta Cryst., B28, 2682-2686.

Monoclinic, $P2_1/c$, a = 7.22, b = 11.88, c = 7.00 Å, β = 98°40', D_m = 1.44, Z = 4. Cu radiation, R = 0.095 for 491 reflexions.

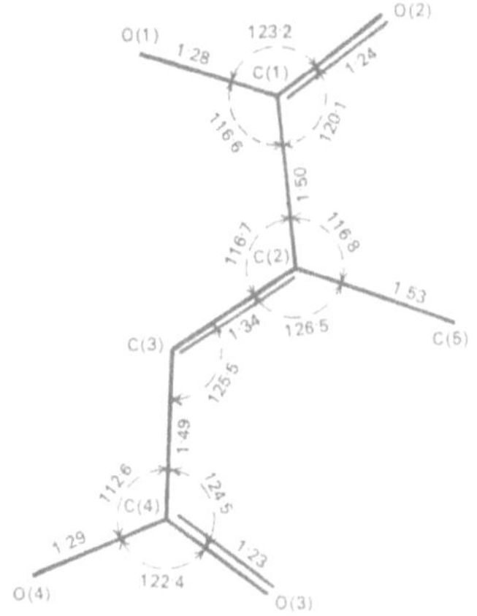

Fig. 1. Bond lengths and angles in mesaconic acid.

The bond lengths and angles in the molecule are normal and are shown in Fig. 1. The molecules are linked together in infinite chains parallel to <u>a</u> by hydrogen bonding. Adjacent molecules in the chain are joined by O(1)···O(3) (2.63 Å) and O(2)···O(4) (2.74 Å) hydrogen bonds. In addition O(4) is also linked with a centrosymmetrically related oxygen atom in an adjacent chain by a hydrogen bond of 2.88 Å. The hydrogen bonds involving O(4) are therefore bifurcated.

The C=O bonds in the two end carboxyl groups of the molecules are in the <u>cis</u> configuration.

trans,trans-MUCONIC ACID

$C_6H_6O_4$ $[-CH{=}CH-CO_2H]_2$

J. BERNSTEIN and L. LEISEROWITZ, 1972. Israel J. Chem., <u>10</u>, 601-612.

Triclinic, P$\bar{1}$, a = 8.982, b = 9.895, c = 3.787 Å, α = 103.68°, β = 75.27°, γ = 101.58°, Z = 2. Mo radiation, R = 0.078 for 1671 reflexions.

The asymmetric unit contains two independent half molecules located about inversion centres and the geometries of the carbon chains of the two molecules are identical to within 0.003 Å. Mean bond lengths are: central C-C = 1.456, C=C = 1.333, C-C(carboxyl) = 1.472(3) Å. The carboxyl group of one molecule is statistically disordered (C-O = 1.257, 1.275(3) Å) whereas in the other molecule the C-O lengths differ significantly (1.236, 1.294(3) Å) indicating a greater degree of order. Both molecules each form hydrogen bonded chains (2.605, 2.614 Å) along the <u>b</u>-direction.

CITRIC ACID MONOHYDRATE

$C_6H_{10}O_8$ $C_6H_8O_7 \cdot H_2O$

G. ROELOFSEN and J.A. KANTERS, 1972. Cryst. Struct. Comm., 1, 23-26.

Orthorhombic, $P2_12_12_1$, a = 6.297, b = 9.319, c = 15.398 Å, Z = 4.
R = 0.044 for 1133 reflexions.

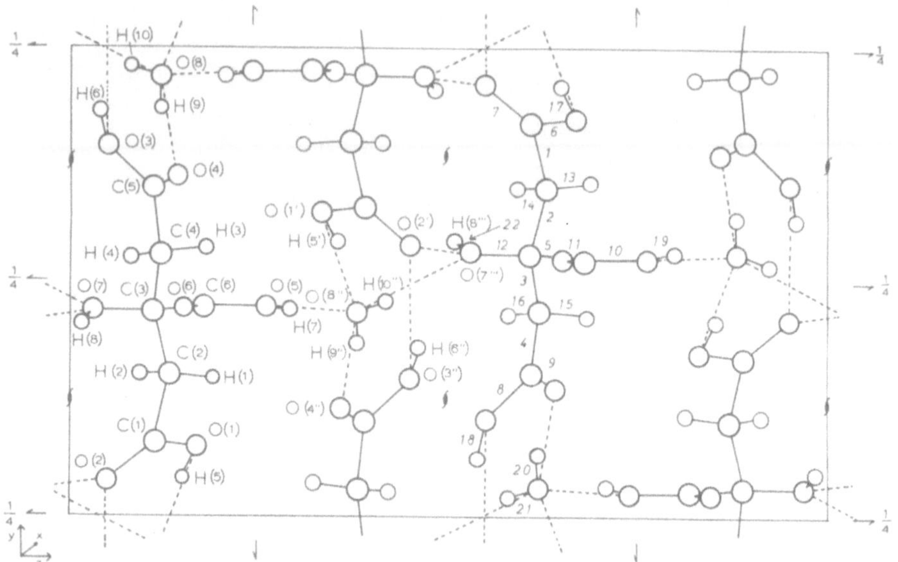

Fig. 1. The crystal structure of citric acid monohydrate.

Only the conformation around the C(1)-C(2) bond in the hydrate is different from that in anhydrous citric acid (1), the dihedral angle between the plane C(1)-C(2)-O(2) and C(1)-C(2)-C(3) being 116.8° in the former and 8.2° in the latter compound. The carboxyl group C(1)-O(1)-O(2) contributes to the hydrogen bond scheme in such a way that O(2) is twice an acceptor (Fig. 1). From the remaining two carbonyl oxygen atoms one acts once as an acceptor, and the other is not involved in a hydrogen bond. The water molecule is tetrahedally surrounded by hydrogen bonds. O-H···O distances are 2.62 - 2.77 Å.

1. C.E. NORDMAN, A.S. WELDON and A.L. PATTERSON, 1960. Acta Cryst., 13, 418; J.P. GLUSKER, J.A. MINKIN and A.L. PATTERSON, 1969. Ibid., B25, 1066.

3,3-DIMETHYLGLUTARIC ACID

$C_7H_{12}O_4$ $CO_2H \cdot CH_2 \cdot C(CH_3)_2CH_2 \cdot CO_2H$

E. BENEDETTI, R. CLAVERINI and C. PEDONE, 1972. Cryst. Struct. Comm., 1, 27-30.

Monoclinic, $P2_1/c$, a = 5.650, b = 11.530, c = 12.849 Å, β = 91.9°, Z = 4. Cu radiation, R = 0.057 for 1237 reflexions.

Bond lengths in the molecule are normal but bond angles are somewhat altered from tetrahedral to release steric strain arising from intramolecular interactions between methyl groups and other atoms in the molecule (angles C(1)-C(2)-C(3) and C(3)-C(4)-C(5) are 118.1(1) and 118.2(1)° respectively, while C(2)-C(3)-C(4) is 114.4(1)°). All the internal rotation angles involving carbon atoms are close to the values of the staggered conformations (180°, ± 60°). Long chains of molecules running along the c axis are formed (chain axis = c/2) by means of hydrogen bonds between carboxylic groups of units related by the glide plane c. This feature is quite rare for dicarboxylic acids which, in the solid state, usually form chains of molecules by hydrogen bonds between carboxylic groups across a centre of symmetry.

ETHYLENEDIAMINETETRA-ACETIC ACID

$C_{10}H_{16}N_2O_8$

M. COTRAIT, 1972. Acta Cryst., B28, 781-785.

Monoclinic, C2/c, a = 13.286, b = 5.578, c = 16.120 Å, β = 96.30°, D_m = 1.65, Z = 4. Cu radiation, R = 0.053 for 1203 reflexions.

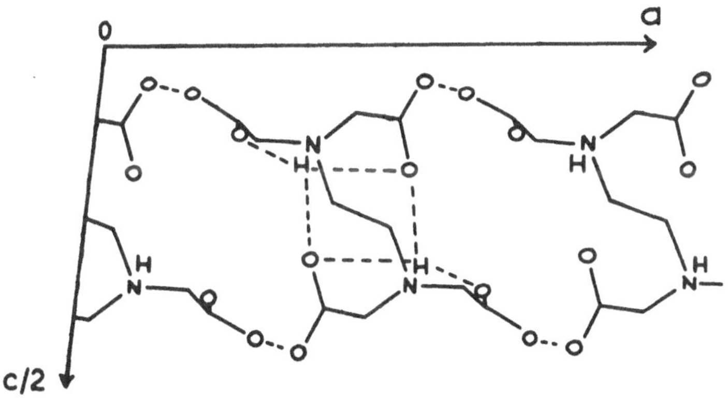

Fig. 1. The EDTA crystal structure viewed along b.

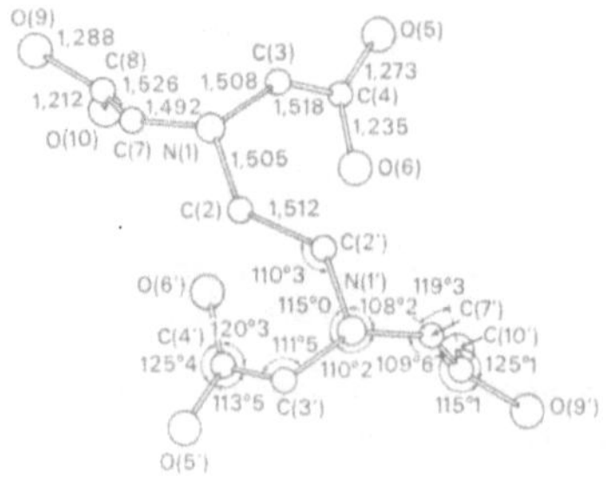

Fig. 2. Bond lengths (σ = 0.002 Å) and angles in EDTA.

In the EDTA molecule the nitrogen atoms are protonated and two carboxyl groups ionised. In the crystal structure neighbouring molecules are linked by short hydrogen bonds (2.46 Å) between COOH and COO^- groups (Fig. 1) as well as by NH···O (2.68 - 2.79 Å) hydrogen bonds. The bond lengths and angles in the molecule (which has twofold crystallographic symmetry) are shown in Fig. 2.

9-KETO-trans-2-DECENOIC ACID

$C_{10}H_{16}O_3$

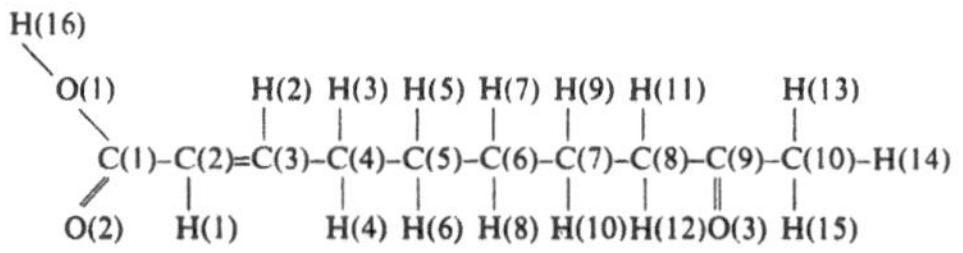

```
H(16)
  \
   O(1)          H(2) H(3) H(5) H(7) H(9) H(11)        H(13)
    \             |    |    |    |    |    |             |
     C(1)-C(2)=C(3)-C(4)-C(5)-C(6)-C(7)-C(8)-C(9)-C(10)-H(14)
    //     |         |    |    |    |    |    ||    |
   O(2)   H(1)      H(4) H(6) H(8) H(10)H(12)O(3) H(15)
```

D.T. CROMER and A.C. LARSON, 1972. Acta Cryst., B28, 2128-2132.

Monoclinic, $P2_1/c$, a = 9.584, b = 8.642, c = 13.371 Å, β = 96°58', Z = 4. Mo radiation, R = 0.087 for 855 reflexions.

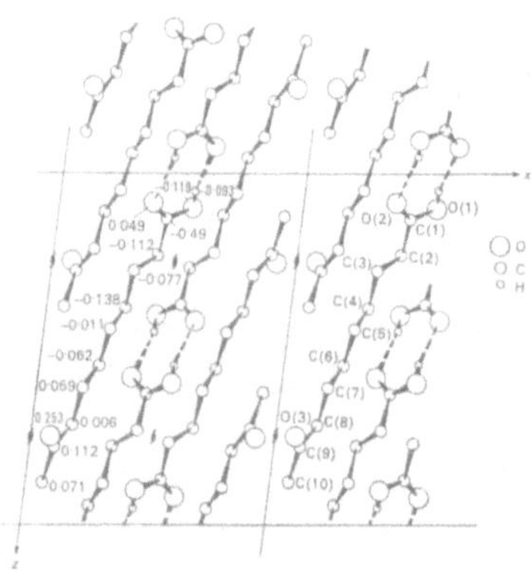

Fig. 1. View of the 9-keto-trans-2-decenoic acid structure looking down b.

The crystal structure (Fig. 1) contains hydrogen-bonded dimers linked via the carboxyl groups about centres of symmetry (O-H···O = 2.628(7) Å). The observed conformation of the molecule can be described in terms of two planes rotated by 32°; plane 1 contains C(1), C(2), C(3), C(4), O(1), O(2), and H(16) with a maximum deviation of 0.04 Å while plane 2 contains C(5) through C(10), O(3), and H(15) with maximum deviation 0.06 Å. H(15) on the methyl group is in the eclipsing position with respect to the carboxyl oxygen O(3). There is a small bend in the molecule between C(4) and C(5) and the angle of twist about the double bond is very small (0.75°). The atoms have unusually large anisotropic thermal parameters and possibly a small amount of disorder caused by occasional reversal of molecules in the unit cell. The bond lengths are shorter than expected (e.g. C=C = 1.177(16) Å) and corrections are needed but could not be made for lack of a suitable model of correlated motion.

MYCOPHENOLIC ACID

$C_{17}H_{20}O_6$

W. HARRISON, H.M.M. SHEARER and J. TROTTER, 1972. J. Chem. Soc., Perkin II, 1542-1544.

Triclinic, $P\bar{1}$, a = 9.555, b = 11.639, c = 7.341 Å, α = 90.75, β = 90.77, γ = 102.70°, D_m = 1.33, Z = 2. Cu radiation, R = 0.076 for 1548 reflexions.

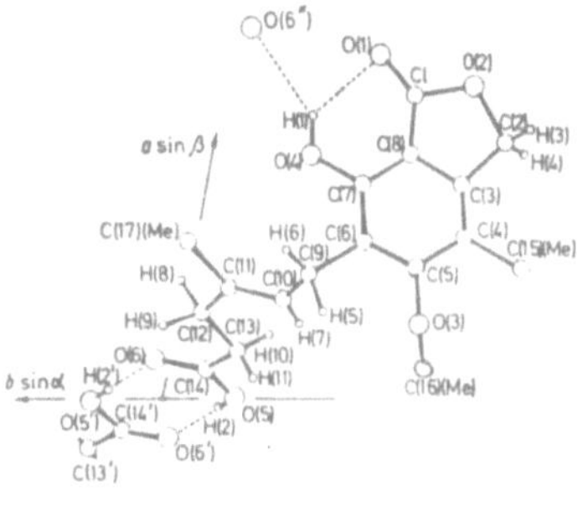

Fig. 1. The structure of mycophenolic acid viewed along c.

The molecule (Fig. 1) contains two approximately planar sections, the ring system and the extended side-chain, at an angle of 79°. The configuration of the C=C double bond in the side chain is trans, and bond lengths and angles are normal. The molecules are joined by one normal O-H···O hydrogen bond between carboxy-groups, and by one bifurcated hydrogen bond, which is partly intra- and partly inter-molecular.

9,10-trans-β-IONYLIDENE-γ-CROTONIC ACID

$C_{17}H_{24}O_2$

B. KOCH, 1972. Acta Cryst., B28, 1151-1159.

Triclinic, P$\bar{1}$, a = 10.931, b = 13.481, c = 7.546 Å, α = 108.12°, β = 127.81°, γ = 68.01°, D_m not given, Z = 2. Cu radiation, R = 0.07 for 1342 reflexions.

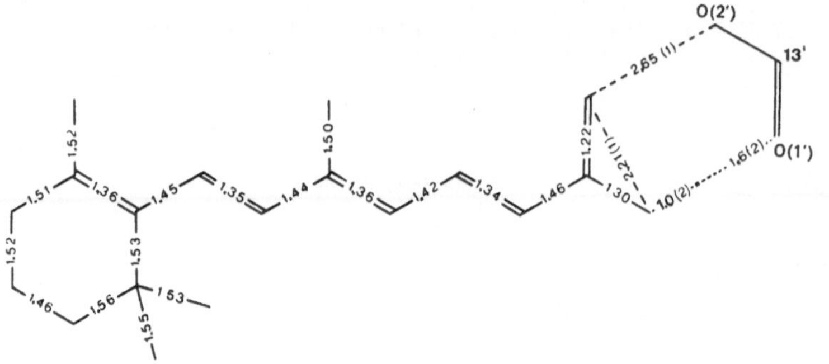

Fig. 1. The 9,10-trans-β-ionylidene-γ-crotonic acid molecule showing bond lengths (σ = 0.007 - 0.012 Å).

Except for C(2)-C(3) (1.46 Å) the bond lengths in the molecule are normal (Fig. 1). This distance has frequently been found to be too short, possibly due to conformational disorder. The thermal parameters suggest some evidence for disorder in the cyclohexene ring, with one conformation favoured perhaps by intramolecular steric hindrance. There are no unusually short intramolecular H···H contacts. The dihedral angle between the ring plane and the overall chain plane is 6.2°. The molecules are paired in the crystal by hydrogen bonding between two carboxylic acid groups across a centre of symmetry, O(1)···H-O(2') = 2.65 Å. All other intermolecular contacts are normal van der Waals interactions.

13-OXOISOSTEARIC ACID

$C_{18}H_{34}O_3$ $(CH_3)_2CH(CH_2)_2CO(CH_2)_{11}CO_2H$

B. DAHLÉN, 1972. Acta Cryst., B28, 2555-2562.

Triclinic, P$\bar{1}$, a = 4.933, b = 5.623, c = 34.460 Å, α = 95.65, β = 94.01, γ = 103.60°, D_m = 1.071, Z = 2. Cu radiation, R = 0.065 for 2300 reflexions.

The molecules are arranged in hydrogen bonded dimers (O-H···O = 2.669 Å) in which the chain axes are tilted 44° to the end-group planes to allow the methyl branches to be accommodated between the methyl ends of the chains. The keto-oxygen atom is directly incorporated into the chain packing and this leads to a bend of the chain axis in the zigzag plane of the molecule. The carboxyl group forms an angle of 13° with the chain plane. The mean $C(sp^3)$-$C(sp^3)$ bond length is 1.512(4) Å and the mean C-C-C angle is 113.9(2)°, but the angles at even-numbered carbon atoms are all greater than those of odd-numbered atoms.

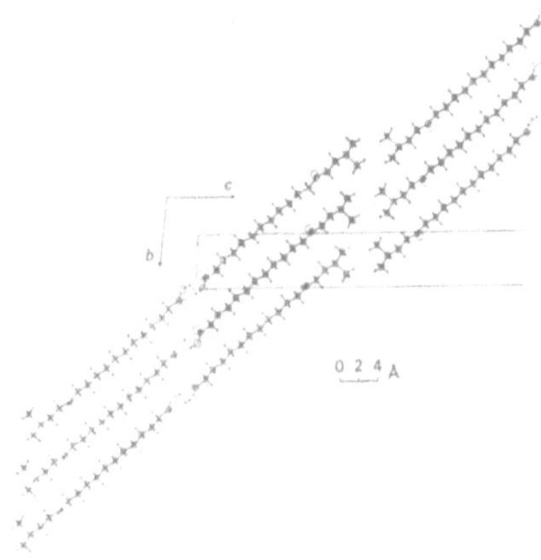

Fig. 1. Molecular arrangement of 13-oxoisostearic acid viewed along the a axis.

ISOSTEARIC ACID

$C_{18}H_{36}O_2$ $(CH_3)_2CH(CH_2)_{14}CO_2H$

S. ABRAHAMSSON and B.-M. LUNDÉN, 1972. Acta Cryst., B28, 2562-2567.

Triclinic, $P\bar{1}$, a = 4.9356, b = 5.6522, c = 34.408 Å, α = 95.22, β = 95.21, γ = 103.62°, Z = 2. Cu radiation, R = 0.066 for 680 reflexions.

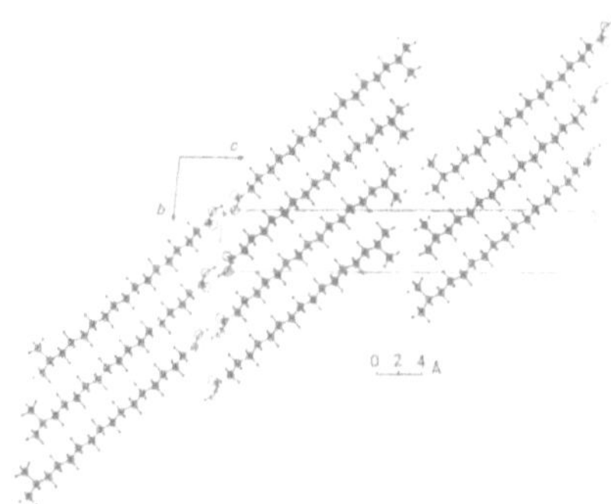

Fig. 1. Molecular packing of isostearic acid viewed along the a axis.

The molecules form hydrogen bonded dimers (O-H···O = 2.702 Å) and are maximally extended (Fig. 1). The molecular packing is dominated by the space requirements of the methyl branches, which are accommodated between the ends of carbon chains. The chain axes are tilted at 44° to the end-group planes and the zigzag chain forms an angle of 11.5° with the plane through the carboxyl group. The mean C-C distance in the chain is 1.515(13) Å and the average bond angle is 113.3(7)°.

POTASSIUM HYDROGEN OXALATE

C_2HKO_4

H. EINSPAHR, R.E. MARSH and J. DONOHUE, 1972. Acta Cryst., B28, 2194-2198.

Monoclinic, $P2_1/c$, a = 4.3043, b = 12.8334, c = 7.6322 Å, β = 102.01°, Z = 4. Mo radiation, R = 0.027 for 1266 reflexions.

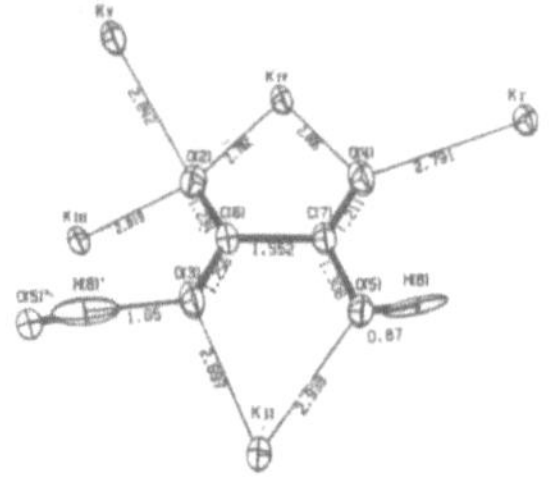

Fig. 1. Projection of the potassium hydrogen oxalate structure.

All features of the crystal structure are nearly identical with those found in a previous X-ray study (1). The structure is composed of layers of anions and cations parallel to ($10\bar{2}$). The binoxalate ions are hydrogen bonded together in chains with an asymmetrical hydrogen bond (O-H···O = 2.523(2) Å). Adjacent binoxalate chains within layers are held together by sevenfold coordination to potassium ions (K···O = 2.782 - 2.938 Å). The binoxalate anion is found to be twisted about its C-C bond with an angle of 13.9(2)° between carboxylate planes and normal bond distances (Fig. 1).

1. Structure Reports, 33B, 11.

LITHIUM HYDROGEN OXALATE MONOHYDRATE

(NORMAL AND DEUTERATED FORMS)

$C_2H_3LiO_5$ $Li \cdot HC_2O_4 \cdot H_2O$

$C_2D_3LiO_5$ $Li \cdot DC_2O_4 \cdot D_2O$

J.O. THOMAS, 1972. Acta Cryst., B28, 2037-2045.

	Hydrogen form Triclinic, P1	Deuterium form Triclinic, P1
a Å	5.056	5.055
b	6.140	6.138
c	3.411	3.410
α°	95.06	95.05
β	98.93	98.62
γ	78.57	78.57
Z	1	1
	Cu radiation	Cu radiation
R	0.024	0.026
reflexions	1023	1057

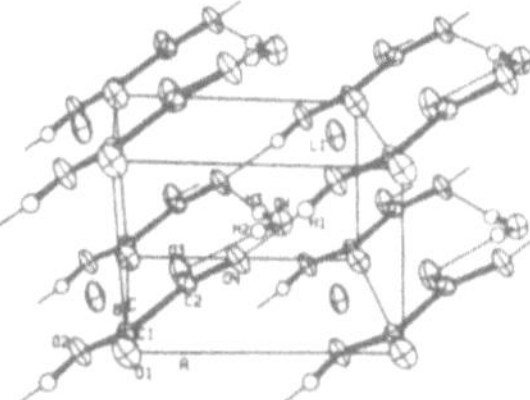

Fig. 1. Illustration of the lithium hydrogen oxalate monohydrate structure.

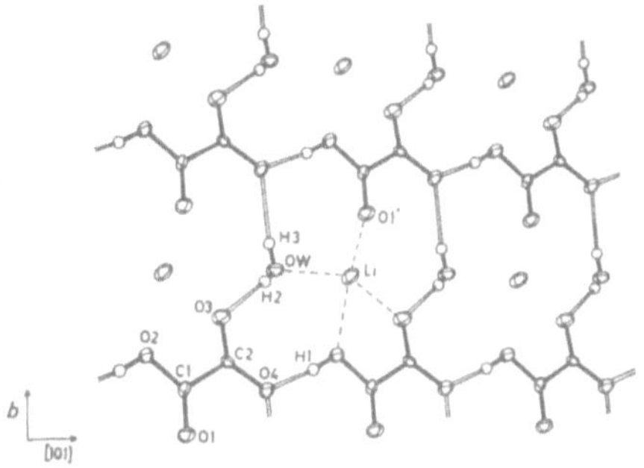

Fig. 2. A view of the lithium hydrogen oxalate monohydrate structure perpendicular to the (10$\bar{1}$) plane. Hydrogen bonds: unfilled lines; ionic bonds: broken lines.

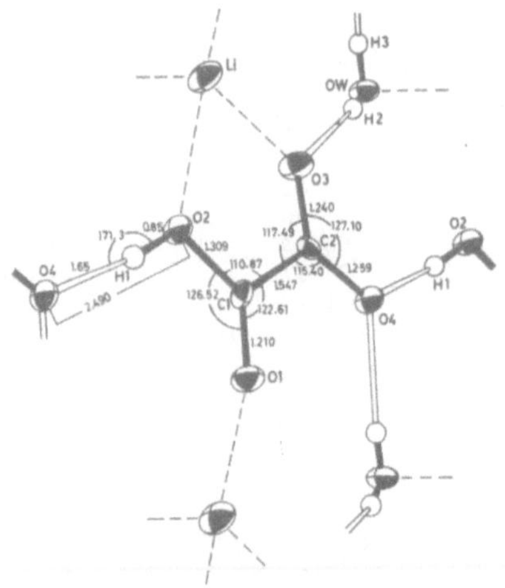

Fig. 3. Bond lengths and angles of the hydrogen oxalate ion.

The distinct layer-like structure (Fig. 1) of the $Li \cdot H_2C_2O_4 \cdot H_2O$ crystals arises as a result of near planar hydrogen oxalate ions linking together via short asymmetric O-H···O hydrogen bonds, 2.490(1) Å, to form infinite chains running in the [101] direction as seen in Fig. 2. These chains are linked transversely by means of O-H···O hydrogen bonds, 2.777(1) and 2.702(1) Å, via the water molecules and Li^+···O electrostatic forces. The infinite planes so formed are held together mainly by the zigzag ···Li^+···O(W)···Li^+··· chains running in the [001] direction. The oxygen atom of the water molecule lies 0.62 Å out of the least-squares plane through the non-hydrogen atoms of the hydrogen oxalate ion and the Li^+ ion lies only 0.3 Å out of this plane in the opposite direction. Of the five near contacts of the Li^+ ion, four lie within the same plane containing the atoms O(1), O(2'), O(3), and O(W); the fifth contact Li^+···O(W") provides the electrostatic interaction between the layers of atoms. The water molecules have a tetrahedral environment comprising two Li^+ ions and two oxygen atoms O(3) and O(4'). The bonding situation around the hydrogen oxalate ion is shown in Fig. 3.

The structure is generally unchanged in the deuterated form except for an increase of 0.016(1) Å in the short asymmetric hydrogen bond O(2)-H(1)···O(4) between the hydrogen oxalate ions, $(O\cdots O)_D$ = 2.506(1) Å. The longer hydrogen bonds from the water molecules O(W)-H(2)···O(3) and O(W)-H(3)···O(4') show insignificant changes of 0.001(2), $(O\cdots O)_D$ = 2.778(1) Å, and -0.002(2) Å, $(O\cdots O)_D$ = 2.700(1) Å, respectively. Small, although possibly significant, changes of 0.29(8) and -0.23(8)° are observed in the O(1)-C(1)-C(2) and O(2)-C(1)-C(2) angles, 122.90(6) and 110.64(6)° respectively. These would indicate that the isotope effect in the O(2)-H(1)···O(4) bonds linking the $HC_2O_4^-$ ions is capable of bringing about a small distortion in the internal geometry of the ions themselves.

AMMONIUM OXALATE MONOHYDRATE

$C_2H_{10}N_2O_5$ $(NH_4)_2C_2O_4 \cdot H_2O$

J. TAYLOR and T.M. SABINE, 1972. Acta Cryst., B28, 3340-3351.

Orthorhombic, $P2_12_12_1$, a = 8.035, b = 10.309, c = 3.795 Å, D_m = 1.50, Z = 2. Cu radiation, R = 0.070 for crystal 1 and 0.040 for crystal 2; neutron radiation, R = 0.039.

Neutron and X-ray studies were performed on protonated crystals, which have been previously studied by X-rays (1), as well as deuterated and null-matrix (63% H, 37% D) crystals where the deuterium atoms were found to show preference for the ammonium ion sites and the protons for the water sites; there is no significant expansion of the hydrogen bonds on deuteration. The hydrogen bonding forces the oxalate ion out of the planar configuration and the carboxyl planes are twisted about 28° relative to each other. O-H···O hydrogen bonds, 2.767(4) Å, link the oxalate ions in the a-direction. The ammonium ion H or D atoms form N-H···O hydrogen bonds, 2.850 - 3.172(4) Å and the ammonium ions as well as the water molecules are both in tetrahedral hydrogen-bonding environments; the hydrogen bonds are non-linear with O-H···O and N-H···O angles lying between 166.9(4) and 177.8(4)°. The ammonium ion is nearly regular with H-N-H angles between 107.5(3) and 112.0(3)°.

1. J.H. ROBERTSON, 1965. Acta Cryst., 18, 410; Structure Reports, 30B.

AMMONIUM OXALATE MONOPERHYDRATE

$C_2H_{10}N_2O_6$ $(NH_4)_2C_2O_4 \cdot H_2O_2$

B.F. PEDERSEN, 1972. Acta Cryst., B28, 746-754.

Orthorhombic, $P2_12_12$, a = 8.531, b = 10.476, c = 3.8228 Å, D_m = 1.51, Z = 2. Mo radiation, R = 0.042 for 1006 reflexions.

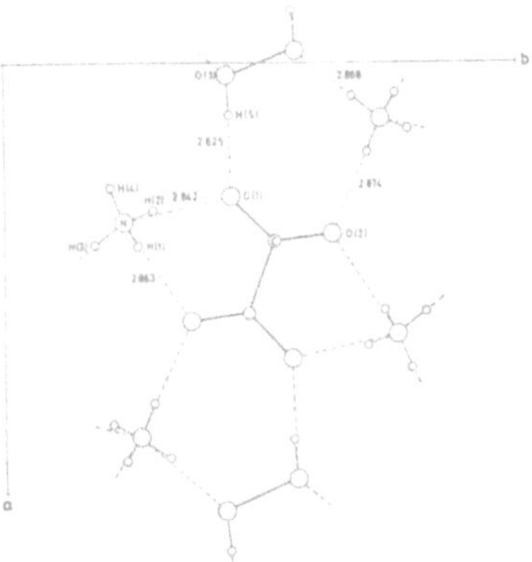

Fig. 1. Hydrogen bonding in ammonium oxalate perhydrate.

Both oxalate ion and H_2O_2 molecule are situated on twofold axes. Interatomic distances are normal but the oxalate ion is non-planar, the angle between the two CO_2 groups of an ion being 28.15(9)°. The hydrogen peroxide molecule has a skew conformation with a dihedral angle of 121(3)°. Approximately 6% of the H_2O_2 sites are occupied by water molecules. All available hydrogen atoms are involved in hydrogen bonding (Fig. 1). The O-H···O distance is 2.625(1) Å and the mean N-H···O distance is 2.860(1) Å. The crystal structure is closely related to that of ammonium oxalate monohydrate (1).

1. J.H. ROBERTSON, 1965. Acta Cryst., 18, 410; and preceding report.

SILVER TRIFLUOROACETATE DIMER

$C_4Ag_2F_6O_4$

R.G. GRIFFIN, J.D. ELLETT, M. MEHRING, J.G. BULLITT and J.S. WAUGH, 1972. J. Chem. Phys., 57, 2147-2155.

Monoclinic, C2/c, a = 12.256, b = 8.003, c = 11.070 Å, β = 128.95°, D_m = 3.61, Z = 4. Mo radiation, R = 0.039 for 486 reflexions.

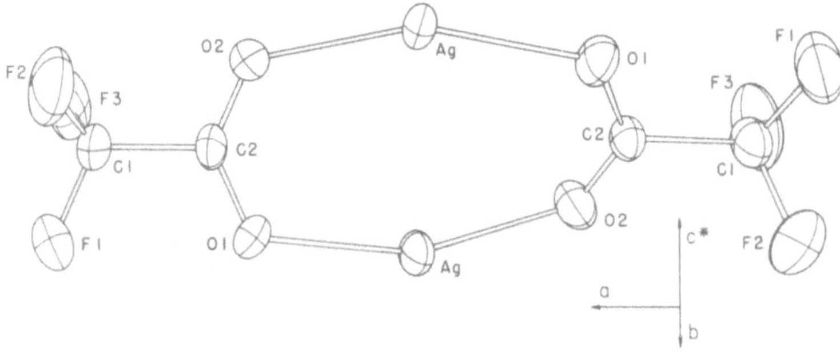

Fig. 1. The silver trifluoroacetate dimer.

Each dimer has (crystallographic) twofold symmetry (Fig. 1). In the eight-membered $(CO_2Ag)_2$ ring bond distances are Ag-O(1) = 2.249(6), Ag-O(2) = 2.232(6), C(2)-O(1) = 1.215(6), C(2)-O(2) = 1.239(8) Å; the Ag···Ag separation is 2.967(3) Å. The C-C and mean C-F distances are 1.55(1) and 1.31(1) Å respectively.

POTASSIUM HYDROGEN BIS(TRIFLUOROACETATE)
POTASSIUM DEUTERIUM BIS(TRIFLUOROACETATE)
(NEUTRON STUDIES)

A.L. MACDONALD, J.C. SPEAKMAN and D. HADŽI, 1972. J. Chem. Soc., Perkin II, 825-832.

$C_4HF_6KO_4$ $KH(CF_3CO_2)_2$

Monoclinic, I2/a, a = 8.773, b = 10.169, c = 9.255 Å, β = 99.85°, D_m = 2.085, Z = 4. Neutron radiation, R = 0.052 for 849 reflexions.

$C_4DF_6KO_4$ $KD(CF_3CO_2)_2$

Monoclinic, I2/a, a = 8.784, b = 10.177, c = 9.264 Å, β = 99.83°, Z = 4. Neutron radiation, R = 0.050 for 1048 reflexions.

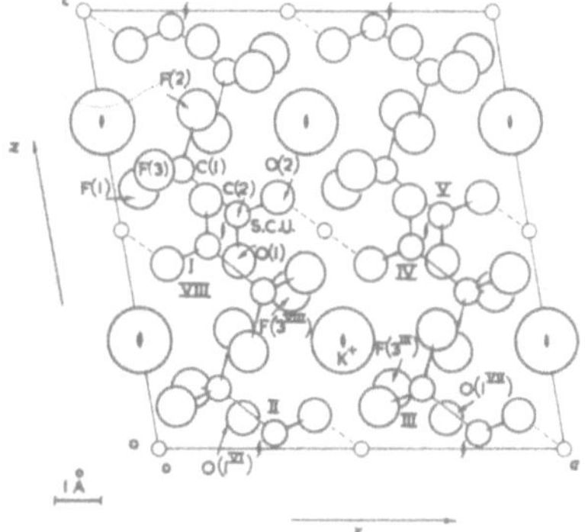

Fig. 1. Potassium hydrogen bis(trifluoroacetate); view of the unit cell in its b-axial projection. (S.C.U. indicates the standard chemical unit.)

The potassium hydrogen bis(trifluoroacetate) structure (Fig. 1) has been studied precisely by X-ray diffraction methods from which it was deduced that two equivalent trifluoroacetate residues were joined by a short hydrogen bond (O···O = 2.435(7) Å) lying across a centre of inversion (1). From the neutron studies the hydrogen bond lengths are O···H···O = 2.437(4) and O···D···O = 2.437(3) Å. The absence of significant isotope effect suggests that this may be a genuinely symmetrical hydrogen bond, with the hydrogen nucleus vibrating (anharmonically) in a single potential-energy well. This is strongly supported by spectroscopic measurements, as well as by analysis of the vibrational motions of the proton or deuteron, within its O···O environment. Some of the problems of studying hydrogen bonds in crystallographically symmetrical sites by diffraction methods are discussed.

1. L. GOLIC and J.C. SPEAKMAN, 1965. J. Chem. Soc., 2530. [Two typographical errors in this 1965 paper are noted in the 1972 neutron study, namely "a for the rubidium salt in Table 1 should be 8.813 Å; y for K in Table 4 should be 28.124(x10^{-5})". Also the β angles given in the 1972 paper are in error; the correct values are given above.]

POTASSIUM HYDROGEN meso-TARTRATE (at -160°C)

$C_4H_5KO_6$

J. KROON and J.A. KANTERS, 1972. Acta Cryst., B28, 714-722.

Triclinic, P$\bar{1}$, a = 9.33, b = 10.29, c = 9.09 Å, α = 80.71, β = 108.77, γ = 130.95°, D_m = 1.99, Z = 4. Mo radiation, R = 0.035 for 5308 reflexions.

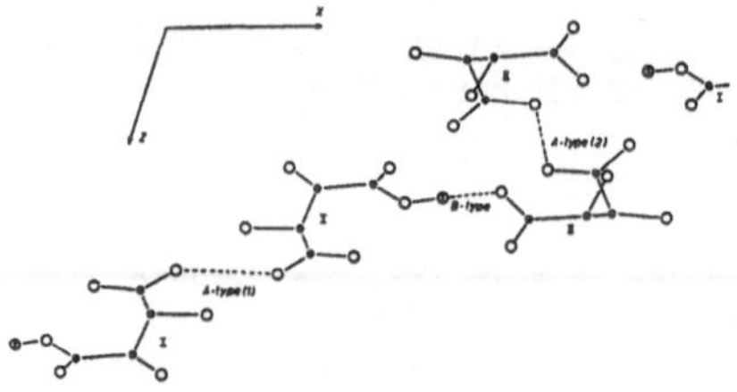

Fig. 1. Projection of part of the structure of potassium hydrogen meso-tartrate along b. Hydrogen bonds are indicated by broken lines. I and II denote the independent meso-tartrate ions in the unit cell.

The molecules occur as enantiomorphic pairs. One carboxyl group has the unusual conformation with the O-H bond in an anti direction with respect to the C=O bond. The same carboxyl group also deviates from the general findings that the α-hydroxyl is in a syn position with respect to the C=O bond. The hydrogen bonding is intermolecular (Fig. 1) and is of a mixed A/B type with three short hydrogen bonds 2.452, 2.483 and 2.543(1) Å (A, A, B) as well as four longer ones O-H···O = 2.720 - 2.877(1) Å. One of the hydrogen atoms is found in an acentric position in one of the A-type bonds that is generally supposed to be symmetric. One K^+ ion has eight and the other seven oxygen contacts in the range 2.73 - 3.16 Å.

POTASSIUM HYDROGEN DIACETATE

$C_4H_7KO_4$ $KH(CH_3CO_2)_2$

M. CURRIE, 1972. J. Chem. Soc., Perkin II, 832-835.

Monoclinic, P2_1/n, a = 4.085, b = 23.84, c = 7.226 Å, β = 97.4°, Z = 4. Mo radiation, R = 0.106 for 849 reflexions.

This acid salt is of type B, i.e. the CH_3CO_2H and CH_3COO^- are crystallographically distinct; these residues are linked through a short unsymmetrical hydrogen bond with O-H···O = 2.476(8) Å (Fig. 1). Bond distances in the acetic acid molecule and acetate ion are normal. The K^+ ion makes contact with seven oxygen atoms. The K^+···O distances are in the range 2.714(6) - 2.942(5) Å. There are no unusual non-bonded intermolecular contacts in the structure.

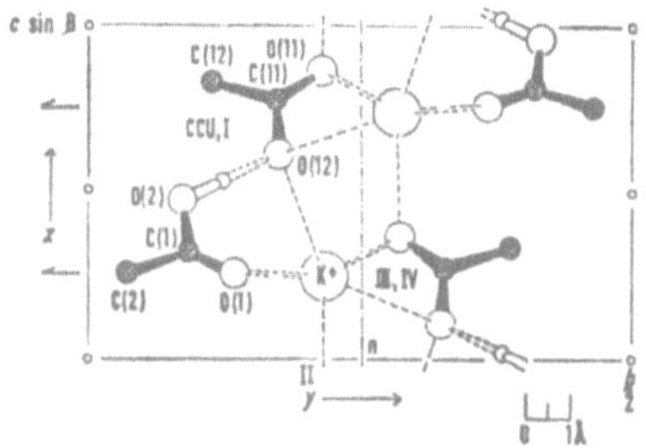

Fig. 1. The crystal structure of potassium hydrogen diacetate in its a axial projection.

CALCIUM FUMARATE TRIHYDRATE

$C_4H_8CaO_7$ $CaC_4H_2O_4 \cdot 3H_2O$

M.P. GUPTA, S.M. PRASAD, R.G. SAHU and B.N. SAHU, 1972. Acta Cryst., B28, 135-139.

Orthorhombic, $Pna2_1$, a = 6.62, b = 17.63, c = 6.97 Å, Z = 4. Cu radiation, R = 0.095 for 424 reflexions.

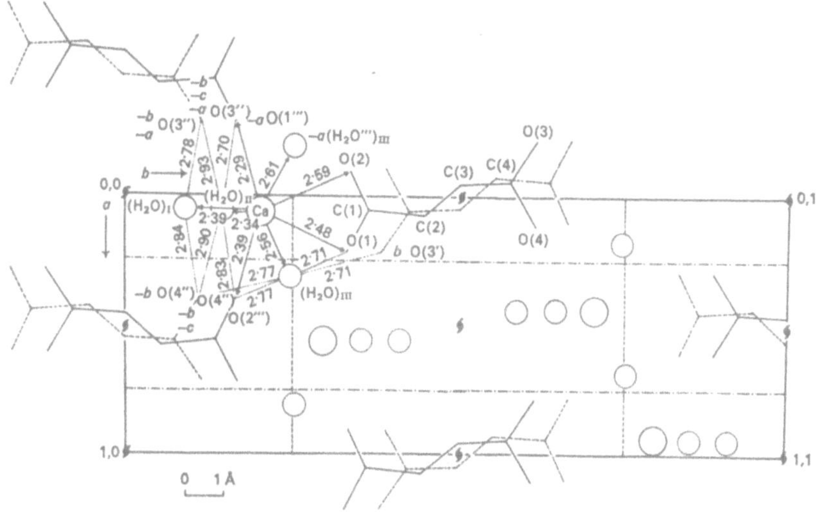

Fig. 1. The crystal structure of $CaC_4H_2O_4 \cdot 3H_2O$ looking down the [001] axis.

Chains of fumarate ions are aligned parallel to b (Fig. 1) and are in two layers at $z \sim 0$ and $z \sim c/2$. The structure is held together by ionic Ca···O linkages with eightfold coördination around Ca (average Ca···O = 2.45 Å) and by ten hydrogen bonds (ranging from 2.70 - 2.93 Å) linking the oxygen atoms of the fumarate ion via the three water molecules. Two of the water molecules take part in bifurcated hydrogen bonds. Bond lengths in the fumarate ion are normal.

AMMONIUM HYDROGEN DI-CHLOROACETATE

$C_4H_9C\ell_2NO_4$ $NH_4^+[H(C\ell CH_2COO)_2]^-$

M. ICHIKAWA, 1972. Acta Cryst., B28, 755-760.

Monoclinic, C2/c, a = 10.521, b = 11.576, c = 8.387 Å, β = 119.48°, D_m = 1.528, Z = 4. Mo radiation, R = 0.068 for 511 reflexions.

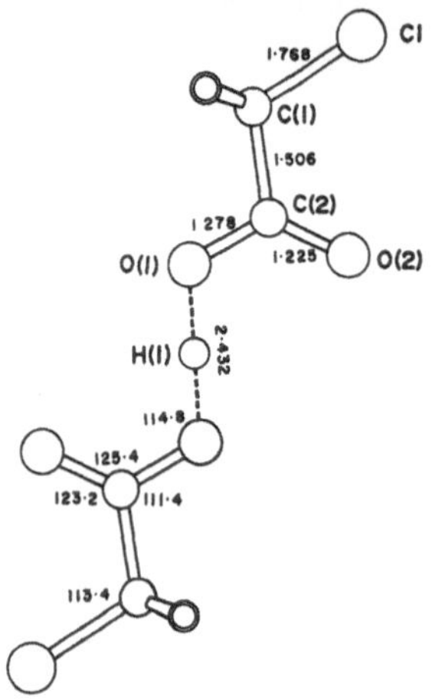

Fig. 1. The hydrogen bonded chloroacetate ions in $NH_4^+[H(C\ell CH_2COO)_2]^-$. For bond lengths, σ = 0.003 - 5 Å.

Each ammonium ion lies on a twofold axis and is surrounded by six oxygens from different chloroacetate ions (N-H···O = 2.880, 2.926 and 3.148(3) Å). Pairs of chloroacetate ions are linked by a very short hydrogen bond (Fig. 1) across a centre of symmetry.

LITHIUM AMMONIUM TARTRATE MONOHYDRATE

$C_4H_{10}LiNO_7$ $LiNH_4C_4H_4O_6 \cdot H_2O$

H. HINAZUMI and T. MITSUI, 1972. Acta Cryst., B28, 3299-3305.

Orthorhombic, $P2_12_12_1$, a = 7.884, b = 14.565, c = 6.409 Å, Z = 4. Mo radiation, R = 0.042 for 1758 reflexions.

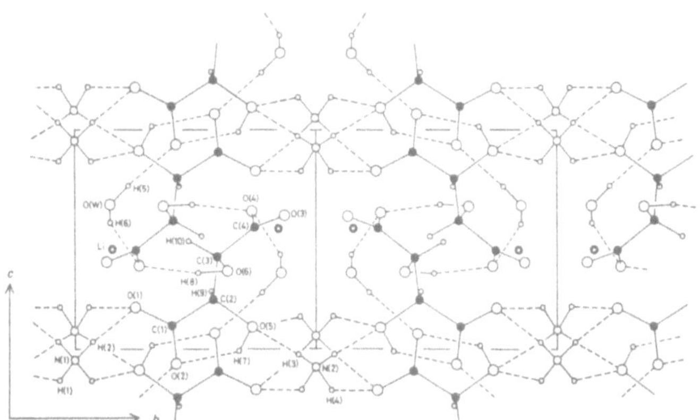

Fig. 1. Lithium ammonium tartrate monohydrate structure projected onto the (100) plane.

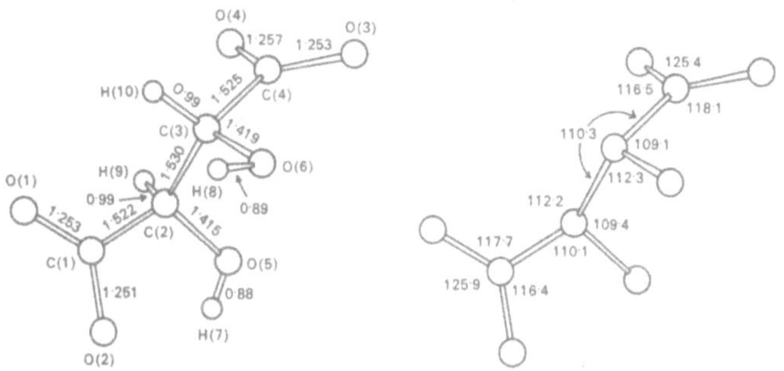

Fig. 2. Bond lengths and angles in the tartrate ion.

The crystal structure (Fig. 1), which has been previously reported (1), is more accurately determined with the hydrogen positions now located. The bond lengths and angles for the tartrate ion are given in Fig. 2; the mean carboxylate C-O, carbinol C-O and C-C bond lengths are 1.254(3), 1.417(2) and 1.526(4) Å, respectively. There are three planar groups in the tartrate molecule, namely, the two carboxylate groups and the C-C chain. The molecules are held together by hydrogen bonds, with the hydroxy group O(5) - H(7) forming an intramolecular hydrogen bond with O(2), O-H···O = 2.583 Å and O(5) linked to N(1) and N(2) by

N-H···O hydrogen bonds, 2.867 and 2.925 Å. The lithium ion is surrounded by five oxygen atoms and the shortest Li···O distance is 1.957 Å. The authors mention that the compound exhibits ferroelectric activity below 100°K, with spontaneous polarization along the b axis in the ferroelectric phase, but do not attempt to discuss the mechanism.

1. A.J.J. SPRENKELS, 1956. Koninkl. Ned. Akad. Wetenschap. Proc., B59, 221; Structure Reports, 20, 480.

POTASSIUM HYDROGEN GLUTARATE

$C_5H_7KO_4$ $KH(C_5H_6O_4)$

A.L. MACDONALD and J.C. SPEAKMAN, 1972. J. Chem. Soc., Perkin II, 942-946.

Orthorhombic, Cmma, a = 18.476, b = 7.282, c = 5.133 Å, D_m = 1.61, Z = 4. Mo radiation, R = 0.034 for 627 reflexions.

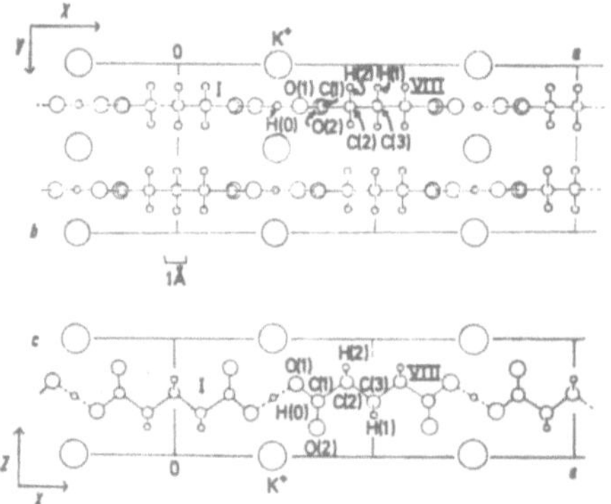

Fig. 1. The crystal structure of potassium hydrogen glutarate, seen in its c and b axial projections. In the lower picture, only the glutarate chain at b/4 is shown.

The crystal structure (Fig. 1) contains infinite chains of glutarate residues (of mm symmetry) linked end-to-end by hydrogen bonds across points of 2/m symmetry. The hydrogen bonds are very short, with O···H···O = 2.445(3) Å. The librational component of the glutarate vibration is large in one direction where it is unconstrained by hydrogen bonding and negligibly small in the other two directions. The K^+ ion is at a site of 222 symmetry; the two independent K^+···O distances (out of eight) are 2.799(1) and 3.000(1) Å.

POTASSIUM HYDROGEN DL-METHYLSUCCINATE

$C_5H_7KO_4$ $KO_2CCH(CH_3)CH_2CO_2H$

Y. SCHOUWSTRA, 1972. Acta Cryst., B28, 2217-2221.

Monoclinic, $P2_1/c$, a = 11.927, b = 6.217, c = 9.731 Å, β = 105.65°, D_m = 1.64, Z = 4. Cu radiation, R = 0.040 for 1296 reflexions.

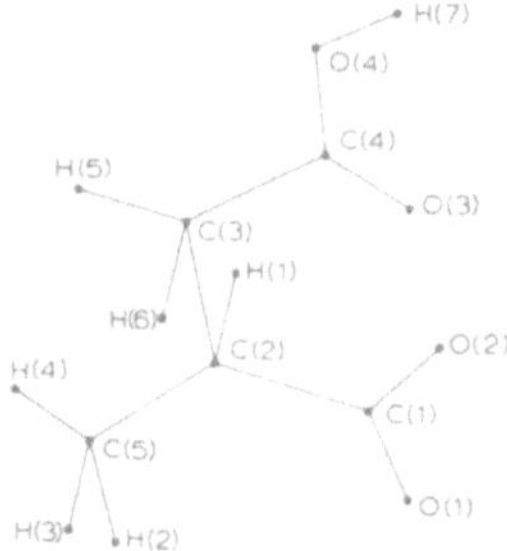

Fig. 1. The structure of the methylsuccinate ion.

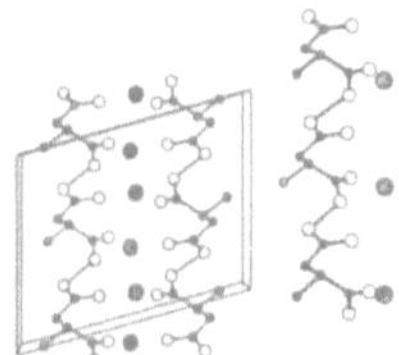

Fig. 2. A view of the crystal structure of potassium hydrogen DL-methylsuccinate along [010].

The acid methylsuccinate ion (Fig. 1) has ionized and non-ionized carboxyl groups which are easily distinguishable; the ionized C-O distances are 1.270 and 1.242(3) Å while the unionized distances are 1.313 and 1.213(3) Å. The methyl group is in the α-position with respect to the ionized carboxyl group. The carboxyl group atoms C(1), C(2), O(1), O(2) and C(3), C(4), O(3), O(4) are individually planar. The carbon chain C(1) - C(4) is bent, but atoms C(4), C(3), C(2), C(5) are roughly coplanar with an average deviation of 0.02 Å, resulting in a staggered conformation about C(2) - C(3). Three linked layers parallel to the (100) plane can be distinguished; two of them contain the DL-methylsuccinate ions, the third, K^+ ions, thus forming a 'superlayer' (Fig. 2). The methylsuccinate layers consist of zigzag chains linked by an intermolecular acentric hydrogen bond (O-H···O = 2.543(3) Å) with the chains fixed in a three-dimensional network by K^+···O interactions. The potassium ion has eightfold coordination consisting of a distorted tetrahedron with four oxygens from the ionized carboxyl group (2.705 - 2.763 Å) and four other weaker interactions (2.956 - 3.085 Å) resulting in an asymmetric arrangement.

DIPOTASSIUM cis-ACONITATE

$C_6H_4K_2O_6$

J.P. GLUSKER, W. OREHOWSKY, C.A. CASCIATO and H.L. CARRELL, 1972. Acta Cryst., B28, 419-425.

Monoclinic, $P2_1/c$, a = 9.417, b = 12.421, c = 7.615 Å, β = 96.27°, D_m = 1.86, Z = 4. Cu radiation, R = 0.025 for 1878 reflexions.

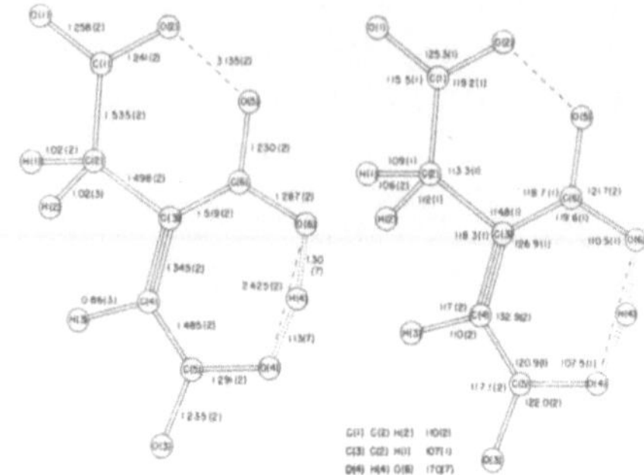

Fig. 1. Bond distances and angles in the cis-aconitate ion.

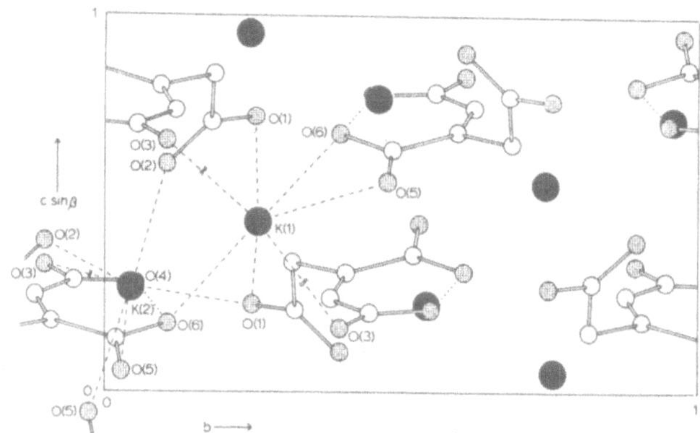

Fig. 2. Projection of the structure of dipotassium cis-aconitate down the c-axis. Oxygen atoms are stippled, potassium ions are black.

Details of the molecular geometry are in Fig. 1. A notable feature is a short intramolecular O···H···O hydrogen bond. The system of atoms C(3) - C(6), O(4), and O(6) is essentially planar. The packing in the crystal (Fig. 2) is determined by packing of the anions with the potassium ions (the only hydrogen available for hydrogen bonding is involved in the intramolecular hydrogen bond). Each potassium ion is surrounded by seven oxygen atoms (K···O = 2.648 - 3.085(1) Å).

POTASSIUM DIHYDROGEN trans-ACONITATE

$C_6H_5KO_6$ $KH_2(C_6H_3O_6)$

J.M. DARGAY, H.M. BERMAN, H.L. CARRELL and J.P. GLUSKER, 1972. Acta Cryst., B28, 1533-1539.

Monoclinic, C2/c, a = 11.861, b = 15.919, c = 8.416 Å, β = 90.90°, D_m = 1.77, Z = 8. Mo radiation, R = 0.043 for 2081 reflexions.

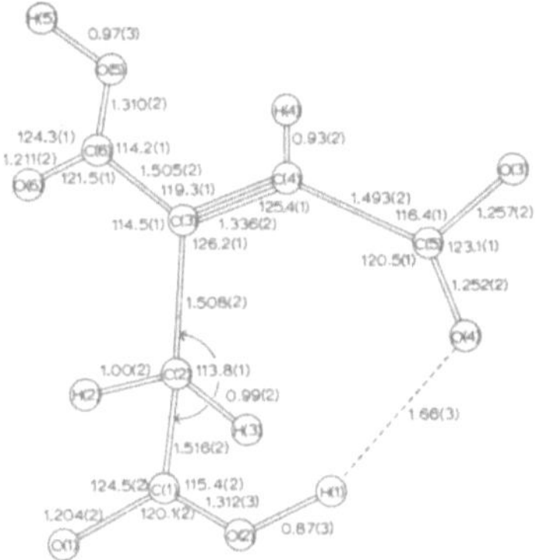

Fig. 1. Bond lengths and interbond angles in the trans-aconitate ion (with e.s.d.'s in parentheses).

Bond lengths and angles in the trans-aconitate ion (Fig. 1) agree with those in the cis ion (1). The C(5)-C(4)=C(3)<C(2), C(6) portion of the anion is approximately planar. The potassium ion lies on a twofold axis and there are two unique potassium positions, K(1) and K(2). Eight oxygen atoms from six anions surround each K^+ ion with $K^+\cdots O$ distances in the range 2.70 - 3.32 Å. The coordination of K(2) is described as a distorted cube while that of K(1) does not conform to any standard geometry. A seven-membered ring is formed by the bidendate chelation (through O(1) and O(6)) of one K^+ ion by the trans-aconitate ion. In addition to the intramolecular hydrogen bond shown in Fig. 1 an intermolecular hydrogen bond O(5)H(5)···O(3') (2.52 Å) links the anions in chains.

1. Preceding report.

DICAESIUM DIMANGANESE TRIOXALATE TRIHYDRATE

$C_6H_6Cs_2Mn_2O_{15}$ $Cs_2Mn_2(C_2O_4)_3 \cdot 3H_2O$

H. SIEMS and J. LÖHN, 1972. Z. anorg. Chem., 393, 97-104.

Triclinic, P$\bar{1}$, a = 7.289, b = 10.525, c = 5.987 Å, α = 92.16, β = 102.74, γ = 92.84°, D_m = 2.63, Z = 1. Mo radiation, R = 0.049 for three dimensional data.

One oxalate ion is situated on a centre of symmetry whereas the other lies in a general position; the Cs and Mn ions are also in general positions in the cell. The Mn atom is six-coordinated in a distorted octahedron with Mn-O in the range 2.13 - 2.21(1) Å. The Cs ion is surrounded by eight oxygen atoms with Cs-O distances in the range 3.13 - 3.39(1) Å; a ninth oxygen is 3.58(1) Å from the Cs^+ ion. Bond lengths in the oxalate ions are normal (C-C = 1.55(1), C-O = 1.22 - 1.26(1) Å).

CALCIUM SALT OF HIBISCUS ACID TETRAHYDRATE

$C_6H_{12}CaO_{11}$ $Ca(C_6H_4O_7)\cdot 4H_2O$

J.P. GLUSKER, J.A. MINKIN and F.B. SOULE, 1972. Acta Cryst., B28, 2499-2505.

Orthorhombic, $P2_12_12_1$, a = 8.409, b = 21.232, c = 6.245 Å, D_m = 1.80, Z = 4. Cu radiation, R = 0.056 for 1383 reflexions.

The absolute configuration of hibiscus acid, the lactone of (+)-allo-hydroxycitric acid is shown in (I), 1S:2R-1,2-hydroxy-1,2,3-propanetricarboxylic

(I) 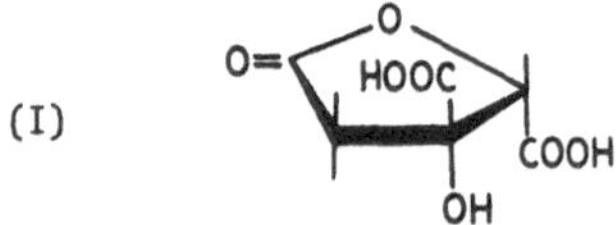

acid. Distances and angles in the lactone ion are in Fig. 1. The two carboxyl groups are trans with respect to the plane of the lactone ring and the two O-C-COO⁻ groupings are each almost planar. The lactone forms a bidentate chelate with the calcium ion (Fig. 2) which is surrounded by nine oxygen atoms in a square antiprism arrangement with an additional square pyramid on one side. Ca-O distances are in the range 2.41 - 2.64. Two of the oxygen atoms of the water molecules are markedly anisotropic indicating possible disorder. These molecules lie in tunnels parallel to the c-axis and are probably lost as the crystal effloresces.

Fig. 1. Distances and angles in the lactone ion of hibiscus acid.

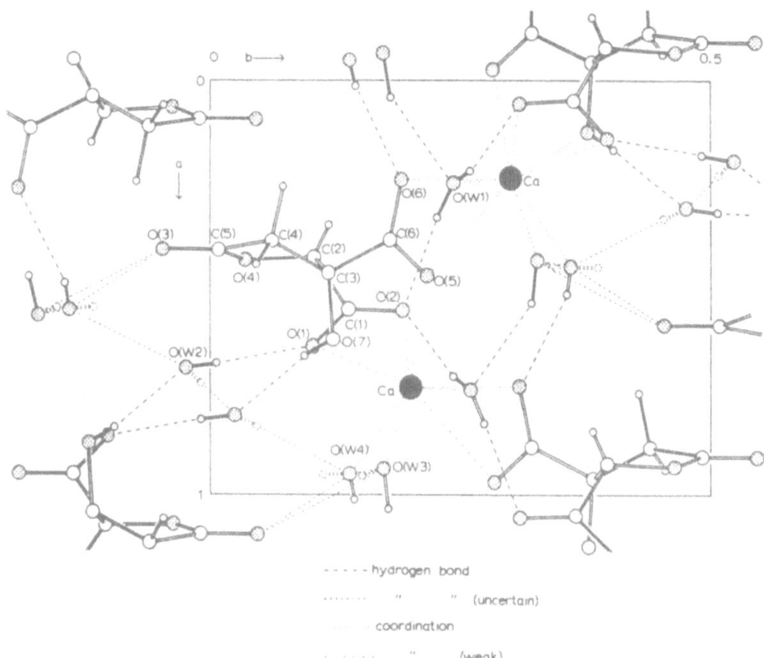

Fig. 2. The packing of the tetrahydrate calcium salt of hibiscus acid viewed down the c axis. Ca ions are black and oxygen atoms are stippled. There are nine independent O-H···O hydrogen bonds with an average O···O distance of 2.82 Å. Uncertain hydrogen atom positions are marked with broken lines.

ACID POTASSIUM HYDROGEN FUMARATE

$C_{12}H_{10}K_2O_{12}$ $2KC_4H_3O_4 \cdot C_4H_4O_4$

M.P. GUPTA and N. PRASAD, 1972. Acta Cryst., B28, 2628-2630.

Triclinic, $P\bar{1}$, a = 8.603, b = 7.475, c = 6.902 Å, α = 78°20', β = 112°20', γ = 107°55', Z = 2. Cu radiation, R = 0.078 for 701 reflexions.

This three-dimensional analysis serves to confirm the findings of an earlier two-dimensional study (1). There are two molecules of potassium hydrogen fumarate in general positions whereas the fumaric acid molecule is on a centre of symmetry at ½, ½, ½. Bond lengths and angles are in good agreement with accepted values.

1. Structure Reports, 35B, 16.

DIMETHYLAMMONIUM CHLORIDE

C_2H_8ClN $(CH_3)_2NH_2Cl$

J. LINDGREN and N. ARRAN, 1972. Acta Chem. Scand., 26, 3043-3052.

Tetragonal, I4/mmm, a = 5.146, c = 9.959 Å, Z = 2. Cu radiation, R = 0.093 for 111 reflexions.

The structure is disordered but contains planar domains which are internally ordered and consist of dimethylammonium and chloride ions linked together by NH···Cl (3.29(3) Å) hydrogen bonds. The choice among several models possible from the X-ray data was made from a study of the infrared spectra.

2-AMMONIOETHYLAMMONIUM TETRACHLOROCUPRATE(II)

$C_2H_{10}Cl_4CuN_2$ $(NH_3(CH_2)_2NH_3)(CuCl_4)$

G.B. BIRRELL and B. ZASLOW, 1972. J. Inorg. Nucl. Chem., 34, 1751.

Monoclinic, $P2_1/b$, a = 8.110, b = 7.187, c = 7.366 Å, γ = 92.46°, D_m = 2.10, Z = 2. Mo radiation, R = 0.11 for 386 reflexions.

The $CuCl_4^{2-}$ ion is square planar within experimental error; Cu-Cl(1) and Cu-Cl(2) are 2.298(3) and 2.292(3) Å, and the angle Cl(1)-Cu-Cl(2) is 89.9(1)°. Chlorine atoms from neighbouring square planar ions at distances of 2.882(3) Å from each Cu atom give rise to a distorted octahedral coordination about the metal, and a two-dimensional sheet-like structure of anions results. The tetrachlorocuprate(II) ions are tipped slightly so that all Cu and Cl(1) are not coplanar. The sheets of anions are held together by linearly extended cations with $-NH_3^+$ groups fitting into cavities lined by eight Cl atoms; N···Cl interatomic distances vary between 3.21 Å and 4.17 Å.

BIS(METHYLAMMONIUM) TETRABROMOFERRATE(III) BROMIDE

$C_2H_{12}Br_5FeN_2$ $(H_3CNH_3)_2[FeBr_4]Br$

G.D. SPROUL and G.D. STUCKY, 1972. Inorg. Chem., 11, 1647-1650.

Orthorhombic, $Pna2_1$, a = 14.989, b = 6.463, c = 28.130 Å, D_m = 2.72, Z = 8. Mo radiation, R = 0.115 for 1135 reflexions.

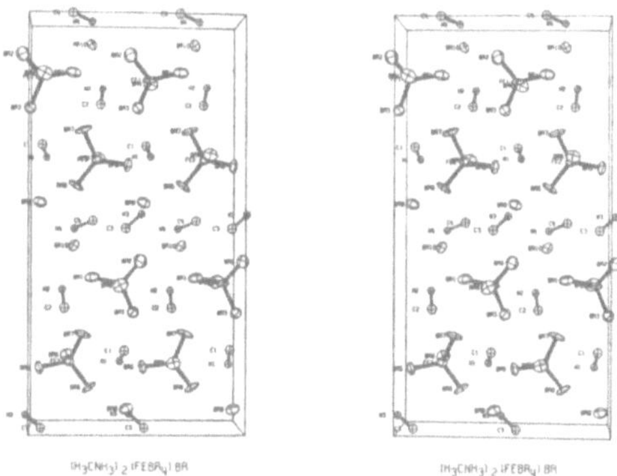

Fig. 1. View of packing in one unit cell of $(CH_3NH_3)_2[FeBr_4]Br$.

The tetrahedral tetrabromoferrate(III) ions occur as two independent forms, with an average Fe-Br interatomic distance of 2.32(10) Å. Methylammonium cations are apparently hydrogen bonded through the nitrogen-hydrogen atoms to both the ligand and nonligand bromide ions. Each unit cell has two independent sets of atoms that are related to one another by a pseudoinversion centre at 1/8, 1/4, ~0 (Fig. 1).

TETRAMETHYLAMMONIUM CADMIUM CHLORIDE

$C_4H_{12}CdCl_3N$ $(CH_3)_4NCdCl_3$

B. MOROSIN, 1972. Acta Cryst., B28, 2303-2305.

Hexagonal, $P6_3/m$, a = 9.138, c = 6.723 Å, Z = 2. Mo radiation, R = 0.031 for 390 reflexions.

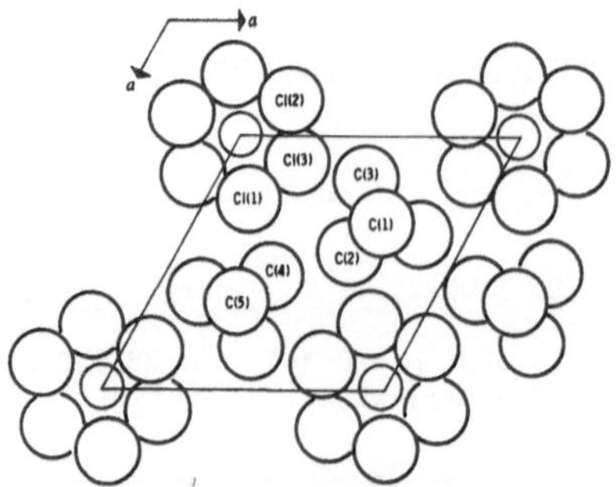

Fig. 1. A representation of the crystal structure of tetramethylammonium cadmium chloride viewed along the c axis. Chains formed by face-sharing of octahedrally coordinated cadmium-chloride ion, -Cd-$C\ell_3$-Cd-, are parallel to the 6_3 axis at the origin, $[N(CH_3)_4]^+$ ions are located between these chains in a disordered manner.

The structure consists of infinite linear chains formed from face-shared cadmium-chloride ion octahedra and of disordered $[N(CH_3)_4]^+$ ions as seen in Fig. 1. There is a Cd-Cℓ separation of 2.641 Å and a Cℓ-Cd-Cℓ angle of 96.05° along the chain. Light atom bond distances do not compare well with standard values due to the disorder in the structure.

TETRA(METHYLAMMONIUM) HEXACHLOROINDIUM CHLORIDE

$C_4H_{24}C\ell_7InN_4$ $[CH_3NH_3{}^+]_4\cdot[InC\ell_6]^{3-}\cdot C\ell^-$

H.-U. SCHLIMPER and M.L. ZIEGLER, 1972. Z. Naturf., 27B, 377-379.

Monoclinic, P2/c, a = 16.113, b = 7.466, c = 20.300 Å, β = 127.9°, Z = 4. Cu radiation, R = 0.076 for 834 reflexions.

The crystal structure (Fig. 1) contains discrete ions $(CH_3NH_3)^+$, $InC\ell_6{}^{3-}$, and $C\ell^-$. The indium atom is octahedrally coordinated with In-Cℓ in the range 2.500 - 2.569 Å. There are two independent In atoms in the crystal structure, each occupying a space group inversion centre. The chloride ions also lie on special positions on twofold axes.

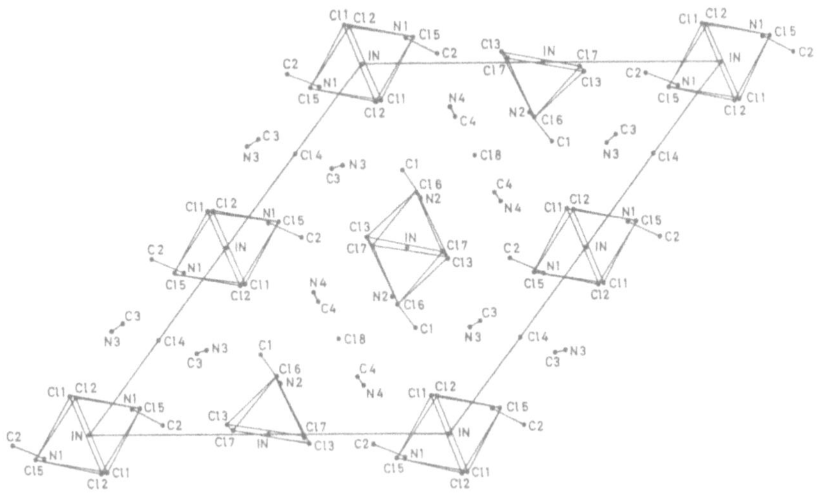

Fig. 1. The crystal structure of $[CH_3NH_3^+]_4 \cdot [InCl_6]^{3-} \cdot Cl^-$ projected onto the ac plane.

BIS(PROPYLAMMONIUM) MANGANESE TETRACHLORIDE

$C_6H_{20}Cl_4Mn$ $(C_3H_7NH_3)_2MnCl_4$

E.R. PETERSON and R.D. WILLETT, 1972. J. Chem. Phys., 56, 1879-1882.

Orthorhombic, Cmca, a = 7.29, b = 25.94, c = 7.51 Å, D_m = 1.50, Z = 4. Cu radiation, R = 0.116 for 421 reflexions.

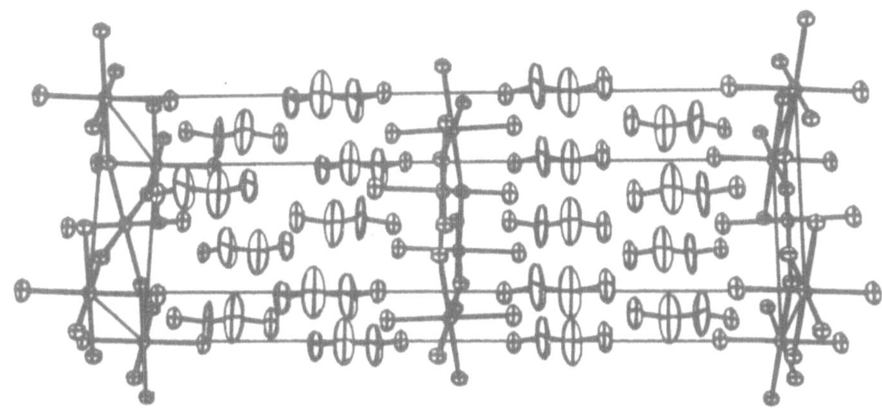

Fig. 1. The structure of $(C_3H_7NH_3)_2MnCl_4$.

The outstanding feature of the structure is the existence of a two-dimensional metal-halogen framework formed by $MnC\ell_6$ octahedra sharing corners. Each metal-halogen sheet is "sandwiched" between layers of propylammonium ions, thus effectively isolating it from adjacent sheets (Fig. 1). This provides a new class of compounds for studying two-dimensional magnetic interactions. The coordination sphere around the Mn(II) ion has approximately D_{4h} symmetry with two Mn-Cℓ bond lengths of 2.475 Å and four bond lengths of 2.629 Å. Existence of N-H···Cℓ hydrogen bonds cause the fourfold axis of the distorted octahedra to be tipped 8.0° away from the crystallographic b axis. This produces a "washboard" effect when each metal-halogen sheet is viewed from the [100] direction and leads to nonlinear Mn-Cℓ-Mn bridges.

BIS(TETRAMETHYLAMMONIUM) HEXACHLORO(DODECA-μ-CHLORO-HEXANIOBATE)

$C_8H_{24}C\ell_{18}N_2Nb_6$ $[(CH_3)_4N]_2[(Nb_6C\ell_{12})C\ell_6]$

F.W. KOKNAT and R.E. McCARLEY, 1972. Inorg. Chem., 11, 812-816.

Trigonal, P$\bar{3}$m1, a = 10.850, c = 8.615 Å, D_m = 2.52, Z = 1. Mo radiation, R = 0.043 for 512 reflexions.

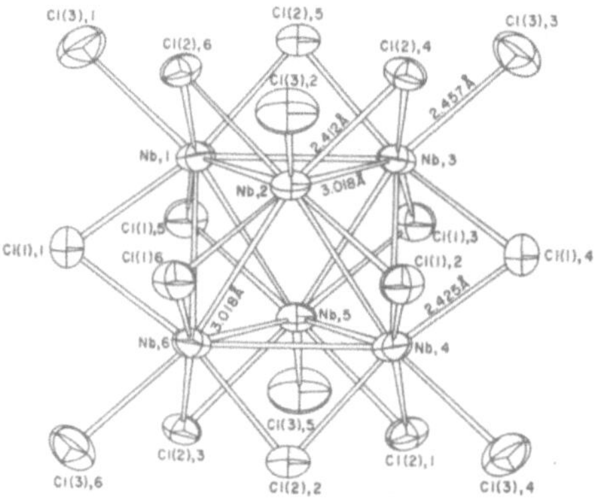

Fig. 1. Structure of the cluster anion showing bond lengths.

In this structure the $[(Nb_6C\ell_{12})C\ell_6]^{2-}$ cluster anions are centred in the eight corners of the trigonal unit cell. The $(CH_3)_4N^+$ groups are centred on threefold axes on top of or below three cluster units with each hydrogen atom H(1) of one methyl group reaching into the space between two cluster units.

Though only D_{3d} symmetry is required, the metal-metal bonded octahedron of Nb atoms is not noticeably distorted from O_h symmetry. Comparison of the anion dimensions (Fig. 1) with those reported for $K_4Nb_6C\ell_{18}$ (1) shows that the Nb-Nb distance becomes longer (by 0.10 Å) and the Nb-Cℓ(terminal) distance shorter (by 0.14 Å) upon removing two electrons from $Nb_6C\ell_{18}^{4-}$ to form $Nb_6C\ell_{18}^{2-}$. These results are taken as confirmation that the two electrons removed are from a bonding MO centred primarily on the metal atoms.

1. A. SIMON, H.G. SCHNERING and H. SCHÄFER, 1968. Z. anorg. Chem., 361, 235.

TETRAMETHYLENE BIS(TRIMETHYLAMMONIUM) DIBROMIDE DIHYDRATE

$C_{10}H_{30}Br_2N_2O_2$ $[(CH_3)_3N\text{-}(CH_2)_4\text{-}N(CH_3)_3]^{2+}\cdot 2Br^-\cdot 2H_2O$

Y. BARRANS, 1972. Acta Cryst., B28, 651-653.

Monoclinic, C2/m, a = 21.240, b = 7.125, c = 5.710 Å, β = 100°40', D_m = 1.460, Z = 2. Cu radiation, R = 0.11 for 657 reflexions.

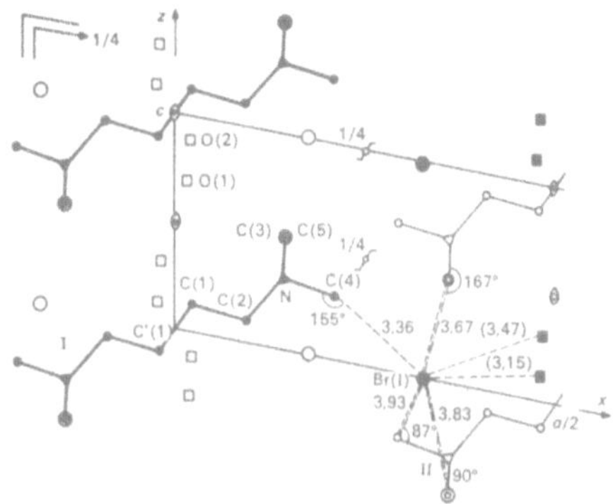

Fig. 1. The structure of $C_{10}H_{30}Br_2N_2O_2$ projected along b. Filled and open circles are bromide ions at y = 0 and y = ½ respectively. Filled and open squares denote the water molecules.

In the crystal each cation has 2/m symmetry and the water molecules are disordered. The structure is of the layer type (Fig. 1) with each bromide anion surrounded by methyl groups. Bond lengths are normal within the accuracy of the analysis.

DL-1-(3,5-DIHYDROXYPHENYL)-2-(ISOPROPYLAMINO)ETHANOL SULPHATE HEMIETHANOLATE

(ALUPENT, ORCIPRENALINE)

$C_{11}H_{17}NO_3\cdot\frac{1}{2}H_2SO_4\cdot\frac{1}{2}C_2H_5OH$

J.P. BEALE, 1972. Cryst. Struct. Comm., 1, 297-300.

Monoclinic, $P2_1/c$, a = 8.440, b = 31.51, c = 11.237 Å, β = 112.84°, Z = 8. Cu radiation, R = 0.067.

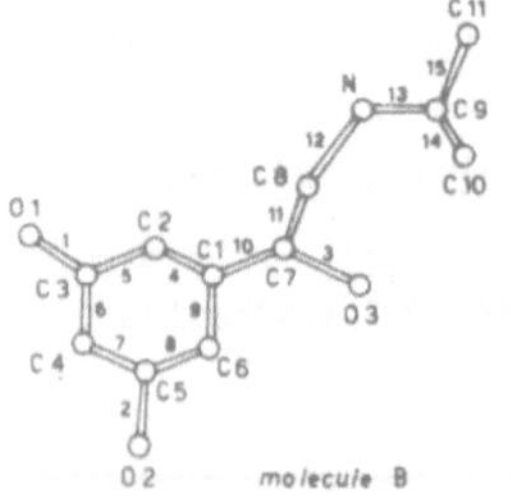

Fig. 1. Conformations of the two independent Alupent molecules.

This structure illustrates the Alupent molecules in a fully extended conformation with the O(3) and nitrogen atoms in a cis configuration (Fig. 1) while the C(7)-O(3) bond is directed out of the plane of the phenyl ring only in the case of the R-isomer. This structure is similar to the structure of isoprenaline (1) which also has two molecules in the asymmetric unit. The bond distances in this structure, except for the O(3)-C(7) distances, are in good agreement with previously reported values.

Complex hydrogen bonding is evident between the oxygens of the sulphate ion and (i) the nitrogen and hydroxyl groups of the Alupent molecules and (ii) the hydroxyl group of the solvated ethanol molecule. The net result is that all the oxygens of the sulphate ion act as receptors for hydrogen bond formation. Finally, molecule B exhibits disordering (about the carbon atom CB(7)) between the bonded oxygen atom OB(3) and the hydrogen atom HB(6).

1. M. MATHEW and G.J. PALENIK, 1971. J. Amer. Chem. Soc., 93, 497.

TRIETHYLAMMONIUM TETRACHLOROCUPRATE

$C_{12}H_{32}C\ell_4CuN_2$ $((C_2H_5)_3NH)_2\ CuC\ell_4$

J. LAMOTTE-BRASSEUR, O. DIDEBERG and L. DUPONT, 1972. Cryst. Struct. Comm., 1, 313-315.

Monoclinic, $P2_1/c$, a = 12.878, b = 13.079, c = 12.227 Å, β = 97.85°, Z = 4. Cu radiation, R = 0.064 for 1655 reflexions.

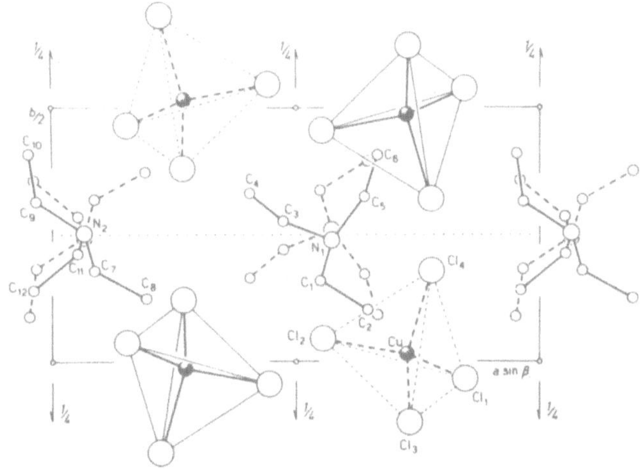

Fig. 1. Projection of the structure of $((C_2H_5)_3NH)_2CuCl_4$.

In this compound, the $CuCl_4^{2-}$ ion may be considered as a tetrahedron distorted by flattening, giving a configuration intermediate between square planar and tetrahedral. Fig. 1 shows that the triethylammonium ions are located in the holes formed by four $CuCl_4^{2-}$ tetrahedra. All interatomic distances and bond angles are within the limits given in the literature, with Cu-Cl = 2.237 - 2.247(3) Å, Cl(1)CuCl(2) = 136.8(1)°, Cl(3)CuCl(4) = 132.9(1)°.

TETRAMETHYLAMMONIUM NONABROMODIANTIMONATE(III)-DIBROMINE

$C_{12}H_{36}Br_{11}N_3Sb_2$ $[(CH_3)_4N]_3Sb_2Br_9 \cdot Br_2$

C.R. HUBBARD and R.A. JACOBSON, 1972. Inorg. Chem., 11, 2247-2250.

Hexagonal, $P6_3/mmc$, a = 9.585, c = 22.667 Å, D_m = 2.55, Z = 2. Mo radiation, R = 0.137 for 289 reflexions.

The structure contains $Sb_2Br_9^{2-}$ anions (two $SbBr_6$ octahedra sharing a face) with each Sb bridged by three Br atoms (Sb-Br(terminal) = 2.625(8), Sb-Br(bridge) = 3.043(8) Å). The anions are themselves linked via Br_2 molecules (with Br···Br_2, 2.89(1) Å, indicative of an important bridging interaction). Only one-third of the terminal bromines are bridged. The Br-Sb-Br···Br-Br···Br-Sb-Br chain is nearly linear. The tetramethylammonium cations fill large holes between anions. The cations and the Br_2 molecules are disordered.

PROCAINE HYDROCHLORIDE

(2-DIETHYLAMINOETHYL p-AMINOBENZOATE HYDROCHLORIDE)

$C_{13}H_{21}C\ell N_2O_2$ $NH_2C_6H_4CO \cdot OCH_2CH_2\overset{+}{N}H(C_2H_5)_2\ C\ell^-$

D.D. DEXTER, 1972. Acta Cryst., B28, 77-82.

Orthorhombic, Pcab, a = 25.017, b = 8.305, c = 14.192 Å, D_m = 1.232, Z = 8. Cu radiation, R = 0.072 for 2486 reflexions.

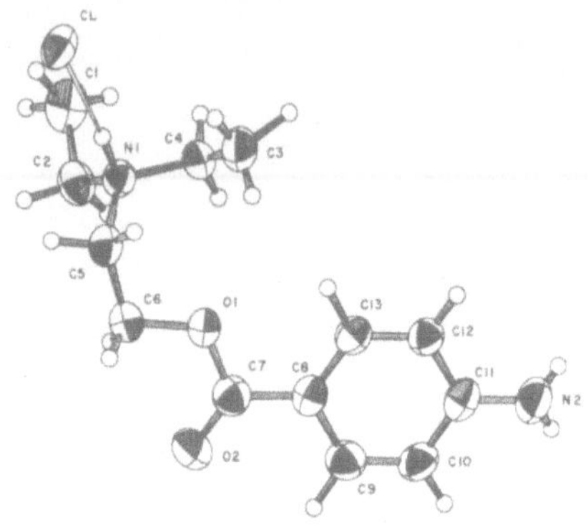

Fig. 1. Perspective drawing of procaine as the hydrochloride.

The molecular structure of procaine has been reported previously as its hydrochloride salt (1) and also as a 1:1 p-nitrophenyl phosphate complex (2). The results of this independent study of procaine hydrochloride are compared with the previous study (1); only five of the 117 parameters differ by more than 2σ. The only discrepancies are associated with the ends of the molecule; three with hydrogen atoms in the ethyl groups, one with a hydrogen atom in the p-amino group, and one with a terminal carbon atom in an ethyl group. An examination of the procaine conformation in the hydrochloride salt and in the complex (2) shows that major conformational features are preserved in the different crystal environments (Fig. 1).

1. E. BEALL, J. HERDKLOTZ and R.L. SASS, 1970. Biochem. Biophys. Res. Comm., 39, 329.
2. M. SAX, J. PLETCHER and R. GUSTAFFSON, 1970. Acta Cryst., B26, 114.

DL-N-t-BUTYL-2(4-HYDROXY-3-HYDROXYMETHYLPHENYL)-2-HYDROXYETHYLAMINE (SALBUTAMOL)

$C_{13}H_{21}NO_3$

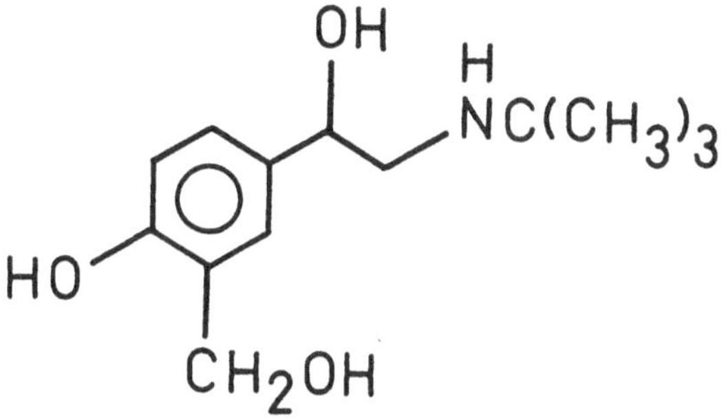

J.P. BEALE and C.T. GRAINGER, 1972. Cryst. Struct. Comm., 1, 71-74.

Orthorhombic, Pbca, a = 21.654, b = 8.798, c = 14.565 Å, D_m = 1.15, Z = 8. Cu radiation, R = 0.048 for 2644 reflexions.

Salbutamol exhibits the same *cis* orientation of amino and hydroxyl groups as observed with another related hydroxylethylamine (designated Th1165) (1). The N···O distance (2.826(2) Å) is in fact a little larger than the corresponding distance (2.766(3) Å) in Th1165. The benzene ring is inclined at 74.5(2)° to the plane of the C(7)-C(8)-N-C(9) atoms, whilst in the Th1165 molecule this angle is 66.9(3)°. The tertiary butyl group is on the *opposite* side of the Salbutamol molecule to the amino and hydroxyl groups. The Th1165 molecule has the substituent groups on the nitrogen atom on the *same* side of the molecule as the amino and hydroxyl groups and therefore in closer proximity to the receptor site.

1. Following report.

DL-N-[1-(4-HYDROXYPHENYL)-PROP-2-YL]-2-(3,5-DIHYDROXYPHENYL)-2-HYDROXYETHYLAMINE HYDROBROMIDE

$C_{17}H_{22}BrNO_4$

J.P. BEALE, 1972. Cryst. Struct. Comm., 1, 67-70.

Monoclinic, $P2_1/c$, a = 8.713, b = 17.347, c = 13.978 Å, β = 123.50°, D_m = 1.48, Z = 4. Cu radiation, R = 0.050 for an unspecified number of three-dimensional data.

The molecule (designated Th1165) and its geometry are depicted in Fig. 1. Each molecule has two asymmetric C atoms and hence four stereoisomers are possible. One racemic pair (Th1165) is 9 to 20 times more biologically active than the other racemate (designated Th1179) (1).

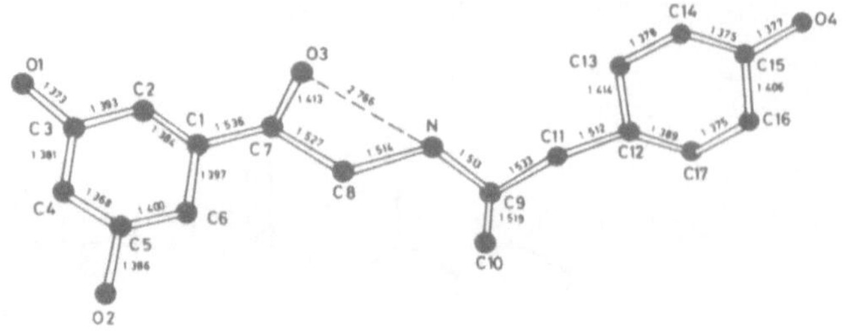

Fig. 1. Bond distances in Th1165.

1. Following report.

DL-N-[1-(4-HYDROXYPHENYL)-PROP-2-YL]-2-(3,5-DIHYDROXYPHENYL)-2-HYDROXYETHYLAMINE HYDROBROMIDE

$C_{17}H_{22}BrNO_4$

J.P. BEALE, 1972. Cryst. Struct. Comm., 1, 223-226.

Monoclinic, $P2_1/c$, a = 14.308, b = 10.425, c = 12.559 Å, β = 111.37°, D_m = 1.47, Z = 4. Cu radiation, R = 0.063 for 3323 reflexions.

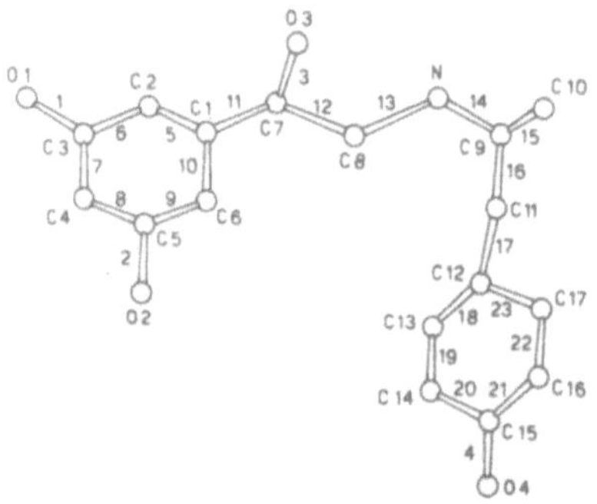

Fig. 1. Conformation of Th1179.

In the crystalline state this molecule (code named Th1179) has a conformation in which the molecular chain and phenyl rings are not fully extended and, in particular, the phenyl rings at either end of the molecule are perpendicular to each other (Fig. 1). By contrast, the diastereoisomer of Th1179 is fully extended possessing a twist type structure (1).

1. Preceding report.

BIS(TETRAETHYLAMMONIUM) OXOPENTACHLOROPROTACTINATE(V)

$C_{16}H_{40}Cl_5N_2OPa$ $[(C_2H_5)_4N]_2PaOCl_5$

D. BROWN, C.T. REYNOLDS and P.T. MOSELEY, 1972. J. Chem. Soc., Dalton, 857-859.

Monoclinic, C2/c, a = 14.131, b = 14.218, c = 13.235 Å, β = 91.04°, Z = 4. Mo radiation, R = 0.067 for 558 reflexions.

The Pa atom is in a distorted octahedral configuration with Pa=O = 1.74(9) Å; the Pa-Cl distances are in the range 2.59 - 2.72(3) Å except for the Pa-Cl bond *trans* to the oxygen (2.42(3) Å).

2-DIETHYLAMINOETHYL-1-PHENYLCYCLOPENTANECARBOXYLATE HYDROCHLORIDE (PARPANIT)

$C_{18}H_{28}ClNO_2$

E.A.H. GRIFFITH and B.E. ROBERTSON, 1972. Acta Cryst., B28, 3377-3384.

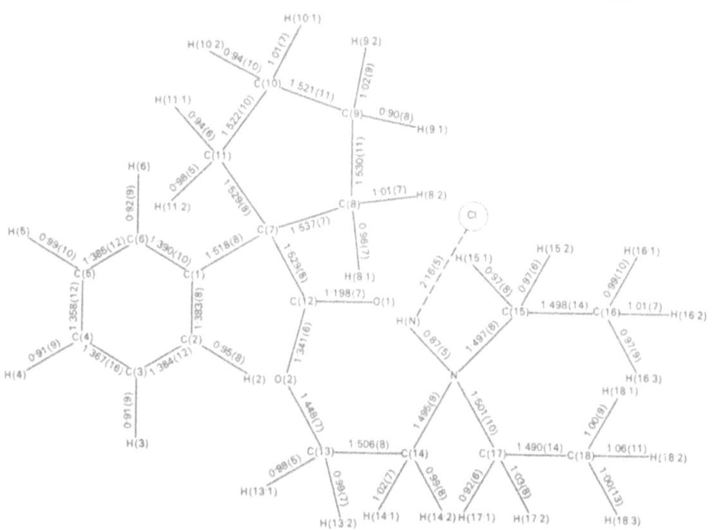

Fig. 1. Bond lengths for the parpanit molecule.

Monoclinic, $P2_1/c$, a = 6.557, b = 23.068, c = 12.512 Å, β = 111.58°, D_m = 1.22, Z = 4. Mo radiation, R = 0.044 for 2711 reflexions.

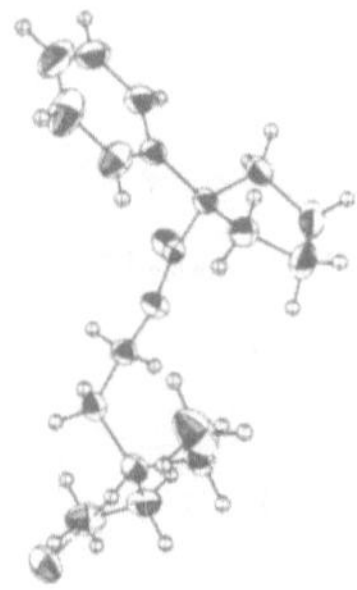

Fig. 2. View of the parpanit molecule.

The molecular structure is shown schematically with bond lengths in Fig. 1 and in a general view in Fig. 2. The $O-C-C-N^+$ moiety has torsion angle ∓85.7° which is the syn-clinal conformation. The torsion angle (±170.16°) about the ester bond C(O)-O-C-C is anti-periplanar as are most mimics of acetylcholine. The ammonium proton is strongly hydrogen bonded to the anion, N-H···Cℓ = 3.023(4) Å, rather than the acyl oxygen atom, as has been inferred to be the case in solution studies. The molecule is involved in no unusually short contacts other than a hydrogen bond to the anion, and the conformation should be reasonably unaffected by the molecular packing. Bond lengths (Fig. 1) appear reasonable.

BIS(NNN'N'-TETRAETHYLETHYLENEDIAMMONIUM) HEXA-μ-CHLORO-μ4-OXO-TETRA[CHLOROCUPRATE(II)]

$C_{20}H_{52}Cl_{10}Cu_4N_4O$ $[(C_2H_5)_2NH(CH_2)_2NH(C_2H_5)_2{}^{2+}]_2[Cu_4OCl_{10}{}^{4-}]$

R. BELFORD, D.E. FENTON and M.R. TRUTER, 1972. J. Chem. Soc., Dalton, 2345-2350.

Monoclinic, $C2/c$, a = 14.46, b = 15.23, c = 17.59 Å, β = 104.55°, D_m = 1.75, Z = 4. Mo radiation, R = 0.063 for 1478 reflexions.

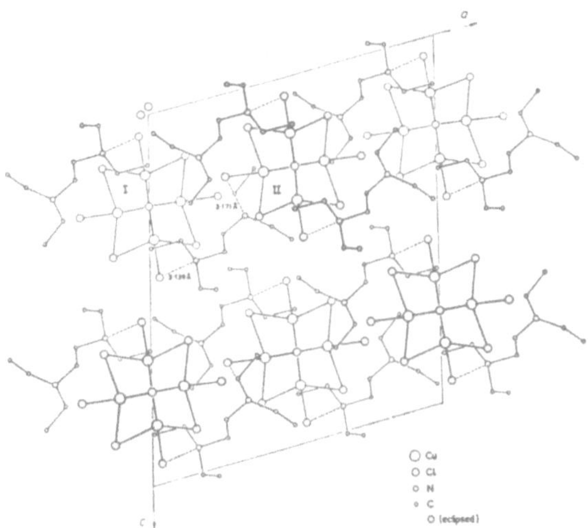

Fig. 1. The crystal structure of $[(C_2H_5)_2NH(CH_2)_2NH(C_2H_5)_2^{2+}]_2[Cu_4OC\ell_{10}^{4-}]$ viewed along b. The broken lines indicate NH···Cℓ hydrogen bonds.

In the crystal structure (Fig. 1) the anion is tetrahedral, with two bridging chlorines and the oxygen lying on the crystallographic twofold axis. The copper atoms have trigonal bipyramidal environments and are surrounded by four chlorines and one oxygen at mean distances Cu-Cℓ(bridge) = 2.401, Cu-Cℓ (terminal) = 2.242, and Cu-O = 1.916 Å. The quaternary nitrogen atoms of the cation, which has a trans conformation about the central C-C bond, are hydrogen-bonded to terminal chlorine atoms, N···Cℓ = 3.139 and 3.171 Å. Intermolecular contacts appear to be normal.

TRIBENZYLAMINE (at -70°C)

$C_{21}H_{21}N$ $(C_6H_5CH_2)_3N$

F. IWASAKI and H. IWASAKI, 1972. Acta Cryst., B28, 3370-3376.

Monoclinic, $P2_1/c$, a = 21.076, b = 9.015, c = 8.917 Å, β = 93.9°, Z = 4. Mo radiation, R = 0.092 for 821 reflexions.

Three benzene rings surround a nitrogen atom so that the molecule has a propeller shape (Fig. 1); three hydrogen atoms H(1), H(8), and H(15) are axially related to the plane formed by the three methylene atoms while H(2), H(9), and H(16) are bonded equatorially. The axial C-H bond is gauche to one of the two N-C bonds and trans to the other, whereas the equatorial C-H bond is gauche to both the N-C bonds. The mean value of the C-C bond distances in the benzene rings is 1.37 Å and the $C(sp^2)-C(sp^3)$ bond lengths average 1.53 Å. The mean C-N distance is 1.47 Å and the C-N-C angles are 108.5 - 111.5(9)°. The molecular conformation is determined by steric hindrance. The intermolecular contacts correspond to normal van der Waals distances.

Fig. 1. Projection of the tribenzylamine molecule parallel to the plane formed by the three methylene carbon atoms C(1), C(8), and C(15).

TRIMETHYLPHENYLAMMONIUM NONACHLORODIRHODATE(III)

$C_{27}H_{42}C\ell_9N_3Rh_2$ $[C_6H_5N(CH_3)_3]_3Rh_2C\ell_9$

F.A. COTTON and D.A. UCKO, 1972. Inorg. Chim. Acta, 6, 161-172.

Orthorhombic, Pnma, a = 32.487, b = 9.274, c = 36.451 Å, Z = 12. Co radiation, R = 0.095 for 1225 reflexions.

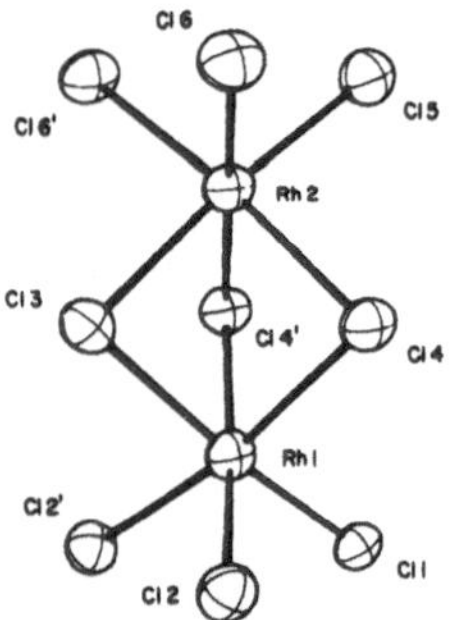

Fig. 1. A projection of the $Rh_2C\ell_9{}^{3-}$ ion.

The $Rh_2C\ell_9{}^{3-}$ ion (Fig. 1) has virtual D_{3h} symmetry and shows symmetry-preserving distortions from the ideal bioctahedron structure which are indicative of Rh-Rh repulsion. The Rh-Rh distance is 3.121(5) Å and the Rh-Cℓ-Rh angles have an average value of 81.3(3)°. When the $Rh_2C\ell_9{}^{3-}$ structure as well as those of 13 other M_2X_9 bioctahedra are analyzed in terms of various structure parameters which reflect the magnitude and sense (i.e., attractive or repulsive) of the M-M interactions, systematic correlations with electronic structures and position in the periodic table are found. Attempts to determine the positions of the atoms of the cations were unsuccessful because of pseudosymmetry-disorder phenomena.

ALKALI METHANE THIOLATES

E. WEISS and U. JOERGENS, 1972. Chem. Ber., 105, 481-486.

Tetragonal, P4/nmm, Z = 2.

	CH_3SK	CH_3SLi	CH_3SNa
a(Å)	4.53	4.36	5.27
c	9.62	8.70	5.67
R	0.15	0.11	0.12

Some 80 or so reflexions for each compound from powder data (Cu radiation).

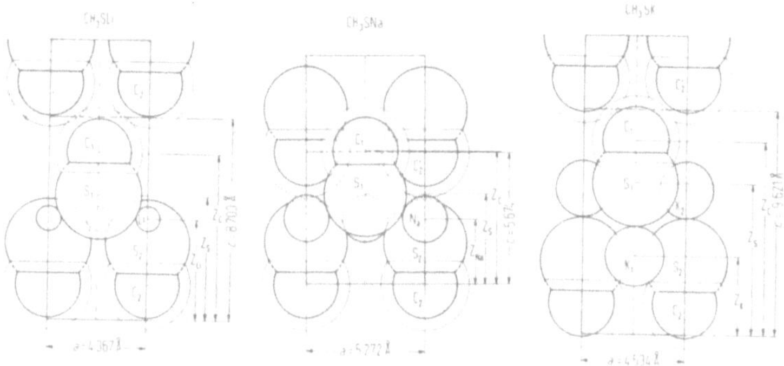

Fig. 1. Projections of the structures of the alkali methane thiolates CH_3SM (M = Li, Na, K).

CH_3SLi and CH_3SNa are of the same type as CH_3OLi and CH_3ONa, whereas CH_3SK is of the CH_3OK-type. Both types represent layer structures with CH_3S groups having an antiparallel arrangement to each other along the c axis. In this way each layer is covered by methyl groups on both sides. The alkali ions are located in the central part of the layer, being coordinated by four (M = Li, Na) or five (M = K) sulphur ions (Fig. 1).

PERDEUTEROMETHANE (NEUTRON STUDY)

CD_4

W. PRESS, 1972. J. Chem. Phys., 56, 2597-2609.

Phase I at 77°K

Cubic, Fm3m, a = 5.96, Z = 4.

Phase II at 24.5°K

Cubic, Fm3c, a = 11.64, Z = 32.

The molecular orientations are completely disordered at 35°K, while an indication of partial order was detected at 77°K. The transition to phase II is accompanied by critical fluctuations with a correlation length of about 24 Å (at 27.7°K) and appears to be of second order (critical exponent β = 0.4 ± 0.1). In phase II six of eight molecules order with a local symmetry $\bar{4}$2m, while the remaining two are orientationally disordered.

The transition to phase III is predominantly of first order. With the data the structure of phase III can not be determined unambiguously. Superlattice reflections can be indexed cubic primitive. The number of molecules per unit cell remains 32, with a = 11.61 Å at 17.5°K. The data suggest that the low temperature structure is to be understood by an ordering within the sublattices of molecules disordered in phase II together with slight distortions of the other sublattices. For phase II C-D distances = 1.033(15), 1.017(12), and 1.079(11) Å are reported as resulting from different treatments of powder and single crystal data.

α-TETRAFLUOROMETHANE (BETWEEN 10 K and 70 K)

CF_4

D.N. BOL'SHUTKIN, V.M. GASAN, A.I. PROKHVATILOV and A.I. ERENBURG, 1972. Acta Cryst., B28, 3542-3547.

Monoclinic, $P2_1/c$, a = 8.435, b = 4.320, c = 8.369 Å, β = 119.40°, Z = 4. Cu radiation, R = 0.15 for 32 reflexions.

Based on a model for CF_4 with $\bar{4}$3m molecular symmetry and with the $\bar{4}$ axis parallel to b, a crystal packing arrangement was derived which gave F···F intermolecular contacts in the range 3.03 - 3.18 Å at 10°K.

BROMOFORM

$CHBr_3$

T. KAWAGUCHI, K. TAKASHINA, T. TANAKA and T. WATANABE, 1972. Acta Cryst., B28, 967-972.

Hexagonal, $P6_3$, a = 6.335, c = 7.214 Å, D_m = 2.8095, Z = 2. Cu radiation, R = 0.147 for 160 reflexions.

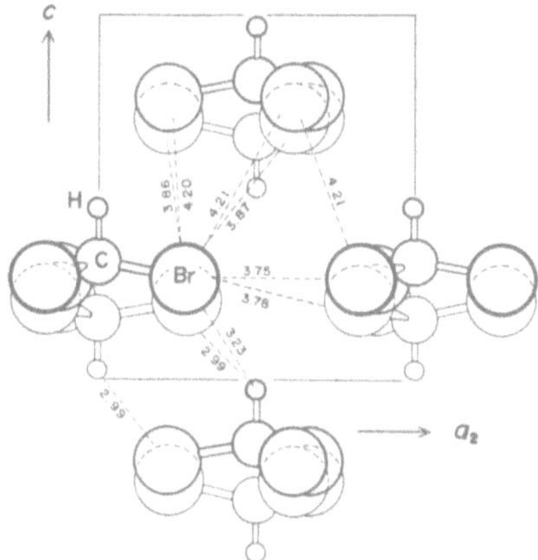

Fig. 1. The crystal structure of bromoform in the (010) plane. The C-Br distance is 1.86 Å and Br-C-Br is 116.5°. Some intermolecular contacts are shown.

The structure (Fig. 1) is statistically disordered with molecules oriented upwards and downwards at random. The order parameter w is almost 0.5 and hence the symmetry of the crystal can be represented approximately by $P6_3/m$. The structure is isomorphous with that of iodoform.

HEXABROMOETHANE

C_2Br_6

G. MANDEL and J. DONOHUE, 1972. Acta Cryst., B28, 1313-1316.

Orthorhombic, Pnma, a = 12.043, b = 10.674, c = 6.705 Å, D_m = 3.823, Z = 4. Cu radiation, R = 0.064 for 607 reflexions.

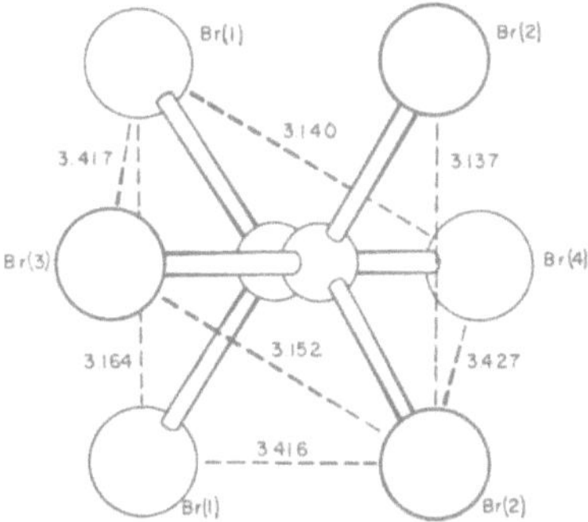

Fig. 1. Intramolecular distances in hexabromoethane.

The molecule has crystallographic m symmetry. The C-C bond length is 1.526(24) Å while the C-Br bond distances range from 1.939 to 1.950 Å, mean 1.944(5) Å. The average bond angle is equal to the tetrahedral value at 109.6(9)°. The intramolecular distances are shown in Fig. 1. The shortest intermolecular Br···Br distance is 3.763 Å, a value 0.14 Å shorter than the sum of the van der Waals radii. For details of an earlier determination see (1).

1. Structure Reports, 9, 316.

MONOSODIUM ACETYLIDE

(NEUTRON STUDY at 293 and 5°K)

C_2HNa $HC{\equiv}C^-Na^+$

M. ATOJI, 1972. J. Chem. Phys., 56, 4947-4951.

At 293°K

Tetragonal, P4/nmm, a = 3.813, c = 8.104 Å, Z = 2. Neutron radiation, R(I) = 0.025 for 17 data.

At 5°K the same space group is found with a = 3.764 and c = 8.062 Å.

The interatomic configuration and the thermal libration modes support the conception that the formal negative charge is virtually located at one carbon, thus $HC{\equiv}C^-Na^+$. The $C{\equiv}C^-$ distance corrected for a large precession-like motion is 1.27(2) Å which is considerably longer than 1.204 Å in acetylene. The $Na^+\cdots C^-$ distance is 2.62(1) Å at 293°K and 2.60(1) and 5°K. For details of an X-ray study see (1).

1. Structure Reports, 33B, 296.

DIIODOACETYLENE

C_2I_2

J.D. DUNITZ, H. GEHRER and D. BRITTON, 1972. Acta Cryst., B28, 1989-1994.

Tetragonal, $P4_2/n$, a = 15.68, c = 4.30 Å, D_m = 3.43, Z = 8. Mo radiation, R = 0.145 for 764 reflexions.

The crystal packing of the linear molecule, C_2I_2, is shown in Fig. 1. The intramolecular I···I distance of 5.16 Å is comparable with distances found in donor-acceptor complexes such as diselenane-diiodoacetylene, 5.167(6) Å (1). The molecules are in general positions, but each molecule points, at both ends, towards the midpoints of two adjacent molecules to give four C···I distances of about 3.4 Å which has been interpreted as donor(C≡C)-acceptor(iodine) interactions.

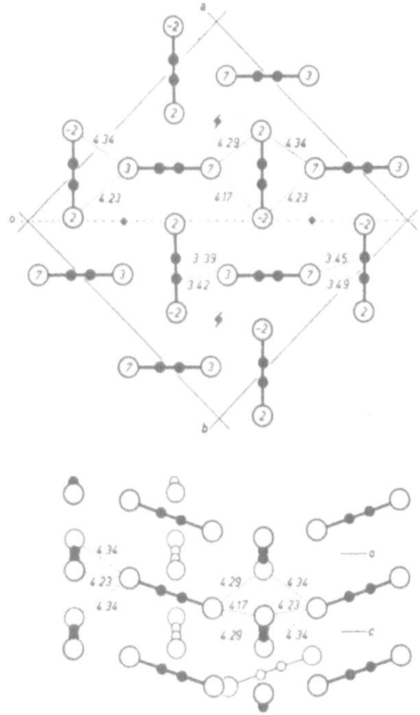

Fig. 1. The crystal structure of diiodoacetylene.

1. O. HOLMESLAND and C. ROMMING, 1966. Acta Chem. Scand., 20, 2601.

1,1,1-TRICHLOROETHANE

(METHYLCHLOROFORM)

$C_2H_3C\ell_3$ $CH_3CC\ell_3$

L. SILVER and R. RUDMAN, 1972. J. Chem. Phys., 57, 210-216.

At -60°C

Orthorhombic, Pnma, a = 11.53, b = 8.000, c = 5.88 Å, Z = 4. Mo radiation, R = 0.053 for 149 reflexions.

At -145°C, a = 11.46, b = 7.890, c = 5.81 Å, R = 0.041 for 145 reflexions.

The results of the structure analyses at the two temperatures rule out any crystallographic phase change at -68°C. At -60° C-Cℓ distances are 1.806(12), 1.806(12), and 1.784(22) Å with C-C = 1.515(28) Å. At -145°C the corresponding values are 1.781(11), 1.781(11), 1.783(22), and 1.548(22) Å. The molecules adopt a staggered conformation. The average angle involving the central carbon atom is 109.5° at -60° and 109.7° at -145°. However, in both cases the average value of the C-C-Cℓ angles (110.2°) is slightly larger than the average value of the Cℓ-C-Cℓ angles (108.8°).

1,2,3,4-TETRACHLOROBUTADIENE

$C_4H_2Cl_4$

Y. OTAKA, 1972. Acta Cryst., B28, 342-345.

Monoclinic, $P2_1/n$, a = 3.872, b = 9.087, c = 10.061 Å, β = 101.22°, D_m = 1.817, Z = 2. Cu radiation, R = 0.11 for 355 reflexions.

The molecule lies on a crystallographic inversion centre, is planar, and the carbon skeleton has a trans-configuration. Bond lengths are C-C = 1.49(1), C=C = 1.34(2), C-Cl = 1.71(1) Å. There are no unusual intermolecular distances.

2,5-DIMETHYL-2,5-HEXANEDIOL TETRAHYDRATE

$C_8H_{26}O_6$ $C_8H_{18}O_2 \cdot 4H_2O$

G.A. JEFFREY and M.S. SHEN, 1972. J. Chem. Phys., 57, 56-61.

Monoclinic, $P2_1/c$, a = 6.158, b = 6.169, c = 17.854 Å, β = 101.8°, D_m = 1.06, Z = 2. Mo radiation, R = 0.057 for 1015 reflexions.

The crystal structure (Fig. 1) consists of layers of hydrogen bonded hydroxyl and water molecules alternating with layers of hydrocarbon chains. The water hydroxyl layers consist of puckered sheets of edge-sharing pentagons analogous to the water structure found in the pinacol and piperazine hexahydrates (1). The O-H···O distances are in the range 2.763 - 2.844 Å. The hydrocarbon chains of the glycol molecules are aligned parallel, analogous to the long-chain hydrocarbons and fatty acids, but spaced further apart. There are no hydrogen bonds between the puckered sheets of pentagons, and therefore the three-dimensional hydrogen bond framework which characterizes the hexahydrates as "semiclathrates" does not occur in this tetrahydrate.

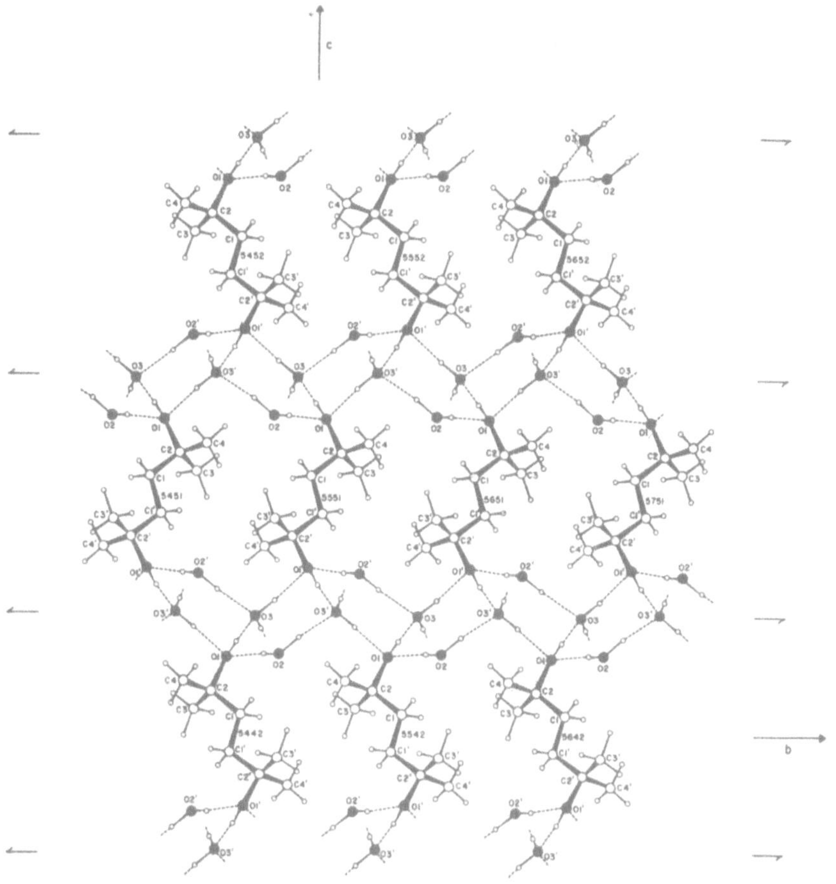

Fig. 1. Crystal structure of 2,5-dimethyl-2,5-hexanediol tetrahydrate; view down a axis. Dotted lines are hydrogen bonds; full lines are covalent bonds; small open circles are hydrogens.

1. Structure Reports, 33B, 113; 35B, 29.

11-BROMOUNDECANOL

$C_{11}H_{23}BrO$ $Br(CH_2)_{11}OH$

L. ROSEN and A. HYBL, 1972. Acta Cryst., B28, 610-617.

Monoclinic, $P2_1$, a = 47.10, b = 5.26, c = 31.14 Å, β = 132.9°, D_m = 1.327, Z = 18. Cu radiation, R = 0.142 for 1353 reflexions.

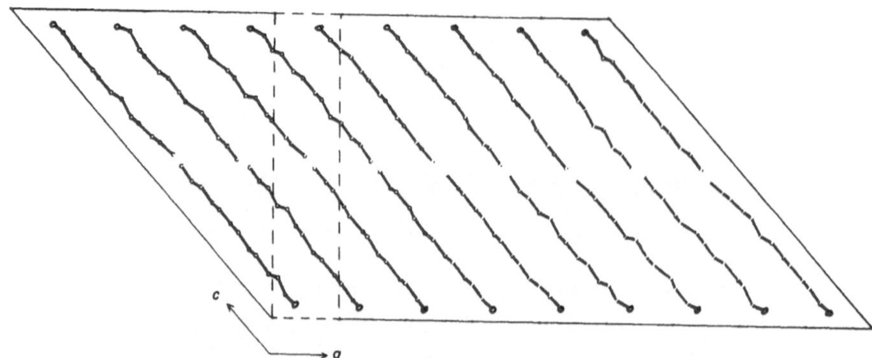

Fig. 1. A drawing of the crystal structure of 11-bromoundecanol with the true cell (heavy outline) and subcell (dashed outline).

There is a prominent subcell with Z = 2 and dimensions a = 5.23, b = 5.26, c = 22.81 Å, β = 90.6°. Both subcell and true cell as well as the crystal structure are shown in Fig. 1. The true unit cell contains a basic pattern of a pair of hydrogen bonded molecules (O-H···O = 2.85 Å) whose coordinates are nearly the same as its neighbours, appropriately translated, but systematically different enough so that nine such pairs must occur before the arrangement truly repeats itself. Molecules pack in infinite sheets with an alteration of molecular tilt approximately 27° to the <u>ac</u> plane occurring from chain to chain.

DDT DERIVATIVES

T.P. DeLACY and C.H.L. KENNARD, 1972. J. Chem. Soc., Perkin II, 2148-2153.

1,1-BIS-(p-CHLOROPHENYL)-2,2,2-TRICHLOROETHANE (p,p'-DDT)

$C_{14}H_9C\ell_5$

Orthorhombic, $Pca2_1$, a = 9.963, b = 19.200, c = 7.887 Å, D_m = 1.55, Z = 4. Cu radiation, R = 0.091 for 900 reflexions.

1-(o-CHLOROPHENYL)-1-(p-CHLOROPHENYL)-2,2,2-TRICHLOROETHANE (o,p'-DDT)

$C_{14}H_9C\ell_5$

Triclinic, $P\bar{1}$, a = 12.016, b = 12.481, c = 10.938 Å, α = 105.43, β = 89.93, γ = 109.03°, D_m = 1.61, Z = 4. Mo radiation, R = 0.048 for 2649 reflexions.

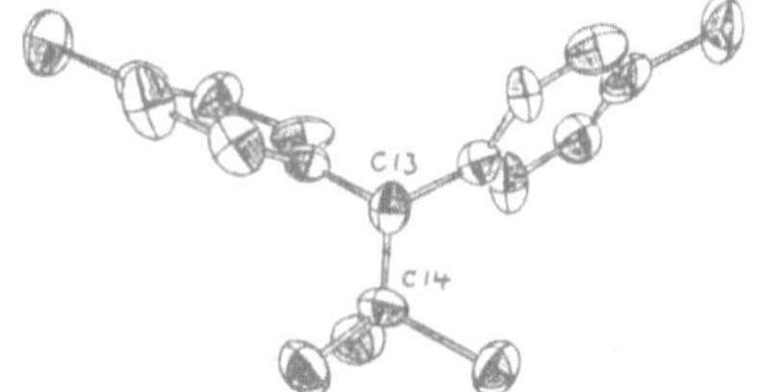

Fig. 1. The structure of p,p'-DDT viewed perpendicular to the C(13)-C(14) bond.

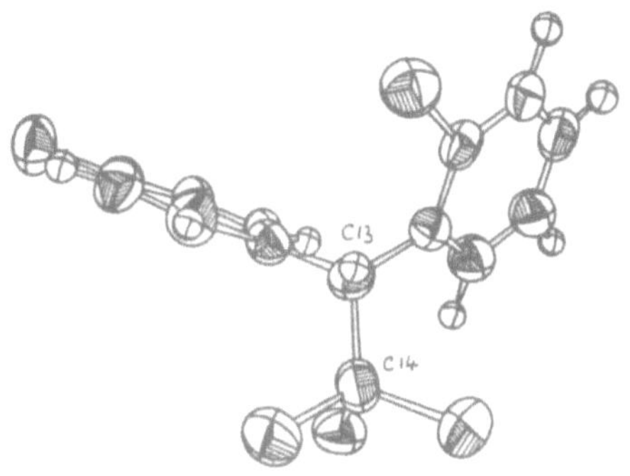

Fig. 2. The structure of o,p'-DDT viewed perpendicular to the C(13)-C(14) bond.

In both molecules the central carbon C(13) joins the chlorophenyl groups to the trihalogenomethyl group. If an ideal 'butterfly' configuration of the molecule is one where the projection in the plane perpendicular to C(13)-C(14) results in the phenyl rings lying 'extended' in the plane, then in the p,p'-DDT molecule (Fig. 1) the phenyl rings have torsion rotations of 4 [anticlockwise with respect to the C(13)-C(1) direction] and 33° [anticlockwise with respect to the C(13)-C(7)]. If viewed down C(13)-C(14), the trichloro-group is rotated slightly away from the staggered configuration with respect to the chlorophenyl groups. Bond lengths and angles in the p,p'-DDT molecule are normal. The o,p'-isomer (Fig. 2) contains two molecules in the asymmetric unit with similar conformations. The butterfly configuration is distorted with phenyl rings having torsional rotations of 4 [clockwise with respect to the C(13)-C(1) direction] and 47° [anticlockwise with respect to the C(13)-C(7)]. The phenyl Cℓ-C bond distances (1.743 and 1.741(5) Å) and paraffinic Cℓ-C bonds are longer than usual with Cℓ(5)-C(14) (1.787 Å) being the longest. Van der Waals interactions appear to govern the packing in both crystal structures.

n-OCTADECANE

$C_{18}H_{38}$

S.C. NYBURG and H. LÜTH, 1972. Acta Cryst., B28, 2992-2995.

Triclinic, $P\bar{1}$, a = 4.285, b = 4.820, c = 24.898 Å, α = 85.15 , β = 67.8 , γ = **72.7°**, Z = 1. R = 0.08 for 291 reflexions.

The previous report (1) of this structure is incorrect because of a mis-indexing error. A new analysis with the earlier data corrected yielded the correct structure. The C-C bond lengths are in the range 1.49 - 1.52(5) Å and the C-C-C angles are 114.2 - 115.6(20)°. The shortest non-bonded intermolecular contacts are of the H···H variety with the shortest distance 2.57 Å.

1. T. HAYASHIDA, 1962. J. Phys. Soc. Japan, 17, 306; Structure Reports, 27, 743.

1,1-BIS-(p-ETHOXYPHENYL)-2,2-DIMETHYLPROPANE

$C_{21}H_{28}O_2$

T.P. DeLACY and C.H.L. KENNARD, 1972. J. Chem. Soc., Perkin II, 2141-2147.

Orthorhombic, $Pca2_1$, a = 22.912, b = 10.424, c = 7.869 Å, D_m = 1.11, Z = 4. Cu radiation, R = 0.035 for 947 reflexions.

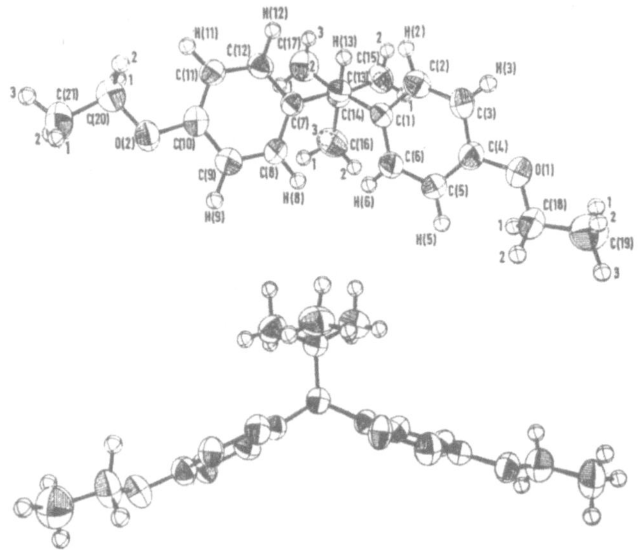

Fig. 1. Two views of the structure of 1,1-bis-(p-ethoxyphenyl)-2,2-dimethylpropane.

The structure of this insecticide is shown in Fig. 1. The central C(13) atom joins the two p-ethoxyphenyls to the neopentane with the two phenyl planes taking up a 'butterfly' configuration. Excluding the ethoxy-groups, the molecule contains an approximate mirror plane, through H(13), C(13), C(14), C(16), and H(163). The ethoxy-groups take up a non-symmetrical 'trans'-configuration and lie almost in the plane of the attached benzene rings. Bond distances for the most part are in the expected range but the C(13)-C(14) bond 1.573(5) is longer than the other C-C bonds in the neopentane grouping (1.524 - 1.534 Å). The packing of the molecules is governed by van der Waals forces.

BIS(p-ACETOXYPHENYL)CYCLOHEXYLIDENEMETHANE

$C_{23}H_{24}O_4$ $C_6H_{10}{=}C(C_6H_4{\cdot}OCOCH_3)_2$

G. PRECIGOUX, B. BUSETTA, C. COURSEILLE and M. HOSPITAL, 1972. Cryst. Struct. Comm., 1, 341-344.

Triclinic, P$\bar{1}$, a = 11.626, b = 11.928, c = 10.899 Å, α = 72.27, β = 130.00, γ = 119.12°, Z = 2. R = 0.073 for 3652 reflexions.

In this molecule one aromatic ring plane is inclined at 83.2° to the other aromatic ring. The ethylenic bond length is 1.338 Å and other bond distances are also normal.

1,1,2,2-TETRAKIS(2-METHOXYPHENYL)ETHANE

$C_{30}H_{30}O_4$ $(CH_3OC_6H_4)_2CHCH(C_6H_4OCH_3)_2$

J.J. DALY, F. SANZ, R.P.A. SNEEDEN and H.H. ZEISS, 1972. J. Chem. Soc., Perkin II, 1614-1618.

Monoclinic, P2_1/c, a = 7.721, b = 8.929, c = 18.025 Å, β = 97.6°, D_m = 1.233, Z = 2. Mo radiation, R = 0.057 for 1326 reflexions.

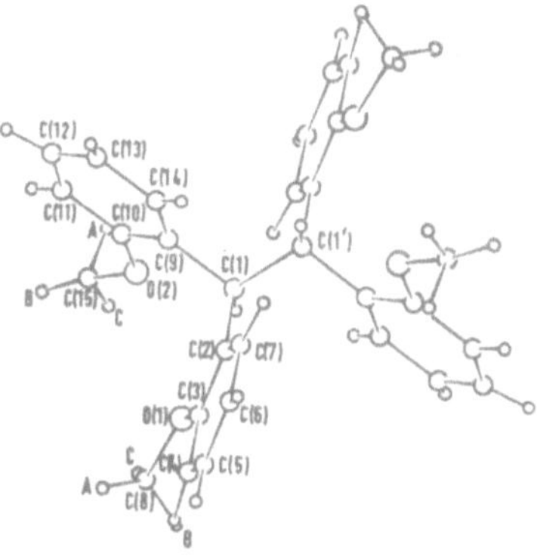

Fig. 1. The conformation of 1,1,2,2-tetrakis(2-methoxyphenyl)ethane.

The molecule centre occupies an inversion centre in the crystal structure and hence a fully staggered trans structure is found (Fig. 1). The central C-C bond length is 1.555(7) Å and a trigonal distortion at the sp^3 carbon atom gives large C-C-C angles near 111.5°. The carbon atoms of the methoxy-groups lie 0.24 and 0.16 Å from their corresponding benzenoid rings, and one of these rings is nearly parallel to the C(sp^3)-H bond. Rotation of the methoxyphenyl groups by 7.5° about the C(sp^2)-C(sp^3) bonds would give the molecule a close approximation to the symmetry 2/m.

PENTAERYTHRITOL TETRACINNAMATE

$C_{41}H_{36}O_8$ $(C_6H_5CH{=}CHCOOCH_2)_4C$

G. TIEGHI and M. ZOCCHI, 1972. Cryst. Struct. Comm., 1, 167-171.

Triclinic, $P\bar{1}$, a = 21.15, b = 16.90, c = 5.75 Å, α = 102°55', β = 89°19', γ = 99°51', Z = 2. Mo radiation, R = 0.122 for 2581 reflexions.

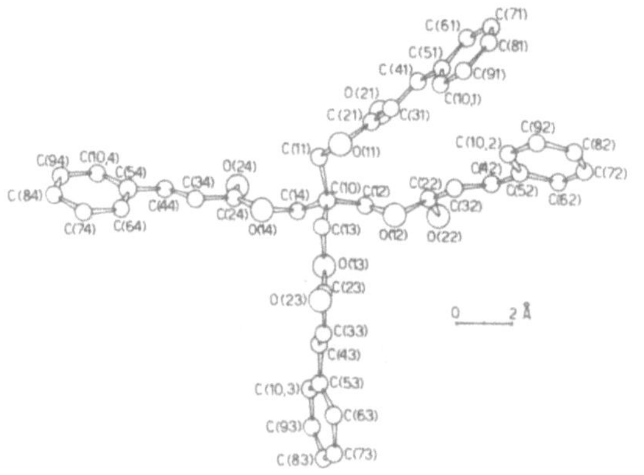

Fig. 1. Pentaerythritol tetracinnamate.

Pentaerythritol tetracinnamate does not retain the $\bar{4}$ molecular symmetry in the crystal because of the high intramolecular rotational freedom and packing requirements (Fig. 1). The conformational differences of the four arms in the molecule are shown by the values of the relevant internal rotation angles:

C(10)-C(11)-O(11)-C(21) = 103.9° C(10)-C(12)-O(12)-C(22) = 171.4°
C(10)-C(13)-O(13)-C(23) = 176.2° C(10)-C(14)-O(14)-C(24) = 144.7°

N-TETRACOSANOYLPHYTOSPHINGOSINE

$C_{42}H_{85}NO_4$

$CH_3-(CH_2)_{13}-CH(OH)-CH(OH)-CH(NH-CO-(CH_2)_{22}-CH_3)-CH_2-OH$

B. DAHLÉN and I. PASCHER, 1972. Acta Cryst., B28, 2396-2404.

Triclinic, P1, a = 6.181, b = 4.929, c = 37.278 Å, α = 90.96 , β = 91.30 , γ = 105.51°, D_m = 1.015, Z = 1. Cu radiation, R = 0.082 for 1271 reflexions.

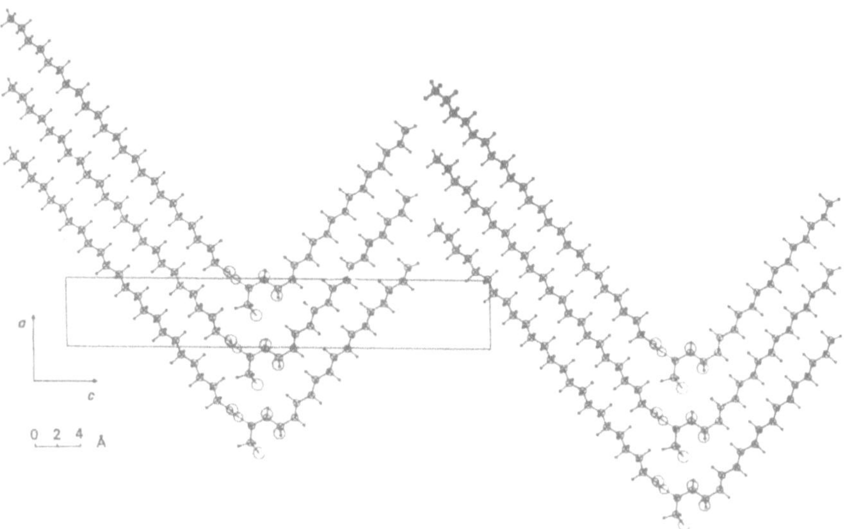

Fig. 1. The N-tetracosanoylphytosphingosine structure viewed in projection on the (010) plane.

The molecular structure and packing are shown in Fig. 1. The molecules cannot pack laterally when fully extended due to the space requirements of the polar groups. The chains form an angle of 101° thereby accommodating the polar groups in the apices of the V. The molecules are linked together in the polar region by one NH···O (2.977(12) Å) and two OH···O (2.687 and 2.755(9) Å) hydrogen bonds. There is also an intramolecular OH···O hydrogen bond (2.724 Å) linking the primary hydroxyl and a secondary hydroxyl group. The mean C-C bond length and C-C-C bond angle are 1.50(4) Å and 116(3)°; the C-C bonds are probably underestimated because of thermal motion.

BENZOYLACETONE

$C_{10}H_{10}O_2$ $C_6H_5C(OH)=CH-CO-CH_3$

D. SEMMINGSEN, 1972. Acta Chem. Scand., 26, 143-154.

Monoclinic, $P2_1/c$, a = 8.244, b = 5.596, c = 19.793 Å, β = 111.77°, D_m = 1.24, Z = 4. Mo radiation, R = 0.053 for 1090 reflexions.

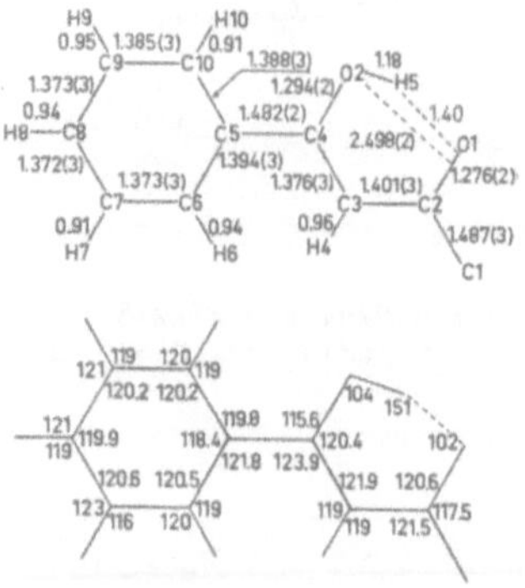

Fig. 1. Bond distances (with standard deviations in parentheses) and angles in benzoylacetone.

β-Diketones exist in a cis-enol form in their crystal structures with a strong intramolecular hydrogen bond. Benzoylacetone conforms to this pattern. The bond lengths in the enol ring (Fig. 1) are asymmetric with the enol hydrogen located on the oxygen adjacent to the phenyl ring; the intramolecular O-H···O distance is 2.498(2) Å and the hydrogen atom (H5) is situated significantly off the line between the two oxygen atoms. The angle O(1)···O(2)-H(5) is 16(1)°. The angle between the phenyl and enol planes is 6.3°.

TRICYANOALKANE DERIVATIVES

J.R. WITT, D. BRITTON and C. MAHON, 1972. Acta Cryst., B28, 950-955.

BROMOTRICYANOMETHANE

C_4BrN_3 $BrC(CN)_3$

Orthorhombic, Pbca, a = 6.09, b = 11.49, c = 17.62 Å, Z = 8. Cu and Mo radiations, R = 0.15 for 475 reflexions.

CHLOROTRICYANOMETHANE

$C_4C\ell N_3$ $C\ell C(CN)_3$

Hexagonal, $P6_3/m$, a = 10.23, c = 9.95 Å, Z = 6. Cu radiation, R = 0.088 for 218 reflexions.

1,1,1-TRICYANOETHANE

$C_5H_3N_3$ $CH_3C(CN)_3$

Hexagonal, $P6_3/m$, a = 10.23, c = 9.95 Å, D_m = 1.18, Z = 6. Cu radiation, R = 0.155 for 245 reflexions.

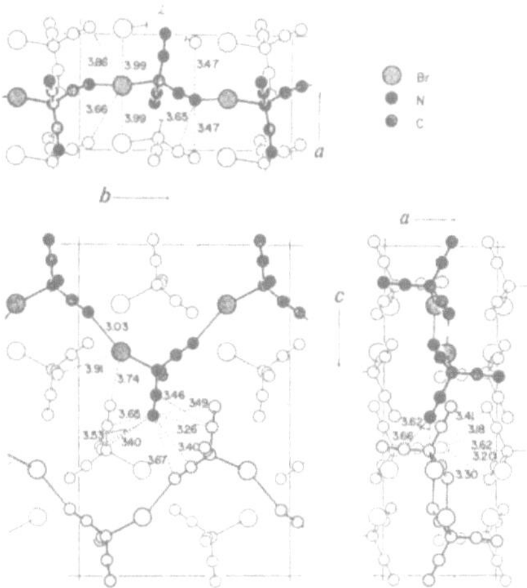

Fig. 1. The crystal structure of $BrC(CN)_3$ showing short intermolecular contacts.

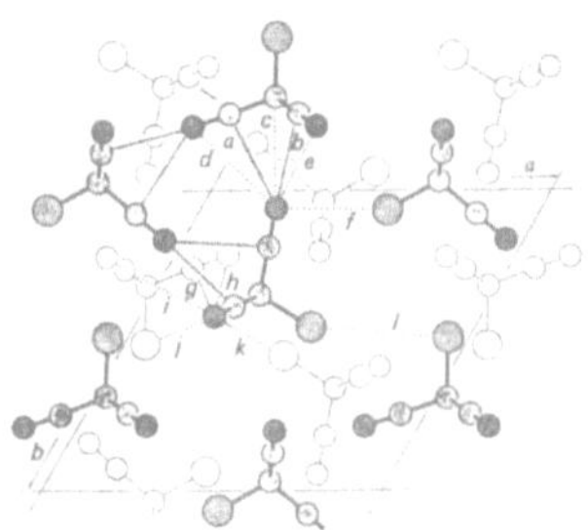

Fig. 2. The $C\ell C(CN)_3$ and $CH_3C(CN)_3$ crystal structure. The intermolecular contacts labelled a, b and g have distances 3.09, 3.09, and 3.14 Å in $C\ell C(CN)_3$ and 3.09, 3.11, and 3.29 Å in $CH_3C(CN)_3$.

The $BrC(CN)_3$ structure contains tetrahedral molecules with normal bond distances. The crystal packing is dominated by a short (3.03 Å) Br···N intermolecular contact and two short N···C distances of 3.18 and 3.20 Å (Fig. 1). The $C\ell C(CN)_3$ and $CH_3C(CN)_3$ structures are isomorphous. They both contain tetrahedral molecules with m symmetry required by the space group and normal bond lengths (e.g. C-Cℓ = 1.78(1), C-CH_3 = 1.57(2), C≡N = 1.15 - 1.29(3) Å). Short intermolecular contacts are found between a nitrogen atom on one molecule and three cyanide carbon atoms of the next (with N···C ∿ 3.10 Å). Weakly bound trimers are thus formed (Fig. 2).

DICHLOROFUMARONITRILE

$C_4C\ell_2N_2$ $NC(C\ell)C{=}C(C\ell)CN$

B. KLEWE and C. RØMMING, 1972. Acta Chem. Scand., 26, 2272-2278.

Monoclinic, C2/c, a = 7.89, b = 9.91, c = 7.66 Å, β = 98.4°, Z = 4. Mo radiation, R = 0.028 for 1010 reflexions.

The crystal structure contains well separated molecules with no unusual intermolecular contacts. Molecules lie on crystallographic centres of symmetry and are hence in the trans-configuration. Bond lengths are C=C = 1.350(4), C-C = 1.427(3), C≡N = 1.147(4) and C-Cℓ = 1.702(2) Å. The molecule is essentially planar (the largest deviation (0.017 Å) involves the cyano-carbon atom).

2-AMINO-1,1,3-TRICYANOPROPENE (FORM B)

(DIMERIC MALONONITRILE)

$C_6H_4N_4$ $NC{\cdot}CH_2{\cdot}C(NH_2){=}C(CN)_2$

B. KLEWE, 1972. Acta Chem. Scand., 26, 317-327.

Monoclinic, $P2_1$, a = 5.28, b = 5.42, c = 11.56 Å, β = 96.2°, D_m = 1.32, Z = 2. Mo radiation, R = 0.048 for 606 reflexions.

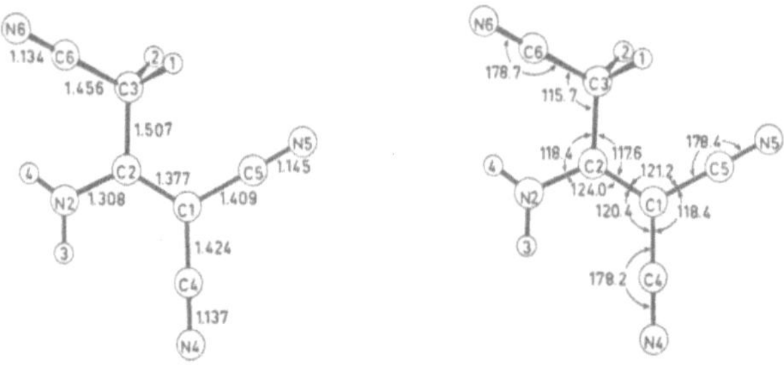

Fig. 1. Dimeric malononitrile showing bond distances (mean σ = 0.005 Å) and angles.

Bond distances and angles (Fig. 1) in this modification are essentially the same as those found in the A-form (1). The major difference between the two forms occurs in the dihedral angle (NH_2)-$\bar{C}$-CH_2-(CN), which has values of 12° in B and 27° in A, and is attributed to packing effects. Intermolecular hydrogen bonds between the amino group and nitrogen atoms of the cyano group form infinite two-dimensional networks parallel to the (001) planes with N-H···N = 2.95 and 3.00 Å.

1. B. KLEWE, 1971. Acta Chem. Scand., 25, 1999; Structure Reports, 37B, 24.

2,4,6-TRIHALOGENOBENZONITRILES

V.B. CARTER and D. BRITTON, 1972. Acta Cryst., B28, 945-950.

2,4,6-TRIBROMOBENZONITRILE

$C_7H_2Br_3N$ $Br_3C_6H_2CN$

Monoclinic, $P2_1/m$, a = 8.82, b = 10.34, c = 4.89 Å, β = 95.8°, Z = 2. Cu and Mo radiation, R = 0.089 for 304 reflexions.

2,4,6-TRICHLOROBENZONITRILE

$C_7H_2Cl_3N$ $Cl_3C_6H_2CN$

Monoclinic, $P2_1/c$, a = 3.97, b = 16.06, c = 12.84 Å, β = 91.0°, Z = 4. Cu radiation, R = 0.134 for 349 reflexions.

The packing of the molecules in both structures is dominated by weak Lewis acid-base interactions, CN···X between cyanide groups and ortho halogen atoms in a manner reminiscent of the dimerization of carboxylic acids. In the bromo compound these interactions lead to infinite chain-like polymers (Fig. 1) with CN···Br = 3.06 Å whereas in the chloro compound only distinct dimers (Fig. 2) are formed (with CN···Cl = 3.22 Å). Bond lengths and angles within the molecules have normal values.

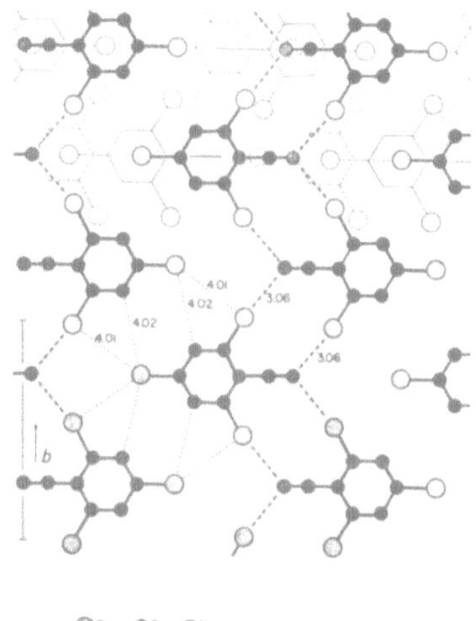

Fig. 1. The tribromobenzonitrile structure seen in projection perpendicular to the layers.

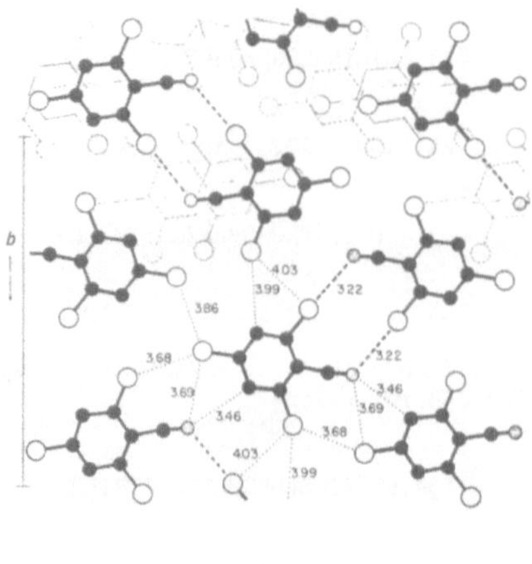

Fig. 2. The trichlorobenzonitrile structure seen in projection perpendicular to the layers.

2,4,5-TRICHLORO-6-(METHYLTHIO)ISOPHTHALONITRILE

$C_9H_3C\ell_3N_2S$

D.R. CARTER, J.W. TURLEY and F.P. BOER, 1972. Acta Cryst., B28, 3430-3434.

Monoclinic, $P2_1/c$, a = 7.705, b = 32.356, c = 8.901 Å, β = 99.86°, D_m = 1.682, Z = 8. Mo radiation, R = 0.046 for 1333 reflexions.

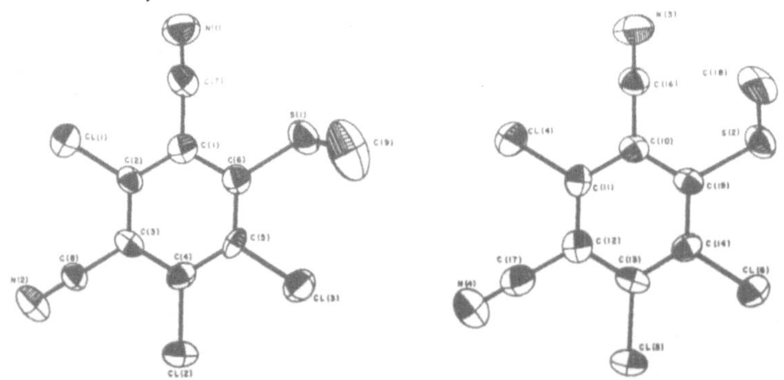

Fig. 1. The two independent molecules of 2,4,5-trichloro-6-(methylthio)-isophthalonitrile.

The crystal structure contains two independent molecules (Fig. 1) which show significant deviations from planarity. Average distances for the six bond types are C-C(ring) = 1.390, C-CN = 1.459, C-Cℓ = 1.772, C≡N = 1.128, C(sp^2)-S = 1.766 and C(sp^3)-S = 1.823 Å. Bond angles at the twelve ring carbons range from 116.7 to 123.0(1.1)°. Although bond lengths and angles in the two molecules are individually very close to normal, a series of cumulative distortions arising from substituent crowding and/or crystal packing forces leads to very significant deviations of individual atoms from the least-squares planes defined by the ring atoms. The shortest intermolecular contacts are Cℓ(2)···N(2) = 3.000(1) and Cℓ(5)···N(4) = 2.995(11) Å and these values suggest that appreciable donor-acceptor interactions can occur in the solid state between the C≡N group and an aromatic chloride.

BENZYLIDENEMALONONITRILE

$C_{10}H_6N_2$

D.A. WRIGHT and D.A. WILLIAMS, 1972. J. Cryst. Mol. Struct., 2, 31-40.

Monoclinic, $P2_1/c$, a = 9.385, b = 3.976, c = 22.152 Å, β = 93.53°, D_m = 1.231, Z = 4. Cu radiation, R = 0.090 for 1412 reflexions (films, visual intensities).

Steric hindrance of the malononitrile group and a hydrogen atom in the *ortho* position of the phenyl ring causes simultaneous expansion of the bond angles between the malononitrile group and the benzylidene group and rotations of these groups relative to each other. These constraints cause the two groups to be non-planar, the dihedral angle between their least-squares planes being 11.0° (Fig. 1). Bond lengths (σ = 0.004 Å) and angles are close to normal values.

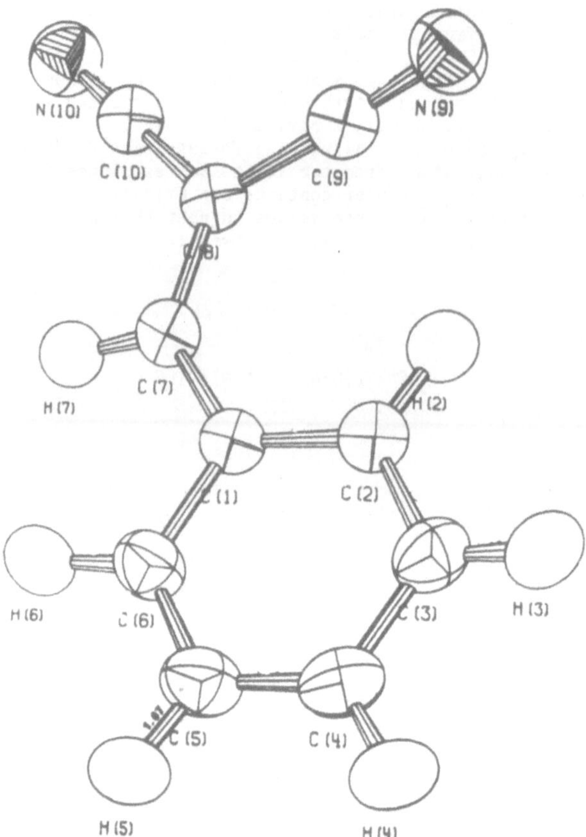

Fig. 1. Perspective drawing of the benzylidenemalononitrile molecule with 70% thermal ellipsoids.

CARBOHYDRAZIDE

CH_6N_2O $(NH_2NH)_2CO$

P. DOMIANO, M.A. PELLINGHELLI and A. TIRIPICCHIO, 1972. Acta Cryst., B28, 2495-2498.

Monoclinic, $P2_1/c$, a = 3.728, b = 8.841, c = 12.659 Å, β = 109.1°, D_m = 1.525, Z = 4. Cu radiation, R = 0.04 for 652 reflexions.

Fig. 1. Bond lengths (σ N-N, N-C, C-O = 0.002 Å) and angles in carbohydrazide.

The molecule (Fig. 1) has the NH-NH_2 groups in <u>cis-gauche</u> and <u>trans-antigauche</u> orientations. The group of atoms N-CO-N is planar with bond distances (see Fig. 1) essentially equal to those in urea. Atoms N(2) and N(4) are +0.153 and -0.059 Å out of the N-CO-N plane. One hydrogen bond (NH···NH_2 = 2.950(3) Å) joins the molecules in chains parallel to [001] and weaker hydrogen bonds (NH···N = 3.068(3), NH···O = 3.007 - 3.095(2) Å) connect the chains.

DITHIOBIURET

$C_2H_5N_3S_2$ $H_2NC(S)NHC(S)NH_2$

W.A. SPOFFORD and E.L. AMMA, 1972. J. Cryst. Mol. Struct., <u>2</u>, 151-158.

Monoclinic, $P2_1/c$, a = 4.081, b = 17.684, c = 8.222 Å, β = 100.56°, D_m = 1.54, Z = 4. Mo radiation, R = 0.033 for 1020 reflexions.

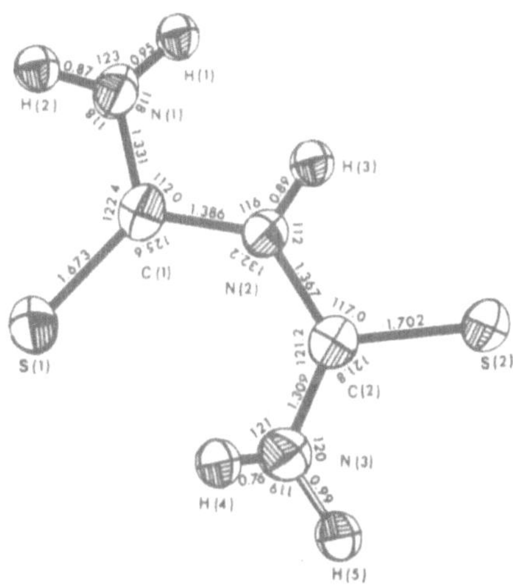

Fig. 1. A view of the dithiobiuret molecule, with bond lengths and angles.

The structure consists of almost planar molecules in the trans configuration with essentially only van der Waals interactions between molecules. Fig. 1 shows the molecule with bond distances and angles; σ(S-C) = 0.003, σ(C-N) = 0.004, σ(N-H) = 0.03 - 0.04 Å.

N-[4-(β-2-METHOXY-5-BROMOBENZAMIDOETHYL)BENZENESULPHONYL]-N'-4-METHYLCYCLOHEXYLUREA

$C_{24}H_{30}BrN_3O_5S$

D. KOBELT and E.F. PAULUS, 1972. Acta Cryst., B28, 3452-3457.

Monoclinic, $P2_1/c$, a = 9.03, b = 23.87, c = 15.48 Å, β = 124.9°, D_m = 1.341, Z = 4. Cu radiation, R = 0.105 for 3305 reflexions.

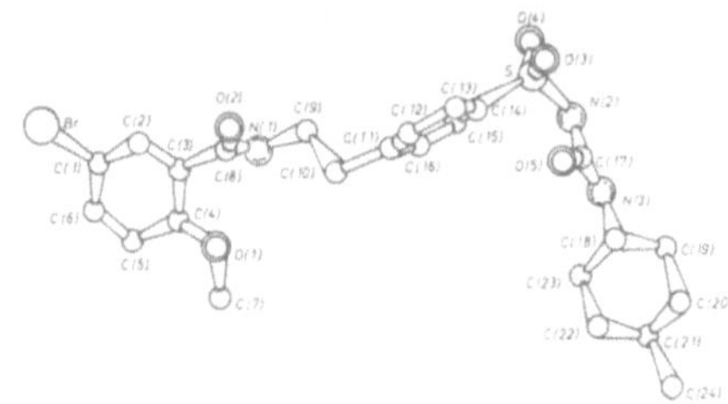

Fig. 1. Structure of N-[4-(β-2-methoxy-5-bromobenzoamidoethyl)benzene-sulphonyl]-N'-methylcyclohexylurea projected on the molecular mean plane.

The molecule (Fig. 1) has a "thickness" of 2.7 Å and a "length" of 19.2 Å (both neglecting hydrogen atoms which could not be found from difference syntheses). The intramolecular geometry is normal. The molecules are hydrogen bonded by NH···O bonds (2.79 Å) between carbonamide groups of neighbouring molecules.

AZIDOFORMAMIDINIUM CHLORIDE

CH_4ClN_5

H. HENKE and H. BÄRNIGHAUSEN, 1972. Acta Cryst., B28, 1100-1107.

Monoclinic, Cc, a = 5.116, b = 10.737, c = 9.707 Å, β = 92.2°, D_m = 1.515, Z = 4. Mo radiation, R = 0.032 for 780 reflexions.

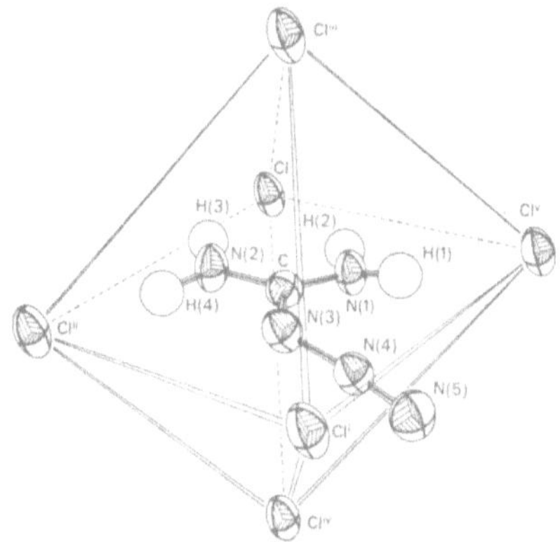

Fig. 1. The Cℓ$^-$ environment around the azido-formamidinium cation.

The structure consists of $[(H_2N)_2C\text{-}N_3]^+$ and Cℓ$^-$ ions connected by hydrogen bonds. The bond lengths in the cation (Fig. 1) are C-N(1) = 1.302, C-N(2) = 1.314, C-N(3) = 1.393, N(3)-N(4) = 1.265, and N(4)-N(5) = 1.110(4) Å. The segment C, N(1), N(2), N(3) is planar with N(4) = 0.33 and N(5) = 0.62 Å above the plane. The long C-N(3) distance suggests that the $(H_2N)_2C$ part and the N_3 part of the cation belong to almost independent π-bonding systems. Each Cℓ$^-$ anion is involved in four N-H···Cℓ hydrogen bonds, 2 from one cation and 1 from each of 2 others. Each cation is similarly involved in 4 hydrogen bonds, 2 to one Cℓ$^-$ and 1 to each of 2 others.

THIOCARBONOHYDRAZIDE SULPHATE

$CH_8N_4O_4S_2$ $[\overset{+}{N}H_3\cdot NH\cdot CS\cdot NH\overset{+}{N}H_3]SO_4{}^{2-}$

F. BIGOLI, A. BRAIBANTI, A.M.M. LANFREDI and A. TIRIPICCHIO, 1972. J. Chem. Soc., Perkin II, 2121-2125.

Orthorhombic, Pbca, a = 10.80, b = 15.44, c = 8.26 Å, D_m = 1.971, Z = 8. Cu radiation, R = 0.047 for 1250 reflexions.

The structure consists of diprotonated ions in a cis-cis conformation with tetrahedral sulphate ions (Fig. 1). Distances and angles have expected values e.g. C=S = 1.663(4), C-N(mean) = 1.344(4), N-N(mean) = 1.412(4), S-O = 1.455 - 1.488(3) Å. The cation has approximate (non-crystallographic) twofold symmetry through the C=S bond. The -NH- and -$\overset{+}{N}H_3$ groups form several N-H···O hydrogen bonds with neighbouring anions as shown in Fig. 1.

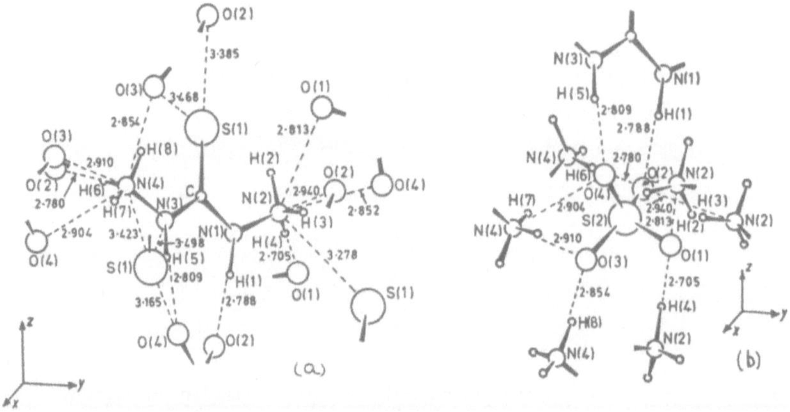

Fig. 1. Environment of (a) the thiocarbonohydrazide ion and (b) the sulphate ion in thiocarbonohydrazide sulphate.

THIOCARBONOHYDRAZIDE HEMIHYDROCHLORIDE

$C_2H_{13}C\ell N_8S_2$

A. BRAIBANTI, A. TIRIPICCHIO and M.T. CAMELLINI, 1972. J. Chem. Soc., Perkin II, 2116-2120.

Monoclinic, C2/c, a = 18.397, b = 9.039, c = 13.847 Å, β = 113.4°, D_m = 1.575, Z = 8. Mo radiation, R = 0.028 for 2111 reflexions..

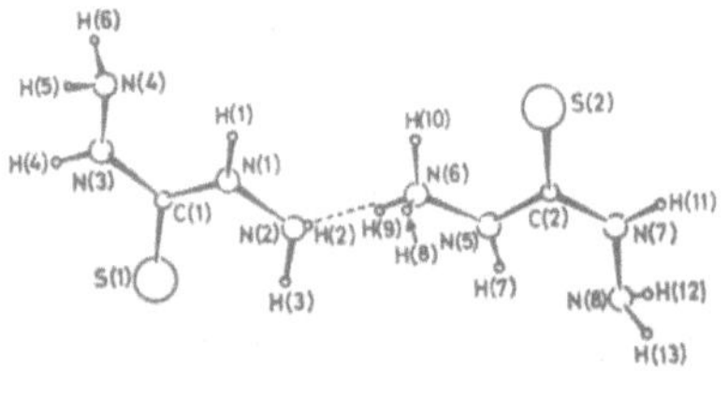

Fig. 1. The thiocarbonohydrazide dimer.

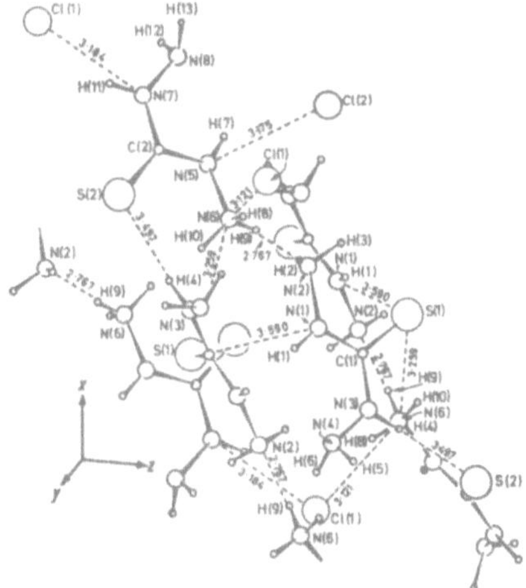

Fig. 2. Clinographic projection of the thiocarbonohydrazide hemihydrochloride structure showing intermolecular hydrogen bonds.

The basic units of the crystal structure are dimers, $H_2N\cdot NH\cdot CS\cdot NH\cdot NH_2\cdots H_3\overset{+}{N}\cdot NH\cdot CS\cdot NH\cdot NH_2$ and chloride anions. The two molecules of the dimer (Fig. 1) are held together by the hydrogen bond (N(6)-H(9)···N(2) = 2.767(3) Å) which is short compared with normal values for hydrogen bonds between nitrogen atoms. Both molecules are in the cis,trans-conformation. The two molecules are almost identical, although there are small differences in distances and angles.

The bond C(1)-S(1) = 1.705(2) Å in the neutral molecule is only slightly longer than C(2)-S(2) = 1.694(2) Å in the protonated species. The N-N bonds do not differ from their mean value 1.416(3) Å. The mean C(1)-N bond length is 1.335(3) Å whereas C(2)-N(5) is 1.349(2) Å and C(2)-N(7) = 1.327(3) Å. Both thio-ureide N-CS-N groups are individually planar, confirming conjugation. In the crystal structure (Fig. 2) the chloride ions lie on twofold axes; one chloride ion (Cℓ(1)) bridges two different dimers whereas the other is bound to one nitrogen only.

NITROGEN-NITROGEN YLIDES

A.F. CAMERON, N.J. HAIR and D.G. MORRIS, 1972. J. Chem. Soc., Perkin II, 1071-1076.

TRIMETHYLAMMONIONITRAMIDATE

$C_3H_9N_3O_2$ $(CH_3)_3\overset{+}{N}-N=\overset{+}{N}(O^-)_2$

Orthorhombic, Pnma, a = 12.245, b = 6.718, c = 7.099 Å, D_m = 1.36, Z = 4. Mo radiation, R = 0.078 for 378 reflexions.

TRIMETHYLAMMONIOBENZAMIDATE

$C_{10}H_{14}N_2O$ $(CH_3)_3\overset{+}{N}\text{-}N{=}C(\bar{O})C_6H_5$

Monoclinic, $P2_1/c$, a = 11.620, b = 7.928, c = 11.454 Å, β = 113.75°, D_m = 1.21, Z = 4. Mo radiation, R = 0.072 for 977 reflexions.

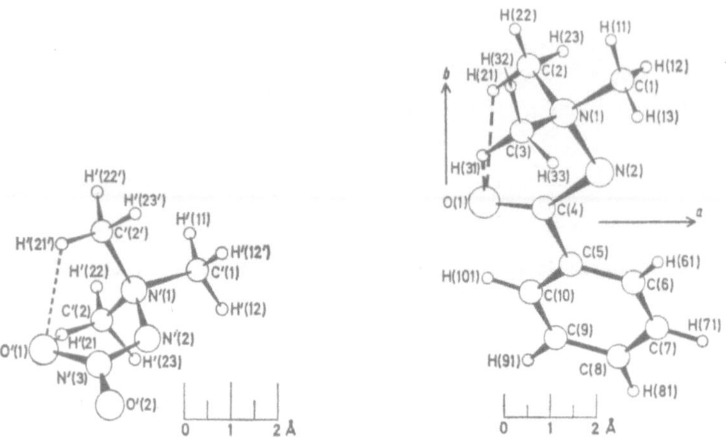

Fig. 1. The structure of trimethylammonionitramidate (left) and trimethylammoniobenzamidate (right).

The solid state conformations of both molecules (Fig. 1) are almost identical and are characterised by syn-planar relationships within the N(1)-N(2)-C(4)-O(1) (benzamidate) and N(1)-N(2)-N(3)-O(1) (nitramidate) systems, the torsion angles about the N(2)-C(4) and N(2)-N(3) bonds being 2° and 0° respectively. In both molecules the bond lengths indicate significant charge delocalisation on the oxygen atoms of the stabilising benzoyl- and nitro-groups as implied by the molecular formulations shown above. Thus in both molecules the N(1)-N(2) bonds (1.470(8) and 1.471(5) Å) correspond to normal single bonds with no evidence of contraction, and the N(2)-C(4) and N(2)-N(3) bonds are both consistent (1.313(6) and 1.323(8) Å) with high double bond character. The C-O and N-O bond lengths (1.243(5) and 1.259(6) Å respectively) are also consistent with the charge delocalisation scheme indicated above. In both compounds packing is determined by van der Waals forces only.

1,4-BIS(DIAZO)-2,3-BUTANEDIONE

$C_4H_2N_4O_2$ $N_2CHCOCOCHN_2$

H. HOPE and K.T. BLACK, 1972. Acta Cryst., B28, 3632-3634.

Monoclinic, $P2_1/n$, a = 3.745, b = 8.326, c = 9.580 Å, β = 93.12°, D_m = 1.537, Z = 2. Cu radiation, R = 0.029 for 614 reflexions.

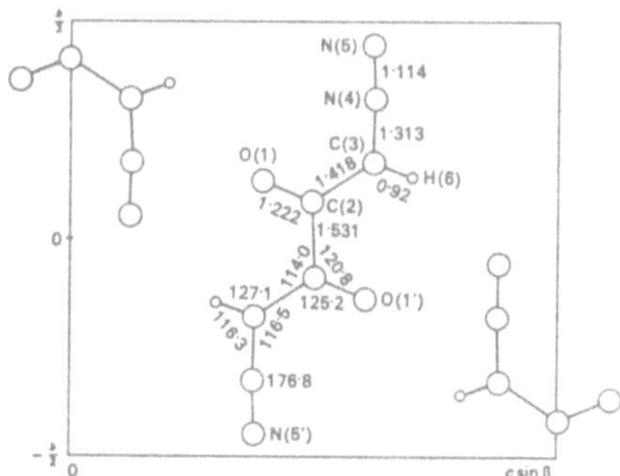

Fig. 1. Molecular geometry and packing of 1,4-bis(diazo)-2,3-butanedione. (σ = 0.001 - 0.002 Å for C-C, C-N, C-O, and 0.1° for C-C-C, N-N-C, C-C-O, and C-C-N.)

The structure (Fig. 1) consists of planar centrosymmetric molecules separated by normal van der Waals distances. The observed molecular geometry is consistent with contributions from both "diazo oxide" ($O=\overset{|}{C}-CH=\overset{+}{N}=N^-$) and "diazo anhydride" ($^-O-\overset{|}{C}=CH-\overset{+}{N}\equiv N$) canonical forms.

p-BENZENEDIAZONIUM SULPHONATE

$C_6H_4N_2O_3S$ $N_2C_6H_4SO_3$

C. RØMMING, 1972. Acta Chem. Scand., 26, 523-533.

Monoclinic, $P2_1/c$, a = 8.02, b = 9.81, c = 11.46 Å, β = 126.7°, D_m = 1.69, Z = 4. Mo radiation, R = 0.052 for 1410 reflexions.

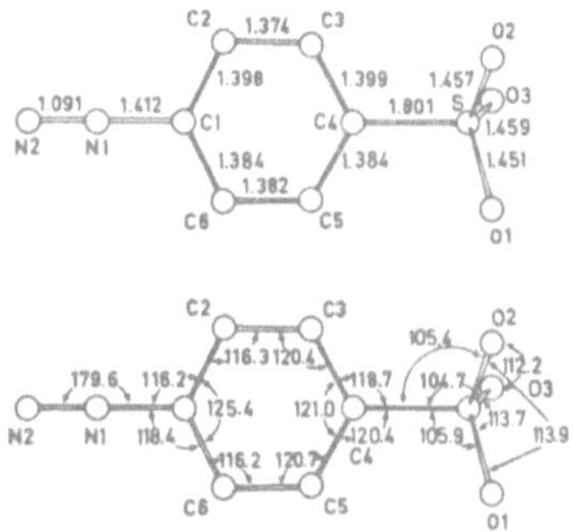

Fig. 1. Benzene diazonium sulphonate: bond lengths (corrected for thermal motion) and angles. Mean standard deviations are 0.003 Å and 0.2°.

This crystal structure has been reported previously (1); the present study is a more precise determination. The molecular geometry is shown in Fig. 1. Except for the oxygen atoms, the molecule is nearly planar with twofold symmetry about the main molecular axis. Atoms S, N(1) and N(2) are 0.003, 0.028 and 0.056 Å respectively from the benzene ring plane. The crystal structure is mainly governed by electrostatic forces between the positively charged diazonium groups and the negatively charged sulphonate groups of neighbouring molecules. Each diazonium group is surrounded by three oxygen atoms from different molecules in a distorted trigonal arrangement in a plane nearly normal to the N-N bond with N···O in the range 2.731 - 3.060(3) Å.

1. C. BUGG, J. LAWSON and R.L. SASS, 1967. Acta Cryst., 17, 767; R.L. SASS and J. LAWSON, 1970. Acta Cryst., B26, 1187.

DIMETHYLGLYOXAL BISGUANYLHYDRAZONE DIHYDROCHLORIDE DIHYDRATE

$C_6H_{20}Cl_2N_8O_2$ $C_6H_{16}N_8^{2+} \cdot 2Cl^- \cdot 2H_2O$

J.W. EDMONDS and W.C. HAMILTON, 1972. Acta Cryst., B28, 1362-1366.

Monoclinic, $P2_1/c$, a = 5.799, b = 9.239, c = 14.447 Å, β = 113.98°, D_m = 1.44, Z = 2. Mo radiation, R = 0.055 for 1997 reflexions.

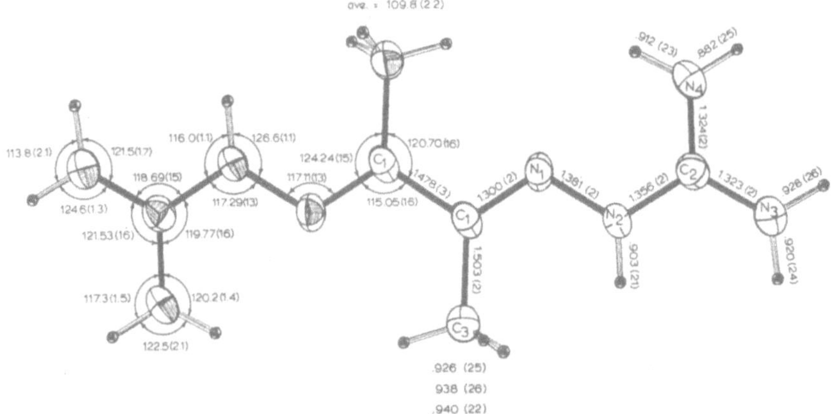

Fig. 1. Distances and angles corrected for rigid-body thermal motion (except those involving hydrogen) in dimethylglyoxal bisguanylhydrazone.

The structure consists of planar dipositive $C_6H_{16}N_8^{2+}$ ions (Fig. 1) hydrogen bonded to both Cl^- and H_2O. The dominant interaction found in the monomethyl compound (1), an interaction involving π-orbitals on adjacent parallel molecules, is less pronounced in the dimethyl structure. The mean interplanar spacing is 3.40 Å compared with 3.19 Å in the monomethyl structure.

1. Structure Reports, 33B, 31.

4-PHENYL-THIOSEMICARBAZIDE

$C_7H_9N_3S$ $C_6H_5NH{\cdot}CS{\cdot}NH{\cdot}NH_2$

A. KÁLMÁN, Gy. ARGAY and M. CZUGLER, 1972. Cryst. Struct. Comm., 1, 375-378.

Monoclinic, $P2_1/c$, a = 11.860, b = 5.607, c = 12.637 Å, β = 92.4°, D_m = 1.321, Z = 4. Cu radiation, R = 0.112 for 1381 reflexions.

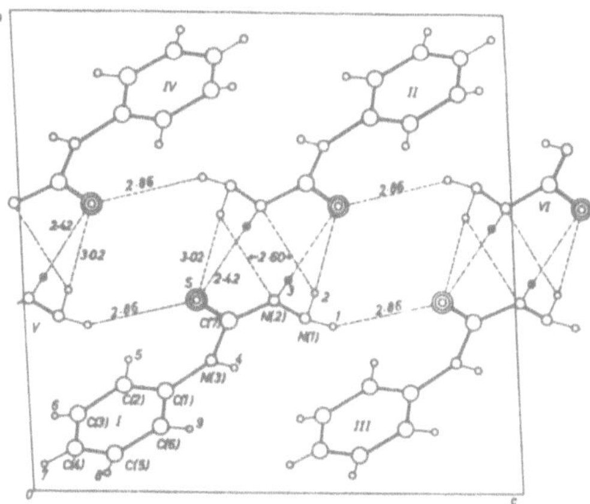

Fig. 1. Packing of the $C_6H_5NH{\cdot}CS{\cdot}NH{\cdot}NH_2$ molecules.

The thiosemicarbazide portion of the molecule is essentially planar (excluding hydrogen atoms) and the molecular packing is shown in Fig. 1. The C-N(2), C-N(3), and N(1)-N(2) distances are 1.445(7), 1.349(7), and 1.431(6) Å respectively; the C-S bond length is 1.685(5) Å.

AZOBIS(CYANO-ALKANES)

A.B. JAFFE, D.S. MALAMENT, E.P. SLISZ and J.M. McBRIDE, 1972. J. Amer. Chem. Soc., 94, 8515-8521.

AZOBISISOBUTYRONITRILE (AIBN)
(AZOBIS(2-CYANO-2-PROPANE))
HYDROGENATED AND DEUTERATED FORMS

$C_8H_{12}N_4$

$C_8D_{12}N_4$

$$\begin{array}{c} \text{R} \quad \text{R} \\ | \qquad | \\ \text{N}{\equiv}\text{CCN}{=}\text{NCC}{\equiv}\text{N} \\ | \qquad | \\ \text{R} \quad \text{R} \end{array} \qquad \begin{array}{l} R{=}CH_3 \\ R{=}CD_3 \end{array}$$

Triclinic, $P\bar{1}$, Z = 1.

	$C_8H_{12}N_4$	$C_8D_{12}N_4$
a (Å)	7.865	7.855
b	5.555	5.549
c	6.201	6.204
α (deg.)	71.33	71.31
β	77.95	77.98
γ	79.19	79.25
radiation	Mo	Mo
R	0.074	0.051
no. of reflexions	1008	993

Monoclinic, $P2_1/c$, Z = 2.

	$C_8H_{12}N_4$	$C_8D_{12}N_4$
a (Å)	5.509	5.501
b	8.225	8.219
c	11.002	10.989
β (deg.)	96.00	95.98
radiation	Mo	-
R	0.046	-
no. of reflexions	803	

AZOBIS-3-CYANO-3-PENTANE (ACP)

$C_{12}H_{20}N_4$

$$N{\equiv}C\overset{R}{\underset{R}{C}}N{-}N\overset{R}{\underset{R}{C}}C{\equiv}N, \quad R{=}C_2H_5$$

Orthorhombic, Cmca, a = 9.059, b = 9.966, c = 15.559 Å, Z = 4. Mo radiation, R_w = 0.026 for 554 reflexions.

The ACP structure is disordered, otherwise bond lengths and angles are consistent among the ACP and AIBN determinations and with those of other azoalkanes. The nitrile and azo groups are nearly eclipsed to give planar, S-shaped N≡CCN=NCC≡N backbones which are arranged into sheets containing a herringbone pattern of nitrile groups (Fig. 1). The deuterated methyl group packs as if it were smaller than the unlabeled methyl.

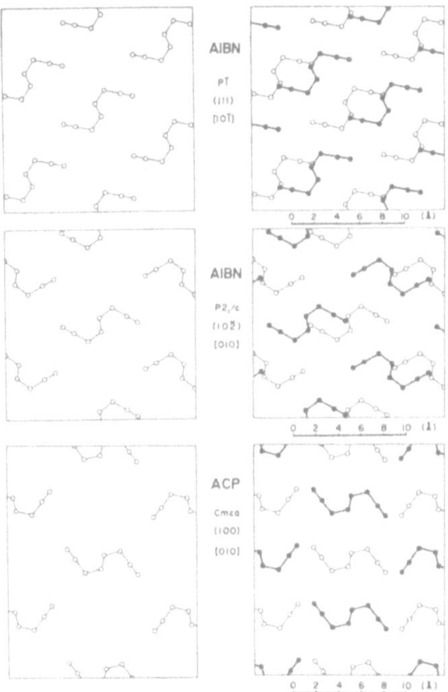

Fig. 1. Azonitrile packings viewed along normal to the plane of N≡CCN=NCC≡N backbones with side chains omitted. Left frame shows a single plane; right frame shows the same plane with the next plane toward the viewer superimposed and darkened. Between frames are given compound, space group, plane of the page, and vertical direction in the page.

ACETONE 4-NITROPHENYLHYDRAZONE

$C_9H_{11}N_3O_2$ $(CH_3)_2C=N-NH-C_6H_4NO_2$

G. MENCZEL, G. SAMAY and K. SIMON, 1972. Acta Chim. Acad. Sci. Hung., 72, 441-450.

Monoclinic, $P2_1/c$, a = 4.053, b = 20.74, c = 11.80 Å, β = 95.5°, D_m = 1.32, Z = 4. Cu radiation, R = 0.109 for 566 reflexions.

The crystal structure contains molecules linked by N-H···O hydrogen bonds (3.13 Å). The N-NH bond length is 1.424(6), C=N is 1.304(13), and N-C(phenyl) is 1.350(6) Å. The phenyl ring is inclined at an angle of 9° to the C-NH-N= plane and the nitro-group plane makes an angle of 5° with the benzene ring.

(+)-3-DIAZOCAMPHOR

$C_{10}H_{14}N_2O$

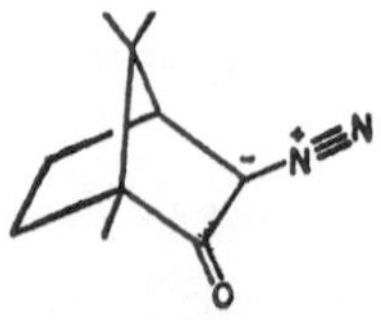

A.F. CAMERON, N.J. HAIR and D.G. MORRIS, 1972. J. Chem. Soc., Perkin II, 1331-1335.

Monoclinic, $P2_1$, a = 7.119, b = 7.466, c = 9.279 Å, β = 91.75°, D_m = 1.18, Z = 2. Mo radiation, R = 0.087 for 661 reflexions.

The analysis was undertaken to investigate the extent to which a resonance structure of the form $R\text{-}C(O^-)\text{=}CH\text{-}N^+\equiv N$ is involved in the stablization of diazoketones. The bond lengths in the region of interest in (+)-3-diazocamphor are C(2)=O = 1.215(8), C(2)-C(3) = 1.448(9), C(3)-N(1) = 1.296(8), and N(1)≡N(2) = 1.125(8) Å. From these data it is inferred that a resonance structure of the type mentioned above plays only a minor role and that stablization of the diazoketone via the diazo function occurs to an extent similar to that which would be expected for diazoalkanes. Other bond lengths, angles and intermolecular contacts have normal values.

α-p-BROMODIAZOAMINOBENZENE

$C_{12}H_{10}BrN_3$ $BrC_6H_4N_2NHC_6H_5$

Ju.A. OMELCHENKO and Ju.D. KONDRAŠEV, 1972. Kristallografija, 17, 947-954 [Soviet Physics-Crystallography, 17, 837-842].

Orthorhombic, $Pbn2_1$, a = 13.40, b = 23.72, c = 7.30 Å, Z = 8. Cu radiation, R = 0.115 for 1993 reflexions.

The molecules pack in the crystals as hydrogen bonded dimers via pairs of N-H···N bonds (3.23, 3.27 Å). Principal mean dimensions in the two independent molecules in the asymmetric unit are C-N = 1.43, N=N = 1.24, N-N = 1.34 Å. The molecules are not themselves coplanar; the dihedral angle between benzene rings is 8° in one molecule and 13° in the other.

β-p-BROMODIAZOAMINOBENZENE

$C_{12}H_{10}BrN_3$ $BrC_6H_4N_2NHC_6H_5$

Ju.A. OMELCHENKO and Ju.D. KONDRAŠEV, 1972. Kristallografija, 17, 947-954 [Soviet Physics-Crystallography, 17, 837-842].

Monoclinic, $P2_1/c$, a = 11.49, b = 4.70, c = 22.35 Å, β = 106°, Z = 4. Cu radiation, R = 0.096 for h0ℓ, h1ℓ, and h2ℓ reflexions.

The structure has been described previously (1). As a result of this refinement the N=N and N-N bond lengths become 1.23 and 1.29 Å.

1. Ju.A. OMELCHENKO and Ju.D. KONDRAŠEV, 1967. Kristallografija, 12, 416.

β-DIAZOAMINOBENZENE

$C_{12}H_{11}N_3$ C_6H_5-NH-N_2-C_6H_5

V.F. GLADKOVA and Ju.D. KONDRAŠEV, 1972. Kristallografija, 17, 33-40 [Soviet Physics-Crystallography, 17, 23-28].

Monoclinic, $P2_1/c$, a = 11.370, b = 5.07, c = 18.404 Å, β = 81°1', D_m = 1.245, Z = 4. Cu radiation, R = 0.12 for 595 reflexions.

In this structure the N-N bond lengths are 1.27 and 1.33(1) Å; C-N distances are 1.41 and 1.43(1) Å. The molecule is almost planar with an angle of 2° between the planes of the benzene rings. The crystal structure contains two kinds (+ and -) spirals formed by hydrogen bonding with N-H···N = 3.22 Å.

p,p-DIBROMOBENZALAZINE

$C_{14}H_{10}Br_2N_2$ $[BrC_6H_4CH{=}N-]_2$

J. MARIGNAN, J.L. GALIGNÉ and J. FALGUEIRETTES, 1972. Acta Cryst., B28, 93-97.

Monoclinic, $P2_1/c$, a = 7.051, b = 4.051, c = 23.39 Å, β = 92.15°, D_m = 1.816, Z = 2. Cu radiation, R = 0.095 for 1018 reflexions.

The molecular centres occupy centres of symmetry; the plane of the benzene ring makes an angle of 4.4° with that of the central C-C=N-N=C=C portion of the molecule. Main bond lengths are N-N = 1.45(2), N=C = 1.28(2) Å; other distances are normal. The disparity between the melting behaviour of anisaldazine (1) and dibromobenzalazine corresponds with a looser packing in the latter system.

1. Structure Reports, 33B, 75.

1-p-TOLYL-3-(α-CYANO)BENZYLIDENETRIAZENE

$C_{15}H_{12}N_4$ $C_6H_5C(CN)=N-N=N-C_6H_4CH_3$

J.W. SCHILLING and C.E. NORDMAN, 1972. Acta Cryst., B28, 2177-2182.

Monoclinic, Pc, a = 4.036, b = 11.109, c = 14.805 Å, β = 93.57°, D_m = 1.23, Z = 2. Cu radiation, R = 0.077 for 1019 reflexions.

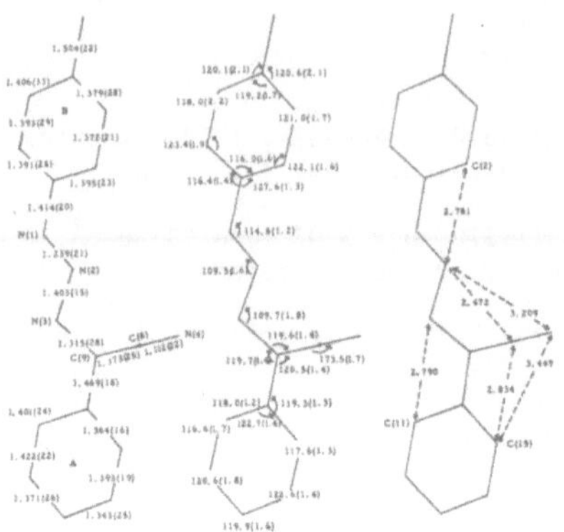

Fig. 1. Bond lengths, angles, and intramolecular contacts in 1-p-tolyl-3(α-cyano)benzylidenetriazene.

In aromatic triazene crystal structures of the form, $R-C_6H_4-NH-N=N-C_6H_4-R$, the two N-N distances are found to be equivalent whereas in this structure the N-N distances are nonequivalent, 1.239(21) and 1.403(15) Å (Fig. 1). The angle N(1)-C(1)-C(2) (Fig. 1) is distorted from the expected 120° by 7.6°; this is probably due to the unsymmetrical intramolecular contact N(2)···C(2) = 2.781 Å, whereas the corresponding angle for ring A is nearly 120° since the intramolecular contacts are balanced. The molecule as a whole is nearly planar with the largest deviation (the CH_3 group) being 0.2 Å. Ring A and B are rotated 8.2 and 8.8° with respect to the plane defined by the three triazene nitrogen atoms. There are no intermolecular contacts to rings A or B closer than 3.8 Å.

4-PHENYLAZOAZOBENZENE

$C_{18}H_{14}N_4$ $C_6H_5N_2C_6H_4N_2C_6H_5$

R.D. GILARDI and I.L. KARLE, 1972. Acta Cryst., B28, 1635-1638.

Monoclinic, $P2_1/c$, a = 13.870, b = 4.617, c = 11.920 Å, β = 113.20°, Z = 2. Cu radiation, R = 0.078 for 1213 reflexions.

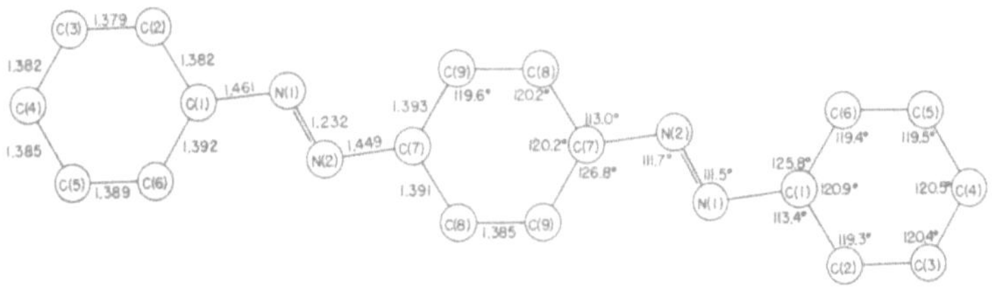

Fig. 1. Bond lengths (σ = 0.0035 - 0.005 Å) and angles (σ = 0.25 - 0.30°) in 4-phenylazoazobenzene.

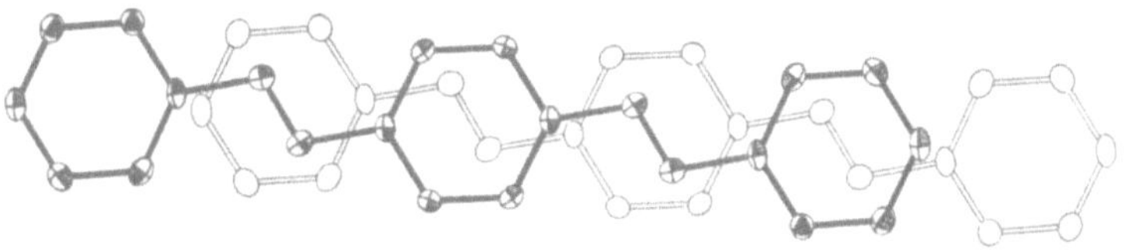

Fig. 2. Two molecules of 4-phenylazoazobenzene viewed along an axis perpendicular to their planes.

The molecule is centred on a crystallographic inversion centre. To a close approximation (±0.05 Å) the entire molecule is planar. Bond lengths and angles are in Fig. 1. Adjacent molecules related by a translation along the b-axis are separated by only 3.325 Å. Fig. 2 illustrates two closely packed molecules. Each azo group is sandwiched between two benzene rings in a manner similar to that found for charge-transfer complexes. In this structure every atom is within 3.6 Å of at least one other molecule.

PHENYL(TRIPHENYLSILYL)DIAZOMETHANE

$C_{25}H_{20}N_2Si$ $(C_6H_5)[(C_6H_5)_3Si]C(N_2)$

C. GLIDEWELL and G.M. SHELDRICK, 1972. J. Chem. Soc., Dalton, 2409-2412.

Orthorhombic, Pbca, a = 26.19, b = 9.17, c = 16.70 Å, Z = 8. Cu radiation, R = 0.117 for 1380 reflexions.

The crystal structure (Fig. 1) contains monomeric molecules each with an essentially linear C-N-N group and planar $CC(N_2)Si$ skeleton. Principal bond lengths and angles are: N-N = 1.130(16), N-C = 1.280(17), mean Si-C = 1.865, mean C-C(aromatic) = 1.373, and C-C(exocyclic) = 1.468(19) Å; Si-$C(N_2)$-C = 125.7(8), Si-C-N = 115.3(1.0), and C-C-N = 118.8(1.2)°. There are no abnormal intermolecular contacts.

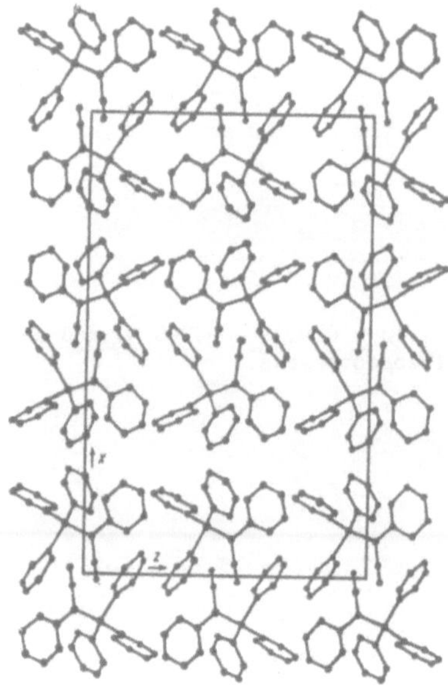

Fig. 1. The molecular packing of $(C_6H_5)[(C_6H_5)_3Si]CN_2$ viewed along the y axis.

1,1'-AZO-2-PHENYLIMIDAZO[1,2-a]PYRIDINIUM DIBROMIDE

$C_{26}H_{20}Br_2N_6$

I. D.J. POINTER and J.B. WILFORD, 1972. J. Chem. Soc., Perkin II, 2259-2262.
II. D.J. POINTER, J.B. WILFORD and D.C. BISHOP, 1972. Nature, 239, 332-333.

Monoclinic, $P2_1/c$, a = 9.15, b = 12.05, c = 12.61 Å, β = 121.90°, D_m = 1.63, Z = 2. Cu radiation, R = 0.112 for 1698 reflexions.

This neuromuscular blocking agent consists of two identical almost planar halves joined by a 2-tetrazene link across a centre of symmetry (Fig. 1). The imidazo-pyridinium groups are nearly coplanar; the molecule as a whole is non-planar because the phenyl and imidazo-pyridinium ring systems are inclined at an angle of 47.5°. The tetrazene chain >N-N=N-N< is in the N-trans-N form with bond lengths 1.39(1) and 1.25(1) Å for the N-N and N=N bonds and the bond angle N(1)-N(10)-N(10') is 111(1)°. Other bond lengths and angles are close to expected values. In the crystal structure the Br^- ion is closer to N(1) than N(4) (N(1)···Br = 3.60, N(4)···Br = 3.78 Å) and suggests that the quaternary centre may be closer to N(1).

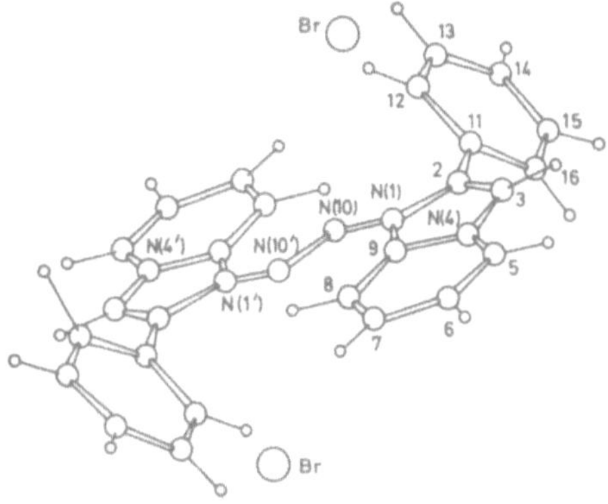

Fig. 1. The structure of the neuromuscular blocking agent $C_{26}H_{20}Br_2N_6$ viewed along the a-axis.

DIPHENYL TRIKETONE HYDRAZONES

D.B. PENDERGRASS, I.C. PAUL and D.Y. CURTIN, 1972. J. Amer. Chem. Soc., **94**, 8730-8737.

DIPHENYL TRIKETONE sym-BENZOYL-p-BROMOPHENYLHYDRAZONE

$C_{28}H_{19}BrN_2O_3$ (I) $(C_6H_5CO)_2C{=}NN\begin{matrix}COC_6H_5\\ C_6H_4Br\end{matrix}$

Orthorhombic, $Pna2_1$, a = 35.876, b = 11.417, c = 5.765 Å, D_m = 1.40, Z = 4. Cu radiation, R = 0.066 for 2005 reflexions.

DIPHENYL TRIKETONE sym-BENZOYLPHENYLHYDRAZONE

$C_{28}H_{20}N_2O_3$ (II) $(C_6H_5CO)_2C{=}NN\begin{matrix}COC_6H_5\\ C_6H_5\end{matrix}$

Orthorhombic, $Pna2_1$, a = 35.750, b = 11.090, c = 5.735 Å, D_m = 1.23, Z = 4. Cu radiation, R = 0.083 for 1321 reflexions.

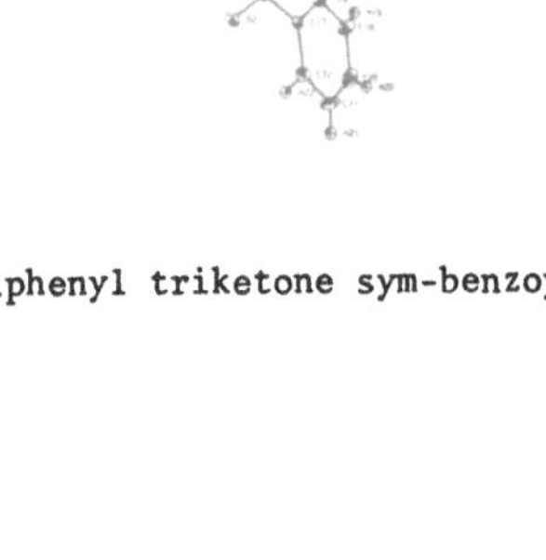

Fig. 1. Diagram of the diphenyl triketone sym-benzoylphenylhydrazone molecule.

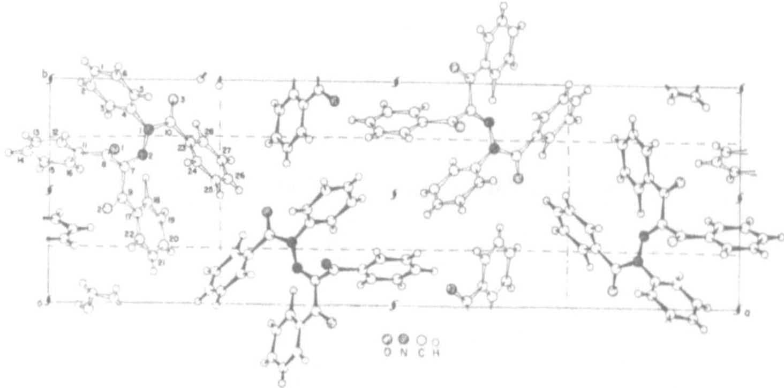

Fig. 2. Packing diagram of diphenyl triketone sym-benzoylphenylhydrazone.

The two compounds, which are isostructural, are the final products of the solid state rearrangement of phenylazotribenzoylmethane and its p-bromophenylazo analogue (1). A view of II is given in Fig. 1 and the packing diagram is seen in Fig. 2. The two compounds are unusual in that the introduction of the p-bromo atom at one of the phenyl rings has had little effect on the crystal structure. The amide groups [C(23), C(10), O(3), N(1), C(4) and N(2)] in I and II are significantly non-planar as a result of pyramidal distortion at nitrogen (N(1) deviates by 0.133 and 0.101 Å from the plane defined by its three bonded neighbours in I and II respectively). The bond lengths and angles found for these molecules confirm the correctness of the formulations shown above and also suggest that a number of other structures (2) referred to in the literature as 'azo' derivatives would be more correctly considered as hydrazones.

1. Following report.
2. H.-C. MEZ, 1968. Ber. Bunsenges. Phys. Chem., 72, 389; C.J. BROWN, 1967. J. Chem. Soc. A, 405; C.T. GRAINGER and J.F. McCONNELL, 1969. Acta Cryst., B25, 1962.

α-p-BROMOPHENYLAZO-β-BENZOYLOXYBENZALACETOPHENONE

$C_{28}H_{19}BrN_2O_3$

$C_6H_5C(=O)O(C_6H_5)C{=}C(COC_6H_5)N{=}NC_6H_4Br$

D.B. PENDERGRASS, D.Y. CURTIN and I.C. PAUL, 1972. J. Amer. Chem. Soc., 94, 8722-8730.

Triclinic, $P\bar{1}$, a = 10.031, b = 6.309, c = 19.387 Å, α = 96.78, β = 103.78, γ = 90.54°, D_m = 1.38, Z = 2. Cu radiation, R = 0.099 for 3510 reflexions.

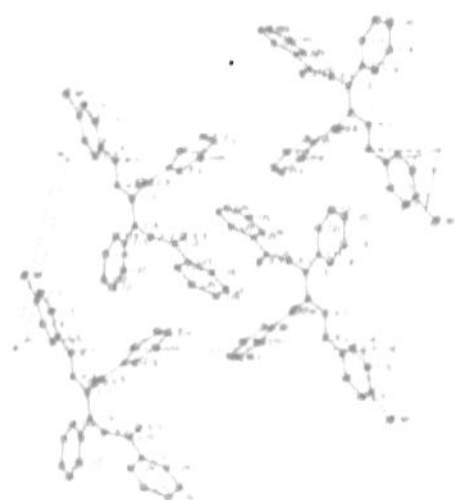

Fig. 1. The molecular packing and conformation of α-p-bromophenylazo-β-benzoyloxybenzalacetophenone.

This molecule isomerises to yield diphenyl triketone sym-benzoyl-p-bromophenylhydrazone, $C_{28}H_{19}BrN_2O_3$ (1); the molecular conformation and packing are shown in Fig. 1. The stereochemistry about the N=N double bond is trans, while the carbonyl group [C(10)-O(3)], which migrates in the isomerization, and the receiving p-bromophenylazo group are trans with respect to the C=C double bond. The central N(2)-C(7)=C(9)-C(17) group is accurately planar while the attached atoms O(2) and C(8) deviate by 0.07 and 0.04 Å, respectively, in opposite directions from the mean plane. The N=N bond length of 1.279(8) Å and the N(2)-C(7) length (1.401(10) Å) imply considerable conjugation between the N(1)-N(2) and C(7)-C(9) double bonds. Both carbonyl groups lie on the same side of the double bond (cis) but point in opposite directions.

The molecules pack in long stacks along the b axis (6.309 Å) with the oxygen atom of the C(8)=O(1) carbonyl group thrust into a cavity formed by C(5), N(2), and C(18) of the adjacent molecule; the distances from O(1) to those three atoms are 3.26, 3.46, and 3.41 Å, respectively. The orientation of the C(10)-C(23)-O(3) benzoyl group places the carbonyl oxygen atom 3.32 Å from C(28) in a molecule related by an inversion centre (Fig. 1). Other close contacts with the benzoyl group include O(3)···C(25) and O(3)···C(24) distances of 3.17 and 3.42 Å, respectively, between stacks of molecules parallel to the b axis. The shortest intermolecular contact between N(1) and C(10) of the migrating benzoyl group is 7.48 Å compared with the intramolecular distance of 5.33 Å.

1. Preceding report.

PHENYLAZOTRIBENZOYLMETHANE

$C_{28}H_{20}N_2O_3$ $(C_6H_5CO)_3CN{=}NC_6H_5$

D.B. PENDERGRASS, D.Y. CURTIN and I.C. PAUL, 1972. J. Amer. Chem. Soc., 94, 8722-8730.

Monoclinic, $P2_1/c$, a = 13.054, b = 10.352, c = 16.560 Å, β = 90.78°, D_m = 1.28, Z = 4. Cu radiation, R = 0.047 for 2309 reflexions.

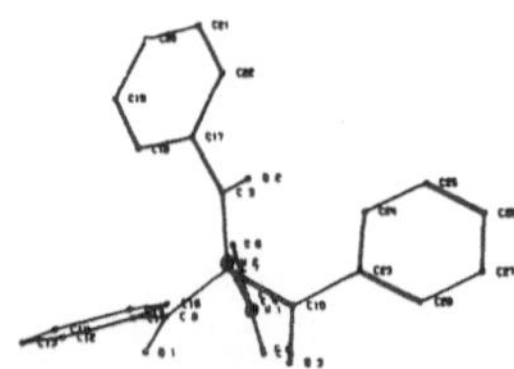

Fig. 1. Drawing of phenylazotribenzoylmethane viewed along the C(7)-N(2) bond.

This molecule (Fig. 1) undergoes a solid state rearrangement to α-phenylazo-β-benzoyloxybenzalacetophenone (1) and to diphenyl triketone sym-benzoylphenylhydrazone (2). While the crystal structure of this molecule is different from that of p-bromophenylazotribenzoylmethane (3), the molecular conformations are almost identical and appear to be influenced by intramolecular C=O···C=O (carbonyl) interactions. Features of the molecular conformation are two C(carbonyl)···O(carbonyl) distances of 2.667(4) and 2.758(4) Å and two C(carbonyl)···N distances of 2.698(4) and 2.998(5) Å. Similar relative orientations of carbonyl groups are found in other di- (and poly-) ketones. There are no intermolecular distances involving migrating centres less than 4.79 Å.

1. For a description of the bromo-derivative of this molecule, see preceding report.
2. This volume, p. 87.
3. R.T. PUCKETT, C.E. PFLUGER and D.Y. CURTIN, 1966. J. Amer. Chem. Soc., 88, 4637.

2-AMINOETHANETHIOSULPHURIC ACID

$C_2H_7NO_3S_2$ $NH_2C_2H_4S_2O_3H$

W.E. KEEFE and J.M. STEWART, 1972. Acta Cryst., B28, 2469-2474.

Triclinic, P$\bar{1}$, a = 8.57, b = 7.39, c = 9.85 Å, α = 80.5°, β = 86.0°, γ = 94.4°, D_m = 1.72, Z = 4. Cu radiation, R = 0.064 for 927 reflexions.

The molecules are zwitterions and in the crystalline state form intramolecular hydrogen bonded seven-membered rings; details of bond lengths in the two independent molecules in the asymmetric unit are in Fig. 1. The two molecules are linked by two equivalent hydrogen bonds (O(33)···H(133) = 1.85 Å and O(22)···H(422) = 2.08 Å) (Fig. 2) to form dimers. Each molecule also has additional intermolecular hydrogen bonds to non-dimer neighbours.

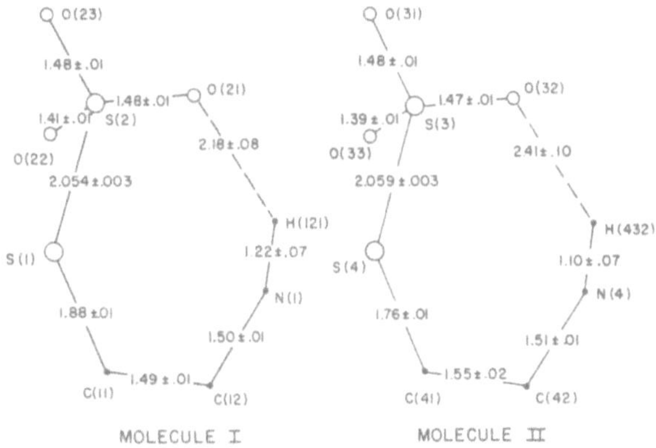

Fig. 1. Bond distances in the $NH_2C_2H_4S_2O_3H$ molecules.

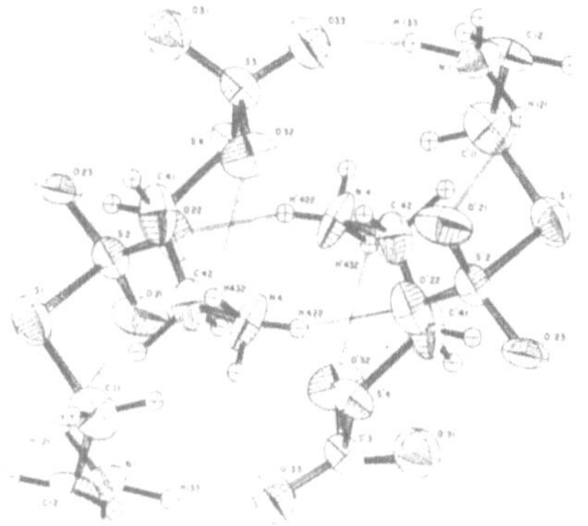

Fig. 2. Hydrogen bonding in $NH_2C_2H_4S_2O_3H$.

α,α'-DITHIOBISFORMAMIDINIUM DICHLORIDE

$C_2H_8C\ell_2N_4S_2$ $[SC(NH_2)_2]_2C\ell_2$

A.C. VILLA, A.G. MANFREDOTTI, M. NARDELLI and M.E.V. TANI, 1972. Acta Cryst., B28, 356-360.

Orthorhombic, Pbca, a = 8.78, b = 10.52, c = 19.69 Å, D_m not given, Z = 8. Cu **radiation,** R = 0.075 for 1211 reflexions.

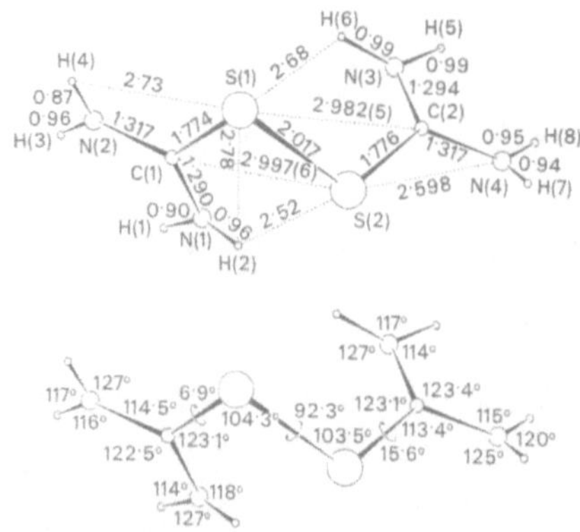

Fig. 1. Distances, angles and torsional angles in $[SC(NH_2)_2]_2^{2+}$, (σ = 0.002 Å for S-S, 0.006 Å for S-C and 0.007 - 0.008 Å for N-C).

Bond distances and angles in the dithiourea cation are in Fig. 1. Both thiourea segments are individually planar; the dihedral angle between their planes is 95.4°. The positions of the chlorine ions are mainly determined by their interactions with NH_2 groups to yield NH···Cℓ hydrogen bonds (in the range 3.120(5) - 3.181(6) Å).

HOMOTAURINE

(3-AMINOPROPANE SULPHONIC ACID)

$C_3H_9NO_3S$ $^{+}NH_3(CH_2)_3SO_3^{-}$

S. UEOKA, T. FUJIWARA and K. TOMITA, 1972. Bull. Chem. Soc. Japan, 45, 3634-3638.

Orthorhombic, $Pmn2_1$, a = 7.059, b = 5.492, c = 7.428 Å, D_m = 1.610, Z = 2. Mo radiation, R = 0.059 for 1143 reflexions.

The molecule is a zwitterion, and has a fully extended configuration; bond lengths and angles are close to expected values. The molecules are joined by N-H···O hydrogen bonds, 2.84 - 2.89 Å, to form a three-dimensional network.

2,5-DIBROMOBENZENESULPHONIC ACID TRIHYDRATE

$C_6H_{10}Br_2O_6S$ $H_7O_3^{+}\cdot C_6H_3Br_2SO_3^{-}$

J.-O. LUNDGREN, 1972. Acta Cryst., B28, 475-481.

Monoclinic, P2/c, a = 7.040, b = 7.656, c = 22.157 Å, β = 96.73°, Z = 4. Cu radiation, R = 0.048 for 1558 reflexions.

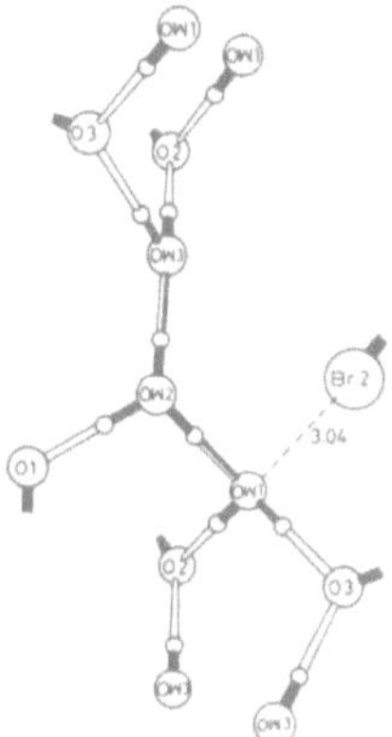

Fig. 1. Hydrogen bond assignment in and around the $H_7O_3^+$ ion (the oxygen atoms of the ion are labelled OW1, OW2, and OW3) in $H_7O_3^+ \cdot C_6H_3Br_2SO_3^-$.

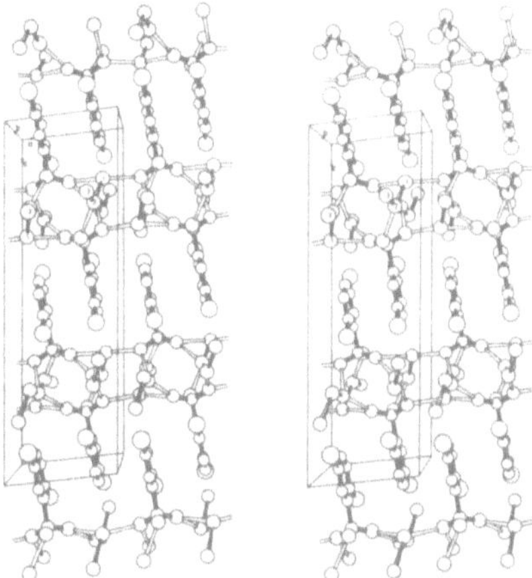

Fig. 2. View of the structure of $H_7O_3^+ \cdot C_6H_3BrSO_3^-$. Oxygen atoms of the $H_7O_3^+$ ion are denoted by 1, 2, and 3. Covalent bonds are filled. Hydrogen bonds within the $H_7O_3^+$ are half filled, other hydrogen bonds are open.

The structure comprises oxonium ions tightly bonded to two water molecules to form $H_7O_3^+$ ions (Fig. 1). The $H_7O_3^+$ ions and the 2,5-dibromobenzenesulphonate ions are hydrogen bonded to form layers (Fig. 2). No hydrogen bonds exist between individual $H_7O_3^+$ ions. Distances and angles involving hydrogen bonds are in Fig. 3. The geometry of the anion is normal.

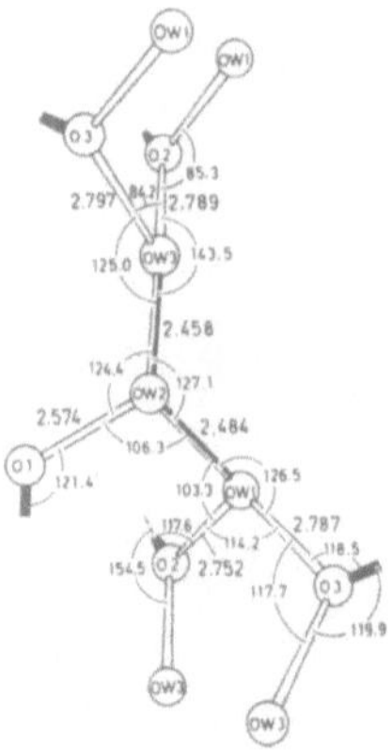

Fig. 3. Distances (σ = 0.009 Å) and angles (σ = 0.3 - 0.4°) involving hydrogen bonds in $H_7O_3^+ \cdot C_6H_3Br_2SO_3^-$.

2,5-DICHLOROBENZENESULPHONIC ACID TRIHYDRATE (AT -188°C)

$C_6H_{10}C\ell_2O_6S$ $H_7O_3^+ \cdot C_6H_3C\ell_2SO_3^-$

J.-O. LUNDGREN and P. LUNDIN, 1972. Acta Cryst., B28, 486-491.

Monoclinic, $P2_1/c$, a = 7.350, b = 23.516, c = 6.862 Å, β = 100.68°, Z = 4. Cu radiation, R = 0.064 for 1341 reflexions.

The structure comprises oxonium ions tightly bonded to two water molecules to form $H_7O_3^+$ ions; very large anisotropy in the thermal parameters of the oxygen atoms indicates some form of disorder. The 2,5-dichlorobenzenesulphonate ions are hydrogen bonded together to form layers (Fig. 1). Distances and angles in and around the $H_7O_3^+$ ion are in Fig. 2; bond lengths and angles in the dichlorobenzenesulphonate ion are normal except for the S-O distances (1.419, 1.426 and 1.448 Å), which are somewhat shorter than corresponding distances in the 2,5-dibromo analogue (1).

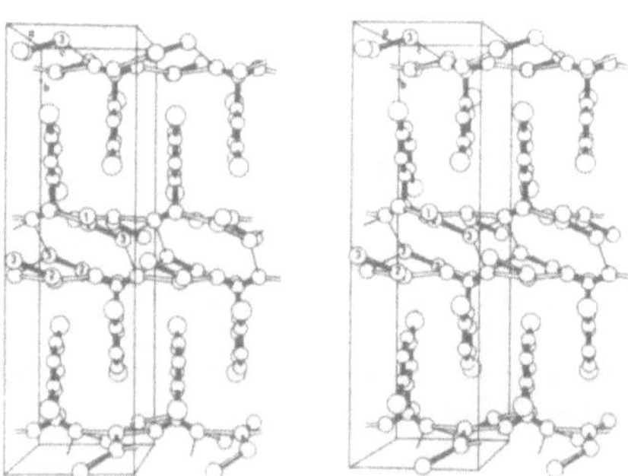

Fig. 1. Drawing of the crystal structure of $H_7O_3^+ \cdot C_6H_3C\ell_2SO_3^-$. Oxygen atoms of $H_7O_3^+$ are denoted by 1, 2, and 3. Covalent bonds are filled. Hydrogen bonds within $H_7O_3^+$ are half-filled, others are open. Other short contacts are represented by a single line.

1. Preceding report.

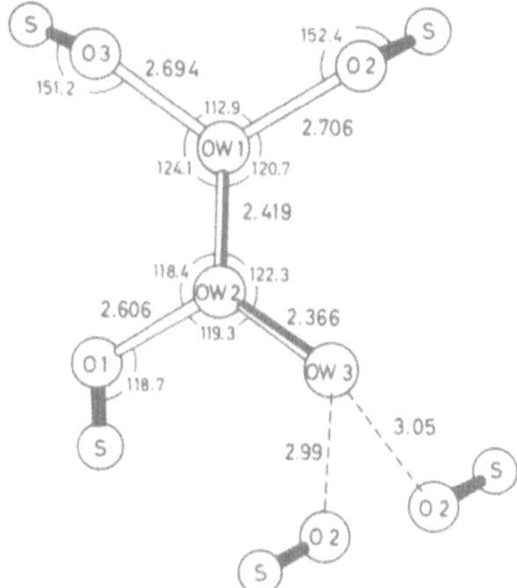

Fig. 2. Distances (σ = 0.007 - 0.009 Å) and angles (σ = 0.03 - 0.04°) involving hydrogen bonds in $H_7O_3^+ \cdot C_6H_3Cl_2SO_3^-$.

2-DIMETHYLAMINOETHYL SELENOLBENZOATE HYDROCHLORIDE

$C_{11}H_{16}ClNOSe$ $C_6H_5 \cdot CO \cdot Se(CH_2)_2 \cdot \overset{+}{N}H(CH_3)_2 \cdot Cl^-$

D.D. DEXTER, 1972. Acta Cryst., B28, 49-54.

Monoclinic, $P2_1/c$, a = 12.387, b = 6.945, c = 15.523 Å, β = 94.57°, D_m = 1.47, Z = 4. Cu radiation, R = 0.051 for 2411 reflexions.

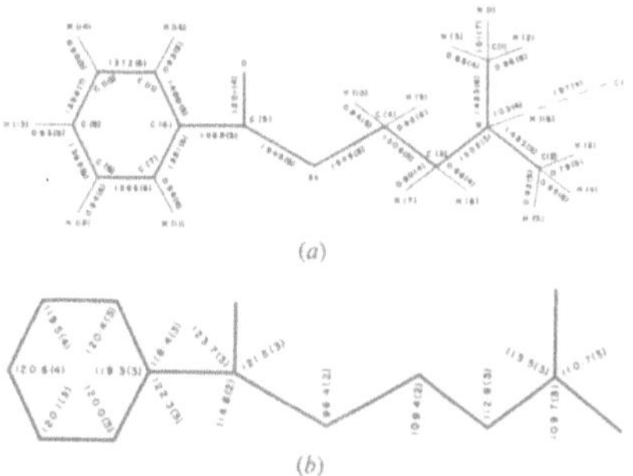

Fig. 1. 2-Dimethylaminoethyl selenolbenzoate hydrochloride. (a) Bond distances. (b) Bond angles (with angles involving H atoms omitted).

Details of the molecular geometry are in Fig. 1. The $\overset{+}{N}$-C-C-Se group adopts the trans conformation with a torsion angle about the C-C bond of 174.4°. The benzene ring is significantly non-planar, with C(6) = 0.035(3) Å below and C(7) = 0.052(4) Å above the plane C(8)···C(11). In the crystal structure (Fig. 2) the protonated nitrogen atoms are hydrogen bonded to Cl^- ions with $N^+ \cdots Cl^-$ = 3.07 Å.

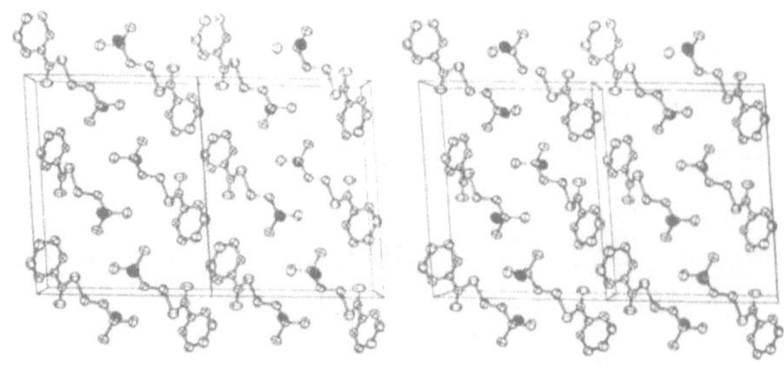

Fig. 2. 2-Dimethylaminoethyl selenobenzoate hydrochloride. Diagram showing molecular packing along b. Hydrogen atoms are omitted and chloride ions shaded.

DIPHENYL DISULPHONE

$C_{12}H_{10}O_4S_2$ $C_6H_5SO_2SO_2C_6H_5$

C.Th. KIERS and A. VOS, 1972. Rec. Trav. Chim. Pays-Bas, 91, 127-132.

Triclinic, P$\bar{1}$, a = 6.331, b = 7.936, c = 8.416 Å, α = 57.63°, β = 112.99°, γ = 116.94°, Z = 1. Mo radiation, R = 0.04 for 2524 reflexions. Molecular symmetry: centre.

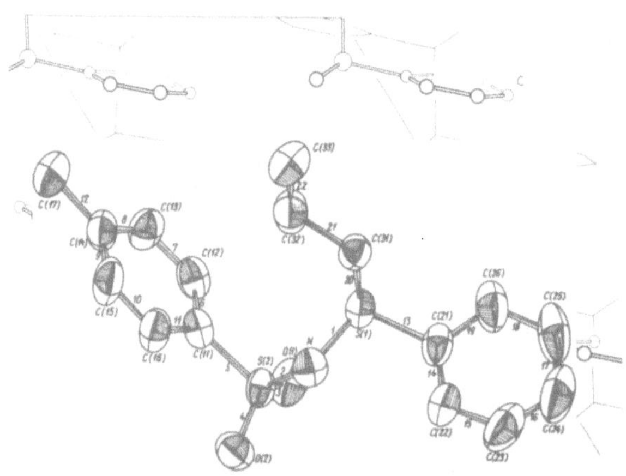

Fig. 1. Structure of diphenyl disulphone projected along a* on to the b-c plane. Molecules lying around the inversion centres at x = -1 are indicated by thin lines. Perpendicular distances between the best S-C_6H_5 planes are: A-B = 3.54, B-C = 3.68 Å.

A view of the crystal structure (1) is given in Fig. 1. Bond lengths are S-O = 1.428(1), S-C = 1.753(1), S-S = 2.193(1) Å. The O-S-O angle is 120.3(1)° and S-S-C is 101.5(1)°. There are no unusual intermolecular contacts.

1. Structure Reports, 11, 681.

BIS(4-BROMOPHENYLSULPHONYL)METHANE

$C_{13}H_{10}Br_2O_4S_2$ $(BrC_6H_4SO_2)_2CH_2$

J. BERTHOU, G. JÉMINET and A. LAURENT, 1972. Acta Cryst., B28, 2480-2485.

Monoclinic, C2/c, a = 30.05, b = 5.015, c = 11.455 Å, β = 118°, Z = 4. Cu radiation, R = 0.06 for 779 reflexions. Molecular symmetry C_2.

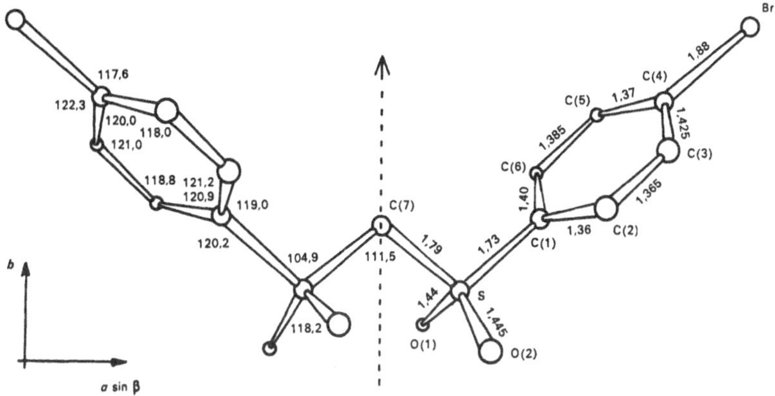

Fig. 1. Bond lengths (σ = 0.02 Å) and angles (σ = 0.7°) in $(Br(C_6H_4)SO_2)_2CH_2$.

The molecular conformation and geometry details are given in Fig. 1. In the crystal structure molecules form layers approximately parallel to (001). There are no unusual intermolecular contacts.

S-(2-METHOXYPHENYL)-N-(2,6-DIMETHYLPHENYL)DITHIOURETHANE OXIDATION PRODUCTS

$C_{16}H_{17}NO_2S_2$

K. KATO, 1972. Acta Cryst., B28, 2653-2658.

Monoclinic, $P2_1/c$, a = 8.165, b = 21.654, c = 20.190 Å, β = 110.79°, Z = 8. Cu radiation, R = 0.053 for 3227 reflexions.

Fig. 1. The S-oxide and sulphenic acid oxidation products of S-(2-methoxyphenyl)-N-(2,6-dimethylphenyl)dithiourethane.

There are two crystallographically independent molecules in the asymmetric unit and the crystal seems to be a solid solution of the S-oxide and sulphenic acid shown in Fig. 1. Principal bond lengths are (in molecule 1 first with those for molecule 2 in square brackets) C-S(phenyl) = 1.764(4) [1.752(4)], C-S(O) = 1.727(4) [1.706(4)], S-O = 1.602(3) [1.568(3)], C-N(phenyl) = 1.296(4) [1.308(4)], S-C(aromatic) = 1.781(4) [1.774(4)] Å.

S-PROPYL,S-PHENYL-N-p-TOLYLSULPHONYL SULPHILIMINE

$C_{16}H_{19}NO_2S_2$

C_6H_5
SN-SO_2-$C_6H_4CH_3$
C_3H_7

A. KÁLMÁN and K. SASVÁRI, 1972. Cryst. Struct. Comm., 1, 243-246.

Monoclinic, $P2_1/a$, a = 12.060, b = 11.902, c = 12.589 Å, β = 111.0°, D_m = 1.276, Z = 4. Cu radiation, R = 0.087 for 1621 reflexions.

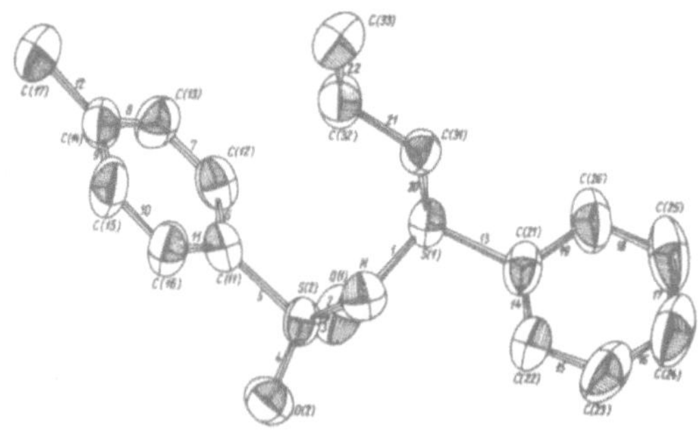

Fig. 1. Molecular conformation of $C_{16}H_{19}NO_2S_2$.

The stereochemistry of the molecule is shown in Fig. 1. Principal bond lengths are S(1)-N = 1.620(7); S(2)-N = 1.618(7); S-O = 1.436, 1.441(7); S-C(phenyl) = 1.763(8), 1.780(9); S-C(methylene) = 1.820(9) Å. Thus the S-N bond lengths are identical.

METHYL 2-PHENYL-(3',4'-DIMETHYL-2-PHENYL)VINYL SULPHOXIDE

$C_{17}H_{18}OS$

D. TRANQUI and H. FILLION, 1972. Acta Cryst., B28, 3306-3315.

Monoclinic, C2/c, a = 30.497, b = 5.951, c = 19.273 Å, β = 121.74°, D_m = 1.25, Z = 8. Cu radiation, R = 0.06 for 2321 reflexions.

The molecular conformation (Fig. 1) of this isomer has phenyl ring orientations which can be explained in terms of the minimization of non-bonding interactions between the ring atoms. The vinyl C(15)-C(16) bond length of 1.390(9) Å is longer than the usual double bond value (Fig. 2) while the S-O distance, 1.521(3) Å, represents a bond character intermediate between a single bond and a double bond; the S-C(16)(sp^2) bond, 1.722(4) Å is shorter than the S-C(17)(sp^3) bond, 1.827(4) Å, which is as expected. The vinyl group, C(16)C(15)SH(9)C(1)C(9), is approximately planar. The molecules are packed together by van der Waals contacts (e.g. C(12)···C(16) = 3.59 Å).

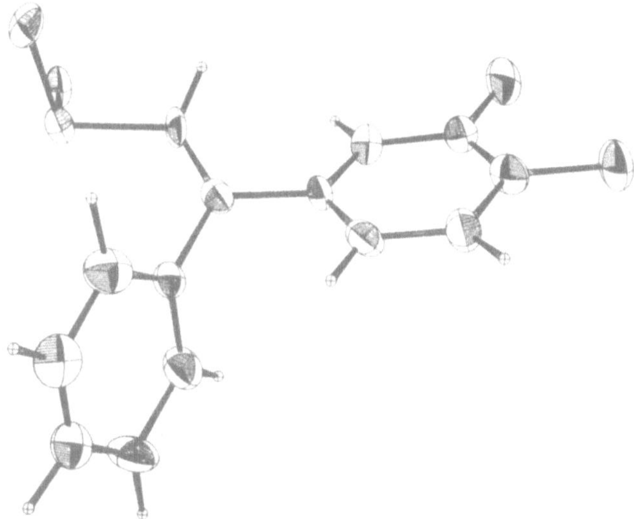

Fig. 1. Conformation of the methyl-2-phenyl-(3',4'-dimethyl-2-phenyl)vinyl sulphoxide molecule.

Fig. 2. Bond lengths (σ = 0.004 - 0.009 Å) for methyl-2-phenyl-(3',4'-dimethyl-2-phenyl)vinyl sulphoxide.

2-PHENYL-4-ACETYLPHENOXY-2,6-DIMETHYLPHENYLIMINOMETHANE-SULPHENIC ACID

$C_{23}H_{21}NO_3S$

K. KATO, 1972. Acta Cryst., B28, 55-59.

Monoclinic, $P2_1/c$, a = 13.461, b = 16.638, c = 11.309 Å, β = 123.549°, D_m not given, Z = 4. Cu radiation, R = 0.049 for 3169 reflexions.

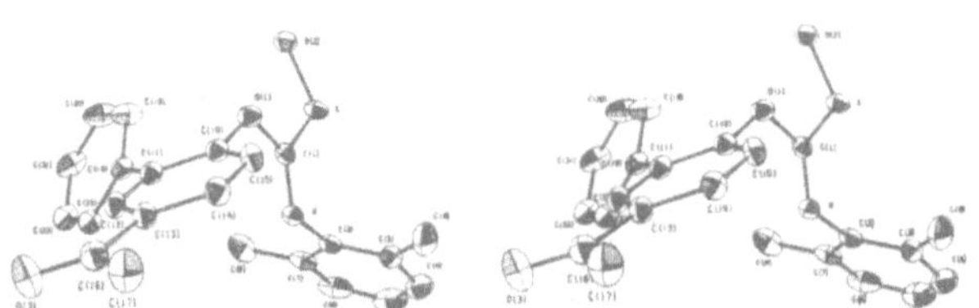

Fig. 1. View of the $C_{23}H_{21}NO_3S$ molecule.

Fig. 2. Bond lengths (σ = 0.002 - 0.005 Å; 0.02 - 0.04 Å for bonds involving H atoms) in $C_{23}H_{21}NO_3S$.

The benzene rings are planar and their orientations with respect to each other are shown in Fig. 1; bond lengths are in Fig. 2. The molecules are hydrogen bonded (O(1)-H···N = 2.794 Å) to form infinite chains parallel to <u>c</u>.

TETRAPHENYL ORTHOTHIOCARBONATE

$C_{25}H_{20}S_4$ $(C_6H_5S)_4C$

K. KATO, 1972. Acta Cryst., <u>B28</u>, 606-610.

Monoclinic, $P2_1/n$, a = 10.266, b = 22.696, c = 10.115 Å, β = 103.185°, D_m not given, Z = 4. Cu radiation, R = 0.046 for 3774 reflexions.

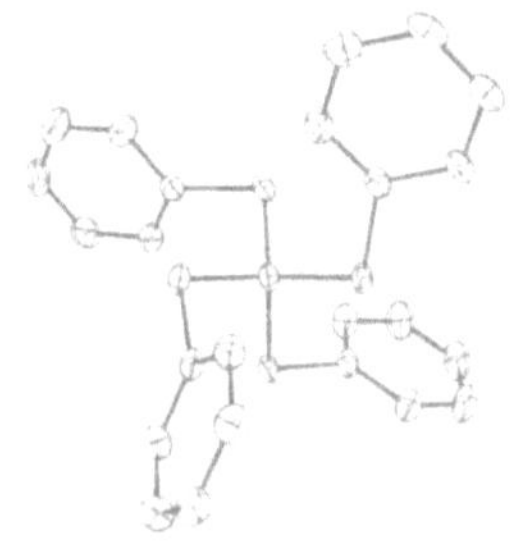

Fig. 1. Diagram of $(C_6H_5S)_4C$.

The molecular conformation is shown in Fig. 1. Mean bond distances are S-C(sp^3) = 1.826(3), S-C(sp^2) = 1.776(3), C-C(aromatic) = 1.376(5) Å; the mean C-S-C angle is 106.1(1)°. In the crystal the shorter intermolecular contacts are between (a) phenyl ring carbon atoms (C···C = 3.523 Å), (b) sulphur atoms (S···S = 3.673 Å), and (c) S and a phenyl carbon (3.667 Å).

DIPHENYLDI(PHENYL-DI(TRIFLUOROMETHYL)METHOXY)SULPHURANE

$C_{30}H_{20}F_{12}O_2S$ $(C_6H_5)_2S[OC(CF_3)_2C_6H_5]_2$

I.C. PAUL, J.C. MARTIN and E.F. PEROZZI, 1972. J. Amer. Chem. Soc., 94, 5010-5017.

Triclinic, $P\bar{1}$, a = 10.026, b = 14.268, c = 10.802 Å, α = 109°50', β = 92°52', γ = 95°20', D_m not given, Z = 2. Cu radiation, R = 0.070 for 3364 reflexions.

Fig. 1. View of the diphenyldialkoxysulphurane molecule.

The molecule has an approximate trigonal bipyramidal geometry about sulphur, with a lone pair considered to occupy one equatorial position (Fig. 1), and two phenyl ligands occupying the other two equatorial positions (C-S-O angles range from 86.4(2) to 91.2(2)°; the C-S-C angle is 104.4(3)°), and with the electronegative alkoxy ligands occupying apical positions (O-S-O angle is 175.1(2)° with the distortion from linearity in the direction of the equatorial phenyl rings). The S-O bond lengths (1.889(4) and 1.916(4) Å) are approximately 0.2 Å longer than the sum of the covalent radii as would be expected from an S-O bond order less than unity.

METHYLOXOCARBONIUM TETRACHLOROALUMINATE

$C_2H_3A\ell C\ell_4O$ $A\ell C\ell_4{}^- CH_3CO^+$

J.-M. Le CARPENTIER and R. WEISS, 1972. Acta Cryst., B28, 1421-1429.

Orthorhombic, Pnma, a = 11.158, b = 7.108, c = 10.796 Å, Z = 4. Mo radiation, R = 0.037 for 454 reflexions.

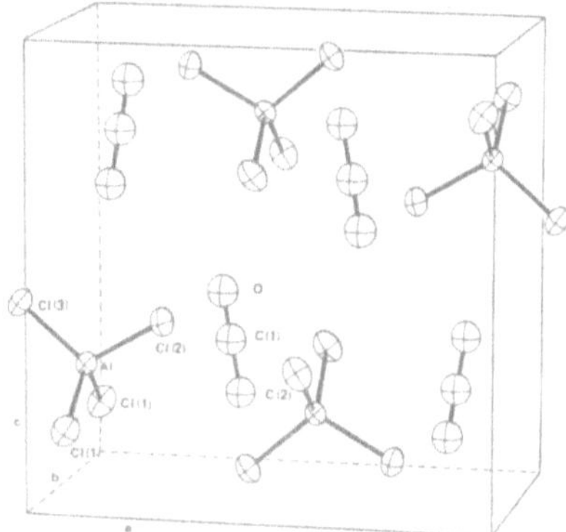

Fig. 1. Arrangement of the ions in the structure of $A\ell C\ell_4^- \cdot CH_3CO^+$.

The crystal structure (Fig. 1) provides direct proof of an ionic structure in the solid state. The methyloxocarbonium ion is linear but disordered with a C(methyl)···O distance 2.56 Å. The $A\ell C\ell_4^-$ ion lies on a mirror plane with mean Aℓ-Cℓ bond length 2.125(2) Å and Cℓ-Aℓ-Cℓ angles in the range 108.3(1) - 110.2(1)°. No association of the oxocarbonium ion with the anions, other than ionic crystal packing is observed in accordance with the infrared spectrum.

The corresponding tetrachlorogallate derivative is isomorphous with the tetrachloroaluminate. The structure $SbC\ell_6^- \cdot CH_3CO^+$ is also described in detail (1).

1. Following report.

METHYLOXOCARBONIUM HEXACHLOROANTIMONATE

$C_2H_3C\ell_6OSb$ $SbC\ell_6^- CH_3CO^+$

J.-M. Le CARPENTIER and R. WEISS, 1972. Acta Cryst., B28, 1421-1429.

Orthorhombic, Imm2, a = 7.144, b = 9.091, c = 7.974 Å, Z = 2. Mo radiation, R = 0.029 for 648 reflexions.

This structure determination establishes that the structure is ionic with well separated $SbC\ell_6^-$ anions and CH_3CO^+ cations (Fig. 1). Both ions have crystallographic C_{2v} symmetry (which demands that the methyl hydrogen atoms be disordered). The C-O distance is 1.109(30) Å while the C-C is 1.452(24) Å. The mean Sb-Cℓ distance in the octahedral anion is 2.364(2) Å. The structure $A\ell C\ell_4^- \cdot CH_3CO^+$ is also described in detail (1).

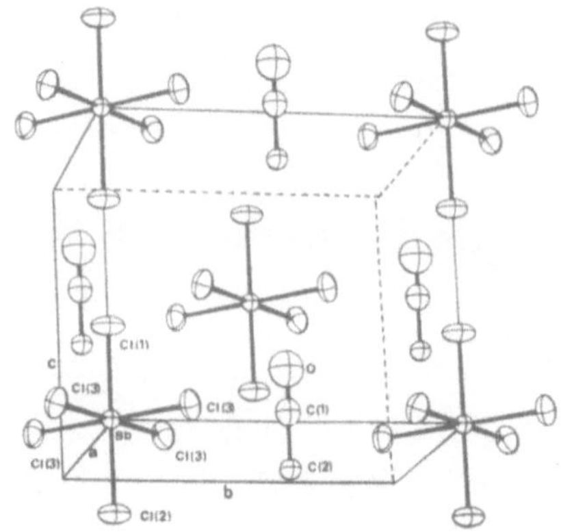

Fig. 1. Arrangement of the ions in the $SbC\ell_6^{-} \cdot CH_3CO^{+}$ structure.

1. Preceding report.

POTASSIUM CYANODINITROMETHANIDE

$C_2KN_3O_4$ $K^{+}[NCC(NO_2)_2]^{-}$

B. KLEWE, 1972. Acta Chem. Scand., 26, 1921-1930.

Orthorhombic, $P2_12_12_1$, a = 6.82, b = 12.26, c = 6.64 Å, D_m = 1.99, Z = 4. Mo radiation, R = 0.049 for 926 reflexions.

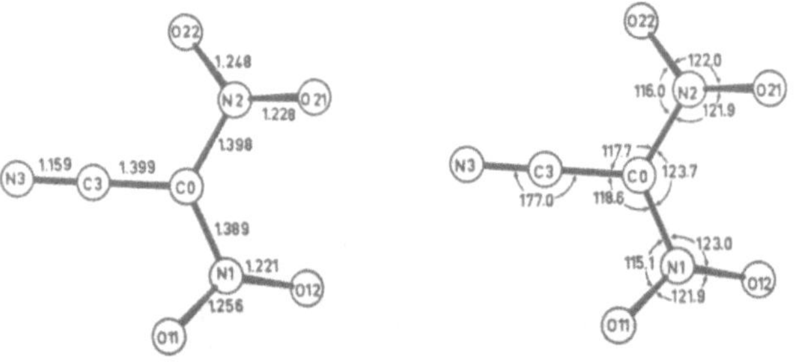

Fig. 1. Bond lengths (σ = 0.004 Å) and bond angles in the anion of the $K^{+}[NCC(NO_2)_2]^{-}$ system.

The anion is propeller-shaped, with the nitro groups twisted 7° from planar conformation, and possesses C_2 (non crystallographic) symmetry. The central carbon atom and its neighbour atoms are coplanar. Bond distances and angles (Fig. 1) are normal. The coordination about the potassium ion is irregular and quite different from that of the rubidium salt (1), although in both structures the metal ion is surrounded by eight oxygen atoms and one cyano nitrogen atom. $K^+\cdots O$ distances are in the range 2.93 - 3.08 Å, $K^+\cdots N$ is 2.81 Å.

1. Following report.

RUBIDIUM CYANODINITROMETHANIDE

$C_2N_3O_4Rb$ $Rb^+[NCC(NO_2)_2]^-$

H.J. BJØRNSTAD and B. KLEWE, 1972. Acta Chem. Scand., 26, 1874-1882.

Orthorhombic, $P2_12_12_1$, a = 7.56, b = 14.96, c = 5.09 Å, D_m = 2.46, Z = 4. Mo radiation, R = 0.063 for 1195 reflexions.

Fig. 1. Bond distances (σ = 0.006 - 0.007 Å) and angles in the anion of the $Rb^+[NCC(NO_2)_2]^-$ system.

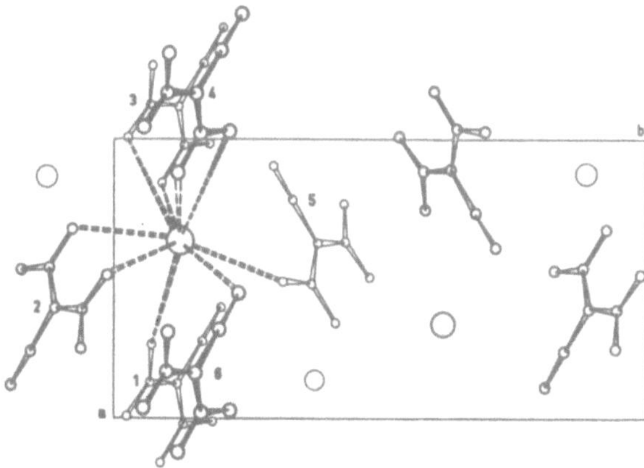

Fig. 2. The crystal structure and coordination around the rubidium in $Rb^+[NCC(NO_2)_2]^-$.

The anion $NCC(NO_2)_2$ is essentially planar with the nitro groups slightly twisted out of the plane. The twists have opposite directions and the anion is thus not propeller-shaped as was found in the potassium derivative (1). Bond distances and angles (Fig. 1) are normal. The O(12)···O(21) intra-ion contact is 2.62 Å. The cation is coordinated to eight oxygen atoms and one cyano nitrogen (Fig. 2), with Rb···O distances in the range 2.91 - 3.20 Å and Rb···N = 3.20 Å.

1. N.V. GRIGOR'EVA, N.V. MARGOLIS, I.N. ŠOKHOR, I.V. TSELINSKIJ and V.V. MEL'NIKOV, 1969. J. Struct. Chem., 10, 834; B. KLEWE, 1972. Acta Chem. Scand., 26, 1921; see preceding report.

DIRUBIDIUM 1,1,2,2-TETRANITROETHANEDIIDE

$C_2N_4O_8Rb_2$ $2Rb^+[(NO_2)_2CC(NO_2)_2]^{2-}$

B. KLEWE, 1972. Acta Chem. Scand., 26, 1049-1057.

Orthorhombic, Pccn, a = 7.70, b = 13.35, c = 8.84 Å, Z = 4. Mo radiation, R = 0.077 for 926 reflexions.

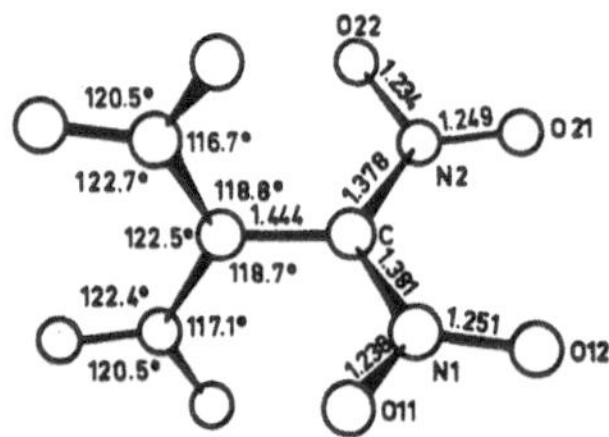

Fig. 1. Bond distances (σ = 0.009, 0.010 and 0.014 Å for N-O, C-N and C-C bonds respectively) and bond angles of the anion $[(NO_2)_2CC(NO_2)_2]^{2-}$.

The anions are centred on twofold rotation axes with an angle of 68° between the planes of the $C(NO_2)_2$ groups. In the potassium salt the angle is 63° (1). The double bond character of the C-C bond is small (Fig. 1). The cation is coordinated to eight oxygens of six different cations with Rb···O distances in the range 2.83 - 3.14 Å. A useful table summarising the results of structure determinations of eleven 1,1-dinitrocarbanion salts is given in the paper.

1. Structure Reports, 33B, 34.

ETHYLOXOCARBONIUM TETRACHLOROGALLATE

$C_3H_5C\ell_4GaO$ $GaC\ell_4^- \cdot CH_3CH_2CO^+$

J.-M. Le CARPENTIER and R. WEISS, 1972. Acta Cryst., B28, 1430-1437.

Monoclinic, $P2_1/c$, a = 7.280, b = 9.623, c = 18.945 Å, β = 132.85°, Z = 4. Mo radiation, R = 0.051 for 802 reflexions.

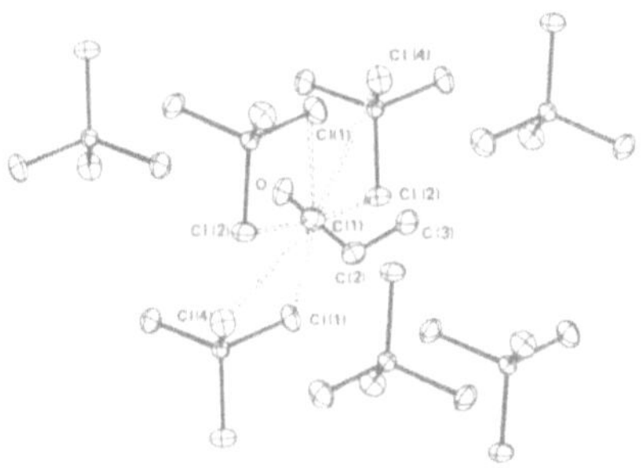

Fig. 1. Environment of the $CH_3CH_2CO^+$ ion in the ethyloxocarbonium tetrachlorogallate structure.

The $GaC\ell_4^-$ and $CH_3CH_2CO^+$ ions occupy general positions in the unit cell and the environment of a $CH_3CH_2CO^+$ ion is shown in Fig. 1. There are six carbonyl carbon···chlorine contacts with C···Cℓ in the range 3.42 - 3.92 Å. Referring to Fig. 1, the C(2)-C(1)-O angle is linear within experimental error (177.5(15)°); C(1)-O = 1.10(2), C(1)-C(2) = 1.44(2) Å. Bond C(2)-C(3) (1.54(3) Å) is typical of a $C(sp^3)$-$C(sp^3)$ single bond. The $GaC\ell_4^-$ ion is in a slightly deformed tetrahedral conformation with Ga-Cℓ(mean) = 2.159 Å and Cℓ-Ga-Cℓ angles in the range 107.2 - 111.1(2)°.

The structure of isopropyloxocarbonium hexachloroantimonate is also described in this paper (1).

1. Following report.

ISOPROPYLOXOCARBONIUM HEXACHLOROANTIMONATE

$C_4H_7C\ell_6OSb$ $SbC\ell_6^- \cdot (CH_3)_2CHCO^+$

J.-M. Le CARPENTIER and R. WEISS, 1972. Acta Cryst., B28, 1430-1437.

Orthorhombic, $P2_12_12_1$, a = 15.432, b = 10.718, c = 7.042 Å, Z = 4. Mo radiation, R = 0.036 for 1286 reflexions.

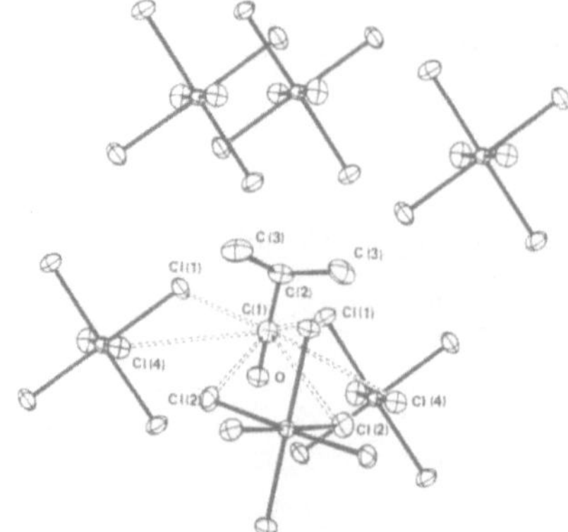

Fig. 1. Environment of the $(CH_3)_2CHCO^+$ ion in $SbC\ell_6^- \cdot (CH_3)_2CHCO^+$.

Both ions have crystallographic m symmetry. Interion contacts are shown in Fig. 1. The six C(carbonyl)···Cℓ contacts are in the range 3.27 - 3.72 Å. The $SbC\ell_6^-$ ion departs very little from ideal octahedral symmetry and the mean Sb-Cℓ distance is 2.370 Å. In the cation the C-C-O angle is 175.7(5)°, the slight departure from linearity being attributed to interion crystal packing forces. The C-O bond length (1.116(10) Å) corresponds to the value found in other oxocarbonium ions. Other distances are C(1)-C(2) = 1.439(12) and C(2)-C(3) = 1.551(11) Å.

POTASSIUM p-CHLOROPHENYLDINITROMETHANIDE

$C_7H_4C\ell KN_2O_4$ $C\ell C_6H_4C(NO_2)_2^-K^+$

B. KLEWE and S. RAMSOY, 1972. Acta Chem. Scand., 26, 1058-1066.

Orthorhombic, Pbca, a = 12.67, b = 20.25, c = 7.41 Å, D_m = 1.80, Z = 8. Mo radiation, R = 0.059 for 827 reflexions.

In the anion the nitro groups are twisted in opposite directions (10° and 7°) from the N-C-N plane; the plane of the phenyl ring makes an angle of 71° with this plane. Bond distances within the anion are normal; the exocyclic C-C bond is 1.464(6) Å. The coordination about the potassium ion is irregular (Fig. 1). K···O distances are in the range 2.87 - 3.02 Å.

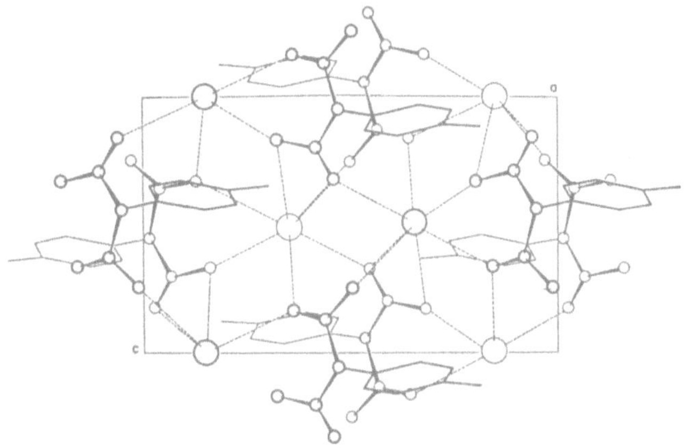

Fig. 1. Schematic drawing of the structure of $K^+[C\ell C_6H_4C(NO_2)_2]^-$ viewed along the b-axis with cation contacts dotted.

2-METHYLPHENYLOXOCARBONIUM HEXACHLOROANTIMONATE

$C_8H_7C\ell_6OSb$ $[CH_3C_6H_4CO]^+[SbC\ell_6]^-$

B. CHEVRIER, J.-M. Le CARPENTIER and R. WEISS, 1972. Acta Cryst., B28, 2673-2677.

Orthorhombic, Pbca, a = 12.38, b = 12.60, c = 19.42 Å, Z = 8. Mo radiation, R = 0.044 for 1797 reflexions.

This salt consists of discrete octahedral $SbC\ell_6^-$ ions (Sb-Cℓ = 2.370 - 2.378(2) Å) and o-methylphenyloxocarbonium ions. The closest interion contacts are between the carbonyl carbon atom and $SbC\ell_6^-$ chlorine atoms (C···Cℓ = 3.31 - 3.46 Å). The C-O bond (1.111(13) Å) is similar to, and the adjacent C-C bond (1.387(14) Å) is much shorter than, the values found in alkyloxycarbonium ions. The exocyclic C-C-O angle is essentially linear (178.7(8)°).

2,2,5,5-TETRAMETHYL-3-CARBAMIDOPYRROLINE-1-OXYL

$C_9H_{15}N_2O_2$

J.W. TURLEY and F.P. BOER, 1972. Acta Cryst., B28, 1641-1644.

Monoclinic, $P2_1/c$, a = 8.039, b = 11.324, c = 11.537 Å, β = 91.75°, D_m not given, Z = 4. Cu radiation, R = 0.052 for 1474 reflexions.

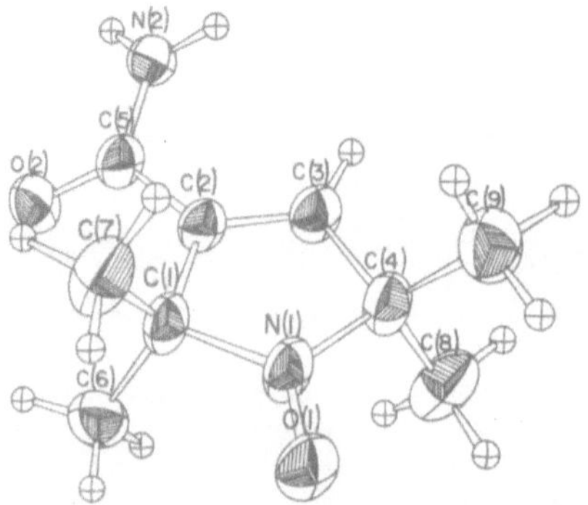

Fig. 1. Molecular structure of the nitroxide free radical 2,2,5,5-tetramethyl-3-carbamidopyrroline-1-oxyl.

The molecular structure is depicted in Fig. 1. The ring system and nitroxide function are planar. The N-O bond distance (1.267(5) Å) is intermediate between the N-O single and double bond length and is consistent with an electronic structure in which bonding σ- and π-orbitals are filled by electron pairs and the anti-bonding π*-orbital contains the lone electron. Bond lengths within the ring are normal and indicate the absence of delocalisation in the ring. The plane of the amide group makes an angle of 29.0° with that of the ring. Two types of N-H···O hydrogen bonds, both linear and fairly weak (N···O = 2.908 and 3.003 Å) exist in the crystal structure.

DIPOTASSIUM cis-HEXACYANOBUTENEDIIDE

$C_{10}K_2N_6$ $K_2((CN)_2CC(CN)C(CN)C(CN)_2)$

E. MAVERICK, E. GOLDISH, J. BERNSTEIN, K.N. TRUEBLOOD, S. SWAMINATHAN and R. HOFFMANN, 1972. J. Amer. Chem. Soc., 94, 3364-3370.

Monoclinic, C2/c, a = 20.077, b = 7.421, c = 17.355 Å, β = 117.8°, D_m = 1.65, Z = 8. Cu radiation, R = 0.072 for 1767 reflexions.

Each K^+ ion is surrounded by seven terminal N atoms from six different anions and is enclosed between planes of anions. In the anion observed bond lengths (Fig. 1) are in agreement with the predictions of M.O. theory. The anion shows considerable deviations from planarity but retains approximately C_2 symmetry. The terminal cyano groups are twisted about the C(1)-C(2) and C(3)-C(4) bonds of the central four-carbon chain (torsion angles C(5)C(1)-C(2)C(3) and C(10)C(4)-C(3)C(2) are -22.2 and -40.7° respectively), and the angles C(1)-C(2)-C(3) and C(2)-C(3)-C(4) are enlarged from the sp^2 angle of 120° to 127.6 and 125.8°, respectively (Fig. 1). These distortions lead to minimum nonbonded C-C distances of 2.69, C(7)-C(8); 2.77, C(6)-C(7); 2.95, C(8)-C(9); and 2.96 Å, C(5)-C(10).

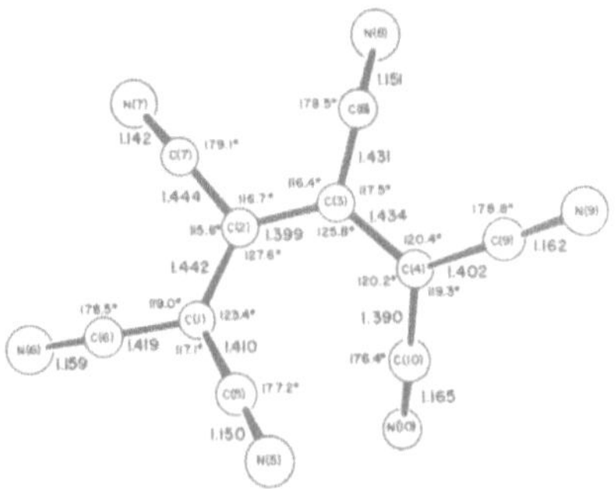

Fig. 1. Bond lengths and bond angles in the $HCBD^{2-}$ anion. Average σ = 0.009 Å for bond lengths and 0.5° for bond angles.

RUBIDIUM 7,7,8,8-TETRACYANOQUINODIMETHANE (AT -160°C)

$C_{12}H_4N_4Rb$

A. HOEKSTRA, T. SPOELDER and A. VOS, 1972. Acta Cryst., B28, 14-25.

Monoclinic, $P2_1/c$, a = 7.187, b = 12.347, c = 13.081 Å, β = 98.88°, D_m = 1.6 at 20°C, Z = 4. Cu radiation, R = 0.066 for 5512 reflexions.

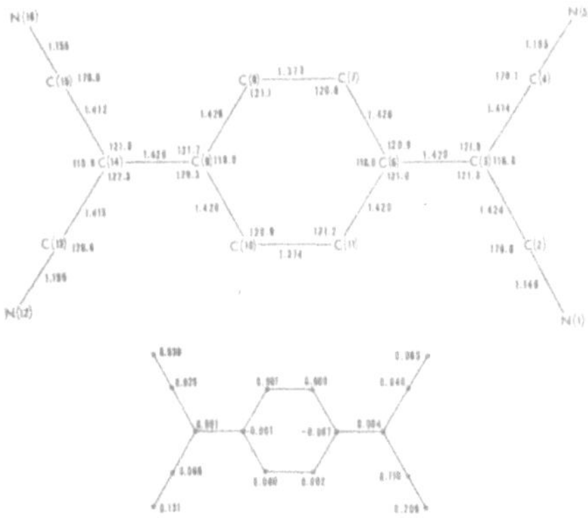

Fig. 1. Molecular geometry of the $TCNQ^-$ ion in Rb-TCNQ. Bond lengths (σ = 0.004 Å) and angles (σ = 0.27°) are shown above and deviations from the best plane through the quinodimethane skeleton are given below.

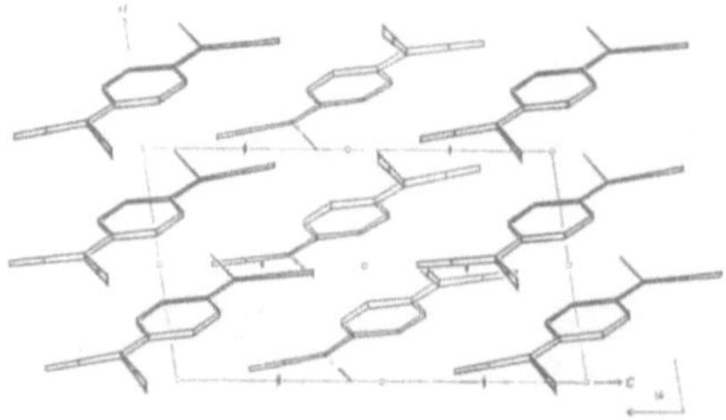

Fig. 2. Projection of the Rb-TCNQ structure along the b axis. The Rb^+ ions coincide with the centres of molecules in this projection and are not shown.

The bond lengths in the $TCNQ^-$ ion (Fig. 1) show that it has quinoid character. The ion is not precisely planar and has the shape of a very shallow boat (see Fig. 1). The TCNQ ions form charge-resonance bounded rows along the a axis (Fig. 2). The interactions between rows is small. Within a row interplanar spacings of 3.159 and 3.484 Å alternate. The Rb^+ ions also form rows along a with alternating distances of 3.483 and 3.726 Å between ions; these distances are much greater than twice the van der Waals radii of Rb^+, ∿3.0 Å. The Rb^+ ion is octahedrally surrounded by eight TCNQ nitrogen atoms with Rb···N in the range 2.982 - 3.108 Å.

TETRAMETHYL-2,2,5,5-AZA-1-CYCLOPENTANONE-3-AZINE-3-OXYL

$C_{16}H_{28}N_4O_2$

B. CHION, A. CAPIOMONT and J. LAJZEROWICZ, 1972. Acta Cryst., B28, 618-619.

Monoclinic, $P2_1/c$, a = 5.865, b = 15.72, c = 10.953 Å, β = 116.358°, D_m not given, Z = 2. Mo radiation, R = 0.11 for 1100 reflexions.

The molecules lie on inversion centres. The molecular dimensions (Fig. 1) show that the central N-N bond is effectively single. The five-membered ring is planar as is the C-NO-C group.

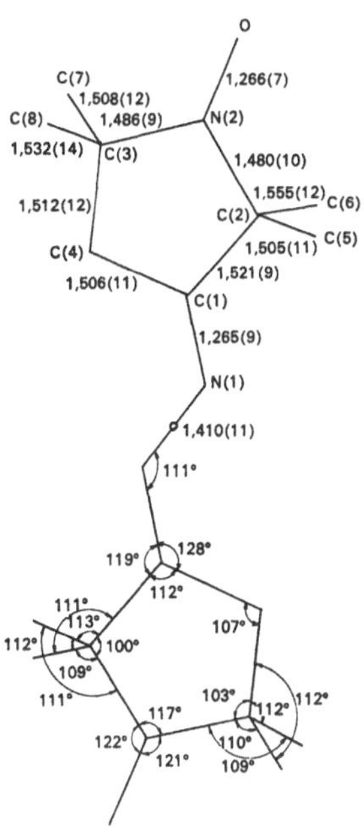

Fig. 1. Bond distances and angles in the nitroxide radical $C_{16}H_{28}N_4O_2$.

BIS[(TETRAMETHYLETHYLENEDIAMINE)LITHIUM(I)]NAPHTHALENIDE

$C_{22}H_{40}Li_2N_4$ $\{Li[(CH_3)_2N(CH_2)_2N(CH_3)_2]\}_2C_{10}H_8$

J.J. BROOKS, W. RHINE and G.D. STUCKY, 1972. J. Amer. Chem. Soc., 94, 7346-7351.

Monoclinic, $P2_1/c$, a = 8.970, b = 15.083, c = 11.474 Å, β = 129.5°, D_m = 1.0, Z = 2. Mo radiation, R = 0.059 for 779 reflexions.

The structure (Fig. 1) consists of the $C_{10}H_8$ moiety bound to two lithium atoms, each of which is coordinated to an N,N,N',N'-tetramethylethylenediamine group. The molecule is situated on an inversion centre so that only half of the naphthalene ring atoms, one lithium atom, and one TMEDA molecule are unique. The coordination sphere of the lithium atom contains two tertiary amine nitrogen atoms and one unsaturated organic group. The location of the lithium atom is not consistent with the simple electrostatic considerations but is explicable in terms of the formation of a multicentre bond between a vacant lithium *p* orbital and the combination of ring-carbon-atom *p* orbitals which is derived from the highest occupied molecular orbital of the isolated carbanion. The bond lengths in the naphthalenide fragment are close to what one would expect from the symmetry of the highest occupied molecular orbital of the dianion of naphthalene. The aromatic group is not planar with atoms RC(1), RC(4) and RC(1'), RC(4') 0.15 - 0.17 Å off the plane through the remaining six carbon atoms of the naphthalenide fragment (RC(1) and RC(4) are displaced in the same direction towards Li (Fig. 1)).

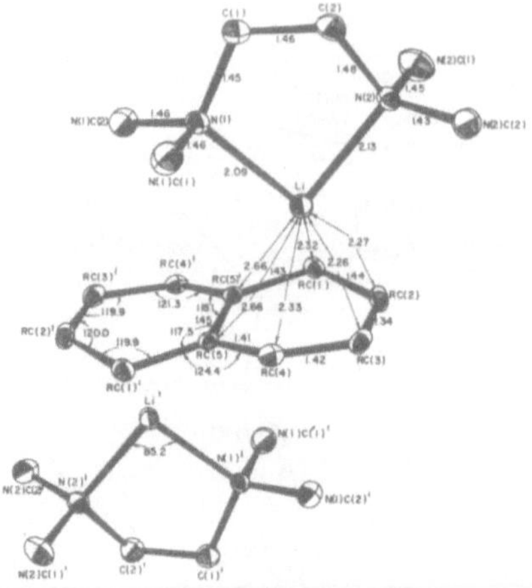

Fig. 1. The molecular geometry of $\{Li[(CH_3)_2N(CH_2)_2N(CH_3)_2]\}_2C_{10}H_8$.

TRIPHENYLMETHYLLITHIUM TETRAMETHYLETHYLENEDIAMINE

$C_{25}H_{31}LiN_2$ $(C_6H_5)_3CLi[(CH_3)_2N(CH_2)_2N(CH_3)_2]$

J.J. BROOKS and G.D. STUCKY, 1972. J. Amer. Chem. Soc., 94, 7333-7338.

Monoclinic, $P2_1/c$, a = 8.546, b = 16.171, c = 16.999 Å, β = 106.75°, Z = 4. Cu radiation, R = 0.063 for 2066 reflexions.

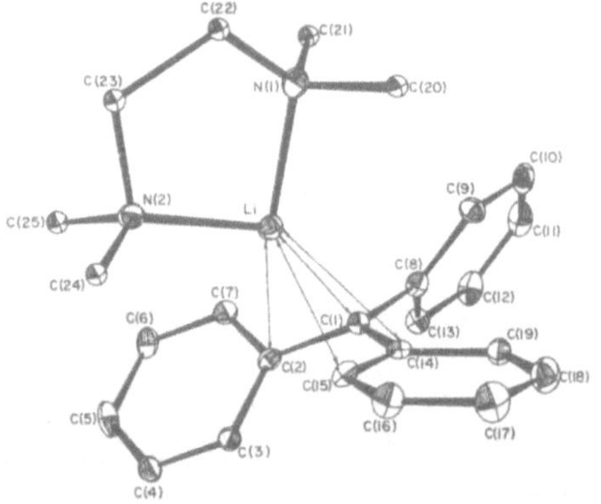

Fig. 1. The molecular geometry of $(C_6H_5)_3CLi[(CH_3)_2N(CH_2)_2N(CH_3)_2]$.

The molecular structure (Fig. 1) is described as a contact ion pair consisting of a triphenylmethyl carbanion and a Li cation coordinated to a tetramethylethylenediamine group. The $(C_6H_5)_3C^-$ group has C_6H_5-C(1)-C_6H_5 bond angles of 117.0(6), 122.8(7), and 118.3(6)°, suggesting predominantly sp^2 hybridization for the central carbon atom, C(1). The twist angles of the phenyl rings depend on their interaction with the lithium atom, with the smallest twist angle (19.7°) being observed for the phenyl ring [C(14)···C(19)] with the closest Li···C contacts. The other twist angles are 30.6° [C(2)···C(7)] and 44.8° [C(8)···C(13)]. Geometrical features are consistent with greater π-electron delocalization between the benzylic carbon atom and the two phenyl groups which exhibit smaller twist angles. The lithium atom is not located directly over C(1) but has four close contacts to the carbanion, 2.23 Å to C(1), 2.49 and 2.51 Å to two carbon atoms [C(14), C(15)] on one phenyl group, and one close contact of 2.54 Å to one carbon atom [C(2)] of a second group.

DI-(2,2,6,6-TETRAMETHYL-4-PIPERIDINYL-1-OXYL) SUBERATE

$C_{26}H_{46}N_2O_6$

A. CAPIOMONT, 1972. Acta Cryst., B28, 2298-2301.

Monoclinic, $P2_1/c$, a = 6.041, b = 21.52, c = 13.62 Å, β = 126.360°, Z = 2. Cu radiation, R = 0.063 for 828 reflexions.

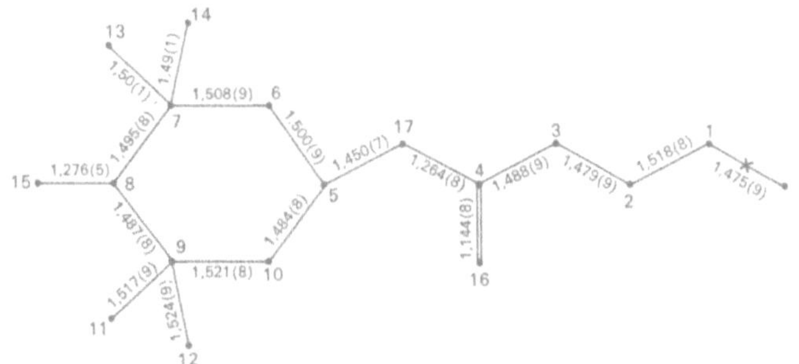

Fig. 1. Bond distances in di-(2,2,6,6-tetramethyl-4-piperidinyl-1-oxyl)suberate. Atom 8 is nitrogen and 15, 16, 17 are oxygen.

The molecule is centrosymmetric and the heterocyclic ring has a chair conformation. The carbon chain $(CH_2)_6$ is planar and is inclined at 13° to the ring plane. Bond distances are shown in Fig. 1. The nitroxide group bond length, N(8)-O(15), is comparable to those found in other nitroxide radicals.

FLUORENYLLITHIUM BISQUINUCLIDINE

$C_{27}H_{35}LiN_2$ $C_{13}H_9Li(NC_7H_{13})_2$

J.J. BROOKS, W. RHINE and G.D. STUCKY, 1972. J. Amer. Chem. Soc., 94, 7339-7346.

Monoclinic, $P2_1/c$, a = 12.376, b = 10.640, c = 19.682 Å, β = 118.13°, D_m = 1.15, Z = 4. Cu radiation, R = 0.073 for 2307 reflexions.

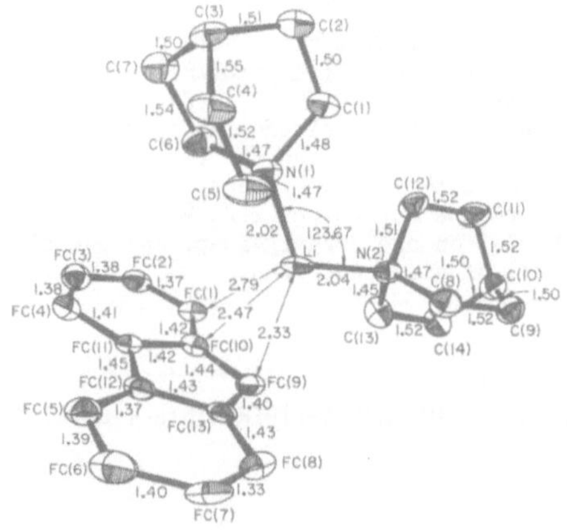

Fig. 1. The molecular geometry of $C_{13}H_9Li(NC_7H_{13})_2$.

The molecular structure is shown in Fig. 1. The Li atom, while it lies above the plane of the carbanion, is not within the periphery of the five-membered ring. The carbanion is not strictly planar although the component rings are planar within experimental error; the benzyl fragment more closely associated with the lithium has atom FC(9) 0.04 Å out of plane in the lithium direction. The molecular geometry is not consistent with a simple electrostatic interpretation of the bonding but can be rationalized in terms of a substantial covalent interaction involving (1) a metal-carbanion three-centre bond which is formed by an empty lithium orbital which is primarily $\underline{p}$ in character and two carbon $\underline{p}_z$ orbitals of the highest occupied molecular orbitals of the carbanion and (2) the σ overlap of a lithium sp^2 orbital with the lower energy bonding molecular orbitals of the carbanion.

p-BROMOBENZOIC ACID

$C_7H_5BrO_2$ $BrC_6H_4CO_2H$

K. OHKURA, S. KASHINO and M. HAISA, 1972. Bull. Chem. Soc. Japan, 35, 2651-2652.

Monoclinic, $P2_1/a$, a = 29.59, b = 6.15, c = 3.98 Å, β = 95.5°, D_m = 1.86, Z = 4. Cu radiation, R = 0.096 for 837 reflexions (films, visual intensities).

The structure resembles that of p-chlorobenzoic acid (1), the crystal containing hydrogen-bonded, centrosymmetrical dimers, O-H···O = 2.65 Å. The carboxyl group is twisted 6° out of the ring plane; bond lengths and angles are close to those in related molecules.

1. Structure Reports, 16, 513.

PENTAFLUOROBENZOIC ACID

$C_7HF_5O_2$ $C_6F_5CO_2H$

V. BENGHIAT and L. LEISEROWITZ, 1972. J. Chem. Soc., Perkin II, 1778-1785.

Triclinic, $P\bar{1}$, a = 7.925, b = 8.637, c = 6.262 Å, α = 97.55°, β = 89.65°, γ = 121.86°, D_m = 1.16, Z = 2. Cu radiation, R = 0.04 for 1598 reflexions.

The molecule is not planar; the angle between the planar pentafluorophenyl group and the carboxy group is 29.8°. Bond lengths are C=O = 1.228(2), C-OH = 1.278(2), $C\text{-}CO_2H$ = 1.500(2), C-F = 1.323 - 1.335 Å. The molecules form centrosymmetric hydrogen-bonded dimers with O-H···O = 2.668 Å.

p-IODOBENZAMIDE

C_7H_6INO $IC_6H_4CONH_2$

K. NAKATA, Y. KATO, Y. TAKAKI and K. SAKURAI, 1971. Mem. Osaka Kyoiku Univ., 20, 93-99.

Monoclinic, $P2_1/a$, a = 31.37, b = 5.30, c = 4.93 Å, β = 107.8°, D_m = 2.10, Z = 4. Cu radiation, two-dimensional analysis, h0ℓ and hk0 zones.

Hydrogen bonds link molecules into centrosymmetrical dimers (NH···O = 2.94 Å). The dimers are further linked by N-H···O hydrogen bonds (2.89 Å) into infinite chains along c. The amide group plane is inclined at 28° to that of the benzene ring.

o-NITROBENZAMIDE

$C_7H_6N_2O_3$ $O_2NC_6H_4CONH_2$

K. FUJIMORI, T. TSUKIHARA, Y. KATSUBE and J. YAMAMOTO, 1972. Bull. Chem. Soc. Japan, 45, 1564-1565.

Monoclinic, $P2_1/a$, a = 12.01, b = 12.79, c = 5.03 Å, β = 97.5°, D_m = 1.43, Z = 4. Cu radiation, R = 0.14 for 1494 reflexions (films, visual intensities).

The ring is planar with the amide and nitro groups rotated out of plane in the same sense by 42 and 47° respectively; the shortest intramolecular contact between the substituent groups is O···O = 2.91 Å. Bond lengths and angles are close to expected values. There are two sets of intermolecular N-H···O hydrogen bonds, the first set (2.94 Å) forming a dimer across a centre of symmetry, and the second set (2.82 Å) linking the dimers into a chain along c.

BENZAMIDE

C_7H_7NO

C.C.F. BLAKE and R.W.H. SMALL, 1972. Acta Cryst., B28, 2201-2206.

Monoclinic, $P2_1/c$, a = 5.607, b = 5.046, c = 22.053 Å, β = 90.66°, D_m = 1.289, Z = 4. Cu radiation, R = 0.068 for 1037 reflexions.

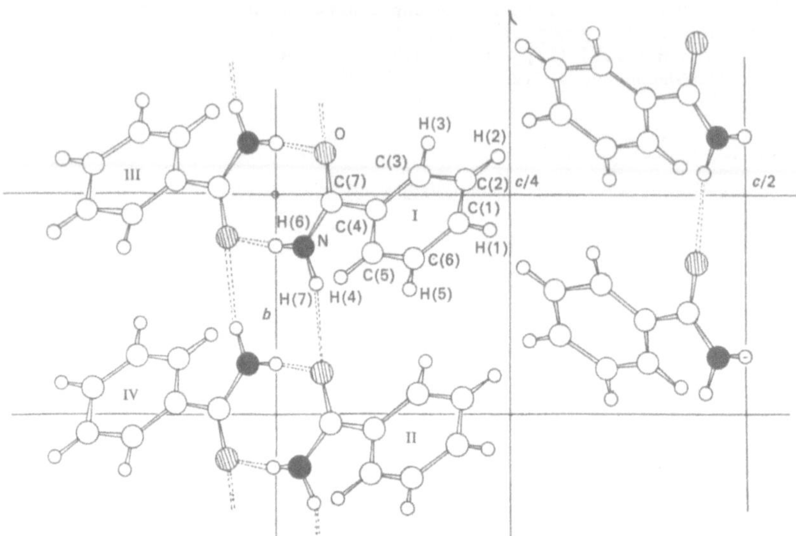

Fig. 1. The molecular arrangement of benzamide projected down a.

The crystal structure of benzamide (1) has been redetermined using three-dimensional X-ray data. The hydrogen-bonding system (N-H···O = 2.907 and 2.933(3) Å) consists of centrosymmetric pairs (Fig. 1) linked side by side to other pairs which are b-translation related. The benzene ring does not deviate significantly from planarity nor regularity and the mean C-C distance is 1.391 Å. Other mean molecular distances are C-O = 1.249(3), C-N = 1.342(3) and C-C(carboamide) = 1.501(4) Å. The deviations of the N (-0.52 Å) and O atoms (0.51 Å) from the plane of the seven carbon atoms indicate a twist of 24.6° of the amide group about the C(4)-C(7) bond relative to the benzene ring; the benzamide molecule is non-planar primarily on account of the repulsion between the amide H(7) and the ortho hydrogen H(4).

1. Structure Reports, 23, 645.

ISOPHTHALIC ACID

$C_8H_6O_4$ $C_6H_4(CO_2H)_2$

R. ALCALA and S. MARTINEZ-CARRERA, 1972. Acta Cryst., B28, 1671-1677.

Monoclinic, $P2_1/c$, a = 3.76, b = 16.42, c = 11.72 Å, β = 90.0°, D_m = 1.52, Z = 4. Cu radiation, R = 0.05 for 809 reflexions.

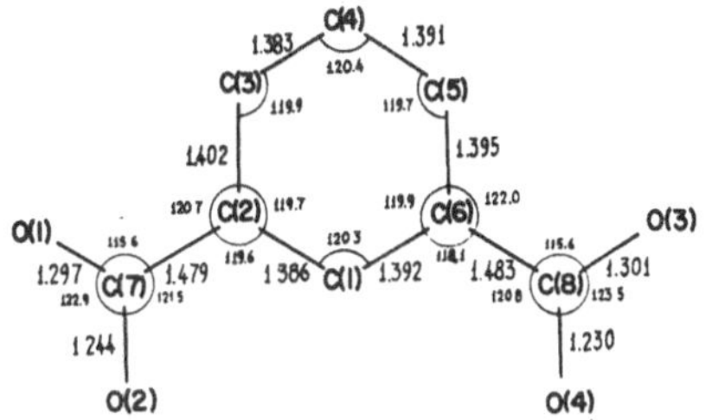

Fig. 1. Bond lengths and angles in isophthalic acid. Mean standard deviations are 0.006 Å and 0.4°.

Details of bond lengths and angles are in Fig. 1. The molecule is basically planar; but the carboxyl groups are rotated about the exocyclic C-C bonds by 2.8 and 4.7° out of the aromatic ring plane. The molecules are hydrogen bonded together to give infinite chains along b with O-H···O = 2.594 and 2.697(6) Å.

TEREPHTHALAMIDE

$C_8H_8N_2O_2$ $C_6H_4(CONH_2)_2$

R.E. COBBLEDICK and R.W.H. SMALL, 1972. Acta Cryst., B28, 2893-2896.

Triclinic, $P\bar{1}$, a = 5.027, b = 5.355, c = 7.165 Å, α = 103.02, β = 100.90, γ = 92.17°, D_m = 1.47, Z = 1. Cu radiation, R = 0.088 for 796 reflexions.

The centrosymmetric molecules are linked together in sheets by NH···O hydrogen bonds 2.90 and 2.94 Å across centres of symmetry (Fig. 1). The plane of the amide group makes an angle of 23° with that of the benzene ring. Corrected bond lengths are C-C(amido) = 1.491(5), C-O = 1.247(5), C-N = 1.305(5), C-C(ring) = 1.380 - 1.409(6) Å.

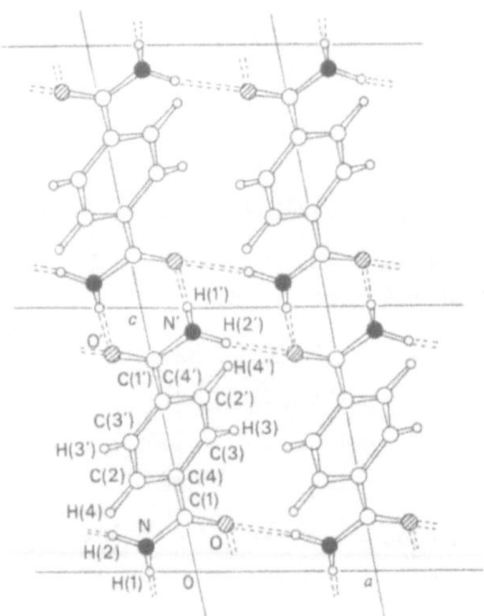

Fig. 1. The crystal structure of terephthalamide projected on to (010).

N-METHYLANTHRANILIC ACID

$C_8H_9NO_2$ $CH_3NHC_6H_4COOH$

N.N. DHANESHWAR and L.M. PANT, 1972. Acta Cryst., B28, 647-649.

Monoclinic, $P2_1/c$, a = 8.15, b = 15.09, c = 7.03 Å, β = 118.3°, D_m = 1.307, Z = 4. Cu radiation, R = 0.101 for 965 reflexions.

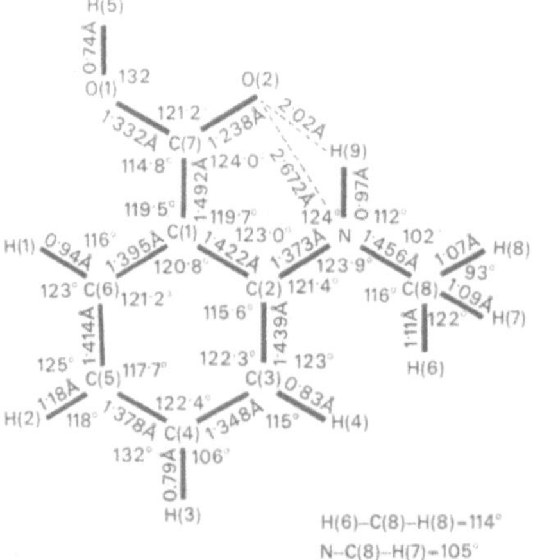

Fig. 1. Bond lengths (σ = 0.006 - 0.011 Å) and angles in N-methylanthranilic acid.

The structure consists of centrosymmetric dimers composed of pairs of molecules linked through O-H···O hydrogen bonds (2.67 Å). Other intermolecular contacts are normal. Bond lengths and angles are in Fig. 1. There is an intramolecular N-H···O hydrogen bond and there is no significant rotation of the carboxyl group about the C(1)-C(7) bond.

o-ETHOXYBENZOIC ACID (AT -10°C)

$C_9H_{10}O_3$ $C_2H_5OC_6H_4CO_2H$

E.M. GOPALAKRISHNA and L. CARTZ, 1972. Acta Cryst., B28, 2917-2924.

Monoclinic, $P2_1/c$, a = 7.28, b = 15.02, c = 16.05 Å, β = 108.27°, D_m = 1.173, Z = 8. Cu radiation, R = 0.09 for 2469 reflexions.

The two independent molecules comprising the asymmetric unit differ slightly in the conformations of the carboxyl and ethoxy groups. In one molecule the carboxyl and ethoxy planes are inclined to the ring plane at angles of 7.2 and 2.7° respectively, whereas in the other molecule the corresponding angles are 1.9 and 16.5°. Unlike many benzoic acid derivatives the carboxyl groups do not hydrogen bond in pairs; instead the carboxyl hydrogen atom forms an intramolecular hydrogen bond with the ethoxy oxygen atom (with H···O = 1.61(8) Å in one molecule and 1.58(5) Å in the other). Bond lengths and angles within the molecules are in good agreement with usually accepted values. There are no abnormal intermolecular contacts.

1,2,3-BENZENETRICARBOXYLIC ACID DIHYDRATE

$C_9H_{10}O_8$ $C_6H_3(COOH)_3 \cdot 2H_2O$

J.M. FORNIES-MARQUINA, C. COURSEILLE, B. BUSETTA and M. HOSPITAL, 1972. Cryst. Struct. Comm., 1, 47-50.

Triclinic, $P\bar{1}$, a = 7.077, b = 8.720, c = 9.114 Å, α = 73.35, β = 79.15, γ = 88.40°, D_m = 1.536, Z = 2. Cu radiation, R = 0.057 for 1989 reflexions.

A view of the crystal structure is given in Fig. 1. All the benzene rings are parallel to the ($\bar{1}$21) plane. The angles which the carboxyl group planes make with that of the benzene ring are +4.3, -87.1, and -9.6° respectively. The molecules are held in the crystal by hydrogen bonding involving the carboxyl oxygen atoms and the water molecules.

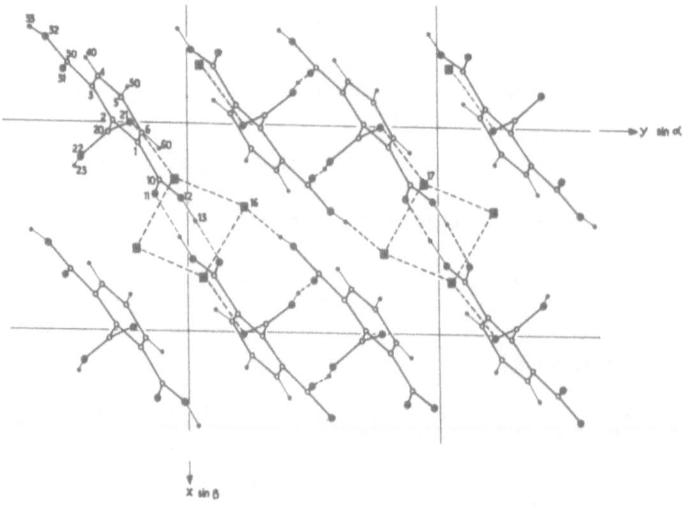

Fig. 1. 1,2,3-Benzenetricarboxylic acid in its c-axis projection.

DIMETHYL 3,6-DICHLORO-2,5-DIHYDROXYTEREPHTHALATE

$C_{10}H_8Cl_2O_6$

S.R. BYRN, D.Y. CURTIN and I.C. PAUL, 1972. J. Amer. Chem. Soc., 94, 890-898.

Yellow Form

Triclinic, P$\bar{1}$, a = 9.595, b = 4.301, c = 7.970 Å, α = 114°19', β = 94°58', γ = 106°9', D_m = 1.75, Z = 1. Cu radiation, R = 0.074 for 906 reflexions.

White Form

Triclinic, P$\bar{1}$, a = 9.842, b = 7.841, c = 10.576 Å, α = 116°23', β = 124°10', γ = 88°59', D_m = 1.71, Z = 2. Cu radiation, R = 0.060 for 1733 reflexions.

The yellow form (Y-1) of the title compound rearranges in the solid state to the white form (W-1). The molecule of Y-1 sits on a centre of symmetry and non-hydrogen atoms are almost coplanar. There is a short intramolecular hydrogen bond (2.532(6) Å) and there are significant exocyclic bond angle distortions resulting from intramolecular overcrowding (Fig. 1). The principal intermolecular forces are van der Waals interactions.

Two crystallographically independent molecules of W-1 sit about the centres of symmetry at 0, 0, 0(molecule A) and 0, ½, 0(molecule B). The planes of the benzene rings in the two molecules are inclined at an angle of 12°4' to each other. The relative tilting of the two molecules is due to the existence of two inter-molecular hydrogen bonds, -O-H···O=C(OCH_3), on one side of the intramolecular Cl-Cl vector (Fig. 2). There appears to be no intramolecular overcrowding on the basis of the molecular geometry (Fig. 1). The major difference in the molecular conformation from that in Y-1 is the rotation of the carbomethoxyl groups out of the plane of the benzene rings by 86°28' and 72°38' in the two molecules. From the X-ray results it follows that the change from Y-1 to W-1 involves not only a change in hydrogen bonding but also the flipping of every other aromatic ring 180°. This process has been shown to involve nucleation followed by a highly anisotropic migration of the reaction front through the crystal. Optical goniometry in conjunction with the X-ray results shows that the migration of the front is rapid in directions normal to [110], the long axis of the crystal, and slow in the [110] direction.

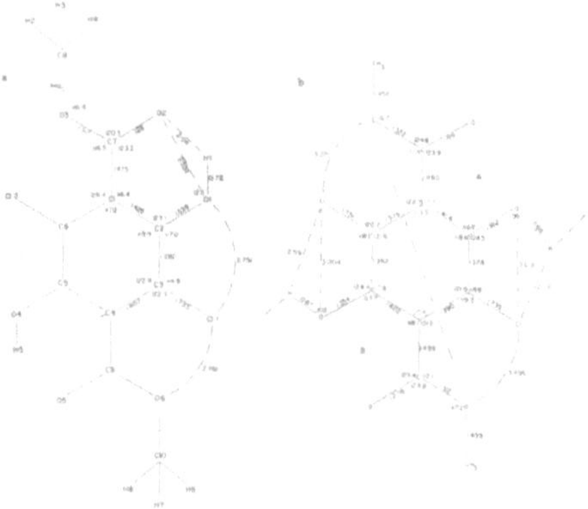

Fig. 1. Bond lengths and angles in Y-1 are shown in (a) while those for the two independent half-molecules of W-1 (separated by the line) are shown in (b). The estimated standard deviations for C-Cℓ, C-C, and C-O bonds range from 0.006 to 0.009 Å for Y-1 and 0.007 - 0.011 Å for W-1, the estimated standard deviations for C-C-C, C-C-Cℓ, and C-C-O angles are from 0.3 to 0.5° for Y-1 and 0.4 to 0.6 for W-1.

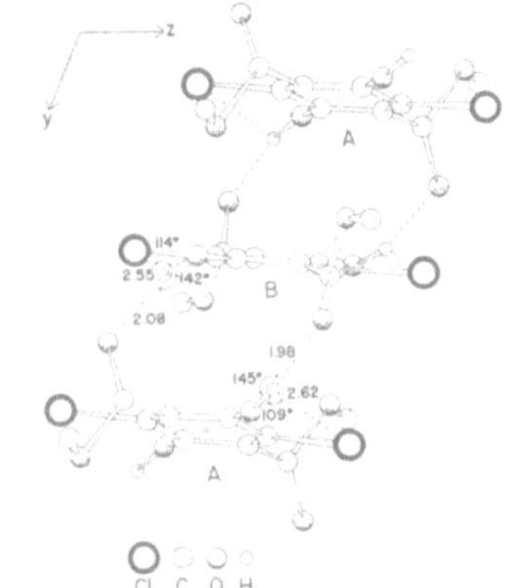

Fig. 2. View of the molecular arrangement within a stack of W-1 molecules showing the dimensions of the hydrogen bonding system. For clarity, the hydrogen atoms on the methyl groups are omitted.

MESITOIC ACID

(2,4,6-TRIMETHYLBENZOIC ACID)

$C_{10}H_{12}O_2$ $(CH_3)_3C_6H_2CO_2H$

V. BENGHIAT and L. LEISEROWITZ, 1972. J. Chem. Soc., Perkin II, 1778-1785.

Monoclinic, C2/c, a = 15.142, b = 7.026, c = 17.439 Å, β = 89.35°, D_m = 1.16, Z = 8. Cu radiation, R = 0.05 for 2060 reflexions.

The overcrowding of the carboxy groups and the o-methyl substituents is relieved primarily by an out-of-plane twist of 48.4°. The carboxyl carbon atom displacement of 0.06 Å from the benzene ring plane is also likely to be due to the overcrowding. The C-O bond lengths are 1.275 and 1.248(2) Å, from which it is inferred that the carboxyl group displays partial orientational disorder in the lattice. The molecules form centrosymmetric hydrogen-bonded dimers with O-H···O = 2.640 Å. An independent account of this structure has appeared previously (1).

1. Structure Reports, 35B, 46.

3'-IODOBIPHENYL-4-CARBOXYLIC ACID

$C_{13}H_9IO_2$ $IC_6H_4C_6H_4CO_2H$

H.H. SUTHERLAND and M.J. MOTTRAM, 1972. Acta Cryst., B28, 2212-2217.

Triclinic, P$\bar{1}$, a = 8.61, b = 15.92, c = 4.36 Å, α = 92.88°, β = 108.82°, γ = 90.00°, D_m = 1.89, Z = 2. Mo radiation, R = 0.06 for 1471 reflexions.

The dihedral angle between the two aromatic ring planes is 30.4(5)°. Principal bond lengths are C-I = 2.091(11), C-O = 1.246 and 1.283(15), C-C(exocyclic) = 1.468 and 1.471(15), C-C(ring) = 1.357 - 1.413(16) Å. The iodine atom is 0.08 Å out of the benzene plane to which it is bonded. The molecules are linked into dimers by carboxyl group hydrogen bonding around a centre of symmetry (O-H···O = 2.61 Å).

DIBENZOYLACETYLENE

$C_{16}H_{10}O_2$ $C_6H_5COC{\equiv}CCOC_6H_5$

R.J. MAJESTE and E.A. MEYERS, 1972. Cryst. Struct. Comm., 1, 231-233.

Monoclinic, $P2_1/c$, a = 7.875, b = 10.863, c = 7.717 Å, β = 115.83°, Z = 2. Mo radiation, R = 0.045 for 425 reflexions.

The molecule lies on a centre of symmetry and is very nearly planar. Bond distances and angles are normal. The carbon-carbon triple bond distance of 1.173(5) Å and the other distances agree well with those found in alkyne systems with delocalized electrons. Intermolecular contacts appears to be normal.

AMMONIUM HYDROGEN TEREPHTHALATE

$C_8H_9NO_4$ $NH_4^+C_6H_4COOHCOO^-$

R.E. COBBLEDICK and R.W.H. SMALL, 1972. Acta Cryst., B28, 2924-2928.

Monoclinic, C2/c, a = 18.913, b = 3.801, c = 11.460 Å, β = 97.44°, D_m = 1.472, Z = 4. Cu radiation, R = 0.076 for 891 reflexions.

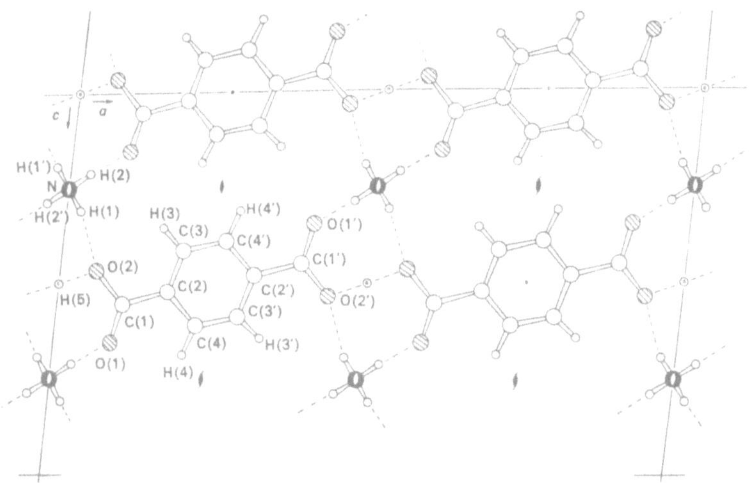

Fig. 1. The crystal structure of ammonium hydrogen terephthalate projected onto (010).

The crystal structure consists of chains of centrosymmetric terephthalate ions linked by short hydrogen bonds across centres of symmetry (O···O = 2.51 Å) and these chains are linked via the ammonium ions to form a three-dimensional hydrogen bonding network (Fig. 1). The carboxyl groups are twisted 9° out of the plane of the benzene ring to form hydrogen bonds with the ammonium ions. Bond lengths for the terephthalate ion (corrected for librational motion) are C-C (carboxylate) = 1.495(4), C(1)-O(1) = 1.237(4), C(1)-O(2) = 1.310(4), C-C(ring) = 1.374 - 1.409(5) Å.

CALCIUM TEREPHTHALATE TRIHYDRATE

$C_8H_{10}CaO_7$ $C_8H_4O_4Ca \cdot 3H_2O$

T. MATSUZAKI and Y. IITAKA, 1972. Acta Cryst., B28, 1977-1981.

Monoclinic, $P2_1/c$, a = 7.11, b = 21.67, c = 6.59 Å, β = 92.3°, D_m = 1.69, Z = 4. Cu radiation, R = 0.135 for 1311 reflexions.

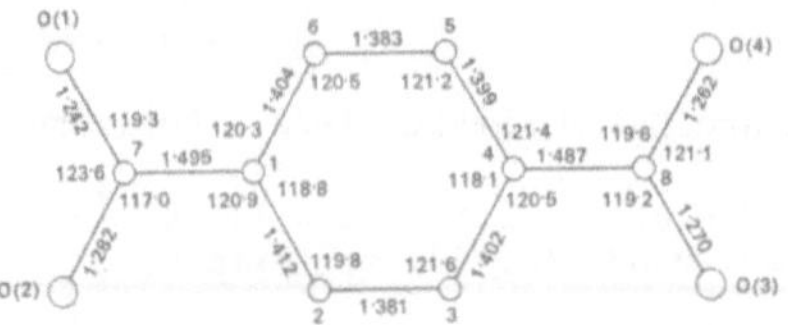

Fig. 1. Bond lengths and angles in calcium terephthalate trihydrate (σ = 0.015 Å and 1.0°).

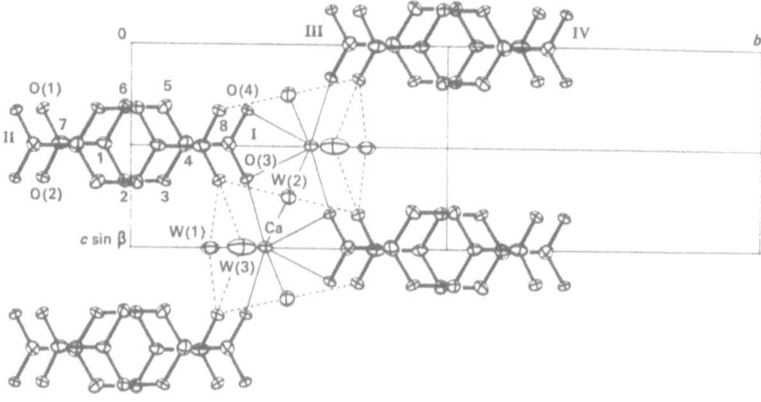

Fig. 2. Projection of the calcium terephthalate trihydrate structure along the a axis. Broken lines indicate hydrogen bonds.

The crystal analysis was performed on a twinned crystal. Each carboxyl group twists by about 5° from the plane of the benzene ring. The bond lengths and angles (Fig. 1) show that the terephthalic acid ion has a small amount of quinoid character; the C-C bonds parallel to the substituent groups are 0.02 Å shorter than the mean of the remaining four bonds in the ring. The terephthalic acid ions (Fig. 2) are parallel to each other in the c-direction and the calcium ion takes eightfold coordination, being surrounded by four oxygen atoms of three terephthalic acid ions and four oxygen atoms of water molecules (Ca···O = 2.350 - 2.642(8) Å). Carboxyl group (II) is coordinated to the calcium ions while carboxyl group (I) is connected to water molecules by hydrogen bonds (O-H···O = 2.73 - 2.94(1) Å).

POTASSIUM HYDROGEN BIS-m-CHLOROBENZOATE

$C_{14}H_9Cl_2KO_4$ $KH(ClC_6H_4CO_2)_2$

A.L. MACDONALD and J.C. SPEAKMAN, 1972. J. Chem. Soc., Perkin II, 1564-1566.

Monoclinic, C2/c, a = 32.55, b = 3.83, c = 11.45 Å, β = 94.6°, D_m = 1.63, Z = 4. Mo radiation, R = 0.059 for 1283 reflexions.

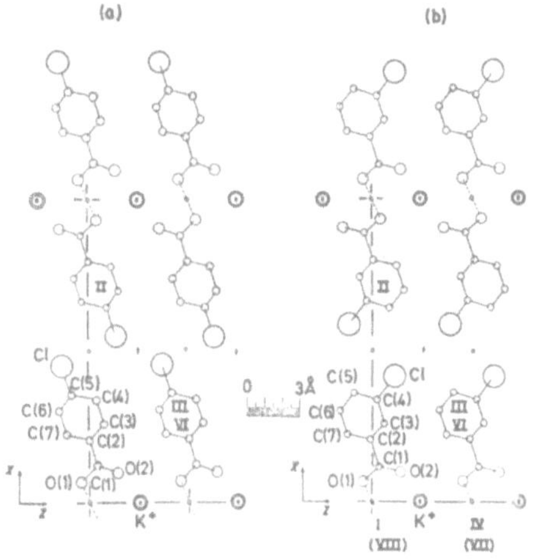

Fig. 1. Crystal structures of the acid salts (KHX_2) of (a) HX = p-chlorobenzoic acid, and (b) HX = m-chlorobenzoic acid (to bring out the similarity of the two structures, the β-angles are drawn here as 90°, though indeed they differ by about 4°).

The structure is shown in Fig. 1. The K^+ ion lies on a twofold axis, the acidic hydrogen atom lies effectively at a centre of symmetry; two chlorobenzoate residues are related by this centre and are linked by a very short (possibly genuinely symmetrical) hydrogen bond, O···H···O = 2.437(6) Å. The anion bond distances have their expected values; the K^+ ion is surrounded by six oxygens (three pairs digonally related belonging to six different anions with K^+···O = 2.750, 2.799, and 2.926(4) Å). The structure is surprisingly similar to that of the corresponding p-chloro derivative (1) (also shown in Fig. 1).

1. H.H. MILLS and J.C. SPEAKMAN, 1963. J. Chem. Soc., 4355.

POTASSIUM HYDROGEN BIS(HOMOPHTHALATE)

$C_{18}H_{15}KO_8$ $K(C_9H_7O_4)_2H$

M.P. GUPTA and D.S. DUBEY, 1972. Acta Cryst., B28, 2677-2681.

Monoclinic, C2/c, a = 32.66, b = 5.51, c = 9.97 Å, β = 95.9°, D_m = 1.48, Z = 4. Cu radiation, R = 0.099 for 594 reflexions.

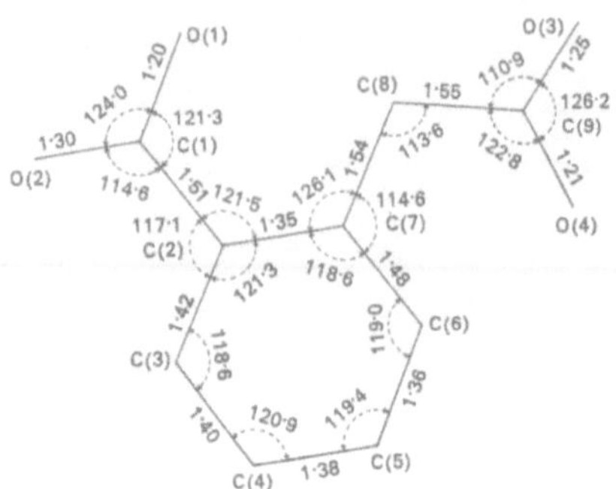

Fig. 1. Bond lengths (σ = 0.02 - 0.03 Å) and angles in the homophthalate moiety.

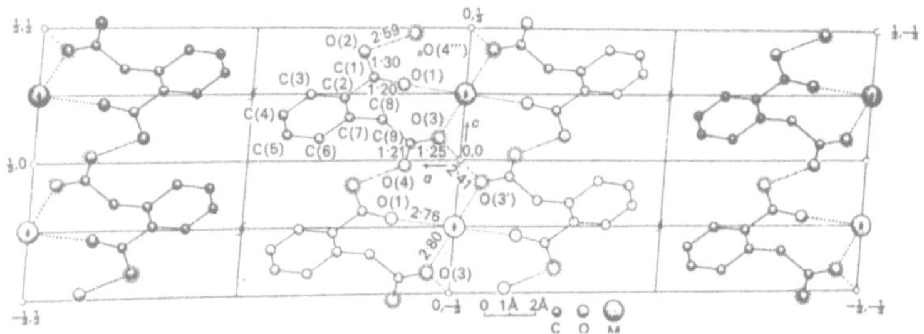

Fig. 2. The $K(C_9H_7O_4)_2H$ structure viewed down [010].

The bond lengths and angles are shown in Fig. 1. The ring C(2)-C(7) is planar and makes dihedral angles of 18.9 and 95.7° with the carboxyl groups C(1)O(1)O(2) and C(9)O(3)O(4) respectively. The potassium ion is surrounded by a tetrahedron of 4 oxygen atoms belonging to 4 different molecules (K-O = 2.76 - 2.80, mean 2.78(1) Å). In addition to these ionic linkages the molecules are joined together in the crystal by hydrogen bonding between O(2) and O(4) of different molecules (2.59 Å), and a short symmetrical hydrogen bond between 2 centrosymmetrically related oxygen atoms (O(3)···O(3') = 2.41 Å) with the proton on a centre of symmetry (Fig. 2).

RUBIDIUM HOMOPHTHALATE

$C_{18}H_{15}O_8Rb$

M.P. GUPTA and D.S. DUBEY, 1972. Z. Kristallogr., 135, 273-286.

Monoclinic, C2/c, a = 32.81, b = 5.47, c = 9.95 Å, β = 95°54', D_m = 1.67, Z = 4. Cu radiation, R = 0.099 for 719 reflexions.

The crystal structure is isomorphous with that of the potassium salt (1). The Rb ion lies on a twofold axis and is coordinated by four oxygen atoms with a mean Rb-O distance 2.87(3) Å. The homophthalate ions are linked by two types of hydrogen bonds. One of these is of the short symmetrical type (O···H···O = 2.45(3) Å) with the hydrogen atom apparently situated on a centre of symmetry; the other is of the normal variety (O-H···O = 2.61(3) Å) with the hydrogen atom in a general position. The carboxyl group C(1)O(1)O(2) is rotated 12.2° about the exocyclic C-C bond from the benzene ring plane; the plane of the other carboxyl group is inclined at an angle of 98.5° to the aromatic plane.

1. Preceding report.

PENTACHLORONITROBENZENE

$C_6Cl_5NO_2$

H.J. ROSSELL and H.G. SCOTT, 1972. Mol. Cryst. and Liq. Cryst., 17, 275-280.

Trigonal, R$\bar{3}$, a = 8.769, c = 11.209 Å, D_m = 1.970, Z = 3. Cu radiation, powder data.

The three molecules in the cell lie flat in close packed planes parallel to (0001). The structure is disordered in that the nitro group is not distinguishable from the chlorine atoms; there is not, however, free rotation of molecules.

POTASSIUM PICRATE

$C_6H_2KN_3O_7$ $C_6H_2(NO_2)_3O^-K^+$

G.J. PALENIK, 1972. Acta Cryst., B28, 1633-1634.

Orthorhombic, Ibca, a = 13.333, b = 19.112, c = 7.118 Å, Z = 8. Mo radiation, R = 0.05 for 559 reflexions.

A comparison is made between three independent determinations of this crystal structure. In the present study, stationary-crystal stationary-counter measurements were made, in the second case densitometer measurement of films was employed (1) and in the third case the intensities were estimated visually from films (2).

The positional parameters from the three sets of data are in good agreement, with some differences in thermal parameters. One point not mentioned in the previous reports is the possible existence of a C-H···O hydrogen bond (H···O = 2.47(2) Å, C-H···O angle 175(2)°) between a benzene ring H atom and a nitro oxygen atom of a neighbouring ion.

1. K. MAARTMANN-MOE, 1969. Acta Cryst., B25, 1452.
2. Structure Reports, 18, 715.

1,3,5-TRINITROBENZENE
(NEUTRON STUDY)

$C_6H_3N_3O_6$

C.S. CHOI and J.E. ABEL, 1972. Acta Cryst., B28, 193-201.

Orthorhombic, Pbca, a = 9.78, b = 26.94, c = 12.82 Å, Z = 16. Neutron data, R = 0.056 for 1096 reflexions.

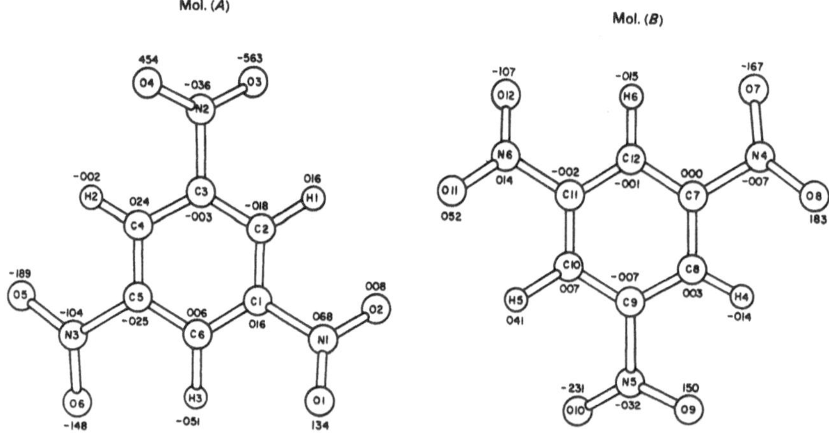

Fig. 1. Deviations of the atoms from the mean plane of the benzene ring (Å x 10^3) of trinitrobenzene.

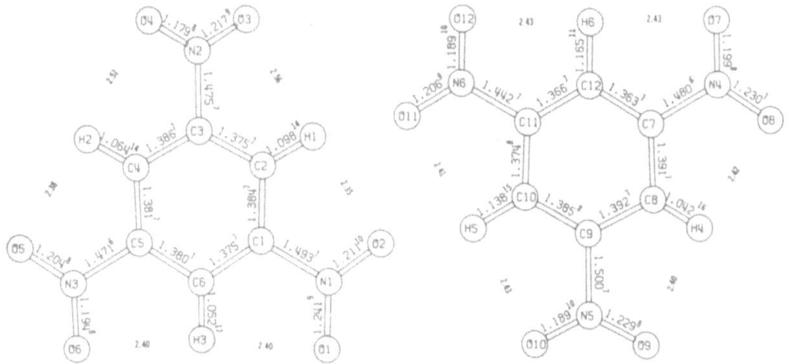

Fig. 2. Bond lengths (Å) in trinitrobenzene. The superscipts are standard deviations (Å x 10^{-3}).

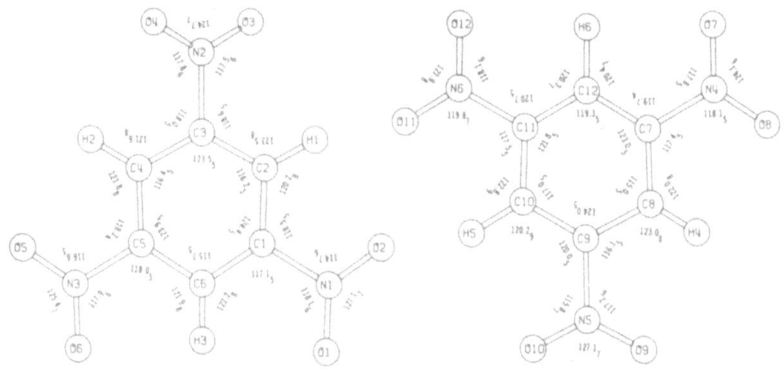

Fig. 3. Bond angles (°) in trinitrobenzene.

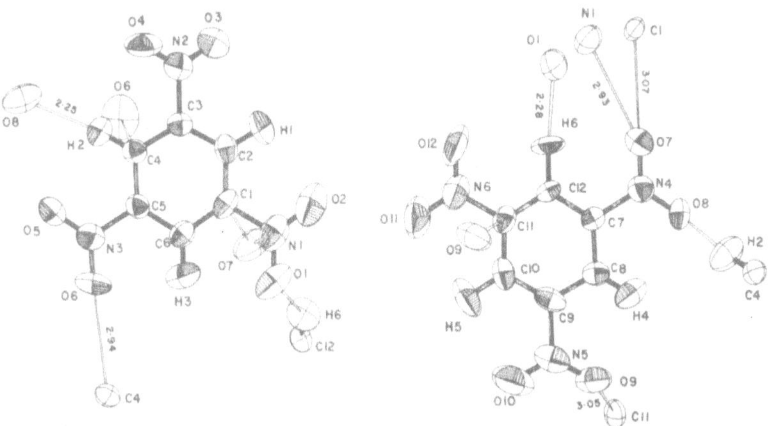

Fig. 4. Short contact distances (Å) in the trinitrobenzene crystal structure.

There are two molecules (designated A and B) in the asymmetric unit, one (A) with a non-planar and the other (B) with an essentially planar benzene ring (Fig. 1). The NO_2 groups are either twisted or bent out of the benzene ring plane. The dihedral angles which each nitro group plane makes with the appropriate benzene ring plane are 3°, 28°, 8°, 10°, 10° and 5° for the N(1)···N(6) nitro groups respectively. Bond lengths and angles (Figs. 2 and 3) agree with those found in trinitrobenzene complexes. Several short O···H distances (intra- and intermolecular) were found (Fig. 4) and appear to indicate C-H···O hydrogen bonding.

2,4,6-TRINITROBENZENESULPHONIC ACID TETRAHYDRATE

(PICRYLSULPHONIC ACID TETRAHYDRATE)

$C_6H_{11}N_3O_{13}S$ $H_5O_2^+ \cdot C_6H_2(NO_2)_3SO_3^- \cdot 2H_2O$

J.-O. LUNDGREN, 1972. Acta Cryst., B28, 1684-1691.

Triclinic, $P\bar{1}$, a = 8.346, b = 11.367, c = 8.065 Å, α = 97.77, β = 109.32, γ = 83.22°, D_m not given, Z = 2. Mo radiation, R = 0.042 for 2848 reflexions.

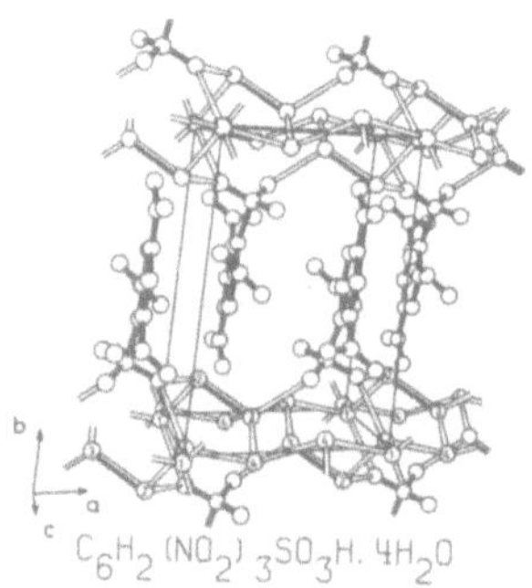

Fig. 1. Drawing of the crystal structure of picrylsulphonic acid tetrahydrate viewed along the c-axis. Covalent bonds are filled. Hydrogen bonds within $H_5O_2^+$ are half-filled and other hydrogen bonds are open. Water oxygen atoms are denoted by numerals 1, 2, 3, and 4 (O(2) and O(3) are in the $H_5O_2^+$ ion).

The structure (Fig. 1) comprises $H_5O_2^+$ ions, picrylsulphonate ions, and water molecules, which are hydrogen bonded together to form layers. The O···O distance within the $H_5O_2^+$ ion is 2.429(2) Å. Other O···O hydrogen bond distances are in the range 2.61 - 2.88 Å. Bond distances in the picrylsulphonate ion are normal and the nitro groups are rotated out of the plane of the aromatic ring by 53, 8, and 66° for the groups at 2, 4, and 6 positions respectively.

syn-p-NITROBENZALDOXIME

$C_7H_6N_2O_3$

L. BREHM and K.J. WATSON, 1972. Acta Cryst., B28, 3646-3652.

Monoclinic, $P2_1/c$, a = 6.252, b = 4.88, c = 24.75 Å, β = 94.66°, D_m = 1.47, Z = 4. Cu radiation, R = 0.100 for 797 reflexions.

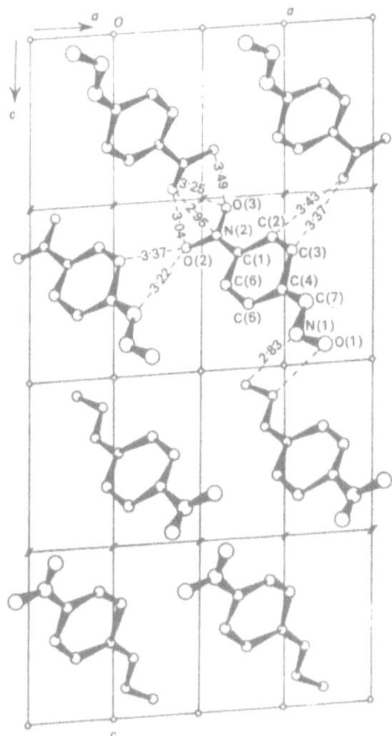

Fig. 1. The syn-p-nitrobenzaldoxime structure viewed along b.

The crystal structure (Fig. 1) is built up of pairs of molecules linked by O-H···N hydrogen bonds (2.83 Å) about centres of symmetry. The bond length data are consistent with a quinoid type contribution to the ground state structure. The benzene ring is planar with the nitro and oxime group planes inclined to it at angles of 4.7 and 9.1° respectively. The shortest (non-hydrogen bond) intermolecular contact (2.95 Å) occurs between N(2) of one nitro group and O(3) of a symmetry related molecule.

p-NITROPHENYL ACETATE

$C_8H_7NO_4$

R. GUTTORMSON and B.E. ROBERTSON, 1972. Acta Cryst., B28, 2702-2708.

Monoclinic, $P2_1/c$, a = 14.115, b = 7.437, c = 16.158 Å, β = 90.83°, D_m = 1.37, Z = 8. Mo radiation, R = 0.082 for 1396 reflexions.

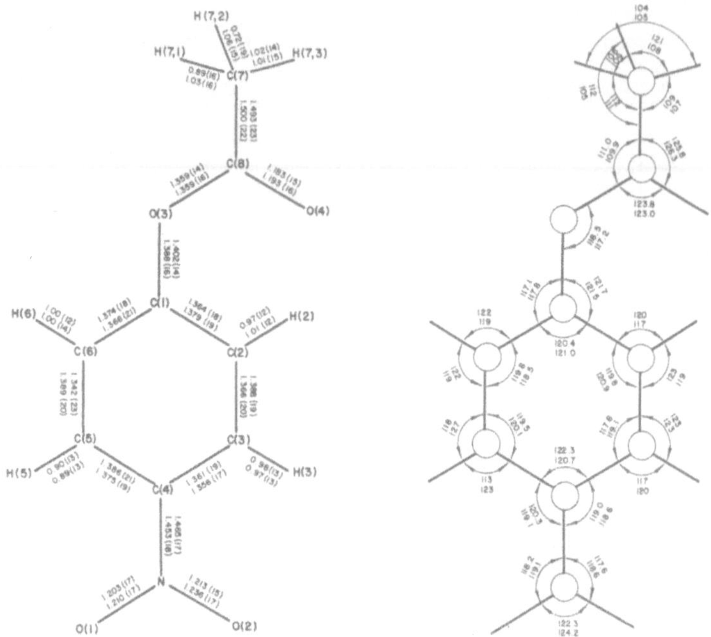

Fig. 1. Bond lengths and angles in the two independent molecules of p-nitrophenyl acetate.

The bond lengths and angles for the two independent molecules in the unit cell are shown in Fig. 1. and indicate the presence of some delocalization in the acyl group. The plane of this group is rotated from that of the phenyl ring in the two molecules by 64.2 and 65.9°, while the dihedral angle between the phenyl ring and the nitro group in the two molecules is 3.3° and 10.8° respectively. These rotations are probably due to steric effects.

2,4,6-TRINITRO-m-XYLENE (TNX)

$C_8H_7N_3O_6$

J.H. BRYDEN, 1972. Acta Cryst., B28, 1395-1398.

Orthorhombic, Pbcn, a = 5.742, b = 14.996, c = 11.384 Å, D_m = 1.615, Z = 4. Cu radiation, R = 0.071 for 732 reflexions.

The four molecules in the unit cell lie on twofold axes in the crystal. The nitro groups in the 4 and 6 positions are rotated by 35.7° out of the aromatic plane while the 2-nitro group is rotated 75.2° out of plane. The structure is isomorphous with that of 1,3-dichloro-2,4,6-trinitro-benzene (1). Bond distances are normal, but bond angles within the benzene ring range from 114.2(3) to 127.2(3)°.

1. J.R. HOLDEN and C. DICKINSON, 1967. J. Phys. Chem., 71, 1129.

o-PHENYLENEDIAMINE DIHYDROBROMIDE

$C_6H_{10}Br_2N_2$ $C_6H_4(NH_3^+)_2 \cdot 2Br^-$

C. STALHANDSKE, 1972. Acta Chem. Scand., 26, 3029-3036.

Orthorhombic, Pmmn, a = 7.534, b = 4.738, c = 12.672 Å, D_m = 1.97, Z = 2. Cu radiation, R = 0.055.

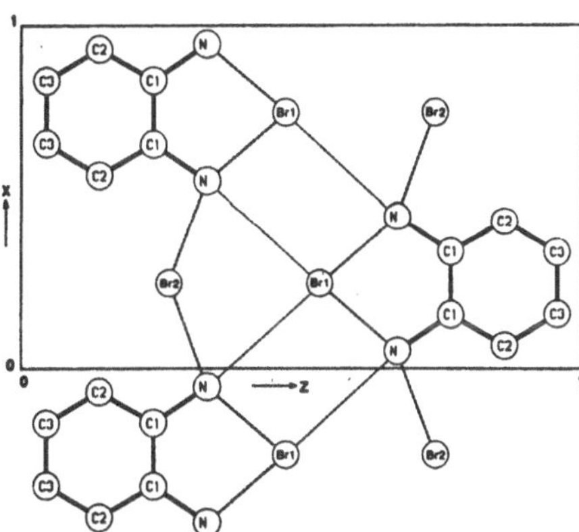

Fig. 1. A projection of the o-phenylenediamine dihydrobromide structure along [010]. Br-N contacts are dotted.

The $C_6H_4(NH_3)_2^{2+}$ cations are, with the exception of the amino hydrogen atoms, planar by symmetry. The C-N distance is 1.445(11) Å. The NH_3^+ groups form three hydrogen bonds with the bromide ions (N-H···Br = 3.308 - 3.402(8) Å) (Fig. 1).

α-p-DIMETHYLAMINOBENZALDOXIME

$C_9H_{12}N_2O$ $(CH_3)_2NC_6H_4 \cdot CH{=}NOH$

F. BACHECHI and L. ZAMBONELLI, 1972. Acta Cryst., B28, 2489-2494.

Monoclinic, $P2_1/c$, a = 7.705, b = 6.190, c = 19.149 Å, β = 95.80°, D_m = 1.20, Z = 4. Mo radiation, R = 0.059 for 980 reflexions.

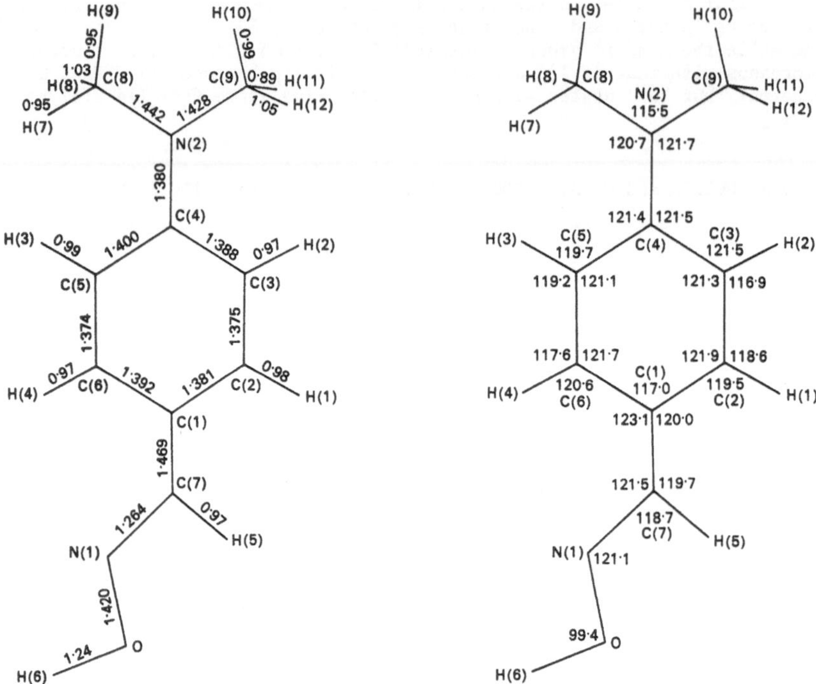

Fig. 1. Bond lengths and angles in $(CH_3)_2NC_6H_4CHNOH$.

In the crystalline state the molecules form centrosymmetrical dimers through two OH···N hydrogen bonds (2.784(4) Å). The bond length data (Fig. 1) are consistent with some contribution from a quinoid canonical form to the ground state structure. The molecule is approximately planar; the dimethylamino group makes an angle of 4.63°, and the aldoxime group 2.52° with the benzene ring plane.

N,N,N',N'-TETRAMETHYL-p-DIAMINOBENZENE PERCHLORATE

(ROOM TEMPERATURE)

$C_{10}H_{17}ClN_2O_4$ $C_{10}H_{17}N_2.ClO_4$

J.L. DE BOER and A. VOS, 1972. Acta Cryst., B28, 835-839.

Orthorhombic, Pnnm, a = 5.956, b = 10.229, c = 10.187 Å, Z = 2. Mo radiation, R = 0.059 for 693 reflexions.

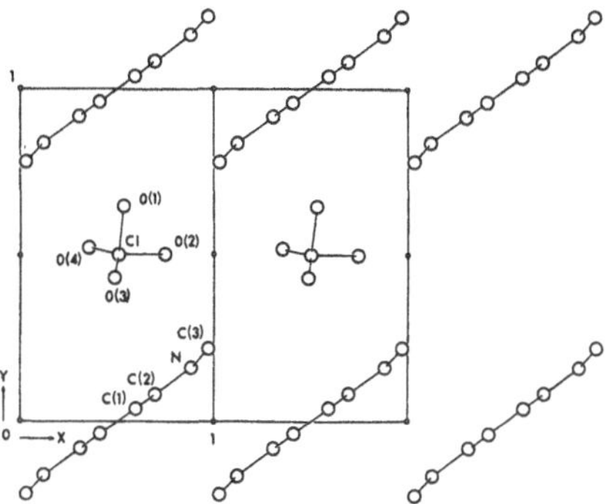

Fig. 1. Projection along [001] of the $TMPD^+ \cdot ClO_4^-$ structure (room temperature form).

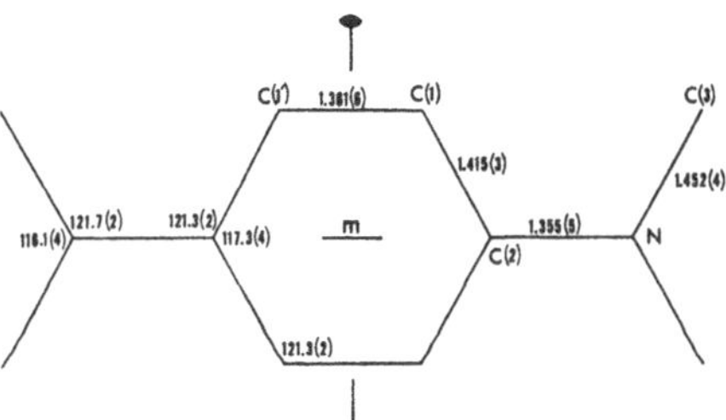

Fig. 2. Bond lengths and angles at room temperature in the TMPD group of $TMPD^+ClO_4^-$.

The TMPD and ClO_4 groups lie at special positions with symmetry 2/m which means that the ClO_4 groups are disordered. In the crystal structure (Fig. 1) the TMPD groups are stacked at equal distances in rows along a. The observed bond lengths (Fig. 2) are in good agreement with those calculated for $TMPD^+$.

N,N,N',N'-TETRAMETHYL-p-DIAMINOBENZENE PERCHLORATE (AT 110°K)

$C_{10}H_{17}C\ell N_2O_4$

J.L. DE BOER and A. VOS, 1972. Acta Cryst., B28, 839-848.

Monoclinic, $B2_1/d$, (C_{2h}^5), a = 11.655, b = 10.147, c = 20.130 Å, β = 92.57°, Z = 8. Mo radiation, R = 0.093 for 3014 reflexions and 0.103 for 2737 reflexions.

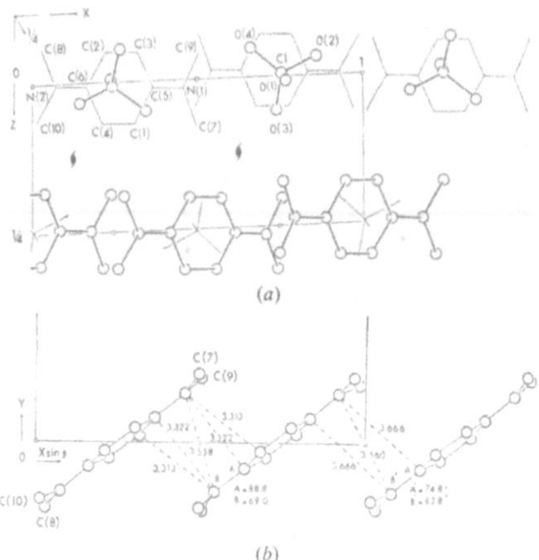

Fig. 1. Structure of the low temperature form of $TMPD^+C\ell O_4^-$. (a) [010] projection, (b) [001] projection.

Fig. 2. Bond lengths (σ = 0.003 Å) and angles in the low temperature form of $TMPD^+ \cdot C\ell O_4^-$.

The a and c axes almost double in length compared with those of the orthorhombic room temperature modification (1). The TMPD groups are arranged in rows (Fig. 1); the distances between benzene planes are alternately 3.10 and 3.62 Å. The bond lengths (Fig. 2) show that the TMPD groups are present as $TMPD^+$ which rules out the "mol-ionic" lattice theory (2).

1. Preceding report.
2. G.T. POTT and J. KOMMANDEUR, 1967. J. Chem. Phys., 47, 395.

BIS-p-NITROPHENYLCARBODI-IMIDE

$C_{13}H_8N_4O_4$ $O_2NC_6H_4$-N=C=N-$C_6H_4NO_2$

A.T. VINCENT and P.J. WHEATLEY, 1972. J. Chem. Soc., Perkin II, 1567-1571.

Orthorhombic, $Pna2_1$, a = 24.738, b = 3.827, c = 13.277 Å, D_m = 1.48, Z = 4. Cu radiation, R = 0.077 for 1205 reflexions.

The molecule adopts a markedly distorted allene type configuration which allows it to pack with a 3.827 Å translational symmetry. Thus the angles C-N=C are 129.6 and 134.4° (compare 120° in an undistorted system), N=C=N is 169.7° (cf. 180°) and the torsion angle C-N···N-C is 113.5° (cf. 90°). The C=N distances are 1.203, 1.219(9) Å. The N-C(phenyl) distances (1.392(8) and 1.406(8) Å) are much shorter than those in di-p-tolylcarbodi-imide (1). The shortening could be due to a significant contribution of a quinoidal structure to the ground state.

1. This volume, p. 143.

N-(p-CHLOROBENZYLIDENE)-p-CHLOROANILINE

$C_{13}H_9Cl_2N$ ClC_6H_4N=CHC_6H_4Cl

J. BERNSTEIN and G.M.J. SCHMIDT, 1972. J. Chem. Soc., Perkin II, 951-955.

Triclinic, $P\bar{1}$, a = 5.986, b = 3.933, c = 12.342 Å, α = 87.38°, β = 78.40°, γ = 89.53°, D_m = 1.42, Z = 1. Mo radiation, R = 0.062 for 1458 reflexions.

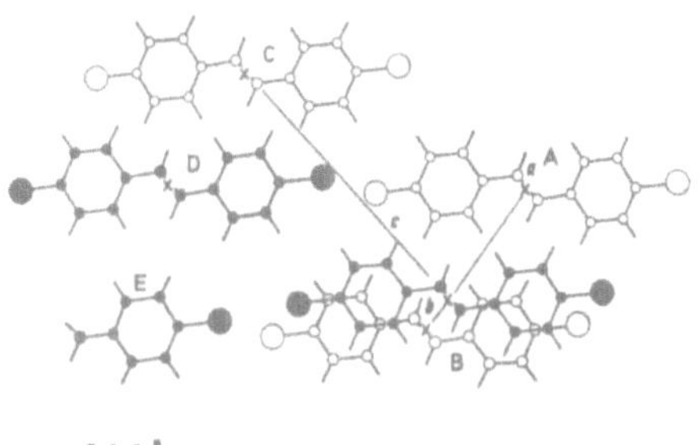

Fig. 1. Molecular packing of N-(p-chlorobenzylidene)-p-chloroaniline. Elevations of other molecules (Å) with respect to the plane of the origin molecule: A -2.54, B -3.56, C -1.24, D 1.36, and E, 0.40. Molecular pseudo-centres of symmetry are marked with a cross.

The structure is disordered about a crystallographic inversion centre which requires that the two phenyl rings must be parallel. The analysis establishes that molecules are essentially planar to within 0.02 Å in striking contrast to the conformations of other benzylideneanilines. The disordered C=N bond length is 1.262 Å. The packing arrangement is dominated by a short Cl···Cl intermolecular contact (3.42 Å) (Fig. 1).

N-(2,4-DICHLOROBENZYLIDENE)ANILINE

$C_{13}H_9C\ell_2N$ $C_6H_5 \cdot N{=}CH \cdot C_6H_3C\ell_2$

J. BERNSTEIN, 1972. J. Chem. Soc., Perkin II, 946-950.

Orthorhombic, $P2_12_12_1$, a = 11.202, b = 3.934, c = 27.077 Å, D_m = 1.45, Z = 4. Mo radiation, R = 0.082 for 1524 reflexions.

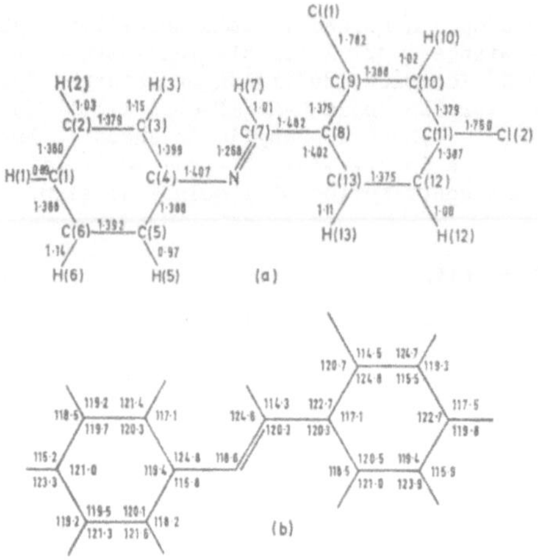

Fig. 1. N-(2,4-Dichlorobenzylidene)aniline: (a) bond lengths; mean σ (Å): C-C = 0.012, C-N = 0.011, C-Cℓ = 0.008, and C-H = 0.08; (b) bond angles; mean σ (°): C-C-C and C-N-C = 0.8, Cℓ-C-C = 0.6, and angles involving hydrogen 4.3.

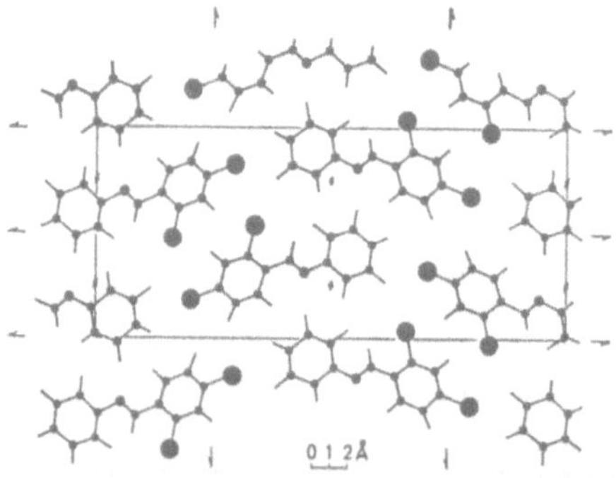

Fig. 2. The crystal structure of N-(2,4-dichlorobenzylidene)aniline viewed down b.

The molecular structure, bond lengths and angles are shown in Fig. 1. The angle of twist of the aniline ring about the N-C(4) bond is 37.9° whereas that of the benzylidene ring about the C(7)-C(8) bond is 6.1°. These values are much smaller than in other benzylideneanilines; the decrease in the twist angles is attributed to packing effects. The packing (Fig. 2) is efficient with no unusally short intermolecular distances.

4-NITRO-4'-METHOXY-BENZALANILINE

$C_{14}H_{12}N_2O_3$ $CH_3O\text{-}C_6H_4\text{-}N{=}CH\text{-}C_6H_4NO_2$

J.M. PIRET, P. PIRET, G. GERMAIN and M. van MEERSSCHE, 1972. Bull. Soc. Chim. Belges, 81, 533-538.

Monoclinic, $P2_1/c$, a = 12.888, b = 7.098, c = 14.066 Å, β = 102.82°, D_m = 1.35, Z = 4. Mo radiation, R = 0.138 for 621 reflexions.

The C-N=CH-C grouping has trans configuration with C-N = 1.42(1) and C=N = 1.24(1) Å. Angles N=C-C and C=N-C are 124 and 117° respectively. The nitrophenyl ring is rotated 16° out of the plane of the C-N=CH-C moiety and the methoxy-phenyl ring is rotated by 37° in the opposite sense. There are only the usual van der Waals contacts between molecules.

2-DIETHYLAMINO-2',6'-ACETOXYLIDIDE HYDROHEXAFLUOROARSENATE

(LIDOCAINE HYDROHEXAFLUOROARSENATE)

$C_{14}H_{23}AsF_6N_2O$

CH_3
$-NH\ CO\ CH_2\ NH^+\ (CH_2CH_3)_2\ ,\ AsF_6^-$
CH_3

A.W. HANSON, 1972. Acta Cryst., B28, 672-679.

Monoclinic, C2/c, a = 22.82, b = 9.19, c = 19.89 Å, β = 120.79°, D_m = 1.58, Z = 8. Cu radiation, R = 0.044 for 2582 reflexions.

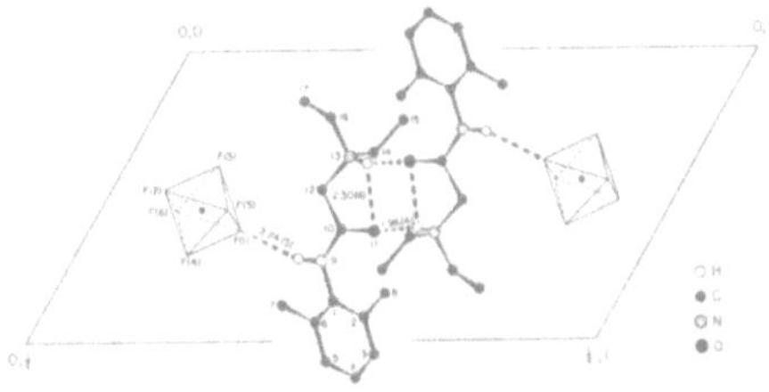

Fig. 1. A part of the structure of lidocaine hexafluoroarsenate viewed along b, showing hydrogen bonding. The only hydrogen atoms shown are those that participate in hydrogen bonding.

The lidocaine moiety is in the physiologically active cation form with the amino nitrogen [N(13)] protonated. The arrangement of the ions in the crystal is shown in Fig. 1. The phenyl ring is planar and the attached nitrogen and methyl carbon atoms lie close to its plane. The side chain from N(9) to C(12) lies in a plane inclined at 66° to that of the ring. The protonated nitrogen, N(13), is strongly hydrogen bonded to the oxygen atom of a neighbouring cation. Pairs of such bonds join pairs of cations across centres of symmetry (Fig. 1) with N(13)H···O(11) = 2.822(4) Å. There is also a weak intramolecular hydrogen bond between N(13)H and O(11). The amido nitrogen [N(9)] is weakly hydrogen bonded to a fluorine atom of a hexafluoroarsenate ion, N(9)H···F(1) = 3.040(5) Å. Infrared spectra of crystalline lidocaine salts (1) had been interpreted in terms of a structure completely free of hydrogen bonding. Bond lengths in the cation are in accord with expected values. The configuration of the hexafluoroarsenate ion is nominally octahedral. The severe thermal motion of the anion results in the As-F distances being unreliable.

1. G.A. NEVILLE and Z.R. REGNIER, 1969. Canad. J. Chem., 47, 4229.

LIDOCAINE HYDROCHLORIDE MONOHYDRATE

$C_{14}H_{25}ClN_2O_2$

CH_3-, NH-CO-CH_2-NH^+-$(CH_2CH_3)_2$, Cl^-, H_2O

A.W. HANSON and M. RÖHRL, 1972. Acta Cryst., B28, 3567-3571.

Monoclinic, $P2_1/c$, a = 8.490, b = 7.110, c = 27.58 Å, β = 106.87°, D_m = 1.21, Z = 4. Cu radiation, R = 0.11 for 2385 reflexions.

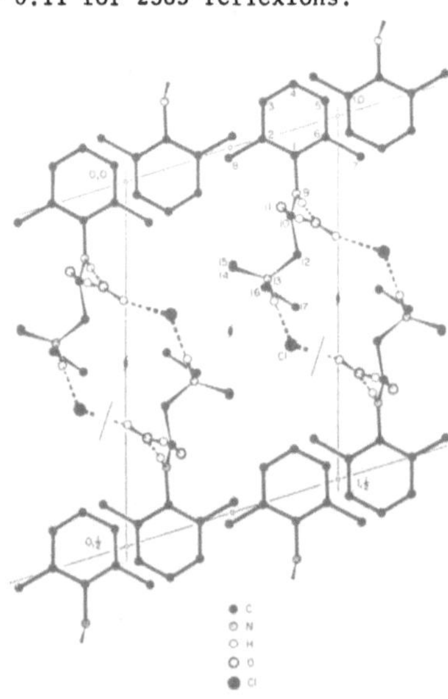

Fig. 1. Hydrogen bonding of lidocaine hydrochloride monohydrate viewed along b.

In the crystal structure (Fig. 1) adjacent lidocaine cations are linked by water molecules to form chains parallel to b. Adjacent chains are joined in pairs by hydrogen bonds involving chloride ions, $\bar{N}H^+$ groups and water molecules (NH···O = 2.85 Å, OH···Cℓ⁻ = 3.19 Å, NH···Cℓ = 3.07 Å, OH···O = 2.84 Å). The structure utilizes all potential donor and acceptor sites. There is disorder, with an alternate uncharacterized structure present at about 5% occupancy. The bond lengths and angles, within the necessarily large estimated uncertainties, are normal.

DI-p-TOLYLCARBODI-IMIDE

$C_{15}H_{14}N_2$ $CH_3C_6H_4$-N=C=N-$C_6H_4CH_3$

A.T. VINCENT and P.J. WHEATLEY, 1972. J. Chem. Soc., Perkin II, 687-690.

Orthorhombic, $P2_12_12_1$, a = 11.310, b = 14.799, c = 7.690 Å, D_m = 1.15, Z = 4. Cu radiation, R = 0.063 for 1511 reflexions.

The molecule has approximately the stereochemistry expected on classical grounds. The angles at the N atoms (127.2, 128.4(3)°) are close to those expected for trigonal hybridisation and the -N=C=N- moiety has an allene-type configuration. The N=C=N angle is 170.4(4)°. The C-N single bond lengths are short and similar (1.428, 1.432(4) Å). The C=N bonds are quite different in length (1.223 and 1.204(4) Å); the differences can be correlated with a difference in the torsion angles (19.8 and 7.9°) at the C-N bonds. The molecules are separated by van der Waals distances in the crystals.

(-)-N-ETHYL-N-METHYLANILINE OXIDE (+)-3-BROMOCAMPHORSULPHONATE

$C_{19}H_{28}BrNO_5S$ $C_9H_{14}NO^+C_{10}H_{14}BrO_4S^-$

R.L. MUNTZ, W.H. PIRKLE and I.C. PAUL, 1972. J. Chem. Soc., Perkin II, 483-488.

Monoclinic, $P2_1$, a = 14.620, b = 7.062, c = 11.534 Å, β = 114°42', D_m = 1.42, Z = 2. Cu radiation, R = 0.050 for 1669 reflexions.

The arrangement of cation and anion in the crystal structure is shown in (Fig. 1); a hydrogen bond O-H···O = 2.522(10) Å links the ions. The absolute configuration of the N-oxide is established as S by internal comparison to the known absolute configuration of the anion. Principal bond lengths in the cation are N-O = 1.409(10), N-C(methyl) = 1.521(14), N-C(ethyl) = 1.473(15), N-C(phenyl) = 1.515(10) Å. Bond distances and angles in the camphorsulphonate anion are normal; the conformation of the camphor ring system can be described as contra (-, +).

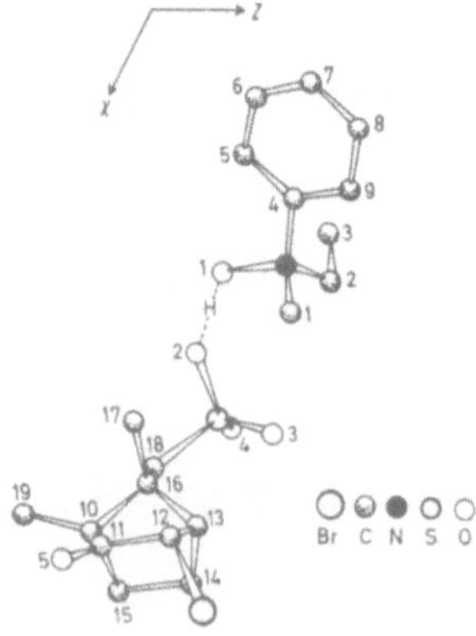

Fig. 1. View of the (-)-N-ethyl-N-methylaniline oxide (+)-3-bromocamphorsulphonate complex looking along b.

DIETHYL ETHER (AT 128°K)

$C_4H_{10}O$

D. ANDRÉ, R. FOURME and K. ZECHMEISTER, 1972. Acta Cryst., B28, 2389-2395.

Orthorhombic, $P2_12_12_1$, a = 11.81, b = 8.07, c = 10.85 Å, Z = 8. Cu radiation, R = 0.052 for 907 reflexions.

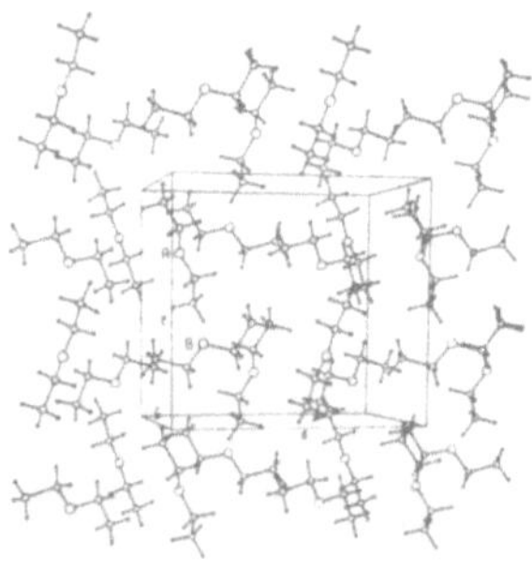

Fig. 1. Drawing of the packing of diethyl ether at 128°K.

Both crystallographically independent molecules in the asymmetric unit possess nearly perfect C_{2v} symmetry. The molecules behave as rigid bodies and distances corrected for libration are mean C-O = 1.433 Å, mean C-C = 1.512 Å. The packing of these essentially planar molecules is very loose (Fig. 1); all intermolecular distances between non-hydrogen atoms exceed the sum of van der Waals radii.

o-CRESOL (AT -50°C)

$C_6H_6O_2$ $C_6H_4(OH)_2$

C. BOIS, 1972. Acta Cryst., B28, 25-31.

Trigonal $P3_1$, a = 16.43, c = 5.94 Å, D_m = 1.12, Z = 9. R = 0.08 for 1373 reflexions.

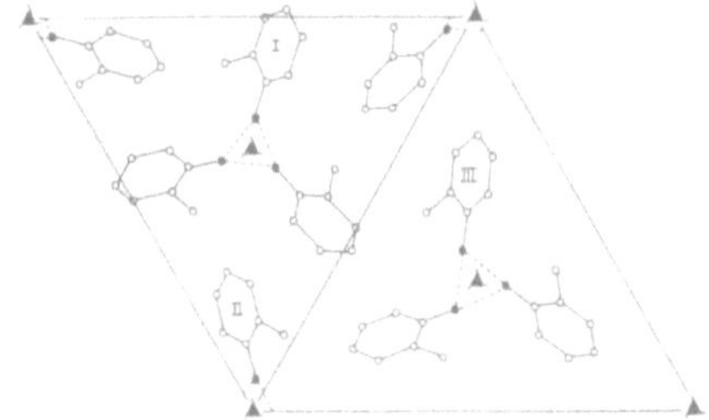

Fig. 1. Projection of the o-cresol structure along c. Dotted lines represent hydrogen bonds.

There are three independent molecules in the asymmetric unit and their orientation in the lattice is shown in Fig. 1. Molecules related by 3_1 screw axes form three independent spirals via hydrogen bonding (OH···O = 2.74, 2.75 and 2.77(1) Å). Bond lengths and angles within the molecules have normal values.

5-HYDROXYDOPAMINE HYDROCHLORIDE

$C_8H_{12}ClNO_3$ $(HO)_3C_6H_2(CH_2)_2\overset{+}{N}H_3 \cdot Cl^-$

A.M. ANDERSON, A. MOSTAD and C. RØMMING, 1972. Acta Chem. Scand., 26, 2670-2680.

Orthorhombic, Pbca, a = 14.58, b = 16.34, c = 7.83 Å, D_m = 1.46, Z = 8. Mo radiation, R = 0.039 for 1310 reflexions.

Details of the molecular geometry are shown in Fig. 1. The pyrogallol part is nearly planar. The C(1) to C(6) carbon atoms deviate less than 0.009 Å from a least squares benzene ring plane whereas the deviation for C(7) is 0.104 Å and for the oxygen atoms -0.015 Å (O1), 0.085 Å (O2), and 0.015 Å (O3). The geometry of the carbon frame is quite normal (compare 6-hydroxydopamine hydrochloride (1)). The angle between the benzene ring plane and the plane defined by C(1), C(7), and C(8) is 57.7°, the C(7)-H(72) bond is thus situated in the plane of the benzene ring. The dihedral angle C(1)-C(7)-C(8)-N is 189.6° and the conformation about the C(8)-N bond is staggered. This extended structure is always found in hydrochloric salts of phenylethylamines.

The crystal structure is mainly stabilized by hydrogen bonds of the types N-H···O, N-H···Cl^- and O-H···Cl^-. The phenol oxygen atoms act as hydrogen donors in hydrogen bonds to chlorine ions, the lengths being 3.100 Å (O1), 3.051 Å (O2), and 3.181 Å (O3). The H···Cℓ⁻ distances are 2.30 Å, 2.24 Å, and 2.42 Å, respectively. The nitrogen atom is hydrogen donor in a corresponding bond, the N···Cℓ⁻ distance being 3.285 Å and HN(1)···Cℓ⁻ distance 2.48 Å. O(1) and O(3) are hydrogen acceptors in hydrogen bonds to nitrogen atoms of neighbouring molecules; the hydrogen bond lengths are 2.850 Å and 2.886 Å, respectively.

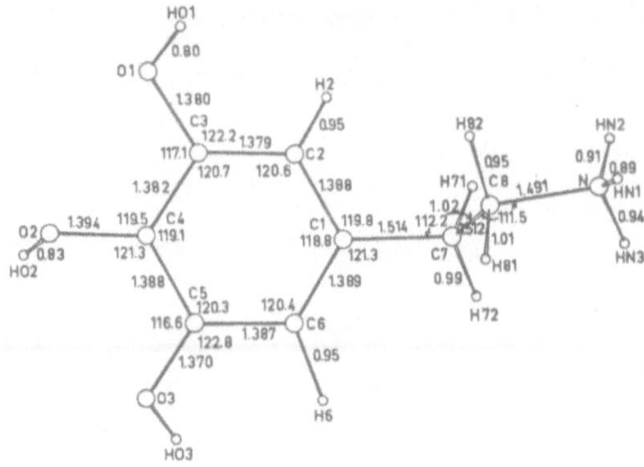

Fig. 1. Bond lengths (corrected for librational effects) and bond angles in 5-hydroxydopamine hydrochloride. Standard deviations are C-C C-N = 0.003 Å, C-O = 0.002 Å; angles involving C, N and O, = 0.2°.

1. Following report.

3,4,6-TRIHYDROXYPHENYLETHYLAMINE HYDROCHLORIDE

(6-HYDROXYDOPAMINE HYDROCHLORIDE)

$C_8H_{12}C\ell NO_3$ $C_6H_2(OH)_3(CH_2)_2\overset{+}{N}H_3\cdot C\ell^-$

M. KOLDERUP, A. MOSTAD and C. RØMMING, 1972. Acta Chem. Scand., 26, 483-493.

Monoclinic, Pc, a = 4.80, b = 7.23, c = 13.12 Å, β = 97.3°, D_m = 1.50, Z = 2. Mo radiation, R = 0.042 for 1781 reflexions.

The bond lengths and angles in the molecule (Fig. 1) are normal except for the C(7)-C(8) bond, which at 1.521(3) Å is shorter than normally found for $C(sp^3)$-$C(sp^3)$ bonds but a similar value was also found in 5-hydroxydopamine (1). The crystal structure is stabilised through a three-dimensional network of hydrogen bonds; each cation is linked to four cations and three chlorine ions by N-H···O (2.87 - 2.94 Å), N-H···Cℓ⁻ (3.17 Å) and O-H···Cℓ⁻ (3.09 - 3.32 Å) hydrogen bonds.

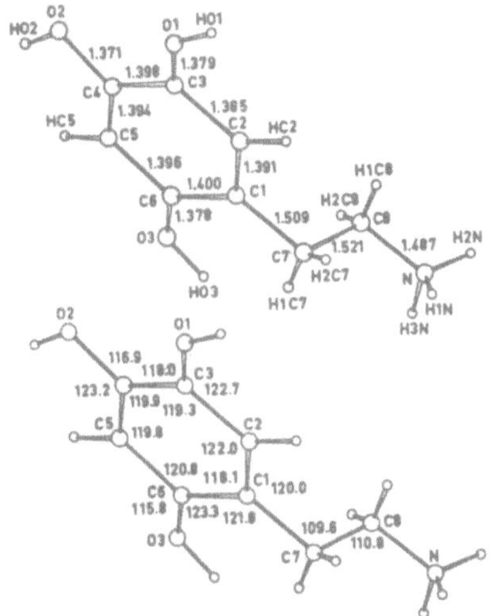

Fig. 1. Hydroxydopamine hydrochloride: bond lengths (Å) (corrected for thermal vibration effects) and angles (°). Mean standard deviations are 0.002 Å and 0.2°.

1. Preceding report.

2-(2'-CARBOMETHOXY-4'-NITROPHENOXY)-1,3,5-TRICHLOROBENZENE

$C_{14}H_8C\ell_3NO_5$

E.A.H. GRIFFITH, W.D. CHANDLER and B.E. ROBERTSON, 1972. Canad. J. Chem., 50, 2979-2988.

Triclinic, P$\bar{1}$, a = 8.862, b = 15.164, c = 13.028 Å, α = 90.35°, β = 116.28°, γ = 97.54°, D_m = 1.58, Z = 4. Mo radiation, R = 0.047 for 4526 reflexions.

Fig. 1. View of the molecular packing in 2-(2'-carbomethoxy-4'-nitrophenoxy)-1,3,5-trichlorobenzene. The origin is in the lower left front corner. The a axis points back into the page, the b axis points up, and the c axis points to the right. Molecule 1 is nearer the origin.

The two independent molecules in the unit cell (Fig. 1) show a conformation similar to that found in related ethers. The molecular conformation appears to be determined partly by steric effects since conjugation between the nitro-substituted ring and ether oxygen is less pronounced than in other derivatives. The C-O(ether) distances do not differ significantly from the mean value, 1.38(1) Å; the mean angle at the ether oxygen is 118.6(13)°. [Figures 5 and 6 in the initial paper are in error. Corrected versions appear in 1.]

1. E.A.H. GRIFFITH, W.D. CHANDLER and B.E. ROBERTSON, 1974. Canad. J. Chem., 52, 182.

4-BROMO-2,6-DI-t-BUTYL-PHENOL

$C_{14}H_{21}BrO$

G. FILIPPINI, C.M. GRAMACCIOLI, A. MUGNOLI and T. PILATI, 1972. Cryst. Struct. Comm., 1, 305-308.

Orthorhombic, $P2_12_12_1$, a = 15.606, b = 10.320, c = 8.926 Å, D_m = 1.318, Z = 4. Mo radiation, R = 0.073 for 1566 reflexions.

The refinement of a structure determined previously (1). The benzene ring and its adjacent atoms are essentially coplanar (within 0.04 Å). The C-O distance is 1.37(2) Å. There is no evidence for any O-H···O hydrogen bonding; all O···O contacts are greater than 4 Å.

1. M. MAZE and C. RERAT, 1964. C. R. Acad. Sci., Paris, 259, 4612.

PENTAMETHYL-o-HYDROXYHYDROCINNAMYL ALCOHOL

$C_{14}H_{22}O_2$

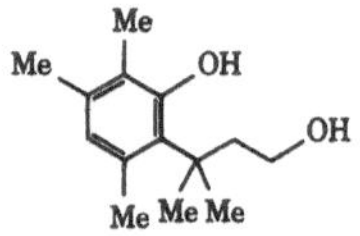

J.M. KARLE and I.L. KARLE, 1972. J. Amer. Chem. Soc., 94, 9182-9189.

Triclinic, $C\bar{1}$, a = 11.615, b = 25.987, c = 18.150 Å, α = 93.66, β = 90.38, γ = 98.36°, Z = 16. Cu radiation, R = 0.095 for 3837 reflexions.

All four independent diol molecules in the asymmetric unit have nearly the same conformation; a representative is shown in Fig. 1. The conformation of the molecules is restricted by the presence of a "trialkyl lock". Mean molecular dimensions are summarised in Fig. 2. The bond lengths are normal but the angles at C(4) and C(5) are distorted considerably from 120° values to accommodate the C(11) and C(12) methyl groups. The phenyl ring is planar but the exocyclic groups are off the phenyl plane. In the crystal structure intermolecular O-H···O bonds (2.68 - 2.94 Å) link the molecules; there are no intramolecular hydrogen bonds. [For another example of a molecule whose conformation is also restricted by a trialkyl lock see pentamethylhydrocoumarin, $C_{14}H_{18}O_2$ (1).]

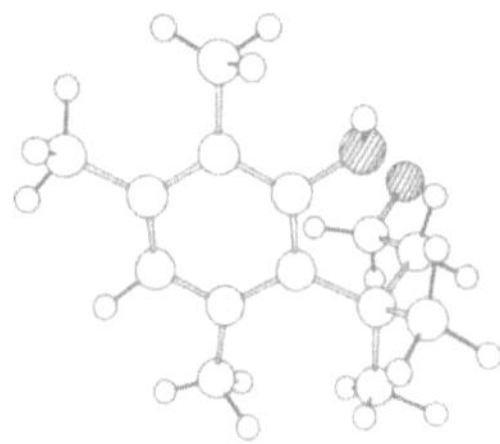

Fig. 1. Diagram showing the conformation of a representative pentamethyl-o-hydroxyhydrocinnamyl alcohol molecule.

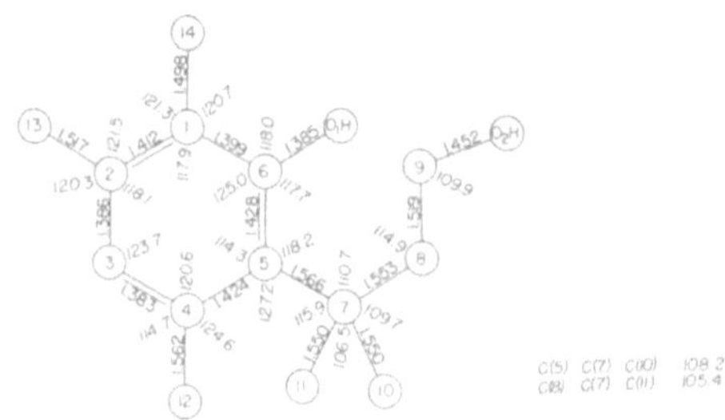

Fig. 2. Bond lengths and angles for the $C_{14}H_{22}O_2$ diol. The figures represent the average values obtained from the four independent molecules in the asymmetric unit.

1. This volume, p. 336.

BUTEIN MONOHYDRATE

$C_{15}H_{14}O_6$ $C_{15}H_{12}O_5 \cdot H_2O$

N. SAITO, K. UENO and Y. SASADA, 1972. Bull. Chem. Soc. Japan, 45, 2274-2277.

Monoclinic, $P2_1/n$, a = 10.89, b = 28.88, c = 4.29 Å, β = 96.7°, D_m = 1.42, Z = 4. Cu radiation, R = 0.17 for 1063 reflexions (film data).

Butein is 3,4,2',4'-tetrahydroxychalcone (Fig. 1); bond lengths and angles are normal. There is an intramolecular hydrogen bond, and the molecules are held together by intermolecular O-H···O hydrogen bonding.

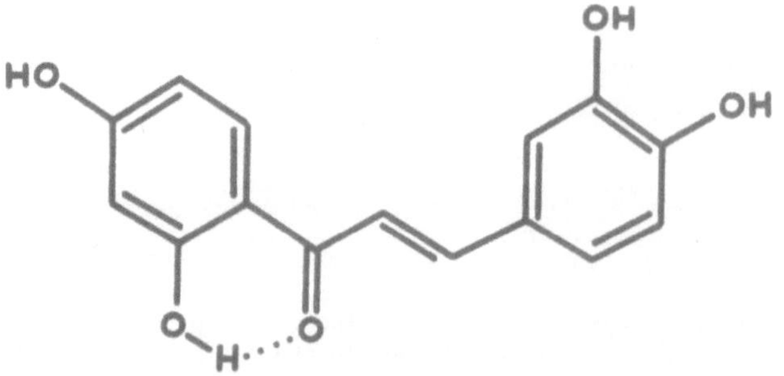

Fig. 1. Butein.

DIENESTROL

$C_{18}H_{18}O_2$

J.M. FORNIES-MARQUINA, C. COURSEILLE, B. BUSETTA and M. HOSPITAL, 1972. Acta Cryst., B28, 655-656.

Orthorhombic, Pbca, a = 18.969, b = 14.296, c = 5.380 Å, D_m not given, Z = 4. Cu radiation, R = 0.059 for 1282 reflexions.

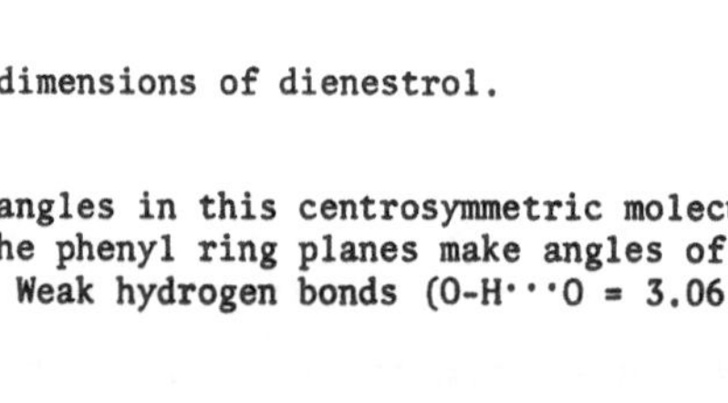

Fig. 1. Molecular dimensions of dienestrol.

Bond lengths and angles in this centrosymmetric molecule (Fig. 1) correspond to expected values. The phenyl ring planes make angles of 85.9° with that of the central dienic chain. Weak hydrogen bonds (O-H···O = 3.06 Å) are reported between molecules.

p-BROMOBENZOPHENONE OXIME O-PICRYL ETHERS

$C_{19}H_{11}BrN_4O_7$

J.D. McCULLOUGH, I.C. PAUL and D.Y. CURTIN, 1972. J. Amer. Chem. Soc., 94, 883-889.

syn Form

Monoclinic, $P2_1/c$, a = 14.67, b = 7.64, c = 37.25 Å, β = 107°10', D_m = 1.62, Z = 8. Cu radiation, R = 0.15 for 4931 reflexions.

anti Form

Monoclinic, $P2_1/c$, a = 12.56, b = 7.96, c = 21.49 Å, β = 113°43', D_m = 1.64, Z = 4. Cu radiation, R = 0.062 for 3038 reflexions.

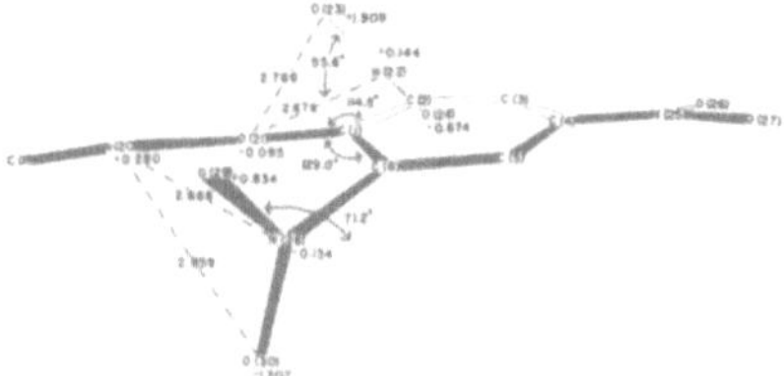

Fig. 1. Diagram showing certain dimensions in or around the picryl ring in 1c. The distances with positive and negative signs refer to the deviations of the atoms from the best plane through the six carbon atoms of the picryl ring.

The benzophenone oxime O-picryl ethers (1a-c) undergo Beckmann rearrangements in solution to form via the intermediates 2a-c the respective N-picrylanilides (3a-c). Crystals of 1a-c were observed to undergo reaction upon heating and the

p-X—C$_6$H$_4$(p-Y—C$_6$H$_4$)C=N—O—Pic (1) → [Pic—O(p-X—C$_6$H$_4$)C=N—C$_6$H$_4$-p-Y] (2) → p-X—C$_6$H$_4$—C(=O)—N(Pic)—C$_6$H$_4$-p-Y (3)

a, X = Y = H
b, X = Br; Y = H
c, X = H; Y = Br

products were shown to be 3a-c (1). A crystallographic investigation of compounds 1a-c and 3a-c was undertaken to provide information on the lattices and molecular geometries of the reactants and products of this potentially solid-state reaction. Unit cell dimensions and space groups are given for all six compounds and full three-dimensional structure analyses were carried out on compounds 1b and 1c.

In the crystal structure of syn-isomer 1b there are two independent molecules per asymmetric unit. There are no major differences in the molecular geometry of the two conformers of 1b and of the molecule of the anti-isomer 1c. Because of this similarity and the greater accuracy of the structure determination of 1c, it is discussed in detail.

In the crystal of 1c, the oxygen and the two ortho nitrogen substituents on the picryl ring (Fig. 1) are displaced from the plane of the picryl ring by -0.095, 0.144, and -0.134 Å. There is considerable evidence for intramolecular overcrowding in this region of the molecule. The phenyl ring syn to the O-picryl group is rotated by 60.5° out of the best plane through the central five-atom ($C_2C{=}N(-O)$) group in the molecule, while the other phenyl ring is only rotated by 18.6°. The C-N(oxime), N-O(ether), and C(picryl)-O(ether) lengths are 1.291(6), 1.441(5), and 1.349(5) Å. The intramolecular distances between the migrating oxygen atom and the carbon atom to which it migrates are 2.213(5), 2.25(2), and 2.23(2) Å in 1c and the two molecules of 1b; there are no intermolecular contacts between these atoms less than 4 Å.

1. J.D. McCULLOUGH, D.Y. CURTIN and I.C. PAUL, 1972. J. Amer. Chem. Soc., 94, 874.

2-(3'-METHYL-4'-NITROPHENOXY)-1,3-DIISOPROPYLBENZENE

$C_{19}H_{23}NO_3$

E.A.H. GRIFFITH, W.D. CHANDLER and B.E. ROBERTSON, 1972. Canad. J. Chem., 50, 2963-2971.

Triclinic, P$\bar{1}$, a = 12.422, b = 13.180, c = 11.241 Å, α = 89.99°, β = 95.81°, γ = 105.46°, D_m = 1.15, Z = 4. Mo radiation, R = 0.060 for 3701 reflexions.

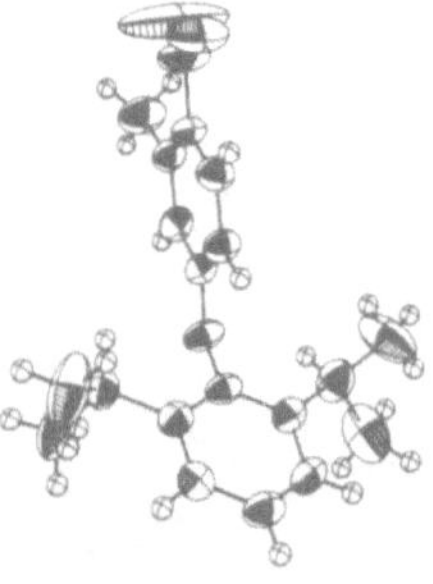

Fig. 1. A molecule of 2-(3'-methyl-4'-nitrophenoxy)-1,3-diisopropylbenzene.

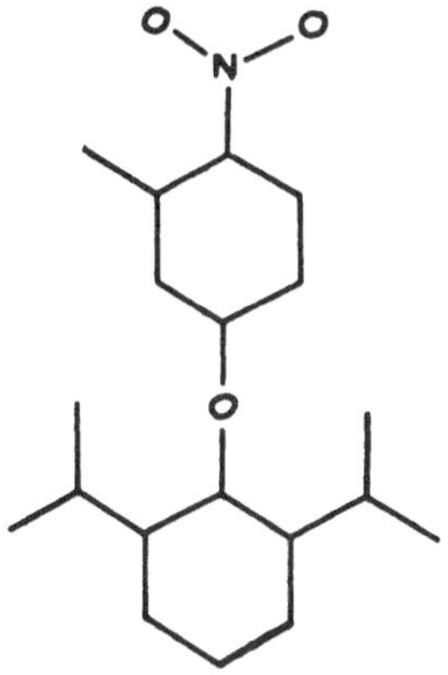

C-C (rings)	1.36 - 1.39(1) Å
Other C-C	1.49 - 1.52(2)
C-O	1.37, 1.41(1)
C-N	1.45(1)
N-O	1.18(2)

Fig. 2. Bond lengths for the two independent molecules of 2-(3'-methyl-4'-nitrophenoxy)-1,3-diisopropylbenzene.

The two independent molecules in the asymmetric unit have similar conformations; that of one molecule is shown in Fig. 1. Bond lengths are shown in Fig. 2. The ether oxygen bond angle in each of the two independent molecules is 120°. The two phenyl rings are nearly orthogonal in each case, while the ring containing the electron withdrawing nitro group is conjugated with the ether oxygen atom. The rings show significant deviations from hexagonal symmetry and the nitro-substituted ring is bent away from the other ring.

2-(2',4'-DINITROPHENOXY)-1,3,5-TRI-t-BUTYLBENZENE

$C_{24}H_{32}N_2O_5$

E.A.H. GRIFFITH, W.D. CHANDLER and B.E. ROBERTSON, 1972. Canad. J. Chem., 50, 2972-2978.

Triclinic, P$\bar{1}$, a = 12.160, b = 10.564, c = 10.684 Å, α = 119.23°, β = 96.06°, γ = 96.55°, D_m = 1.1, Z = 2. Mo radiation, R = 0.053 for 2471 reflexions.

The conformation of the molecule is shown in Fig. 1, and bond lengths are in Fig. 2. The bond angle at the ether oxygen atom is 117°. The presence of the two electron-withdrawing nitro groups on one ring increases the conjugation between the ring and the ether oxygen to a greater degree than was found in a mono-nitro derivative (1). The plane of the tri-t-butyl substituted benzene ring is inclined at 84° to that of the dinitro substituted ring. The p-nitro group is rotated 5.2° from the ring to which it is bonded and the o-group 15.8° from the ring.

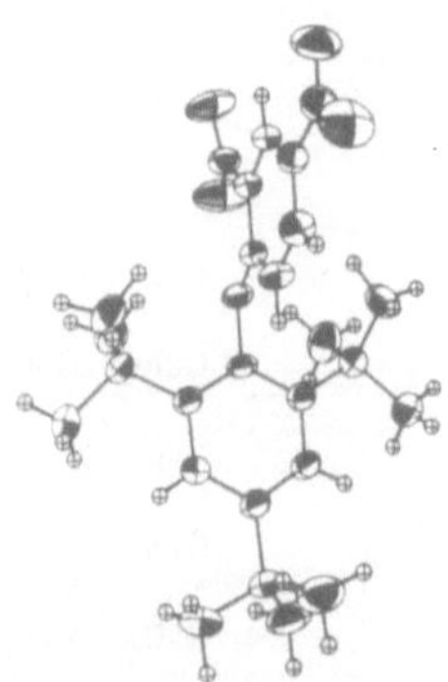

Fig. 1. View of 2-(2',4'-dinitrophenoxy)-1,3,5-tri-t-butylbenzene.

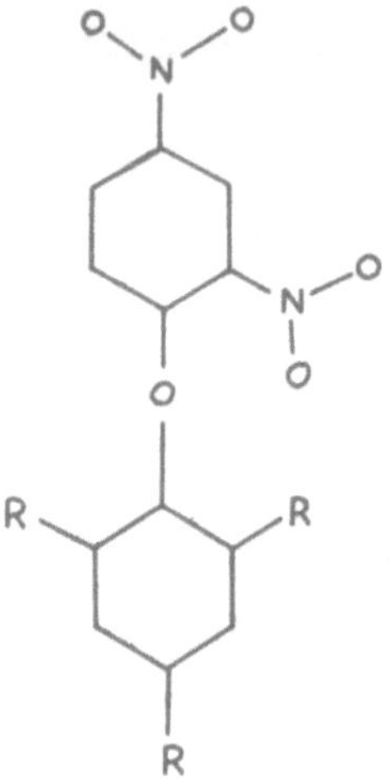

C-C (rings)	1.37 - 1.41(1) Å
Other C-C	1.51 - 1.56(2)
C-O	1.35, 1.43(1)
C-N	1.47(1)
N-O	1.21(1)

R = t-butyl

Fig. 2. Bond lengths (with standard deviations in parentheses) in 2-(2',4'-dinitrophenoxy)-1,3,5-tri-t-butylbenzene.

1. Preceding report.

TETRACHLORO-p-BENZOQUINONE (AT 110°K)

(CHLORANIL)

$C_6C\ell_4O_2$

K.J. van WEPEREN and G.J. VISSER, 1972. Acta Cryst., B28, 338-342.

Monoclinic, $P2_1/a$, a = 8.664, b = 5.658, c = 8.532 Å, β = 105.95°, D_m not given, Z = 2. Cu radiation, R = 0.056 for 4277 reflexions.

The structure of this molecule at room temperature is known (1), and in Fig. 1 the geometries of the molecule at the two temperatures are compared.

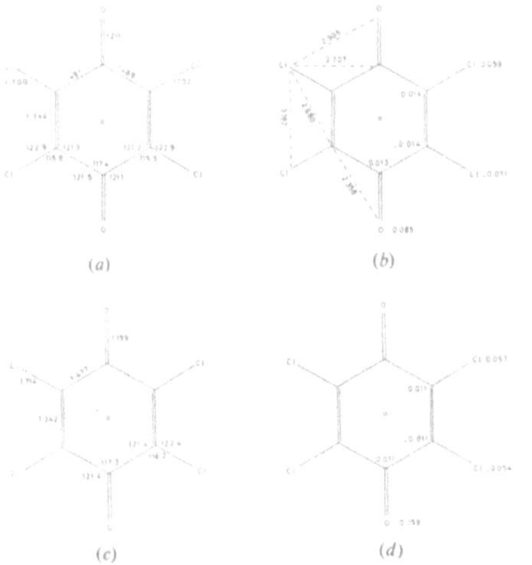

Fig. 1. Geometry of the chloranil molecules, (a) 110°K. Bond lengths (σ(C-Cℓ) = 0.0012 Å, σ(remainder) = 0.0015 - 0.0017 Å) and angles (σ = 0.09 - 0.12°) not corrected for libration. (b) Deviations from the carbon atom best plane and non-bonded distances. (c) and (d) Corresponding data from reference 1. (σ(C-Cℓ) = 0.008 Å, σ(others) = 0.011 Å, σ(angles) = 0.6 - 0.7°).

1. Structure Reports, 27, 899.

1,2,3-TRICHLOROBENZENE

(NEUTRON STUDY)

$C_6H_3C\ell_3$

R.G. HAZELL, M.S. LEHMANN and G.S. PAWLEY, 1972. Acta Cryst., B28, 1388-1394.

Monoclinic, $P2_1/c$, a = 12.66, b = 8.258, c = 15.05 Å, β = 114.42°, D_m not given, Z = 8. Neutron diffraction data, R = 0.07 for 1194 reflexions.

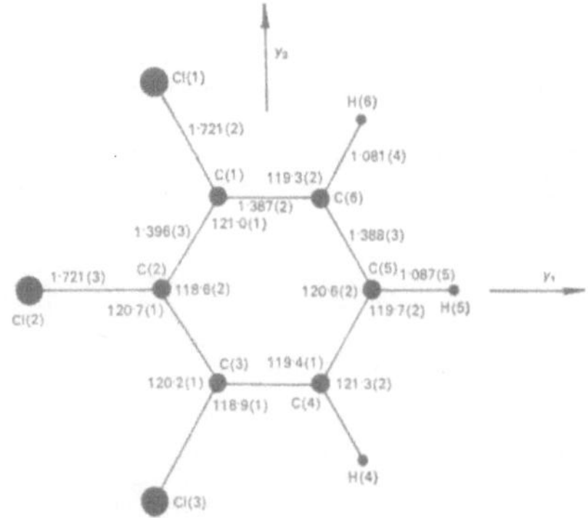

Fig. 1. Molecular parameters from the constrained refinements of 1,2,3-trichlorobenzene.

There are two symmetry unrelated molecules in the asymmetric unit which were constrained to have the same shape and C_{2v} symmetry during refinement. The resulting molecular geometry is in Fig. 1. When constraints were relaxed one of the two molecules took up a distorted conformation (a movement of C(2) by 0.011 Å and Cℓ(2) by 0.007 Å) whereas the other closely retained the free-state symmetry imposed in the constrained refinement. The thermal motion is excellently described by rigid body molecular motion when account is taken of the internal modes on the hydrogen atoms.

p-IODOTOLUENE

C_7H_7I $CH_3C_6H_4I$

C.-T. AHN, S. SOLEDO and G.B. CARPENTER, 1972. Acta Cryst., B28, 2152-2157.

Orthorhombic, $P2_12_12_1$, a = 7.46, b = 16.50, c = 6.11 Å, D_m = 1.918, Z = 4. Mo radiation, R = 0.059 for 585 reflexions.

In spite of some disorder, the molecular structure is as expected, with a mean C-C distance of 1.389 Å in the benzene ring and C-I = 2.06 Å. The molecule is essentially planar with the ring atoms deviating by less than 0.032 Å and the iodine atom lying 0.143 Å from the ring plane. The crystal structure consists of a pattern of C-I···I groups linked into a planar zigzag chain around a twofold screw axis along the c axis with a short I···I distance of 4.06 Å and a C-I···I angle of 158.3°.

2-AMINO-5-BROMOTOLUENE

C_7H_8BrN

H. van der MEER, 1972. Acta Cryst., B28, 3098.

Orthorhombic, $P2_12_12_1$, a = 15.265, b = 8.756, c = 5.578 Å, Z = 4. Cu radiation, R = 0.081 for 542 reflexions.

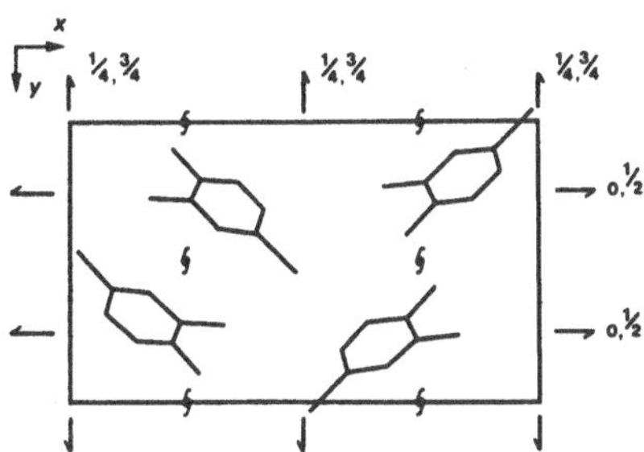

Fig. 1. Projection of the structure of 2-amino-5-bromotoluene along [001].

The crystal structure (Fig. 1) contains well separated molecules with no unusual intermolecular contacts. Bond lengths and angles are normal within the accuracy of the analysis.

2,3,5,6-TETRACHLORO-4-(METHYLTHIO)BENZONITRILE

$C_8H_3Cl_4NS$

D.R. CARTER and F.P. BOER, 1972. J. Chem. Soc., Perkin II, 2104-2107.

Triclinic, $P\bar{1}$, a = 7.559, b = 8.332, c = 8.912 Å, α = 78.63°, β = 80.95°, γ = 80.60°, D_m = 1.77, Z = 2. Mo radiation, R = 0.036 for 2172 reflexions.

The crystal structure contains discrete molecules; all atoms except those of the methyl group lie in the (210) planes of the crystal. The methylthio group forms a dihedral angle of 87.1° with respect to the aromatic ring which is itself slightly non-planar. The pattern of displacements of the ring conforms to a very shallow boat. Principal bond lengths are C-Cl = 1.717 - 1.725(2), $C(sp^2)$-S = 1.763(2), $S-CH_3$ = 1.826(3) Å; the C-S-C angle is 102.0(1)°.

TETRACHLORO-p-XYLENE

$C_8H_6C\ell_4$ $C\ell_4C_6(CH_3)_2$

T.L. KHOCJANOVA, T.A. BABUSHKINA, S.I. KUZNECOV and G.K. SEMIN, 1972. Kristallografija, 17, 552-556 [Soviet Physics-Crystallography, 17, 480-484].

Monoclinic, $P2_1/c$, a = 8.21, b = 3.87, c = 16.95 Å, β = 117°40', Z = 2, R = 0.104.

The crystal structure contains disordered molecules situated on symmetry centres. The molecule is planar and the length of the C-"X" bond is 1.67 Å. The compounds pentachlorotoluene and tetrachloro-m-xylene are isostructural with tetrachloro-p-xylene.

CINNAMIC ACID DERIVATIVES

S.E. FILIPAKIS, L. LEISEROWITZ, D. RABINOVICH and G.M.J. SCHMIDT, 1972. J. Chem. Soc., Perkin II, 1750-1758.

β-CHLORO-trans-CINNAMIC ACID

(I)

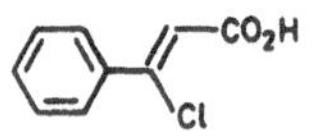

$C_9H_7C\ell O_2$

Monoclinic, A2/a, a = 19.588, b = 5.374, c = 16.683 Å, β = 103.62°, Z = 8. Cu radiation, R = 0.054 for 1873 reflexions.

β-CHLORO-cis-CINNAMIC ACID

(II)

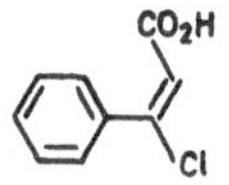

$C_9H_7C\ell O_2$

Monoclinic, $P2_1/c$, a = 5.797, b = 24.247, c = 7.240 Å, β = 121.56°, Z = 4. Cu radiation, R = 0.046 for 1917 reflexions.

β-METHYL-cis-CINNAMIC ACID

(III)

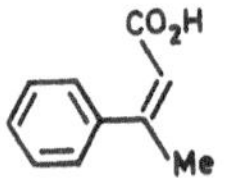

$C_{10}H_{10}O_2$

Monoclinic, $P2_1/c$, a = 7.504, b = 7.442, c = 17.056 Å, β = 107.00°, Z = 4. Cu radiation, R = 0.066 for 1987 reflexions.

Bond lengths and angles in the three molecules are shown in Fig. 1. The three C(5)-C(6) bonds are almost equal in the three acids despite angles of twist around this bond of 11.7° (I), 65.4° (II), and 83.3° (III). In the trans-acid (I) the twist around the C(5)-C(6) bond is 11.7° whereas in the cis-acid (II) the twist angle is 65.4°. In the methyl cis-acid (III) the twist around the C(5)-C(6) bond is larger still (83.3°) and in this conformation the atoms O(1), O(2), C(3)-C(6) are coplanar whereas in the cis-chloro acid (II) the plane of the carboxy group makes an angle of 9° with that of atoms C(3)-C(6). All three crystal structures (Fig. 2) contain centrosymmetric dimers linked by O-H···O hydrogen bonds. The electron-density distributions and C-O bond lengths are indicative of disorder in the paired carboxy groups in the cis-acids (II and III) but not in the β-chloro-trans-acid (I). Molecules of (I) are inclined to the ac plane such that the interplanar spacing between molecules is 3.48 Å. The cis-acid (II) has carboxy acid dimers sandwiched between C-Cℓ bonds, and in (III) hydrogen bonded dimers are sandwiched between antiparallel benzene rings.

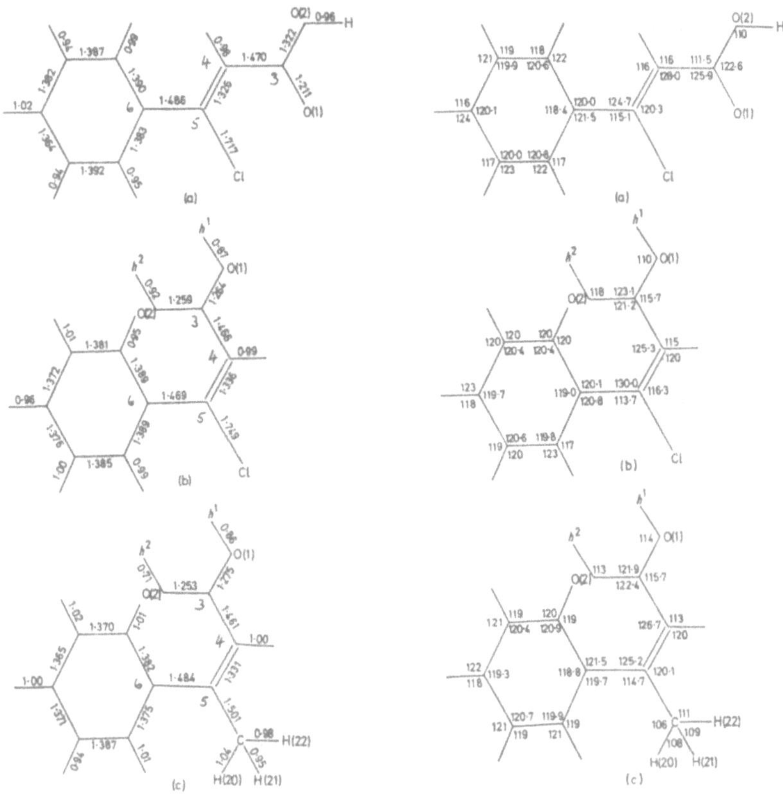

Fig. 1. Bond lengths and angles in (a) β-chloro-trans-, (b) β-chloro-cis- and (c) β-methyl-cis-cinnamic acid. Mean σ values for C-C, C-O, and C-Cℓ bond lengths are 0.004, 0.004, and 0.003 Å respectively.

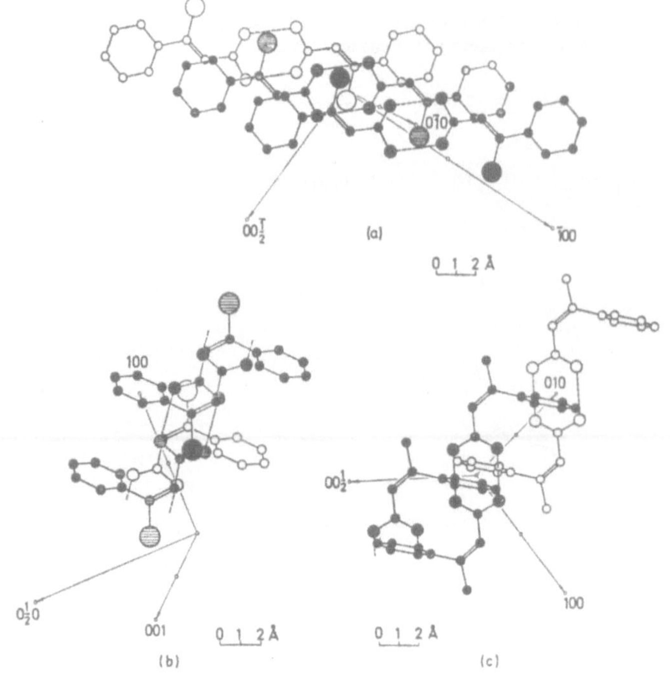

Fig. 2. Molecular packing arrangements seen perpendicular to the plane of the carboxy-group: (a) β-chloro-trans, (b) β-chloro-cis, and (c) β-methyl-cis-cinnamic acid.

1,2,3-TRICHLORO-4,5,6-TRIMETHYLBENZENE (AT 298°K)

$C_9H_9C\ell_3$

R. FOURME, M. RENAUD and D. ANDRE, 1972. Mol. Cryst. Liq. Cryst., 17, 209-221.

Monoclinic, $P2_1/c$, a = 8.22, b = 3.88, c = 17.16 Å, β = 119.33°, Z = 2. Cu radiation, R = 0.08 for 310 reflexions.

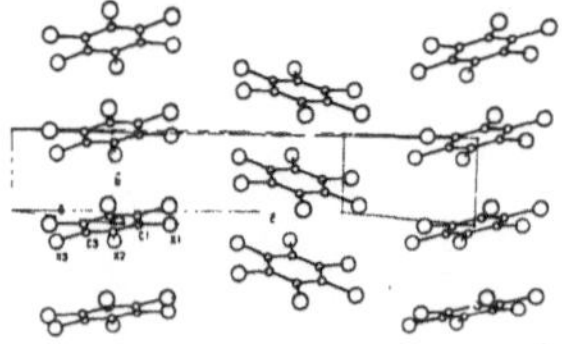

Fig. 1. View of the trichlorotrimethylbenzene structure at 298°K.

The crystal structure (Fig. 1) contains disordered molecules at room temperature and it is not possible to distinguish between chlorine and methyl groups (the molecular centre coincides with a centre of inversion). The structure should be compared with that found at 173°K (1).

1. Following report.

1,2,3-TRICHLORO-4,5,6-TRIMETHYLBENZENE (AT 173°K)

$C_9H_9C\ell_3$

R. FOURME and M. RENAUD, 1972. Mol. Cryst. Liq. Cryst., 17, 223-236.

Triclinic, P$\bar{1}$, a = 8.12, b = 7.62, c = 34.25 Å, α = 90.00, β = 119.33, γ = 90.00°, Z = 8. Cu radiation, R = 0.158 for 500 reflexions.

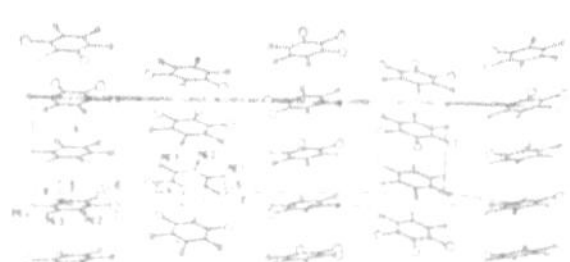

Fig. 1. View of the $C_9H_9C\ell_3$ structure at 173°K.

At 173°K the crystal structure (Fig. 1) contains completely ordered molecules (two independent molecules per asymmetric unit). The effect of lowering the temperature has been to double the b and c axial lengths compared to the 298°K modification (1), and to remove the dynamic disorder present in the 298°K structure.

1. Preceding report.

DIPOTASSIUM MESITYLENE-DISULPHONATE DIHYDRATE

$C_9H_{14}K_2O_8S_2$

M.A.M. MEESTER and H. SCHENK, 1972. Rec. Trav. Chim. Pays-Bas, 91, 213-220.

Monoclinic, P2_1/c, a = 7.341, b = 12.330, c = 15.934 Å, β = 94.60°, D_m = 1.80, Z = 4. Cu radiation, R = 0.042 for 2311 reflexions.

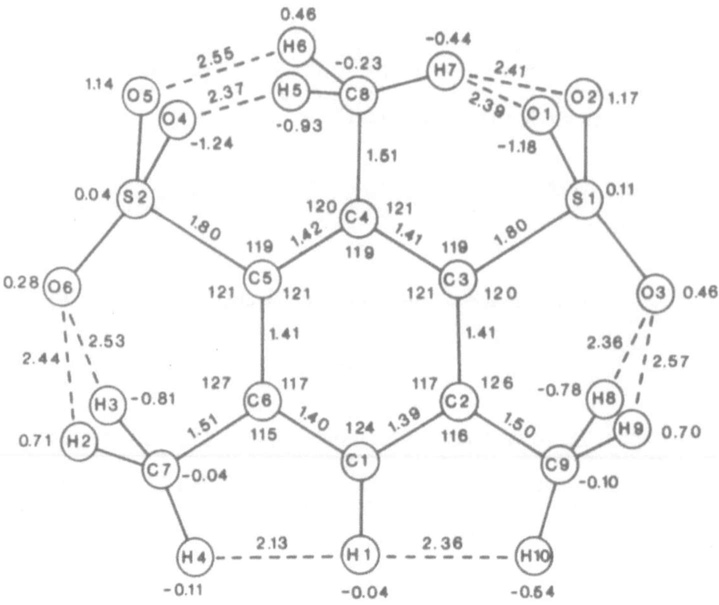

Fig. 1. The mesitylene-disulphonate ion with some bond lengths and all non-bonded distances between outer atoms. The distances of the ring-atoms C(1), C(2), C(3), C(4), C(5), and C(6) from the benzene plane are -0.02, 0.00, 0.02, -0.03, 0.01 and 0.01 Å. The distances of the other atoms from this plane are indicated near each atom.

Details of the molecular structure are shown in Fig. 1. Molecular overcrowding is relieved by bond angle deformation rather than by bond length extension. The benzene ring is nearly planar. Successive substituents are situated on alternate sides of the ring. The water molecules form O-H···O hydrogen bonds 2.81 - 2.93 Å with sulphonate oxygen atoms; the K ions are surrounded by deformed octahedra of oxygen atoms.

PYROMELLITIC DITHIOANHYDRIDE

$C_{10}H_4O_4S_2$ $S(CO)_2C_6H_4(CO)_2S$

I.V. **BULGAROVSKAJA** and Z.V. ZVONKOVA, 1972. Kristallografija, 17, 292-296 [Soviet Physics-Crystallography, 17, 247-250].

Monoclinic, $P2_1/c$, a = 6.326, b = 5.425, c = 13.691 Å, β = 94.75°, D_m = 1.760, Z = 2. Cu radiation, R = 0.153 for 703 reflexions. Molecular symmetry:centre.

The crystal structure contains well resolved molecules whose centres coincide with crystallographic inversion centres. The molecule is essentially planar with dimensions which agree well with accepted values.

p-DIETHYNYLBENZENE

$C_{10}H_6$ $HC{\equiv}C{-}C_6H_4{-}C{\equiv}CH$

N.A. AHMED, A.I. KITAIGORODSKIJ and M.I. SIROTA, 1972. Acta Cryst., B28, 2875-2877.

Monoclinic, $P2_1/c$, a = 4.007, b = 6.018, c = 15.340 Å, β = 91°42', D_m = 1.16, Z = 2. R = 0.092 for 274 reflexions.

In this centrosymmetric molecule the C-C≡ and C≡C bond distances are 1.444 and 1.188 Å. The structure found experimentally is in good agreement with that predicted on theoretical grounds.

DURENE

$C_{10}H_{14}$

C. H. STAM, 1972. Acta Cryst., B28, 2630-2632.

Monoclinic, $P2_1/a$, a = 11.59, b = 5.74, c = 7.04 Å, β = 112.8°, Z = 2. Cu radiation, R = 0.061 for 654 reflexions.

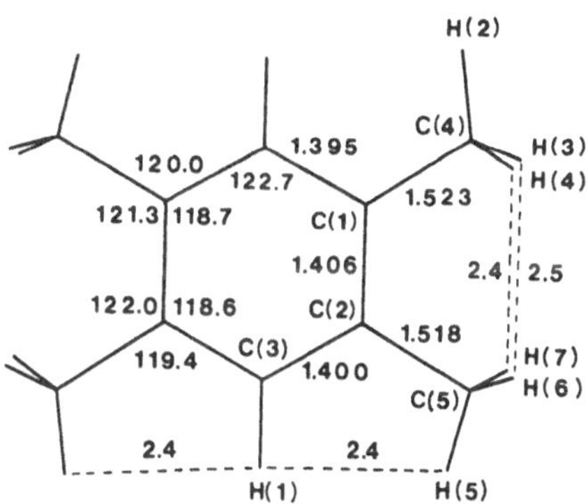

Fig. 1. Bond lengths (corrected for librational motion) and angles in durene.

This three-dimensional analysis confirms the previous two-dimensional study (1) and provides more precise bond lengths and angles (Fig. 1). The carbon skeleton is planar to within 0.001 Å. The methyl groups are symmetrical with respect to the benzene ring such that H(2) and H(5) are coplanar with it.

1. Strukturbericht, 3, 683, 752.

DIPHENYLIODINIUM NITRATE

$C_{12}H_{10}INO_3$ $(C_6H_5)_2INO_3$

W.B. WRIGHT and E.A. MEYERS, 1972. Cryst. Struct. Comm., 1, 95-98.

Orthorhombic, Fdd2, a = 31.906, b = 12.719, c = 12.059 Å, Z = 16. Mo radiation, R = 0.036 for 1584 reflexions.

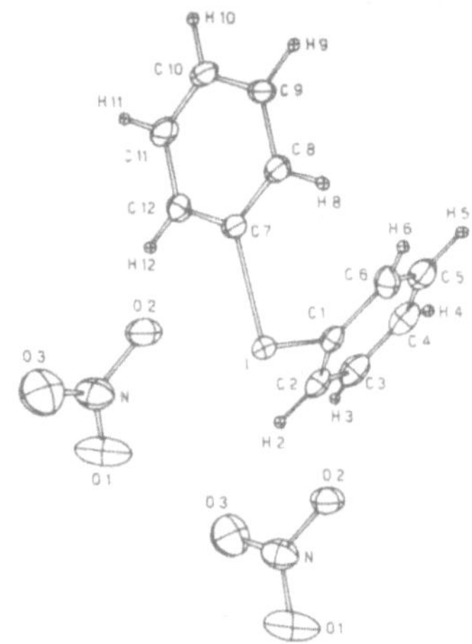

Fig. 1. Diphenyliodinium nitrate.

The iodine has roughly fourfold planar coordination (Fig. 1) and infinite zig-zag chains, -I-O(2)-I'-O(2)'-, extend through the structure. The bond distances and angles are normal with I-C = 2.114(7) Å and C(1)-I-C(7) = 92.0(4)°, I-O(2) = 2.768(8), I-O(2') = 2.877(8) Å.

ETHYL p-CHLORO-α-CYANO-β-METHYL-cis-CINNAMATE

$C_{13}H_{12}ClNO_2$

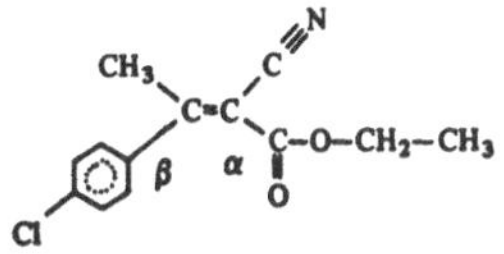

T. HIGUCHI, W. NAGAI, K. NAKATSU, T. MIWA and A. SHIMADA, 1972. Israel J. Chem., 10, 221-228.

Orthorhombic, Pbcn, a = 17.31, b = 8.68, c = 17.20 Å, D_m = 1.295, Z = 8. Cu radiation, R = 0.119 for 1100 reflexions.

The phenyl ring makes an angle of 60° with the ethylenic plane. The conformation of the ethylenic bond and the carbonyl group is approximately cisoid but a twist of 16.5° reduces contact between the carbonyl oxygen and the phenyl ring. Bond lengths and angles are in accord with predicted values. There are no unusual intermolecular contacts.

1-BROMO-2(p-ETHYLPHENYL)-1,2-DIPHENYLETHYLENE

$C_{22}H_{19}Br$

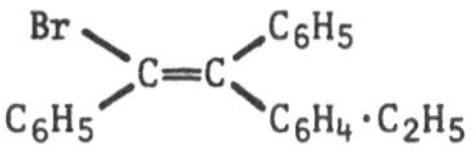

J.M. FORNIES-MARQUINA, C. COURSEILLE, B. BUSETTA and M. HOSPITAL, 1972. Cryst. Struct. Comm., 1, 261-264.

Monoclinic, $P2_1/c$, a = 8.574, b = 35.767, c = 5.767 Å, β = 86.1°, Z = 4. Cu radiation, R = 0.084 for 2943 reflexions.

In this isomer (labelled A) the phenyl rings are trans. Bond distances and angles have normal values. The angles between the planes of the aromatic rings are 78.5° (C(1)phenyl:C(2)phenyl), 61.3° (C(1)phenyl:C(2)p-ethylphenyl) and 61.5° (C(2)phenyl:C(2)p-ethylphenyl).

1,2,4,5-TETRA-t-BUTYLBENZENE

$C_{22}H_{38}$ $(C_4H_9)_4C_6H_2$

C. H. STAM, 1972. Acta Cryst., B28, 2715-2720.

Monoclinic, $P2_1/n$, a = 10.910, b = 9.988, c = 10.051 Å, β = 112.29°, Z = 2. Cu radiation, R = 0.075 for 1446 reflexions.

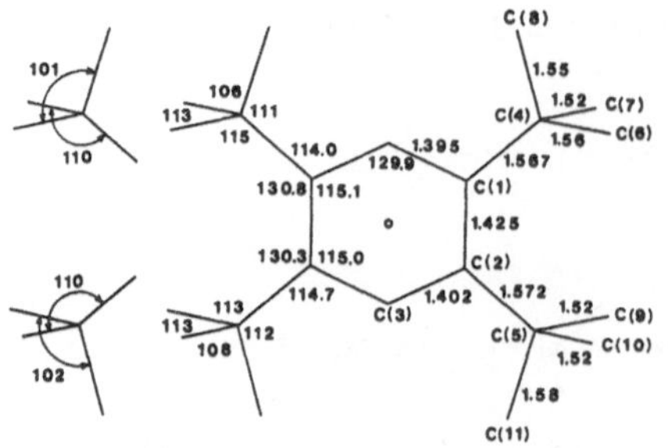

Fig. 1. Bond distances and angles in the 1,2,4,5-tetra-t-butylbenzene molecule.

The bond distances and angles in this centrosymmetric molecule are given in Fig. 1. The considerable strain in the molecule has not affected the planarity of the benzene ring, but is expressed in the bond lengths and angles. The mutual repulsion of the ortho-t-butyl groups results in the C(2)-C(1)-C(4) and C(1)-C(2)-C(5) angles increasing to 130.8(3) and 130.3(2)° respectively. The compensation for this increase is about equally distributed among the other two angles around C(1) and C(2). The strain also results in an increase of the C(1)-C(2) bond length to 1.425(5) Å, and the C(1)-C(4) and C(2)-C(5) bond lengths to 1.567(5) and 1.572(4) Å respectively. The methyl groups of the t-butyl substituents all have the usual staggered conformation, and neighbouring t-butyl groups on the benzene ring are interlocked.

1,2-DICHLORO-1-CYCLOBUTENE-DIONE

(SQUARIC ACID DICHLORIDE)

$C_4C\ell_2O_2$

R. MATTES and S. SCHROEBLER, 1972. Chem. Ber., 105, 3761-3764.

Orthorhombic, $P2_12_12_1$, a = 12.112, b = 7.582, c = 6.175 Å, Z = 4. Cu radiation, R = 0.074 for 327 reflexions.

The molecule is essentially planar. Bond lengths in the four membered ring are C(1)-C(2) = 1.41(2), C(2)-(3) = 1.44(2), C(3)-C(4) = 1.56(2), C(4)-C(1) = 1.48(2), C-Cℓ = 1.66(1), C=O = 1.16 - 1.25(2) Å. The molecules are separated by normal van der Waals distances in the crystal.

POTASSIUM 1,2-DINITRO-3,4,5-TRICHLOROCYCLOPENTADIENIDE

$C_5Cl_3KN_2O_4$

Y. OTAKA, F. MARUMO and Y. SAITO, 1972. Acta Cryst., B28, 1590-1594.

Monoclinic, $P2_1/c$, a = 13.684, b = 8.765, c = 8.065 Å, β = 93.2°, D_m = 2.043, Z = 4. Cu radiation, R = 0.10 for 723 reflexions.

In the cyclopentadienide ion, the three chlorine atoms are in the plane of the five-membered ring but the two nitrogen atoms are displaced on the opposite sides of the plane by about 0.3 Å. From the bond lengths (Fig. 1) it is concluded that the π-electrons in the anion are delocalised as in sandwich type compounds. The two planar $C\text{-}NO_2$ groups are tilted about lines through C(1)···O(1) and C(2)···O(4) through angles of 23 and 25° respectively. The dihedral angle N(1)C(1)C(2)N(2) is 32°. The K^+ ion is surrounded irregularly by six oxygen atoms (Fig. 2) and the unit negative charge of the molecule seems to be distributed on the oxygen atoms.

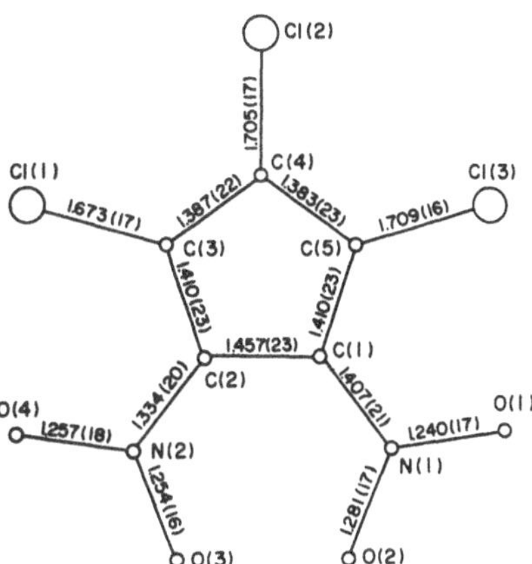

Fig. 1. Bond lengths in the anion 1,2-dinitro-3,4,5-trichlorocyclopentadienide.

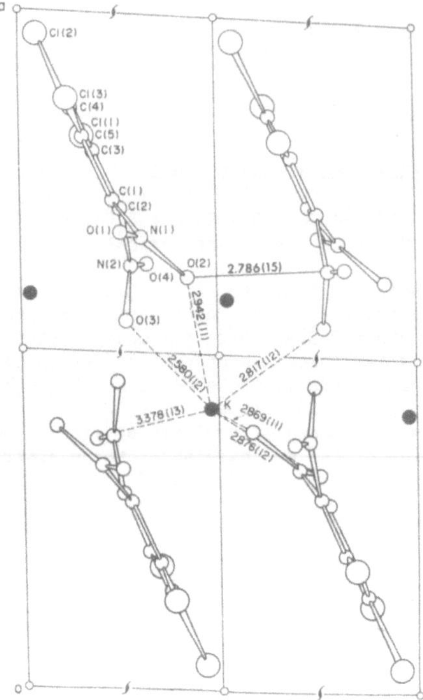

Fig. 2. A projection of the potassium 1,2-dinitro-3,4,5-trichlorocyclopentadienide structure along b.

anti,cis,cis-2,2'-DIBROMOBICYCLOPROPYL

$C_6H_8Br_2$

C. SCHRUMPF and P. SÜSSE, 1972. Chem. Ber., 105, 3041-3049.

Monoclinic, $P2_1/a$, a = 8.638, b = 8.596, c = 5.167 Å, β = 100.75°, D_m = 2.0, Z = 2. Mo radiation, R = 0.102 for 703 reflexions.

The molecule occupies a centre of symmetry. Details of bond lengths and stereochemistry are in Fig. 1. In the crystal the molecules are separated by normal van der Waals distances.

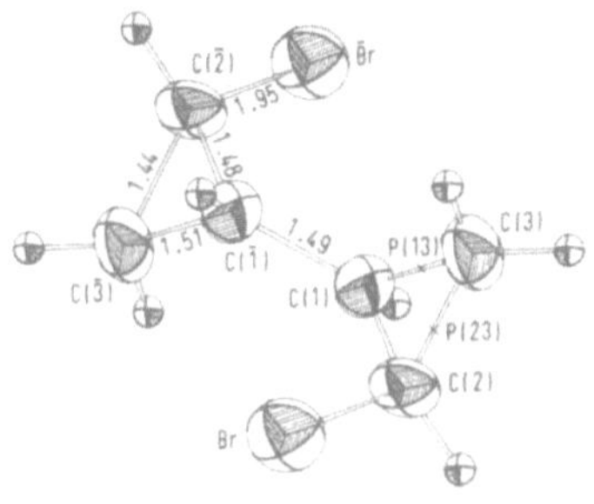

Fig. 1. Molecular conformation and bond lengths (σ(C-C, C-Br) = 0.02, 0.01 Å) in the dibromobicyclopropyl molecule.

cis-1,2-CYCLOBUTANEDICARBOXYLIC ACID

$C_6H_8O_4$

D. van der HELM, I-NAN HSU and J.M. SIMS, 1972. Acta Cryst., B28, 3109-3114.

Monoclinic, $P2_1/c$, a = 10.710, b = 8.559, c = 7.343 Å, β = 95.03°, D_m = 1.423, Z = 4. Cu radiation, R = 0.056 for 1256 reflexions.

In the crystal structure (Fig. 1), the molecules form zigzag chains parallel to the a-axis. The carboxylic acid groups related to each other by a centre of symmetry form O-H···O hydrogen bonds which are not coplanar. The cyclobutane ring is puckered; the dihedral angle between the planes C(4)C(5)C(6) and C(4)C(6)C(7) is 156°. The bond distances and angles are shown in Fig. 2. The cis conformation of the two carboxylic acid groups results in considerable steric hindrance which is relieved by angular deformations. There is an additional twist of C(3) with respect to C(8) compared to C(5) with respect to C(6) of 16.5°. The sum of the three C-C-C bond angles at C(4) is increased to 332.8° whereas at C(7) it is decreased to 312.5°. The carboxylic acid groups are rotated in the same directions around the C(3)-C(4) and C(7)-C(8) bonds, away from the preferred orientation in which the C=O bond of the acid group is syn-planar with the C(4)-C(7) bond.

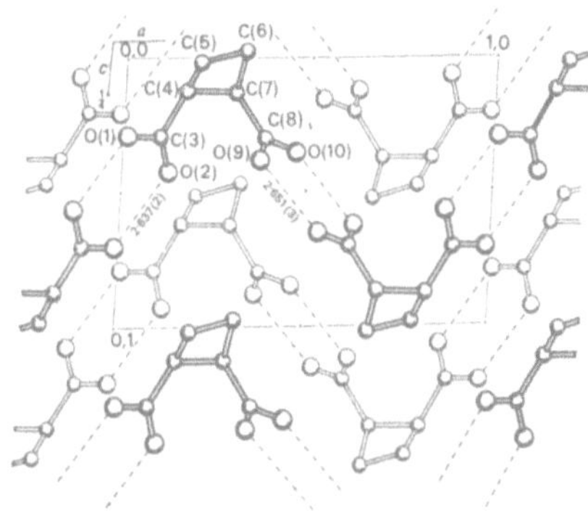

Fig. 1. Projection of cis-1,2-cyclobutanedicarboxylic acid down the b axis.

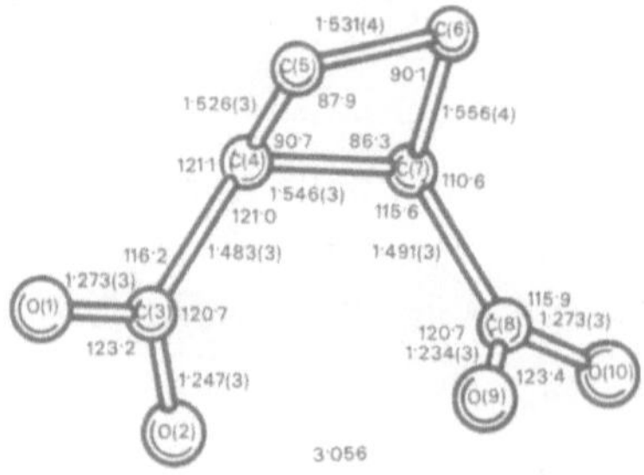

Fig. 2. Bond distances and angles (σ = 0.2°) in cis-1,2-cyclobutanedicarboxylic acid.

trans-2,trans-3-DIMETHYLCYCLOPROPANECARBOXYLIC ACID

$C_6H_{10}O_2$

P. A. LUHAN and A. T. McPHAIL, 1972. J. Chem. Soc., Perkin II, 2372-2375.

Monoclinic, $P2_1/c$, a = 6.39, b = 10.98, c = 9.76 Å, β = 105.6°, D_m = 1.14, Z = 4. Cu radiation, R = 0.085 for 638 reflexions.

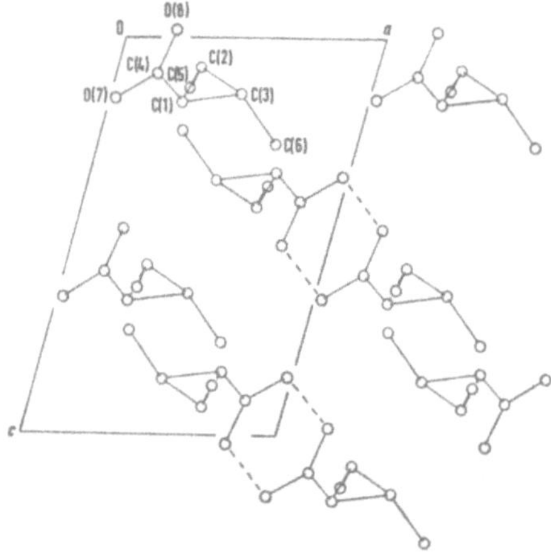

Fig. 1. The crystal structure of trans-2,trans-3-dimethylcyclopropanecarboxylic acid viewed along b.

The crystal structure contains centrosymmetric hydrogen bonded dimers (O-H···O = 2.65 Å) (Fig. 1). Bond lengths and angles have values close to those expected. The dihedral angle between the planar carboxy group and the cyclopropane ring is 85°. The carbonyl oxygen atom lies over the cyclopropane ring but is not equidistant from C(2) and C(3).

1-AMINOCYCLOPENTANE CARBOXYLIC ACID MONOHYDRATE

$C_6H_{13}NO_3$ $C_6H_{11}O_2N{\cdot}H_2O$

M. MALLIKARJUNAN, K. K. CHACKO and R. ZAND, 1972. J. Cryst. Mol. Struct., 2, 53-66.

Monoclinic, $P2_1/c$, a = 11.24, b = 6.27, c = 11.22 Å, β = 97.6°, D_m = 1.23, Z = 4. Cu radiation, R = 0.121 for 1047 reflexions (films, visual intensities).

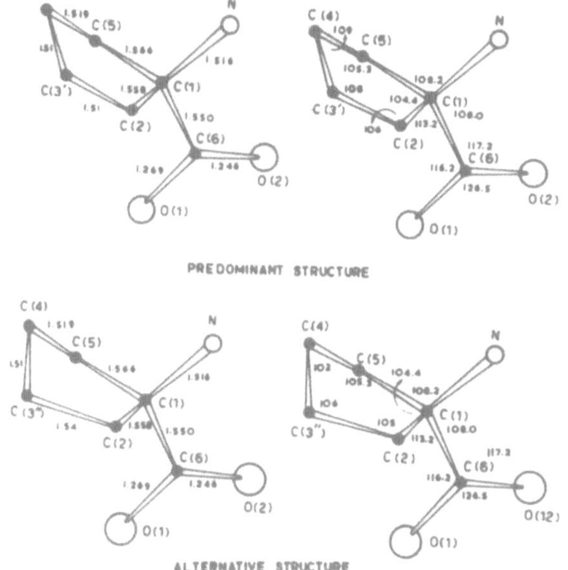

Fig. 1. Bond lengths and angles for the 'predominant' and 'alternate' conformations of $C_6H_{11}O_2N{\cdot}H_2O$.

The cyclopentane ring is disordered; one of the carbon atoms exists in two alternative sites, leading to two possible conformations both of which are of the envelope type. The geometries of the two probable conformations are shown in Fig. 1, σ(C-O, C-N) = 0.007, σ(C-C) = 0.009 Å. There are five hydrogen bonds, N-H···O distances ranging from 2.85 to 2.93, and O-H···O distances of 2.79 and 2.92 Å.

(±)- and (+)-1,2-trans-CYCLOPENTANEDICARBOXYLIC ACID

$C_7H_{10}O_4$

E. BENEDETTI, P. CORRADINI and C. PEDONE, 1972. J. Phys. Chem., 76, 790-797.

(±) form

Monoclinic, C2/c, a = 9.252, b = 6.659, c = 12.185 Å, β = 100°38', D_m = 1.34, Z = 4. Cu radiation, R = 0.063 for 625 reflexions. Molecular symmetry C_2.

(+) form

Monoclinic, $P2_1$, a = 5.957, b = 6.645, c = 12.041 Å, β = 129°13', D_m = 1.34, Z = 2. Cu radiation, R = 0.063 for 602 reflexions.

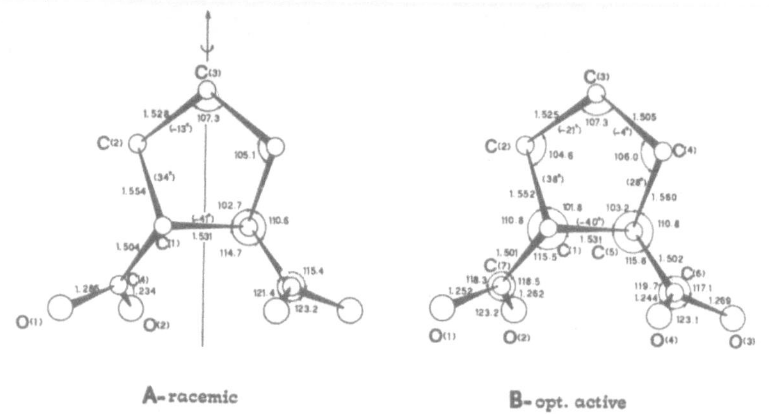

Fig. 1. Bond lengths, angles and torsion angles in 1,2-trans-cyclopentane-dicarboxylic acids.

Details of molecular geometry for both the racemic and optically active forms are shown in Fig. 1 and are for the most part in very good agreement. The C-O and C=O distances are clearly differentiated only in the racemic molecule. In both structures the molecules are linked into infinite chains by pairs of hydrogen bonds (2.66 Å in the centrosymmetric system, and 2.65 and 2.66 Å in the optically active form). In both systems the carboxyl groups are as close to equatorial conformations as the ring puckering will allow.

1,2,3-TRISDIMETHYLAMINOCYCLOPROPENIUM PERCHLORATE

$C_9H_{18}ClN_3O_4$

A. T. KU and M. SUNDARALINGAM, 1972. J. Amer. Chem. Soc., 94, 1688-1692.

Orthorhombic, Pnma, a = 14.232, b = 10.304, c = 9.196 Å, D_m = 1.319, Z = 4. Cu radiation, R = 0.076 for 829 reflexions.

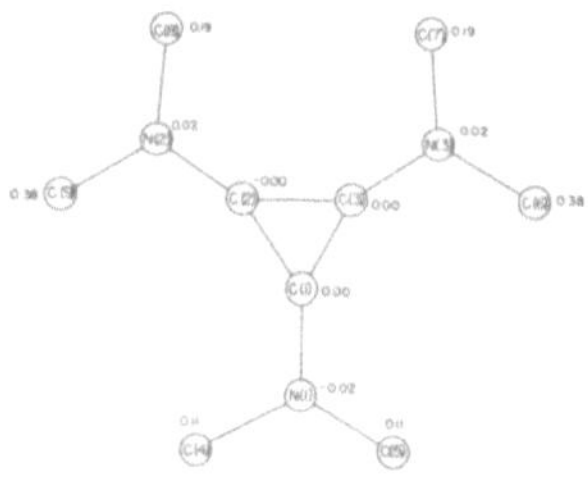

Fig. 1. The deviations of the atoms from the plane through C(1)-C(2)-C(3) in the 1,2,3-trisdimethylaminocyclopropenium cation.

The molecule is bisected by a crystallographic mirror plane. The N-C(methyl) distances average 1.457(7) Å whereas the N-C(ring) distances average 1.333(7) Å. The cyclopropenium ring is symmetrical with mean C-C = 1.363(7) Å. The geometry is accounted for by resonance forms I→VI.

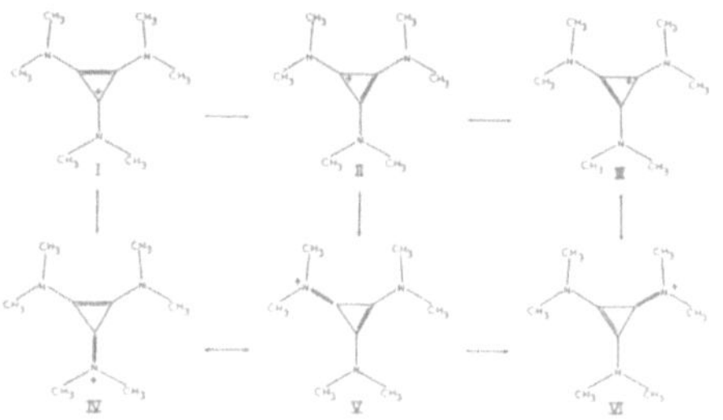

The deviation of the atoms from the cyclopropenium ring are in Fig. 1. The planes of the symmetry related dimethylamino groups are twisted at angles of 20.8° to the plane of the cyclopropenium ring, while the plane of the third amino group makes a dihedral angle of 9.9°. The nonplanarity of the molecule as a whole is due to the hydrogen-hydrogen interactions between the methyl groups on adjacent nitrogen atoms. In the crystal, carbonium ions are sandwiched by the ClO_4^- groups, while the ClO_4^- are held tightly by pairs of cations.

E-2-(p-NITROPHENYL)-CYCLOPROPYL METHYL KETONE

$C_{11}H_{11}NO_3$ $NO_2C_6H_4C_3H_4COCH_3$

J. BORDNER, L. A. JONES and R. L. JOHNSON, 1972. Cryst. Struct. Comm., 1, 389-391.

Monoclinic, $P2_1/c$, a = 7.917, b = 18.78, c = 8.252 Å, β = 121.39°, Z = 4. Cu radiation, R = 0.092 for 951 reflexions.

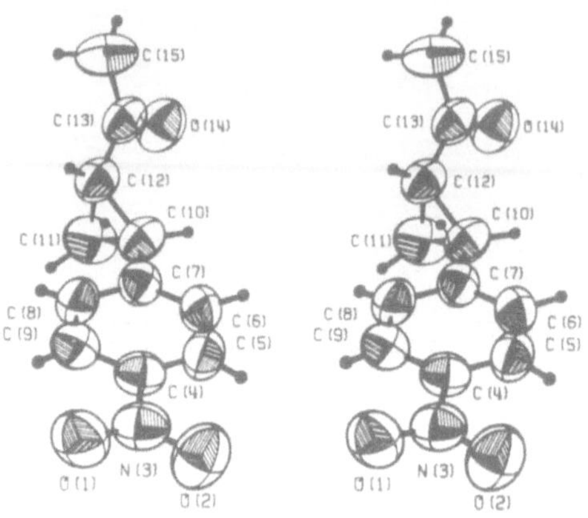

Fig. 1. Diagram of the $NO_2C_6H_4C_3H_4COCH_3$ molecule.

A view of the molecule is given in Fig. 1. The conformation of the molecule is described in terms of the following dihedral angles: 8-7-10-11 = 32°; 8-7-10-12 = 40°; 10-12-13-14 = 43°; 11-12-13-14 = 25°. Bond lengths and angles are normal.

1,1-BIS-(p-CHLOROPHENYL)-2,2-DICHLOROCYCLOPROPANE

$C_{15}H_{10}C\ell_4$

T. P. DeLACY and C. H. L. KENNARD, 1972. J. Chem. Soc., Perkin II, 2141-2147.

Monoclinic, $P2_1/a$, a = 17.351, b = 15.701, c = 11.111 Å, β = 97.09°, D_m = 1.47, Z = 8. Cu radiation, R = 0.061 for 2019 reflexions.

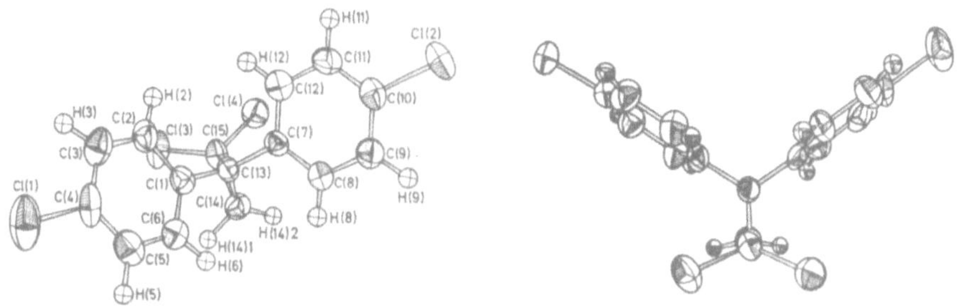

Fig. 1. Two views of the 1,1-bis-(p-chlorophenyl)-2,2-dichlorocyclopropane molecule.

There are two crystallographically independent molecules in the asymmetric unit. Two views of one molecule are shown in Fig. 1; the other molecule has essentially the same conformation. The molecules have pseudo m-symmetry with the mirror plane in the plane of the cyclopropane ring. Important bond lengths and angles in this insecticide are C(cyclopropane)-Cℓ = 1.721 - 1.776(5) Å, C-C(cyclopropane) = 1.48 - 1.54(1) Å, C(1)-C(13)-C(7) (mean) = 114.9(5), C(1)-C(13)-C(14) type = 116.9 - 118.8(5)°. The molecules are separated by van der Waals distances in the crystal structure.

2,4,6-TRICHLOROCYCLOHEXANONE

$C_6H_7Cℓ_3O$

A. LECTARD, F. METRAS and J. GAULTIER, 1972. Cryst. Struct. Comm., 1, 415-418.

Orthorhombic, Pbca, a = 20.634, b = 10.532, c = 7.631 Å, Z = 8. Cu radiation, R = 0.054 for 1136 reflexions.

The molecule has a chair conformation with the 2- and 6-chlorines equatorial and the 4-chlorine axial. All C-C-C angles lie close to their mean value (112°) and all C-C bond lengths are within 1σ of their mean value (1.520(6) Å). The C-Cℓ distances do vary slightly (C(6)-Cℓ = 1.767(5), C(2)-Cℓ = 1.781(5), and C(4)-Cℓ = 1.808(5) Å). The mean C-C=O angle is 124°.

cis-1,3,5-TRICHLOROCYCLOHEXANE

$C_6H_9Cℓ_3$

K. HUML and J. HAŠEK, 1972. Acta Cryst., B28, 2852-2856.

Monoclinic, C2/c, a = 22.131, b = 8.803, c = 9.033 Å, β = 101.3°, D_m = 1.42, Z = 8. Mo radiation, R = 0.063 for 1399 reflexions.

The molecule is in chair conformation with bond lengths and angles shown in Fig. 1. The chloro substituents are equatorial. In the crystals the molecules are separated by normal van der Waals contacts.

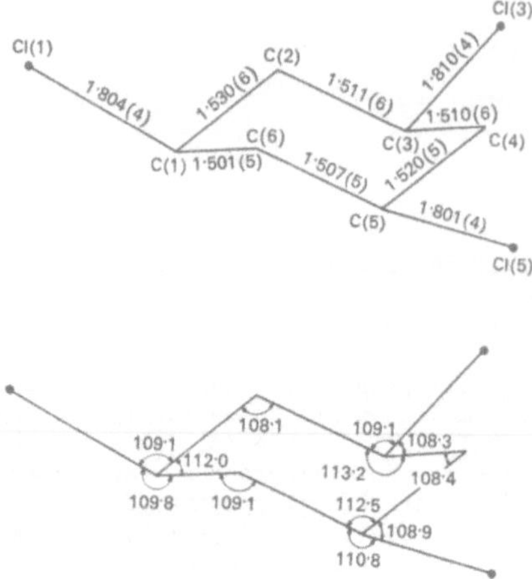

Fig. 1. Bond lengths and angles (σ = 0.3°) in cis-1,3,5-trichlorocyclohexane.

trans-CYCLOHEXANE-1,4-DICARBOXYLIC ACID

$C_8H_{12}O_4$

P. LUGER, K. PLIETH and G. RUBAN, 1972. Acta Cryst., B28, 706-710.

Monoclinic, $P2_1/c$, a = 5.605, b = 8.069, c = 9.644 Å, β = 107.24°, D_m = 1.34, Z = 2. Cu radiation, R = 0.074 for 783 reflexions.

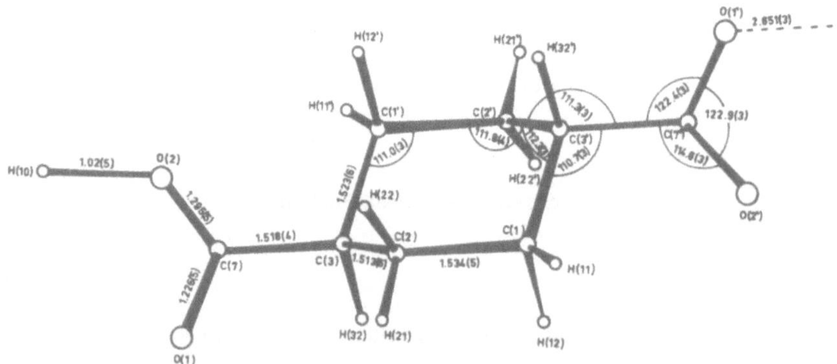

Fig. 1. Bond lengths and angles in trans-cyclohexane-1,4-dicarboxylic acid.

Centrosymmetric molecules are linked across centres of symmetry by O-H···O hydrogen bonds (2.651(3) Å) to produce infinite chains of molecules. The cyclohexane ring is of necessity in a chair conformation; bond distances and angles are shown in Fig. 1. A survey of molecular dimensions of a number of cyclohexane-dicarboxylic acids is also given.

(+)-AEROPLYSININ-I

$C_9H_9Br_2NO_3$

L. MAZZARELLA and R. PULITI, 1972. Gazz. Chim. Ital., 102, 391-394.

Hexagonal, $P6_5$, a = 8.663, c = 25.973, Z = 6. Mo radiation, R = 0.043 for 1038 reflexions.

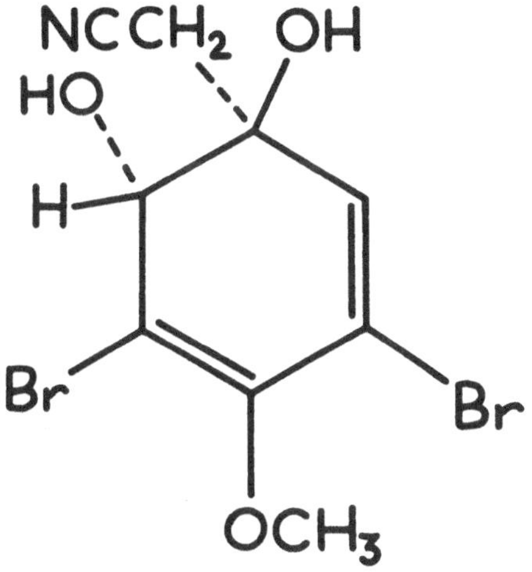

Fig. 1. (+)-Aeroplysinin-I.

The molecular structure is shown in Fig. 1. The two OH groups are in a quasi-axial conformation, the dihedral angle between the mean planes passing through the atoms attached to the two double bonds is about +18°, and an R-diene helix is described. The molecules are joined by O-H···O (2.85 Å) and O-H···N (2.91 Å) hydrogen bonds. The results agree with those for (-)-aeroplysinin (1).

1. D. B. COSULICH and F. M. LOVELL, 1971. Chem. Comm., 397.

cis-2-CHLORO-4-t-BUTYLCYCLOHEXANONE

$C_{10}H_{17}C\ell O$

R. A. G. de GRAAFF, M. Th. GIESEN, E. W. M. RUTTEN and C. ROMERS, 1972. Acta Cryst., B28, 1576-1583.

Triclinic, $P\bar{1}$, a = 8.294, b = 10.563, c = 6.366 Å, α = 100.8, β = 97.4, γ = 79.5°, D_m = 1.15, Z = 2. Mo radiation, R = 0.037 for 1468 reflexions.

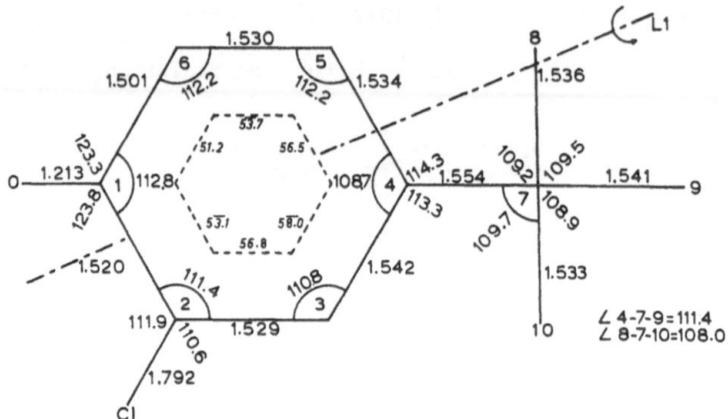

Fig. 1. Bond lengths (Å), valency and torsional angles in the cis-2-chloro-4-t-butylcyclohexanone molecule. The direction of the largest libration axis L_1 is indicated.

The cyclohexane ring has a chair conformation with the Cℓ and t-butyl group in the equatorial positions. The staggering of the latter group with respect to the ring atoms is -1.5°. Molecular geometry details are shown in Fig. 1. The ring is slightly more puckered than cyclohexane. The molecular packing is governed by (i) dipole-dipole interactions between carbonyl groups, (ii) contacts between Cℓ and H atoms, and (iii) H···H interactions. The distances involved are (i) C···O = 3.06, C···C = 3.28, and O···O = 3.29 Å, (ii) H···Cℓ = 3.14 - 4.12 Å, and (iii) H···H = 2.54 - 3.20 Å.

4α-t-BUTYLCYCLOHEXANE-1β,2β-DIOL

$C_{10}H_{20}O_2$

M. D. BRICE, B. R. PENFOLD and W. T. ROBINSON, 1972. Aus. J. Chem., 25, 2117-2124.

Triclinic, $P\bar{1}$, a = 12.268, b = 15.921, c = 6.322 Å, α = 82.53°, β = 114.45°, γ = 111.13°, Z = 4. Cu radiation, R = 0.063 for 1199 reflexions.

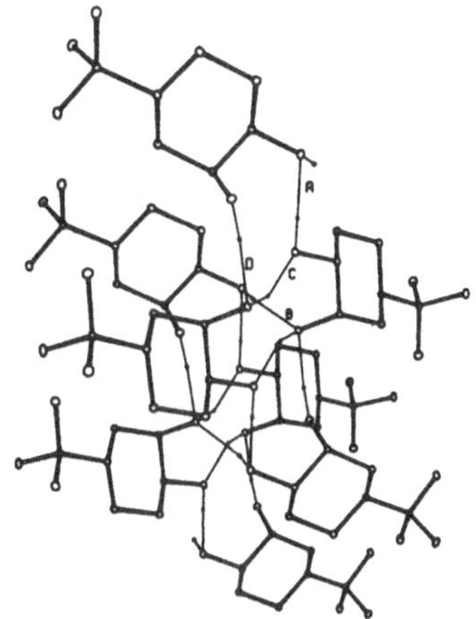

Fig. 1. A view of the contents of two unit cells of t-butylcyclohexanediol looking approximately down the c-axis. The four unique hydrogen bonds are labelled A, B, C, and D.

The crystal structure consists of two sets of crystallographically independent molecules hydrogen bonded (O-H···O = 2.77 - 2.90 Å) to form discrete chains parallel to c (Fig. 1). The cyclohexane rings adopt chair conformations and are slightly flattened (mean C-C torsion angle 55.9°). Bond lengths and angles are normal.

cis-4-t-BUTYLCYCLOHEXANE-1-CARBOXYLIC ACID

$C_{11}H_{20}O_2$ $(CH_3)_3C-C_6H_{10}-CO_2H$

H. van KONINGSVELD. Acta Cryst., B28, 1189-1195.

Monoclinic, $P2_1/c$, a = 17.132, b = 6.271, c = 10.380 Å, β = 92.17°, D_m = 1.08, Z = 4. Cu radiation, R = 0.057 for 1542 reflexions.

The CCO_2H in the molecule is planar and is hydrogen bonded across a centre of symmetry to the carboxyl group of a second molecule thus forming a dimer: O(1)···O(2') = 2.640 Å. The bond angles and distances in the t-butyl group are normal. This group as well as the individual CH_3 groups are twisted away from their perfectly staggered position; this is reflected in high anisotropic thermal parameters for the atoms concerned. The mean C-C bond length in the cyclohexane ring is close to normal (1.529 Å) although they range from 1.521(4) to 1.536(4) Å. Starting with the C(1)-C(2) bond the observed torsion angles along the successive bonds are: -51.4, +54.4, -55.9, +57.2, -57.1, and +52.9°. The ring is puckered at the substituted C(4) atom, and the deviation from the mean C-C-C bond angle of the ring (111.4°) is large: C(3)C(4)C(5) = 108.4°; C(3)C(4)C(7) = 114.0°; and C(5)C(4)C(7) = 114.5°.

3-(1,1,5-TRIMETHYL-5-CYCLOHEXENE-6-YL)-PROPENOIC ACID

$C_{12}H_{18}O_2$

CO_2H

H. SCHENK, 1972. Cryst. Struct. Comm., 1, 143-146.

Monoclinic, $P2_1/c$, a = 9.043, b = 9.812, c = 15.290 Å, β = 123.14°, Z = 4. Cu radiation, R = 0.067 for 1772 reflexions.

Most compounds related to the title compound have an s-cis ring-chain coupling. Here the ring-chain coupling is s-trans. Bond C(2)-C(3) is 1.478(6) Å, too short for a normal C-C bond; a possible explanation is conformational disorder or high thermal movement. Two molecules related by a centre of symmetry form dimers through hydrogen bonds, 2.64 Å. The cyclohexene ring is in a somewhat deformed half-chair conformation and the conjugated system is almost planar.

1,2,3,4,5,6-HEXACHLORO-1-PHENYLCYCLOHEXANE

$C_{12}H_{10}Cl_6$

C. A. de MEY, A. J. de KOK, J. LUGTENBURG and C. ROMERS, 1972. Rec. Trav. Chim. Pays-Bas, 91, 383-393.

Monoclinic, $P2_1$, a = 16.33, b = 6.48, c = 6.96 Å, β = 91.03°, D_m = 1.5, Z = 2. R = 0.143 for 2220 reflexions.

The chlorine atoms are equatorial and the phenyl ring axial. The cyclohexane ring is in a chair conformation with C-C-C-C torsion angles 52 - 58°. The mean C-Cl distance is 1.79(3) Å; other distances and angles are also normal.

3-(p-CHLOROPHENYL)-3,5,5-TRIMETHYLCYCLOHEXANONE

$C_{15}H_{19}ClO$

R. L. R. TOWNS and B. L. SHAPIRO, 1972. Cryst. Struct. Comm., 1, 151-154.

Monoclinic, $P2_1/c$, a = 10.371, b = 15.072, c = 12.856 Å, β = 137.72°, D_m = 1.22, Z = 4. Cu radiation, R = 0.09 for 1246 reflexions.

The molecular conformation is shown in Fig. 1. The cyclohexanone ring is in a chair-like conformation with the dihedral angles (relative to the C(2), C(3), C(5), and C(6) plane) at each end of the ring equal to 135°. No unusual intermolecular contacts were observed.

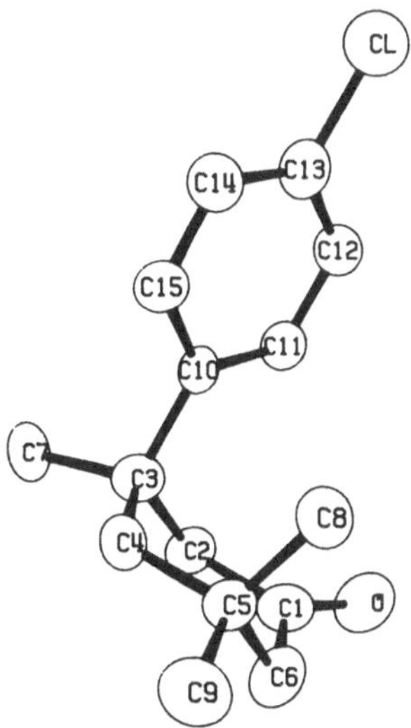

Fig. 1. Molecular conformation of 3-(p-chlorophenyl)-3,5,5-trimethylcyclohexanone.

trans-4-t-BUTYLCYCLOHEXANOL p-BROMOBENZOATE

$C_{17}H_{23}BrO_2$

J. OHRT and R. PARTHASARATHY, 1972. J. Cryst. Mol. Struct., 2, 213-223.

Triclinic, $P\bar{1}$, a = 9.813, b = 13.451, c = 6.564 Å, α = 102.99, β = 98.88, γ = 89.99°, D_m = 1.34, Z = 2. Cu radiation, R = 0.058 for 3165 reflexions.

The bond distances and angles in the molecule are shown in Fig. 1. The study has shown that no flattening is induced into the cyclohexane ring and there is no elongation of the C(t-Bu)-C(cyclohexane) bond. The steric strain introduced by the t-butyl group is minimized by a flattening of the C(4), C(3), C(7), C(5) pyramid. The axial hydrogen atom on C(1) nearly eclipses the carbonyl oxygen.

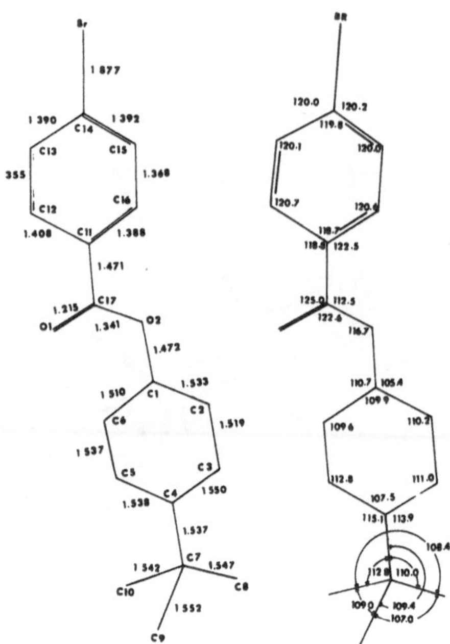

Fig. 1. Bond distances and angles in trans-4-t-butylcyclohexanol p-bromobenzoate; σ(C-O, C-C) = 0.007 Å.

cis-4-t-BUTYLCYCLOHEXYL-p-TOLUENESULPHONATE

$C_{17}H_{26}O_3S$

V. J. JAMES and C. T. GRAINGER, 1972. Cryst. Struct. Comm., 1, 111-114.

Orthorhombic, Pbca, a = 15.7075, b = 34.1172, c = 6.4149 Å, Z = 8. Cu radiation, R = 0.044.

The material is dimorphous. The structure of Type A is described elsewhere (1), that of Type B is described here. The molecule of Type B (Fig. 1) shows the same evidence of strain in the t-butyl group as that reported for Type A. The flattening of the cyclohexane ring and the pattern of the individual torsional angles of the ring also compare favourably with those of Type A, the flattening in both instances being localized about C(8). The computer prediction (2) of a twisted t-butyl group was not observed confirming the conclusion, based on Type A and the trans-isomer of the compound (3), that such a distortion is not experienced with the addition of the t-butyl group to the fourth carbon of the cyclohexane ring.

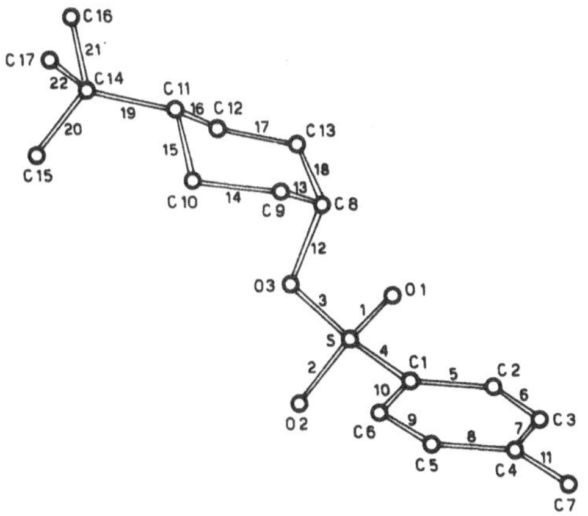

Fig. 1. cis-4-t-Butylcyclohexyl-p-toluenesulphonate. Angles C(14)-C(11)-C(12) and C(14)-C(11)-C(10) are 113.4 and 115.2(1)° respectively.

1. Structure Reports, 39B, 99.
2. C. ALTONA and M. SUNDARALINGAM, 1970. Tetrahedron, 26, 925.
3. P. L. JOHNSON, C. J. CHEER, J. P. SCHAEFER, V. J. JAMES and F. H. MOORE, 1972. Tetrahedron, in press.

p-TERCYCLOHEXANE-2

$C_{18}H_{32}$ $C_6H_{11}\cdot C_6H_{10}\cdot C_6H_{11}$

K. SASVARI, 1972. Cryst. Struct. Comm., 1, 163-164.

Monoclinic, $P2_1/n$, a = 6.32, b = 9.64, c = 26.14 Å, β = 90.3°, Z = 4. Cu radiation, R = 0.089 for 2402 reflexions.

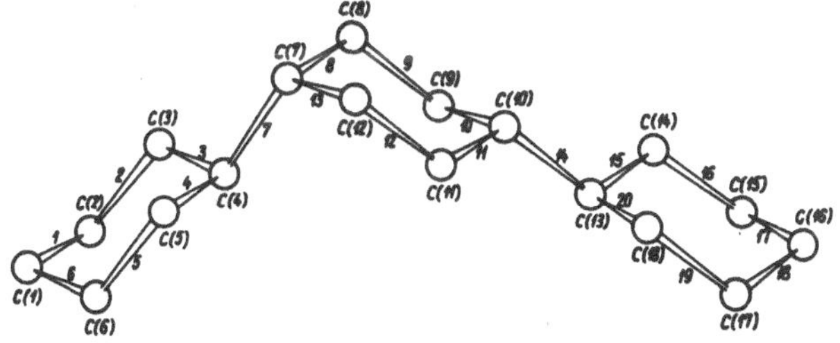

Fig. 1. Conformation of p-tercyclohexane-2.

The molecule of p-tercyclohexane-2 (Fig. 1) is composed of three cyclohexane rings all of which maintain the chair conformation of the free cyclohexane ring. The rings are linked by C-C single bonds forming an L shaped molecule chain. The four-atom midparts of all three rings are planar with a maximum deviation of 0.009 Å. Considering these planes the middle ring is almost parallel to one end ring (6.7°) and nearly perpendicular to the other end ring (92.1°). The six carbon atoms with indices 1, 4, 7, 10, 13, and 16, four of which are linking the rings, form a plane with a maximum deviation of 0.04 Å. This plane is almost perpendicular to the previous three planes of the rings and with rough approximation it can be taken as a non-crystallographic mirror plane of the molecule.

1-BROMO-1-BENZOYL-2-PHENYLCYCLOHEXANE

$C_{19}H_{19}BrO$

A. DUCRUIX and C. PASCARD-BILLY, 1972. Acta Cryst., B28, 1848-1852.

Monoclinic, $P2_1/c$, a = 8.67, b = 22.13, c = 9.64 Å, β = 120°, D_m = 1.40, Z = 4. Cu radiation, R = 0.074 for 1589 reflexions.

The cyclohexane ring is in a chair conformation with the bromine and phenyl substituents axial. The bond distances and angles are normal and there are no unusual intermolecular contacts.

trans-CYCLOHEXANE-1,4-DICARBOXYLIC ACID POTASSIUM SALT

$C_{24}H_{34}K_2O_{12}$ $C_8H_{10}O_4K_2 \cdot 2C_8H_{12}O_4$

P. LUGER, K. PLIETH and G. RUBAN, 1972. Acta Cryst., B28, 699-706.

Triclinic, $P\bar{1}$, a = 10.439, b = 10.569, c = 6.283 Å, α = 97.793, β = 100.090, γ = 87.460°, D_m = 1.44, Z = 1. Cu radiation, R = 0.063 for 2478 reflexions.

This crystal structure analysis establishes that the above formulation is correct rather than $2(C_8H_{10}O_4K) \cdot C_8H_{12}O_4$. In the crystal structure (Fig. 1) all three cyclohexane moieties are centred on inversion centres. Molecule 2 in the figure is the di-ionised form and it is hydrogen bonded to the other two independent diacid molecules (O(2)H···O(5') = 2.555 and O(9')H···O(6') = 2.558 Å). The potassium ion is in a severely distorted octahedral coordination with $K^+ \cdots O$ distances in the range 2.727 - 2.873 Å.

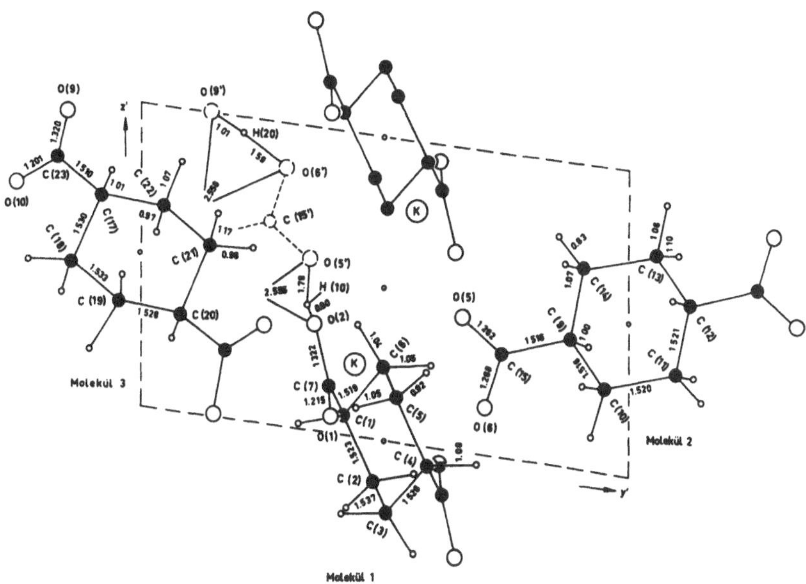

Fig. 1. Projection of the unit cell of $C_8H_{10}O_4K_2 \cdot 2C_8H_{12}O_4$ in the x-direction. Molecule 2 is the dianion. For the bond lengths, $\sigma = 0.005$ Å.

syn-3,7-DIBROMO-cis,cis-CYCLO-OCTA-1,5-DIENE

$C_8H_{10}Br_2$

R. K. MACKENZIE, D. D. MACNICOL, H. H. MILLS, R. A. RAPHAEL, F. B. WILSON and J. A. ZABKIEWICZ, 1972. J. Chem. Soc., Perkin II, 1632-1638.

Orthorhombic, Fdd2, a = 19.64, b = 19.36, c = 4.78 Å, D_m = 1.94, Z = 8. Cu radiation, R = 0.112 for 470 reflexions. Molecular symmetry C_2.

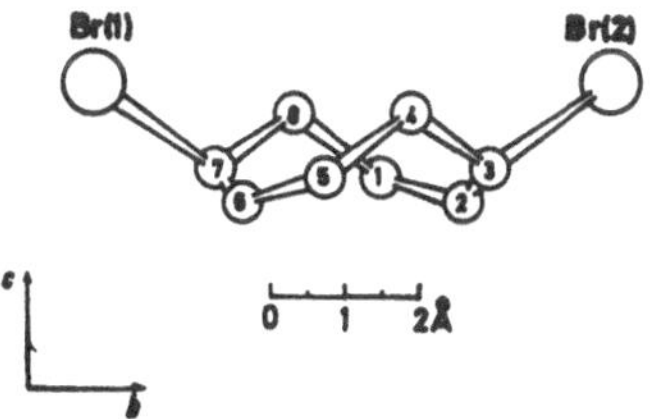

Fig. 1. The conformation of syn-3,7-dibromo-cis,cis-cyclo-octa-1,5-diene as viewed along the a-axis.

The conformation (Fig. 1) adopted is that of a twist-boat with a dihedral angle of about 74° between the mean molecular planes containing the olefinic double bonds. The dihedral angle between the planes defined by C(2), C(3), C(4) and C(3), C(4), C(5) is 65°. The C(3)-Br(2) bond is inclined at an angle of 57° to the c axis (C_2 symmetry axis of the molecule). The bond lengths are in agreement with values found in related molecules. The molecules are separated in the crystal by normal van der Waals distances.

CALCIUM 2,4,6,8-CYCLOOCTATETRAENE-1,2-DICARBOXYLATE DIHYDRATE

$C_{10}H_{10}CaO_6$ $CaC_8H_6(COO)_2 \cdot 2H_2O$

D. A. WRIGHT, K. SEFF and D. P. SHOEMAKER, 1972. J. Cryst. Mol. Struct., 2, 41-51.

Triclinic, $P\bar{1}$, a = 6.270, b = 7.767, c = 12.141 Å, α = 91.10, β = 100.21, γ = 99.76°, D_m = 1.520, Z = 2. Cu radiation, R = 0.077 for 2453 reflexions (films, visual intensities).

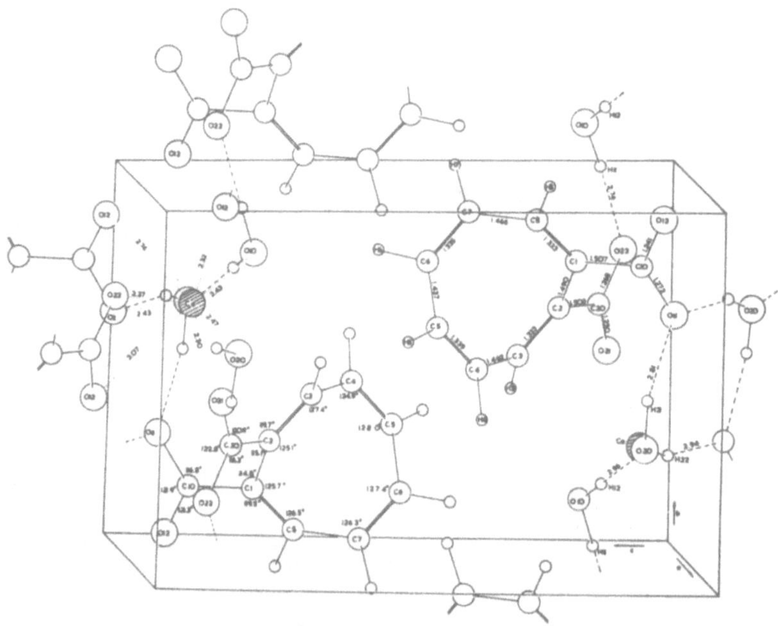

Fig. 1. Crystal structure of calcium 2,4,6,8-cyclooctatetraene-1,2-dicarboxylate dihydrate.

The anion has an approximate 2-fold axis of symmetry to which bond lengths, angles and dihedral angles conform within experimental error. The cyclooctatetraene ring has a tub shape, similar to that found in other cyclooctatetraene structures, with its average bond lengths of 1.46 Å and 1.34 Å for single and double bonds respectively, in good agreement with earlier values for similar molecules. The Ca^{2+} cation is coordinated by eight oxygen atoms (six carboxylate, two water) in a distorted tetragonal antiprism with distances ranging from 2.30 to 3.07 Å. All hydrogen atoms were located; the water hydrogens form a layer-like network of hydrogen bonds linking the water and carboxylate oxygens approximately perpendicular to the $\underline{z}$ axis. The O-H···O distances range from 2.74 to 2.96 Å. The C-O distances range from 1.241 to 1.272 Å, σ(C-C, C-O) = 0.003 - 0.008 Å. Fig. 1 shows the structure with bond distances and angles.

4-ISOPROPYLTROPOLONE

(β-THUJAPLICIN)

$C_{10}H_{12}O_2$

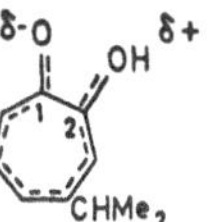

J. E. DERRY and T. A. HAMOR, 1972. J. Chem. Soc., Perkin II, 694-697.

Monoclinic, I2/c, a = 17.063, b = 6.469, c = 15.970 Å, β = 90.77°, D_m = 1.21, Z = 8. Cu radiation, R = 0.083 for 1272 reflexions.

The π-electron system in the molecule is partially delocalised as implied in the formulation above. Within the ring the C(1)-C(2) bond is essentially a single bond (1.469(6) Å) and the means of the other formal single and double C-C bonds are 1.416 and 1.362 Å. The C⋯O(δ-) bond length is 1.261(5) whereas the C⋯OH(δ+) bond length is 1.349(5) Å. The seven-membered ring deviates slightly from planarity (maximum deviation 0.025 Å). The hydroxy-oxygen atom lies in the ring plane but the other oxygen is 0.036 Å off the plane. Bond angles within the ring range from 122.8 - 131.1° (mean value 128.6°). Hydrogen bonding (O···O = 2.80 Å) appears to link the molecules in chains parallel to $\underline{b}$ (the pertinent hydrogen atom could not be located but the C-O···O angle is 140.7°).

HEPTAFULVALENE

$C_{14}H_{12}$

I. R. THOMAS and P. COPPENS, 1972. Acta Cryst., B28, 1800-1806.
II. P.-T. CHENG and S. C. NYBURG, 1973. Ibid., B29, 1358-1359.
III. P. COPPENS, 1973. Ibid., B29, 1359-1360.

Monoclinic, P2$_1$/c, a = 9.688, b = 7.730, c = 6.971 Å, β = 98.03°, D_m not given, Z = 2. Cu radiation, R = 0.041 for 407 reflexions.

Fig. 1. Side view of the heptafulvalene molecule showing the S-shape.

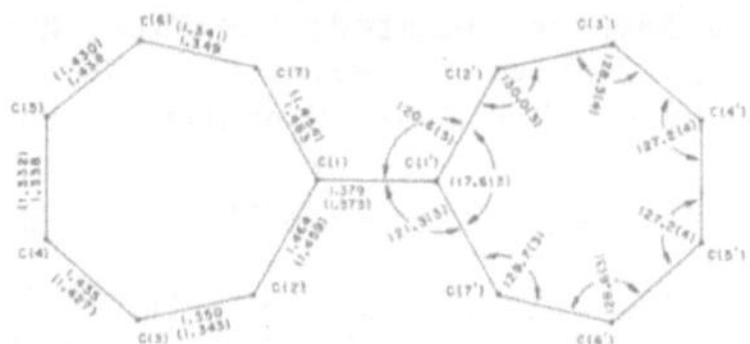

Fig. 2. Bond lengths and angles in heptafulvalene. Values in parentheses are before correction for apparent shortening due to thermal motion. Standard deviations are 0.004 - 0.005 Å for the bond lengths.

Paper I reports the complete analysis. This strained or overcrowded molecule is approximately S-shaped (Fig. 1); largest deviations from the best molecular plane are 0.350 Å. There is an alternation of almost single and double bonds in the seven-membered ring (Fig. 2), in fair agreement with M.O. calculations on a planar idealised molecule (1). In paper II it is argued that the cause of non-planarity in the heptafulvalene molecule is more probably due to conflict of two pairs of hydrogen atoms within the molecule rather than the need to relieve bond angle strain within the ring. Paper III is a response to II and argues that steric hindrance is not severe and that ring strain is a significant contributor to the deformation of the molecule.

1. M. J. S. DEWAR and G. J. GLEICHER, 1965. J. Amer. Chem. Soc., 87, 685.

OCTAMETHYLCYCLOOCTATETRAENE

$C_{16}H_{24}$

J. BÖRDNER, R. G. PARKER and R. H. STANFORD, 1972. Acta Cryst., B28, 1069-1075.

Orthorhombic, Pbcn, a = 12.774, b = 9.339, c = 12.217 Å, D_m not given, Z = 4. Cu radiation, R = 0.086 for 765 reflexions.

The molecule is situated on a crystallographic twofold axis and is shown in Fig. 1. The molecule adopts the tub conformation with alternating single and double C-C bonds. The methyl substituents result in a significant flattening of the ring. In particular the bond angles decrease and the torsion angles increase compared with the unsubstituted ring. The thermal vibrational parameters of the ring carbon atoms were interpreted in terms of rigid-body motion.

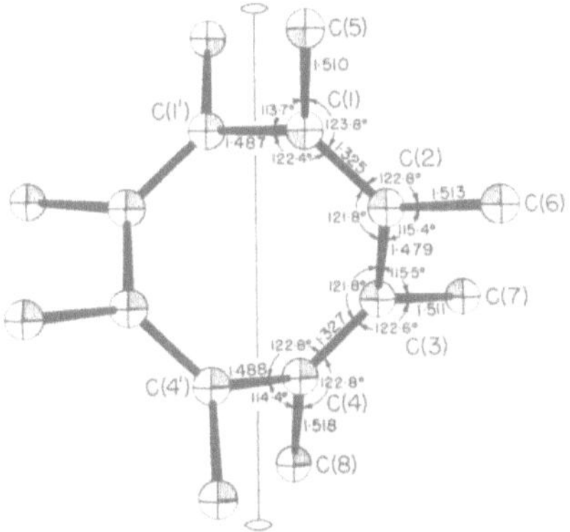

Fig. 1. Bond distances and angles in octamethylcyclooctatetraene.

[14]ANNULENE

$C_{14}H_{14}$

C. C. CHIANG and I. C. PAUL, 1972. J. Amer. Chem. Soc., 94, 4741-4743.

Monoclinic, $P2_1/c$, a = 8.640, b = 4.376, c = 14.996 Å, β = 106°5', Z = 2. Cu radiation, R = 0.054 for 412 reflexions.

The molecular structure of [14]annulene in the crystal is shown in Fig. 1. Some molecular dimensions are shown in Fig. 2. The molecule is nonplanar with C(1) and C(5) deviating by -0.206 and 0.209 Å from the best plane through all the carbon atoms in the molecule; H(1) and H(5) deviate by -0.78 and 0.77 Å from this plane. In the crystal, the molecule is centrosymmetric but approaches C_{2h} symmetry, with the twofold axis running through atoms C(3) and C(10). The cause of the significant nonplanarity is the steric overcrowding of the four internal hydrogen atoms in the centre of the molecule. The H(1)···H(5), H(1)···H(12), and H(1)···H(8) distances are 2.02(7), 2.04(7), and 2.91(7) Å, respectively. The C-C bond lengths range from 1.350(11) to 1.407(11) Å with an average value of 1.378 ± 0.019 Å. While the spread of the individual lengths implies possible significant differences, there is no obvious chemically significant pattern to the values obtained. The distortions from planarity are distributed throughout the molecule. With the exception of the C(6)-C(7) bond, the torsion angles around the crystallographically independent half of the ring lie in the range of 10-20° (for cis bonds) and 160-170° (for trans bonds); the substituents on the C(6) and C(7) atoms are almost completely eclipsed (torsion angle = -1.5°).

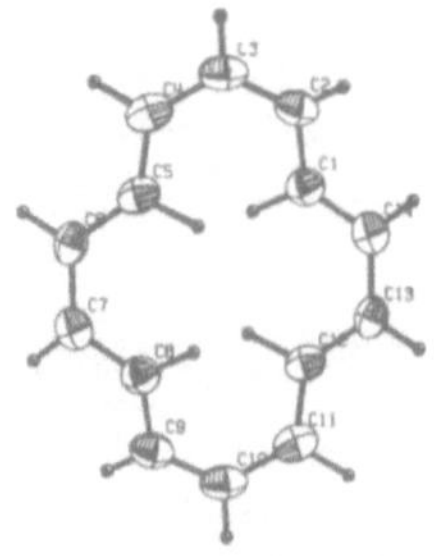

Fig. 1. View of a molecule of [14]annulene.

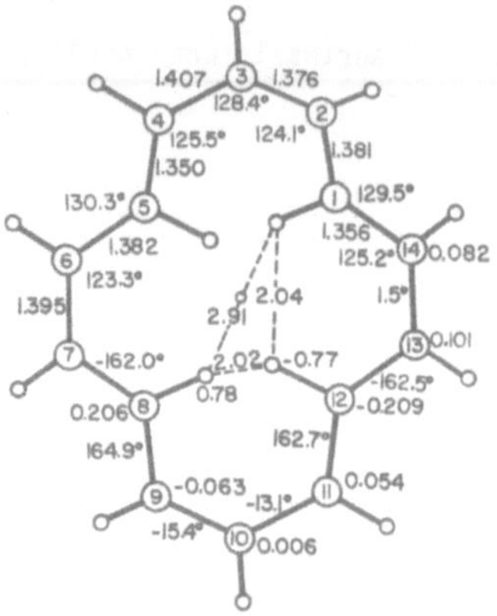

Fig. 2. Molecular dimensions in [14]annulene. The upper half of the drawing shows the bond lengths (Å) and bond angles (°). The lower part of the drawing (related by a centre of inversion) shows the deviations of atoms (angströms) from the best plane through the carbon atoms, and torsion angles around the C-C bonds (degrees). The standard deviations of the C-C bonds are 0.010 Å and of the C-C-C angles are 0.5°.

9,10,15,16-TETRADEHYDRO-11,12,13,14-TETRAHYDRODIBENZO[a,c]CYCLODODECENE

$C_{20}H_{16}$

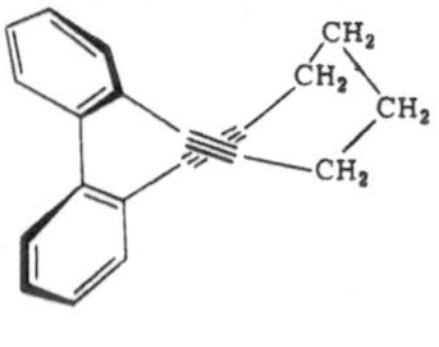

H. IRNGARTINGER, 1972. Chem. Ber., <u>105</u>, 1184-1202.

Orthorhombic, Pbcn, a = 15.117, b = 7.587, c = 24.996 Å, Z = 8. Mo radiation, R = 0.053 for 2183 reflexions.

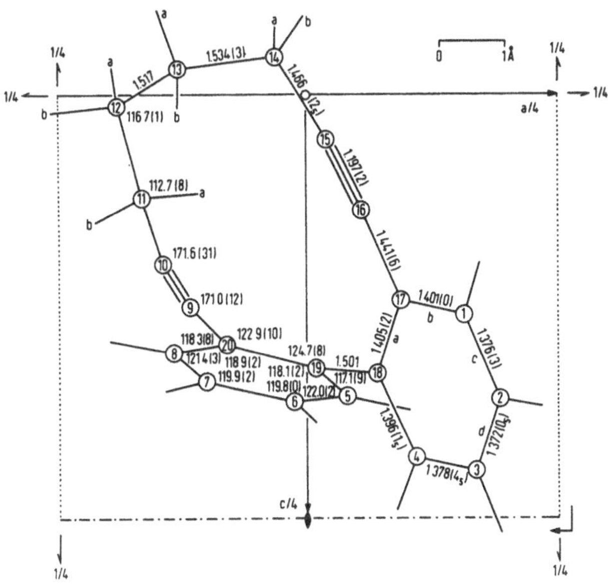

Fig. 1. The $C_{20}H_{16}$ molecule seen in projection along b. Bond lengths and angles have been averaged over equivalent portions of the molecule.

The molecule adopts the conformation implied above with an angle of 47.6° between the triple-bond axes. The molecular geometry is summarized in Fig. 1; the C-C≡C angles deviate by 8.7° from linearity and the mid-points of the two triple bonds are separated by 3.034 Å. Within the 12-ring the remaining bond angles are enlarged because of transannular contacts. The CH_2-groups have gauche-conformation (dihedral angles -79.7°, +56.2°, +50.1°). The molecule has symmetry C_2 with exception of the methylene groups. The distortion of the 12-ring causes twisting of the benzene rings too.

5-cis-15-trans-TETRABENZO[a,c,g,i]CYCLODODECENE

$C_{28}H_{20}$

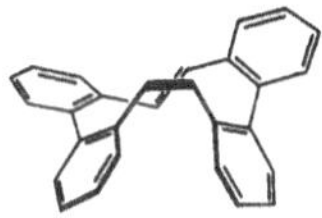

H. IRNGARTINGER, 1972. Chem. Ber., 105, 2068-2084.

Monoclinic, $P2_1/c$, a = 13.981, b = 7.786, c = 19.981 Å, β = 120.25°, Z = 4. Mo radiation, R = 0.052 for 3195 reflexions.

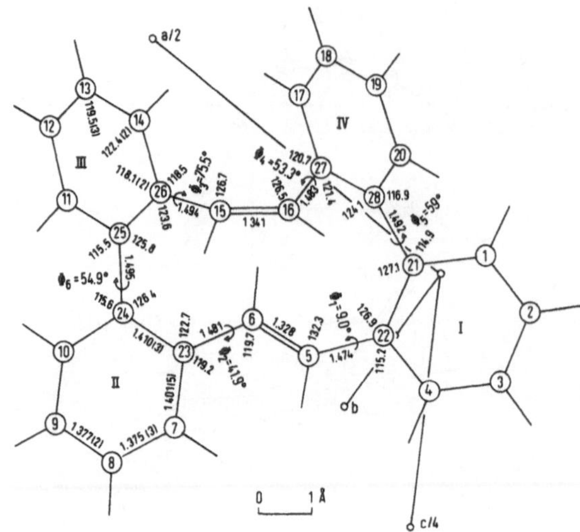

Fig. 1. Bond lengths (σ = 0.003 Å) and angles (σ = 0.2°) in $C_{28}H_{20}$; torsion angles are also shown.

Bond lengths and angles are in Fig. 1. The bond lengths between the phenyl rings and the double bonds depend on the twist-angle of both groups. The cis- and trans-double-bonds are crossed with an angle of 50.5°; the dihedral angle between the planes of the π-orbitals is 31°. The midpoints of these bonds have a distance of 3.086 Å. As a result of their transannular contact they have deviations of 4° out of their planes in addition to the torsion-angle of 4° about the trans-double-bond. For the same reason all the angles are enlarged within the 12-ring.

A rotation mechanism is suggested to account for the identity (n.m.r.) of the inner and outer proton on the trans-double-bond.

HEXA-o-PHENYLENE

$C_{36}H_{24}$

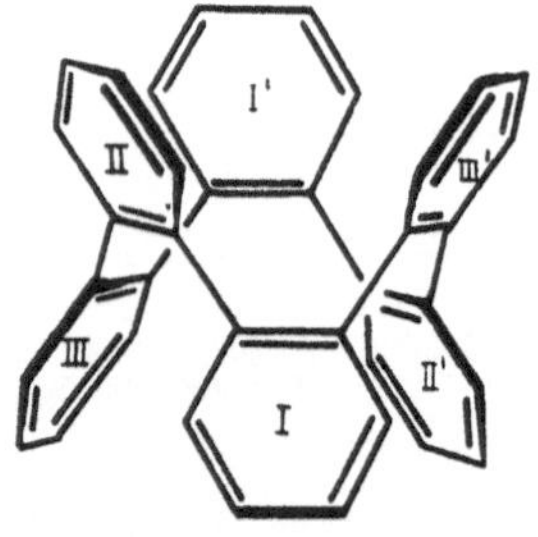

H. IRNGARTINGER, 1972. Israel J. Chem., 10, 635-647.

Triclinic, $P\bar{1}$, a = 8.767, b = 8.830, c = 9.058 Å, α = 106.51°, β = 71.97°, γ = 111.92°, Z = 1. Mo radiation, R = 0.046 for 1869 reflexions.

The molecule has a crown-like conformation as has the inner 12-membered ring. The mean length of the bond linking the benzene rings is 1.498 Å and the angle of twist between the planes of adjacent benzene rings is 77.4°. Bond lengths and angles in the benzene rings show trends around the ring similar to those found in other o-disubstituted aromatic molecules.

β-FLUORONAPHTHALENE

$C_{10}H_7F$

N. B. CHANH and Y. HAGET-BOUILLAUD, 1972. Acta Cryst., B28, 3400-3404.

Monoclinic, $P2_1/a$, a = 7.796, b = 5.965, c = 9.955 Å, β = 122.87°, D_m = 1.240, Z = 2. Cu radiation, R = 0.095 for 290 reflexions.

The β-fluoronaphthalene has a disordered structure at room temperature with the fluorine atom statistically distributed over the four β-positions of the naphthalene ring with equal probability of 0.25. The C-F distance, 1.27(3) Å, has a value between normal C-H (1.00 Å) and C-F (1.33 Å) bond lengths. The C-C distances (1.36 - 1.39 Å) are much shorter than those of naphthalene. The β-substituents have the closest intermolecular contacts (X(6)···X(7) 2.50, 2.52 Å).

RACEMIC trans-1a,7b-DIHYDRO-OXIRENO[a]NAPHTHALENE-3-SPIRO-2'-OXIRAN-2(3H)-ONE

$C_{11}H_8O_3$

B. M. GATEHOUSE and D. J. LLOYD, 1972. J. Chem. Soc., Perkin II, 932-934.

Orthorhombic, Pbca, a = 16.35, b = 12.39, c = 8.44 Å, D_m = 1.45, Z = 8. Cu radiation, R = 0.138 for 312 reflexions.

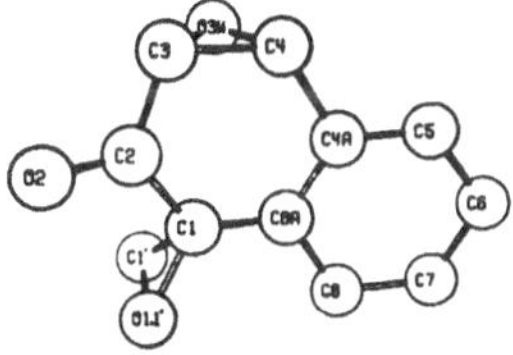

Fig. 1. A view of the $C_{11}H_8O_3$ molecule.

This analysis has determined the positions of the epoxide oxygens in the compound to be as shown in Fig. 1. Bond lengths and angles are normal within the accuracy of the determination. There are no unusual intermolecular contacts.

4,5-DINITRONAPHTHALIC ANHYDRIDE

$C_{12}H_4N_2O_7$

J. BORDNER and L. A. JONES, 1972. J. Cryst. Mol. Struct., 2, 79-87.

Monoclinic, $P2_1/c$, a = 8.096, b = 8.813, c = 15.11 Å, β = 92.47°, D_m = 1.73, Z = 4. Cu radiation, R = 0.054 for 1076 reflexions.

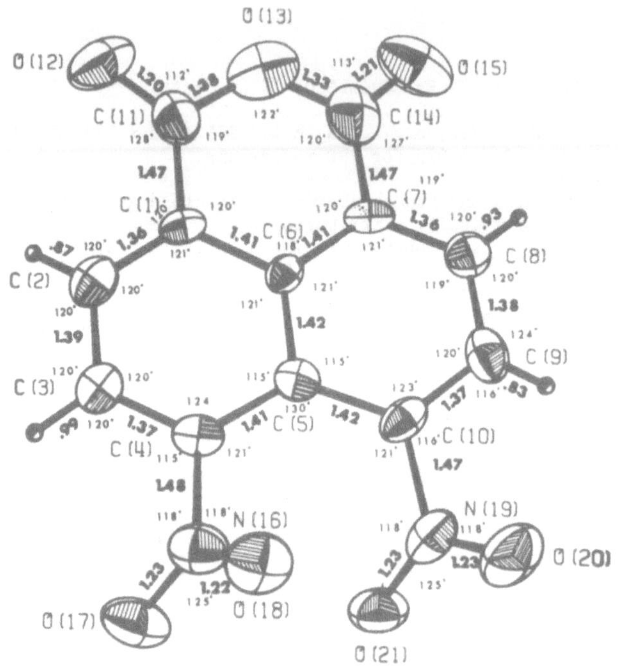

Fig. 1. Bond lengths and angles in 4,5-dinitronaphthalic anhydride.

The molecule is non-planar the carbonyl oxygen atoms being out of the naphthalene plane by -0.29 and 0.26 Å while the anhydride oxygen lies in the naphthalene plane. The bond distances and angles are shown in Fig. 1 , σ(bonds not involving H) = 0.008 Å, σ(C-H) = 0.04 Å.

1,5-DIMETHYLNAPHTHALENE

(NEUTRON STUDY)

$C_{12}H_{12}$

G. FERRARIS, D. W. JONES, J. YERKESS and K. D. BARTLE, 1972. J. Chem. Soc., Perkin II, 1628-1632.

Monoclinic, $P2_1/c$, a = 6.18, b = 8.91, c = 16.77 Å, β = 101.4°, Z = 4.
Neutron radiation, R = 0.041 for 437 reflexions.

The mean bond lengths found by this study are C(1)-C(2) = 1.408(10), C(2)-C(3) = 1.419(9), C(3)-C(4) = 1.386(10), C(4)-C(10) = 1.404(10), C(1)-C(9) = 1.413(8), C(9)-C(10) = 1.444(10), C(1)-CH_3 = 1.514(9) Å. These values compare well with those found by an X-ray study (1). The largest difference is in the C(9)-C(10) bond which is 0.064 Å longer in the neutron study than in the X-ray one. The carbon framework of the molecule is planar. The H(methyl) atoms are locked in fixed orientations in which one hydrogen of each is in the ring plane; closest approaches of out-of-plane H(methyl) atoms to adjacent α-hydrogen atoms are 2.35 - 2.40(4) Å. Aromatic C-H and methyl C-H bond lengths do not differ significantly from their means (1.05 and 1.06 Å).

1. J. BEINTEMA, 1965. Acta Cryst., 18, 647.

1,8-BIS(PROP-1-YNYL)NAPHTHALENE

$C_{16}H_{12}$

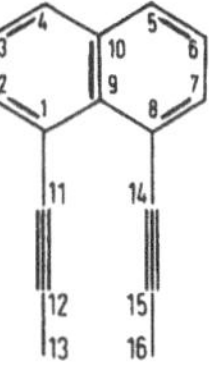

A. E. JUNGK, 1972. Chem. Ber., 105, 1595-1613.

Monoclinic, $P2_1/c$, a = 15.4306, b = 12.7498, c = 11.9690 Å, β = 97.113°, Z = 8.
Mo radiation, R = 0.057 for 3404 reflexions.

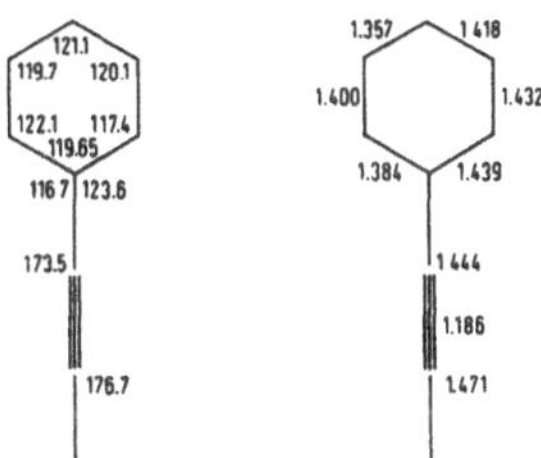

Fig. 1. Mean bond lengths and angles in $C_{14}H_{12}$. (The 1.384 Å bond corresponds to the C(1)-C(2) position in the naphthalene nucleus.)

There are two independent molecules in the asymmetric unit. Mean molecular dimensions averaged over equivalent bonds are in Fig. 1 (bond scatter of equivalent bonds is 0.0027 Å). Overcrowding by the peri-substituents splays the C(11)-C(1)-C(9) and corresponding angles to 123.6 ± 0.6° and the average C(1)-C(9)-C(8) angle to 125.25 ± 0.05°; the distance between the 1,8-substituents reaches 2.874 Å. The nonlinear triple bonds display a trans-configuration due to steric interaction of the triple bonds.

1-ACETYL-2-ETHOXYNAPHTHALENE

$C_{14}H_{14}O_2$

M. P. GUPTA and M. SAHU, 1972. Z. Kristallogr., 135, 262-272.

Triclinic, P$\bar{1}$, a = 8.20, b = 9.07, c = 8.41 Å, α = 103.6°, β = 104.4°, γ = 91.98°, D_m = 1.203, Z = 2. Cu radiation, R = 0.138 for 590 reflexions.

The crystal structure contains discrete molecules separated by van der Waals distances. Bond distances and angles are normal within the accuracy of the analysis. The acetate group plane is rotated through 64° about the exocyclic C-C bond out of the naphthalene plane.

1-(2,5-DICHLOROPHENYLAZO)-2-HYDROXY-3-NAPHTHOIC ACID 4-CHLORO-2,5-DIMETHOXY-ANILIDE

$C_{25}H_{18}Cl_3N_3O_4$

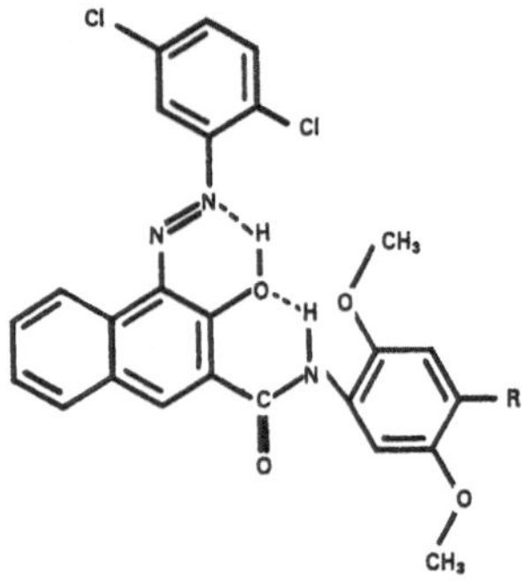

D. KOBELT, E.F. PAULUS and W. KUNSTMANN, 1972. Acta Cryst., B28, 1319-1324.

Triclinic, P$\bar{1}$, a = 12.96, b = 3.963, c = 22.80 Å, α = 93.6, β = 99.6, γ = 99.5°, D_m not given, Z = 2. Cu radiation, R = 0.103 for 3081 reflexions.

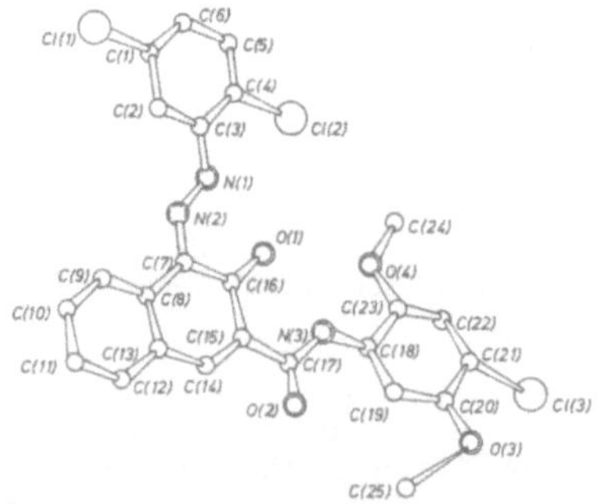

Fig. 1. A view of the molecular structure of $C_{25}H_{18}Cl_3N_3O_4$ projected onto the ac-plane.

The molecular structure is in Fig. 1. Bond lengths and angles are normal and there are N-H···O and O-H···N intramolecular hydrogen bonds. The whole molecule is an almost planar fully conjugated π-system.

1,4,5,8-TETRAPHENYLNAPHTHALENE

$C_{34}H_{24}$

G. EVRARD, P. PIRET and M. van MEERSSCHE, 1972. Acta Cryst., B28, 497-506.

Monoclinic, $P2_1/c$, a = 6.455, b = 24.333, c = 8.017 Å; β = 114.3°, D_m = 1.252, Z = 2. Cu radiation, R = 0.106 for 1726 reflexions.

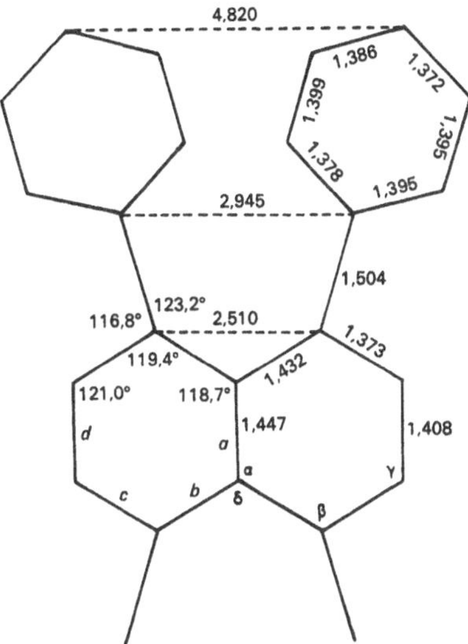

Fig. 1. Interatomic distances (σ = 0.004 - 0.007 Å) and angles (mean values) in tetraphenylnaphthalene.

The molecule lies on a space group centre of symmetry but conforms closely to the idealised 2/m symmetry, with the twofold axis along the direction of the C(9)-C(10) central bond. The naphthalene ring is distorted into a double boat conformation by the intramolecular overcrowding, e.g. C(1) is 0.115 Å out of the naphthalene plane. Further relief from overcrowding is achieved by in-plane splaying-out of the exocyclic C-C bonds (C(10)-C(1)-C(phenyl) = 123.2(3)°), out of plane bending of the exocyclic C-C bonds (7.5°), and rotation of the phenyl rings about the exocyclic C-C bonds (the phenyl ring planes make angles of 58° with that of the naphthalene nucleus). Mean interatomic distances and angles are in Fig. 1. Intermolecular contacts are of the normal van der Waals type.

PERDEUTERIOANTHRACENE

$C_{14}D_{10}$

M. S. LEHMANN and G. S. PAWLEY, 1972. Acta Chem. Scand., 26, 1996-2004.

Monoclinic, $P2_1/a$, a = 8.542, b = 6.016, c = 11.163 Å, β = 124.59°, Z = 2.
Neutron radiation, R = 0.034 for 1001 reflexions.

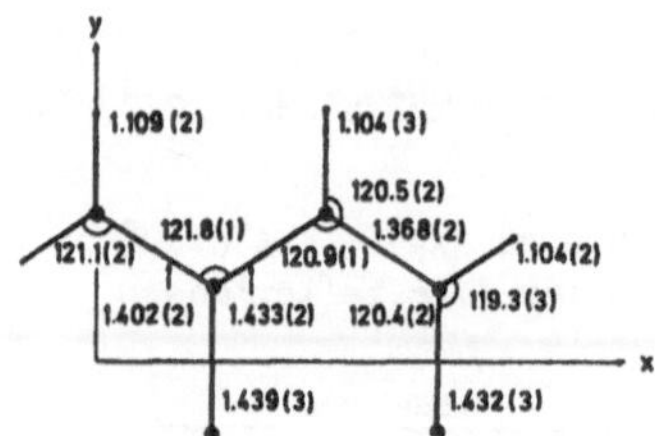

Fig. 1. The best symmetry-averaged molecular geometry, corrected for libration and riding motion effects in perdeuterioanthracene.

The crystal structure is the same as that of hydrogenous anthracene. The best symmetry-averaged molecular geometry is shown in Fig. 1. The thermal motion of the molecule as a rigid unit dominates the overall thermal motion.

7,10-DICHLOROANTHRAQUINONEOXADIAZOLE

$C_{14}H_4Cl_2N_2O_3$

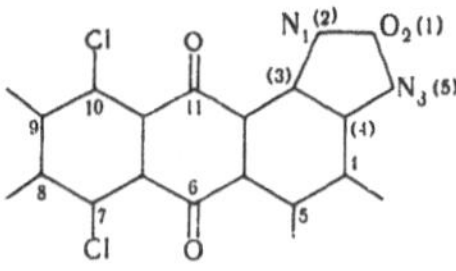

O. A. MIKHNO, L. A. CHETKINA and Z. I. EŽKOVA, 1972. Kristallografija, 17, 297-302 [Soviet Physics-Crystallography, 17, 251-255].

Monoclinic, $P2_1/a$, a = 12.37, b = 25.87, c = 3.81 Å, β = 94.5, D_m = 1.75, Z = 4.
Cu radiation, R = 0.159 for 1595 reflexions.

The molecule is deformed from planarity by intramolecular overcrowding between Cl and O atoms, and the oxygen atoms are displaced by 0.2 and 0.5 Å out of the plane of the central ring. Apart from the central ring, the individual rings of the molecule are flat; the angle between ring I and ring III is 15.5°. The oxadiazole ring is coplanar with ring III to within 1°. Bond angles in ring III are deformed by the hetero ring formation. Bond lengths, within the limits of accuracy, have values close to normal.

1,8-DICHLOROANTHRAQUINONE

$C_{14}H_6C\ell_2O_2$

O. A. MIKHNO, Z. I. EŽKOVA and B. N. KOLOKOLOV, 1972. Kristallografija, 17, 667-670 [Soviet Physics-Crystallography, 17, 580-582].

Orthorhombic, $Pca2_1$, a = 21.43, b = 3.81, c = 13.77 Å, D_m = 1.62, Z = 4. Cu radiation, R = 0.157 for 510 reflexions.

The molecule is overcrowded and steric interaction between adjacent Cℓ and O atoms causes buckling of the molecular framework. The mean Cℓ···O distance is 2.80 Å; the Cℓ atoms are displaced 0.26 - 0.29 Å (and the overcrowded oxygen 0.52 Å in the opposite direction) from the central ring plane. The angle between rings I and II is 7°, and between rings II and III is 6°. The molecule adopts a shallow dish-like conformation. There are no unusual intermolecular contacts.

10-CHLOROMETHYL-2,3,9-TRIMETHYLANTHRACENE

(ICR-489)

$C_{18}H_{17}C\ell$

A. CHOMYN, J. P. GLUSKER, H. M. BERMAN and H. L. CARRELL, 1972. Acta Cryst., B28, 3512-3517.

Monoclinic, $P2_1/c$, a = 11.026, b = 14.387, c = 9.093 Å, β = 103.64°, D_m = 1.25, Z = 4. Cu radiation, R = 0.054 for 2146 reflexions.

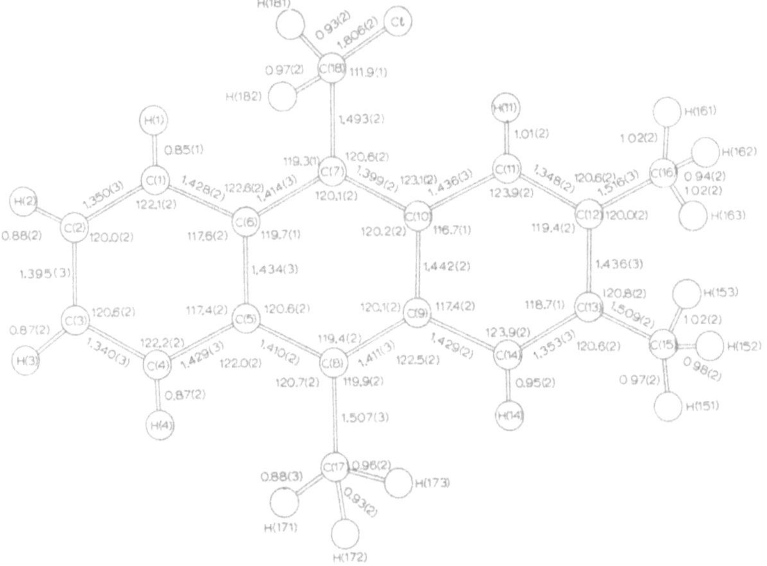

Fig. 1. Bond lengths and angles of 10-chloromethyl-2,3,9-trimethylanthracene.

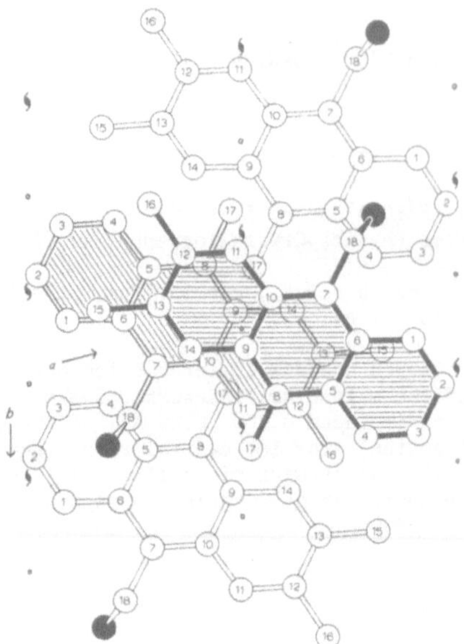

Fig. 2. Molecular overlap in the crystal structure of 10-chloromethyl-2,3,9-trimethylanthracene.

The structure, numbering scheme, bond distances and angles are shown in Fig. 1. The angle between the planes of the outer rings is 1.2°. The molecules stack perpendicular to the y axis in such a manner (Fig. 2) that chlorine atoms pack near those atoms (C(2) and C(4)) of neighbouring molecules expected to be slightly positive.

10-METHYL-9{[(2-CHLOROETHYL)THIO]METHYL}ANTHRACENE

(ICR-358)

$C_{18}H_{17}C\ell S$

J. P. GLUSKER and D. E. ZACHARIAS, 1972. Acta Cryst., B28, 3518-3525.

Triclinic, $P\bar{1}$, a = 9.411, b = 11.984, c = 7.835 Å, α = 107.62°, β = 109.48°, γ = 101.54°, D_m = 1.327, Z = 2. Cu radiation, R = 0.046 for 2296 reflexions.

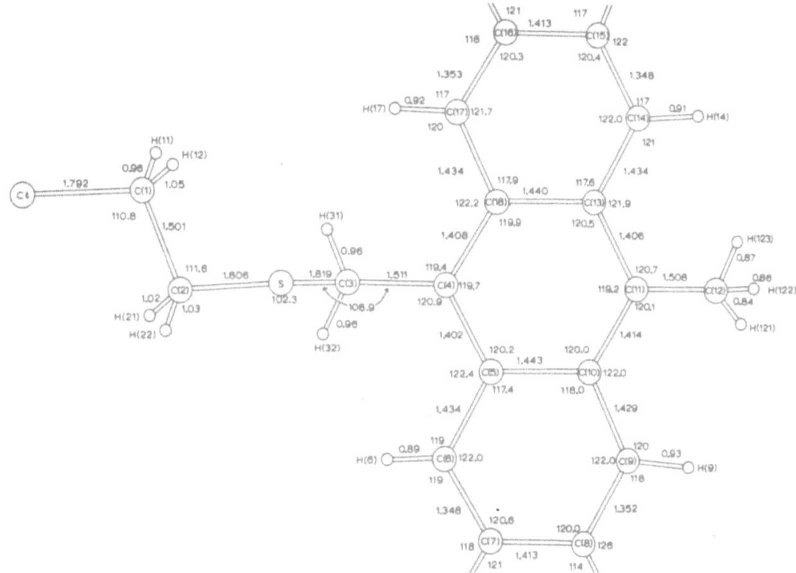

Fig. 1. Structure of 10-methyl-9-{[(2-chloroethyl)thio]methyl}anthracene with intramolecular bond distances and angles (σ = 0.003 Å for C-S and C-Cℓ, 0.004 Å for C-C, 0.03 Å for C-H, 0.1° for C-S-C, 0.3° for C-C-C, and 5° for H-C-H).

The structure and molecular geometry are shown in Fig. 1. There is an angle of 8.7° between the mean planes of the two outer rings. Two of the three rings overlap corresponding rings of neighbouring molecules (3.6 Å separation).

1,3-DICYANOBICYCLO[1,1,0]BUTANE

$C_6H_4N_2$

P. L. JOHNSON and J. P. SCHAEFER, 1972. J. Org. Chem., 37, 2762-2763.

Orthorhombic, Pbcn, a = 10.397, b = 5.813, c = 9.358 Å, D_m = 1.20, Z = 4. Cu radiation, R = 0.057 for 321 reflexions. Molecular symmetry: twofold axis.

The molecular structure and bond lengths are shown in Fig. 1. The dihedral angle between the two C_3 planes is 126.4(4)°. There are no anomalous intermolecular contacts; the C(2)-C(2') intramolecular contact (3.118(5) Å) probably plays a rolé in determining the bicyclobutane geometry.

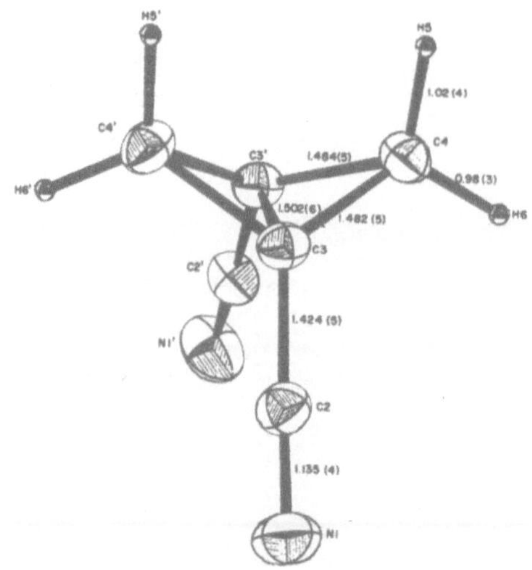

Fig. 1. Bond lengths for 1,3-dicyanobicyclo[1,1,0]butane.

cis-6-CHLORO-trans-7-CHLORO-cis-BICYCLO[3,2,0]HEPTAN-2-ONE

$C_7H_8Cl_2O$

F. P. BOER and P. P. NORTH, 1972. J. Chem. Soc., Perkin II, 416-419.

Monoclinic, $P2_1/c$, a = 6.191, b = 11.172, c = 11.709 Å, β = 94.73°, D_m = 1.47, Z = 4. Mo radiation, R = 0.056 for 1236 reflexions.

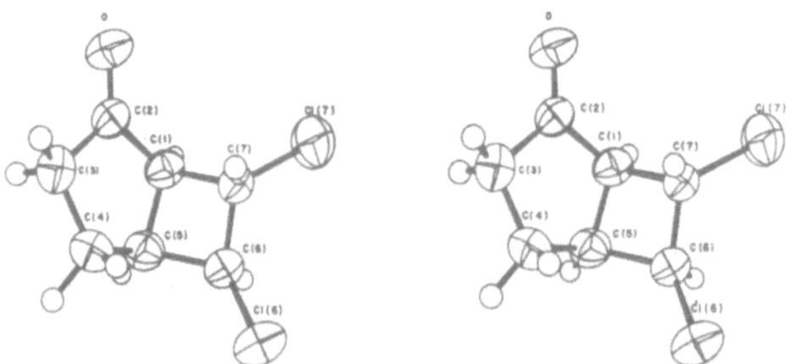

Fig. 1. The bicyclo[3,2,0]heptanone derivative, $C_7H_8Cl_2O$.

The five-membered ring in the molecule (Fig. 1) has an envelope conformation where C(3), C(4), and C(5) form a 'flap' making a dihedral angle of 153.3° with a least-squares plane through C(1), C(2), C(3), C(5), and O (the latter five atoms are nearly coplanar). The envelope is folded so that the flap is bent in the endo-direction. By puckering, the ring relieves eclipsing interactions between hydrogens on C(3) and C(4). The dihedral angle in the four-membered ring is 149.3°, with the ring bent in a direction which relieves a possible interaction between Cl(6) and H(4B) (Fig. 1). The Cl(6)···H(4B) intramolecular contact is 2.816 Å.

Bond angles in the five-membered ring range between 104.8 and 108.1°. The largest of these values, that for C(2), is considerably less than the 120° angle normal for sp^2 hybridized carbon. Bond angles in the four-membered ring are all between 87.3 and 88.2(3)°. On the other hand, the bridgehead angles C(2)-C(1)-C(7) (114.2°) and C(4)-C(5)-C(6) (117.1°) are considerably larger than normal tetrahedral angles, as are the C-C-Cℓ angles (116.7 - 120.1°). Bond distances in the cyclobutane ring (1.512 - 1.550(5) Å) and in the five-membered ring are normal as are the C=O and C-Cℓ distances. The molecules pack with carbonyl dipoles aligned parallel to y. The closest C···O and C···C intermolecular contacts are 3.250 and 3.584 Å respectively.

d-SPIRO[3,3]HEPTANE-2,6-DICARBOXYLIC ACID (at -160°)

(d-FECHT ACID)

$C_9H_{12}O_4$

L. A. HULSHOF, A. VOS and H. WYNBERG, 1972. J. Org. Chem., 37, 1767-1770.

Monoclinic, C2, a = 8.486, b = 7.609, c = 6.928 Å, β = 93.25°, Z = 2. R = 0.056 for 2314 reflexions.

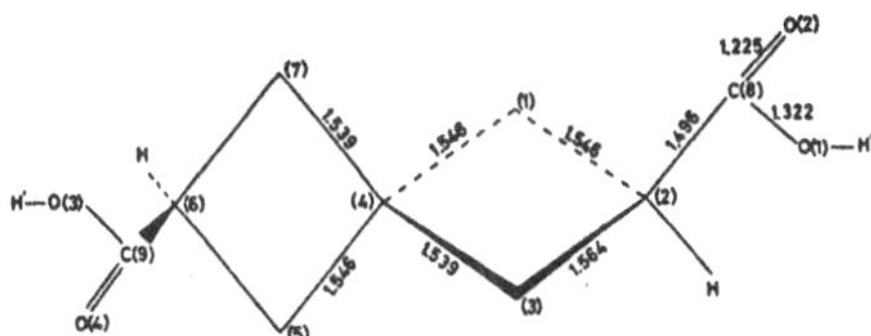

Fig. 1. Absolute configuration and bond lengths (σ = 0.0013 Å) in d-Fecht acid.

The molecule has twofold crystallographic symmetry (Fig. 1). The four-membered rings are puckered with a dihedral angle 152.6° resulting in an approximate equatorial position for the carboxyl groups. The bond angles in the ring range from 88.0 to 88.9(1)°; the C-C-C angles at the spiro carbon atom vary from 114.8 to 126.2°. A strong indication was obtained that the absolute configuration is R as shown in Fig. 1 [later work however shows that the opposite configuration is correct, see J. Org. Chem., 38, 4217 (1973)]. Linear arrays of molecules are formed in the crystal via pairs of OH···O (2.669 Å) hydrogen bonds linking carboxyl groups related by a twofold axis.

SPIRODIENONE II

$C_{11}H_{14}O$

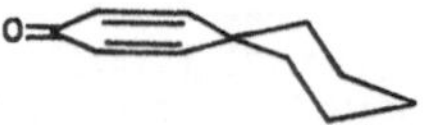

H. KOYAMA and T. IRIE, 1972. J. Chem. Soc., Perkin II, 351-353.

Orthorhombic, $P2_12_12_1$, a = 9.781, b = 9.923, c = 9.428 Å, D_m = 1.165, Z = 4. Mo radiation, R = 0.055 for 878 reflexions.

The molecular conformation is shown above. The cyclodienone ring is planar and the cyclohexane ring has the normal chair conformation. The mean of the angles at the spiro atom is 109.4(3)°. Bond lengths in the cyclodienone ring are normal but all the angles suggest the presence of ring strain. The mean cyclohexane C-C-C angle is 111.3(4)°, and the mean C-C bond (1.530(7) Å) in the cyclohexane ring is also normal.

2-(p-BROMOPHENYL)-1,3-INDANE-DIONE : 0.25 BENZENE SOLVATE

$C_{15}H_9BrO_2,0.25(C_6H_6)$

F. BECHTEL, G. BRAVIC, J. GAULTIER and C. HAUW, 1972. Cryst. Struct. Comm., 1, 159-162.

Triclinic, $P\bar{1}$, a = 18.416, b = 14.655, c = 5.291 Å, α = 86.92, β = 91.76, γ = 103.87°, Z = 4. R = 0.066 for 3795 reflexions.

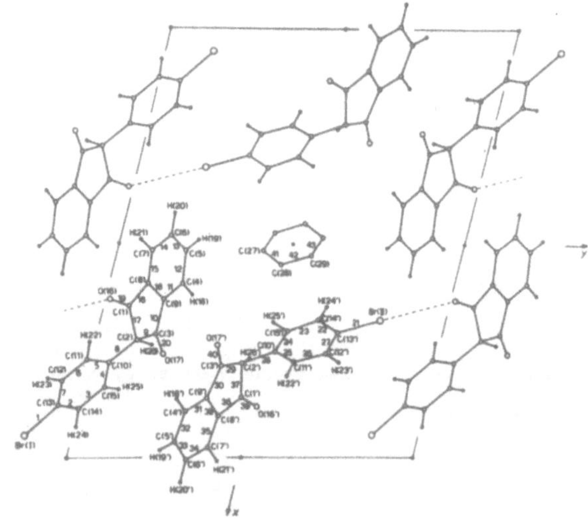

Fig. 1. Structure of the 2-(p-bromophenyl)-1,3-indane-dione benzene solvate.

There are two independent molecules in the asymmetric unit; a molecule of benzene is also found in the unit cell centred on a symmetry centre (Fig. 1). The indane groups are not rigorously planar. Bond lengths in the two molecules are in good agreement with each other but the angles which the bonds C(2)-C(10) make with the appropriate indane planes are 137° and 142°. The shortest contact between the molecules involves a bromine atom and a carbonyl group (Br···O = 3.33 Å, C-Br···O = 176.4°).

1-AMINO-2-p-CHLOROPHENYL-3-INDENONE

$C_{15}H_{10}C\ell NO$

A.E. ŠVETS, Ja.Ja. BLEIDELIS, R.P. ŠIBAEVA and L.O. ATOVMJAN, 1972. Ž. Strukt. Khim., 13, 745-748 [J. Struct. Chem., 13, 700-702].

Orthorhombic, Pcab, a = 7.13, b = 13.52, c = 25.53 Å, D_m = 1.28, Z = 8. Cu radiation, R = 0.13 for 880 reflexions.

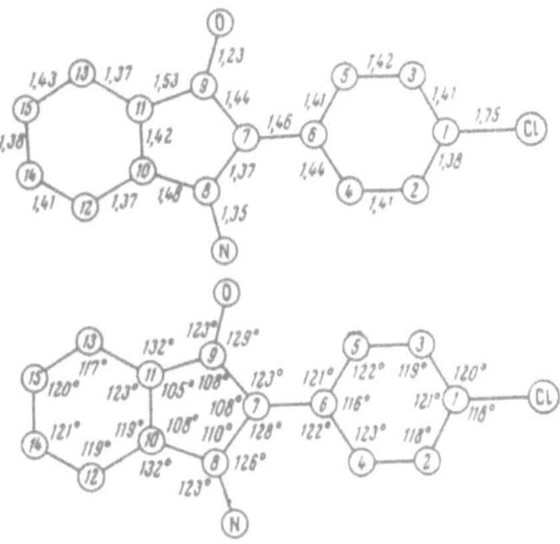

Fig. 1. Geometry of $C_{15}H_{10}C\ell NO$.

The molecule is shown with bond distances (±0.03 Å) and angles (±2°) in Fig. 1. The phenyl ring is inclined 42° with respect to the indene group. Molecules are linked by N(H)···O hydrogen bonds (2.85 Å) to form chains along a.

cis-1,1-BISMETHOXYCARBONYL-4a-METHYL-DECAHYDRONAPHTHALENE

$C_{15}H_{24}O_4$

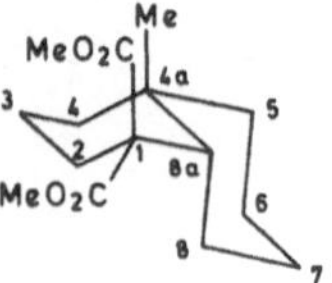

A. T. McPHAIL, P. WONG and D. M. S. WHEELER, 1972. J. Chem. Soc., Perkin II, 2369-2372.

Monoclinic, $P2_1/c$, a = 11.52, b = 7.68, c = 20.10 Å, β = 125.58°, D_m = 1.23, Z = 4. Cu radiation, R = 0.103 for 2156 reflexions.

The molecule adopts the conformation shown above with both rings in chair conformations; the methyl substituent and one of the methoxycarbonyl groups are axially oriented on the same ring. Relief from overcrowding which results from the adoption of this conformation is achieved through a combination of factors: (i) bond-length extension, (ii) bond-angle deformations, and (iii) flattening of the cyclohexane rings in the central region of the molecule. Ring bonds C(1)-C(8a) and C(4a)-C(8a), mean 1.579 Å, are significantly longer than the other $C(sp^3)$-$C(sp^3)$ bonds (mean 1.539 Å), and the C-CO_2Me bonds are also long, 1.548 Å. Torsion angles around the C(4a)-C(8a) bond are smaller than the average (55°) owing to flattening which occurs in the cyclohexane rings in order to reduce the 1,3-diaxial interactions. The shortest intermolecular separations correspond to van der Waals distances.

Δ^{3a}-4-(m-METHOXYBENZYLSULPHONYL)-1,5-DIACETOXY-7a-METHYLHEXAHYDROINDENE

$C_{22}H_{28}O_7S$

C. F. W. van de VEN and H. SCHENK, 1972. Cryst. Struct. Comm., 1, 147-150.

Triclinic, $P\bar{1}$, a = 11.695, b = 22.538, c = 9.391 Å, α = 89.05°, β = 98.42°, γ = 113.75°, Z = 4. Cu radiation, R = 0.12 for 3804 reflexions.

The title compound (Fig. 1) was investigated to establish the mechanism of the synthesis of the corresponding 7-thia steroid (1). The hydrogen atom at position labelled 5 is in the α-position whereas in the 7-thia steroid (1) this hydrogen atom is in the β-position. Ring C is a somewhat distorted half-chair and ring D a nearly ideal half-chair.

Fig. 1. The $C_{22}H_{28}O_7S$ molecule.

1. C. F. W. van de VEN and H. SCHENK, 1972. Cryst. Struct. Comm., 1, 121; this volume, Steroid section.

TETRAETHYL 4-BROMO-8-METHOXYDECAHYDRONAPHTHALENE-2,2,6,6-TETRACARBOXYLATE

$C_{23}H_{35}BrO_9$

H. W. GUIN and S. H. SIMONSEN, 1972. Cryst. Struct. Comm., 1, 397-401.

Monoclinic, $P2_1/c$, a = 13.487, b = 16.823, c = 12.235 Å, β = 97.44°, D_m = 1.291, Z = 4. Cu radiation, R = 0.154 for 2027 reflexions.

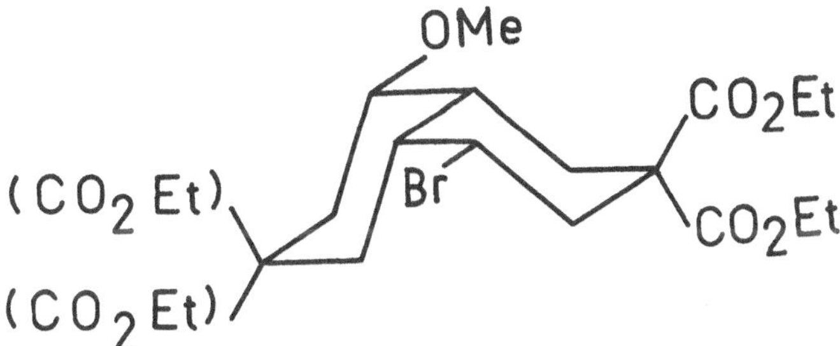

Fig. 1. The $C_{23}H_{35}BrO_9$ molecule.

Despite an inaccurate structure resulting from crystal decomposition, it is shown that the rings are *cis*-fused and that the title compound is a bicyclo-[4.4.0]-decane (Fig. 1). The bromo and methoxy groups are equatorial with respect to the rings indicating that the methoxybromination of tetra-ethyl cis,cis-3,8-cyclodecadiene-1,1,6,6-tetracarboxylate is stereospecific and is analogous to the hypothetical sesquiterpene biogenesis reaction scheme which involves proposed stereospecific double bond cyclization (1).

1. J. B. HENDRICKSON, 1959. Tetrahedron, 7, 82.

trans-1,2-DIHALOGENOACENAPHTHENES

M. T. Le BIHAN and M. C. PERUCAUD, 1972. Acta Cryst., B28, 629-634.

trans-1,2-DIBROMOACENAPHTHENE

$C_{12}H_8Br_2$

Orthorhombic, $P2_12_12$, a = 7.92, b = 7.83, c = 8.47 Å, Z = 2. Cu radiation, R = 0.10. Number of reflexions not given.

This compound is isostructural with the trans-dichloro-derivative if the *a* and *b* cell dimensions given here are interchanged. The *x*-coordinate of the dibromo derivative then becomes the *y*-coordinate of the dichloro derivative (but with a sign change also); the dibromo *y*-coordinate becomes the dichloro *x*-coordinate. [This is nowhere stated in the paper and the two structures are treated as if they are independent.]

trans-1,2-DICHLOROACENAPHTHENE

$C_{12}H_8Cl_2$

Orthorhombic, $P2_12_12$, a = 7.79, b = 7.74, c = 8.38 Å, Z = 2. Cu radiation, R = 0.06. Number of reflexions not given. D_x is said to be 0.989. The correct value for the data given is 1.15.

The structure is isostructural with that of the dibromo-derivative (see above). The molecule lies on a twofold axis of the space group (Fig. 1). Molecular dimensions of the dichloro-derivative are in Fig. 2; those of the dibromo-derivative are essentially identical (C-Br = 1.956(15) Å). Distances between molecules correspond to van der Waals contacts.

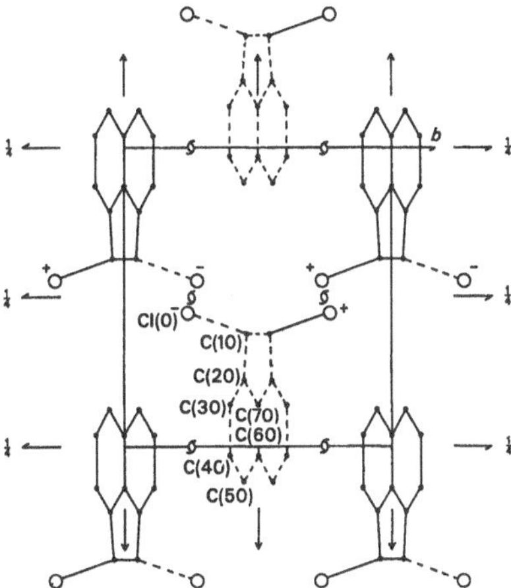

Fig. 1. Projection of the crystal structure trans-1,2-dichloroacenaphthene along a.

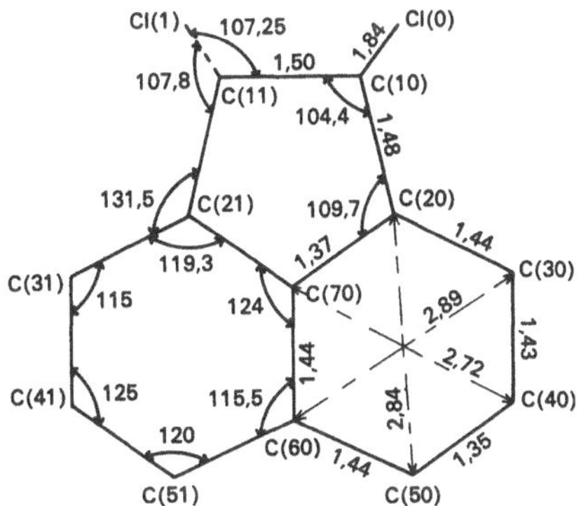

Fig. 2. Bond lengths (σ = 0.02 Å) and angles in trans-1,2-dichloroacenaphthene.

2,4,7-TRINITRO-9-FLUORENONE

$C_{13}H_5N_3O_7$

D. L. DORSET, A. HYBL and H. L. AMMON, 1972. Acta Cryst., B28, 3122-3127 (and B29, 2314 (1973)).

Monoclinic, $P2_1/n$, a = 22.470, b = 5.652, c = 10.278 Å, β = 105.51°, D_m = 1.692, Z = 4. Mo radiation, R = 0.057 for 1001 reflexions.

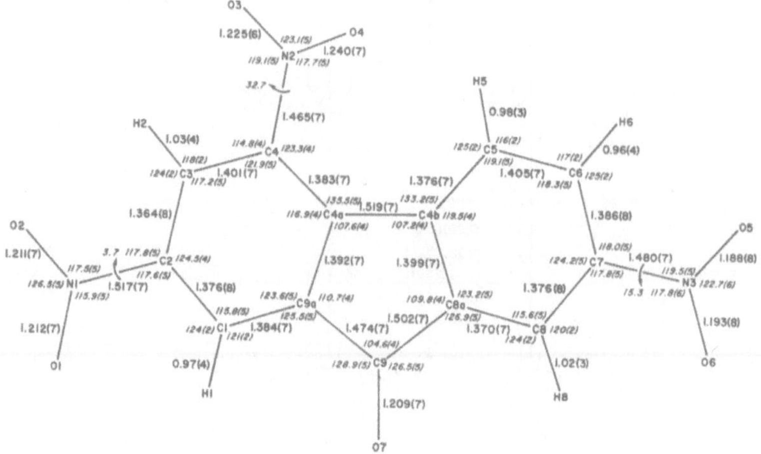

Fig. 1. Bond lengths and angles for 2,4,7-trinitro-9-fluorenone.

Fig. 2. Intermolecular contact diagram viewed along [010] for 2,4,7-trinitro-9-fluorenone.

The fluorene nucleus is planar. The bond lengths given in Fig. 1 indicate that there is little interaction between the benzene rings, or between the rings and the carbonyl group. The nitro group at C(4) is twisted out of the benzene ring plane by 33°. The twist angles for the C(2) and C(7) nitro groups are respectively 3.7 and 15.3°. A packing diagram viewed normal to the ac plane (Fig. 2) shows the shortest intermolecular contacts.

9-FLUORENONE

$C_{13}H_8O$

H. R. LUSS and D. L. SMITH, 1972. Acta Cryst., B28, 884-889.

Orthorhombic, Pcab, a = 16.068, b = 18.650, c = 12.550 Å, D_m = 1.24, Z = 16. Mo radiation, R = 0.056 for 1662 reflexions.

Fig. 1. Averaged bond lengths and angles in 9-fluorenone.

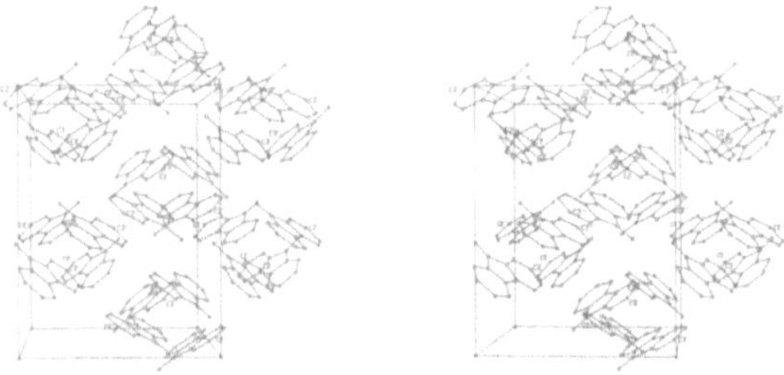

Fig. 2. View of the molecular packing of 9-fluorenone down the b-axis.

Values of equivalent bond lengths and angles for the two independent molecules in the asymmetric unit, averaged assuming C_{2v} symmetry, are shown in Fig. 1. The C-C bonds of the benzo rings average 1.383(1) Å but can be divided into two sets of alternate bonds which average 1.378(1) and 1.388(1) Å. Both molecules are planar to within 0.025 Å. The molecular packing is shown in Fig. 2; the molecules pack end-to-end in layers perpendicular to b.

6,6,12,12-TETRACHLORO-3,3,9,9-TETRAMETHOXYTRICYCLO[9,1,0,$0^{5,7}$]DODECANE

$C_{16}H_{24}C\ell_4O_4$

R. W. BAKER and P. J. PAULING, 1972. J. Chem. Soc., Perkin II, 1451-1453.

Triclinic, $P\bar{1}$, a = 9.988, b = 8.195, c = 6.377 Å, α = 93.07, β = 91.40, γ = 110.70°, D_m = 1.43, Z = 1. Mo radiation, R = 0.043 for 1504 reflexions.

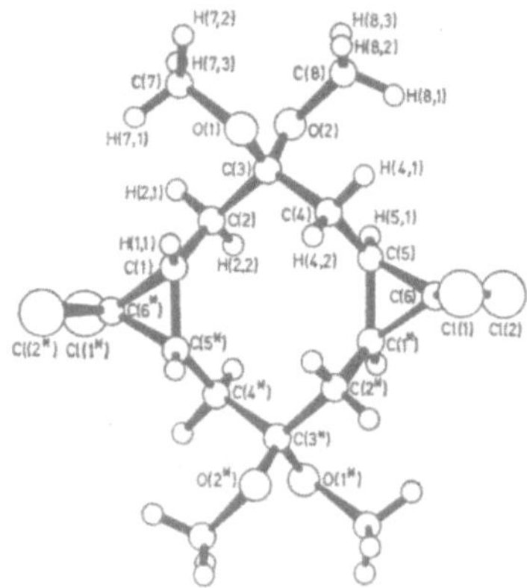

Fig. 1. Perspective drawing of the tricyclo[9,1,0,$0^{5,7}$]dodecane derivative $C_{16}H_{24}C\ell_4O_4$.

The molecule has a centre of symmetry coincident with a crystallographic inversion centre, but possesses approximate 2/m symmetry and its conformation (Fig. 1) is as predicted by Dunitz (1). The (approximate) mirror plane passes through the four Cℓ atoms. The C-C-$\bar{C}$-C torsion angles are 115 and 117° and the methoxy groups are extra-annular. Bond lengths have their expected values.

1. J. D. DUNITZ, 1968. Perspectives in Structural Chemistry, 2, 1.

3a,6-DIMETHYL-6-(trans-3-CHLOROBUT-2-ENE-1-YL)-
2,4,5,6,8,9-HEXAHYDRO-3H-BENZ[e]INDENE-3,7-(3aH)-DIONE

$C_{19}H_{23}C\ell O_2$

C. S. YOO, J. PLETCHER and M. SAX, 1972. Acta Cryst., B28, 2838-2844.

Monoclinic, $P2_1/c$, a = 10.091, b = 11.515, c = 14.709 Å, β = 92.54°, D_m = 1.237, Z = 4. Cu radiation, R = 0.081 for 2912 reflexions.

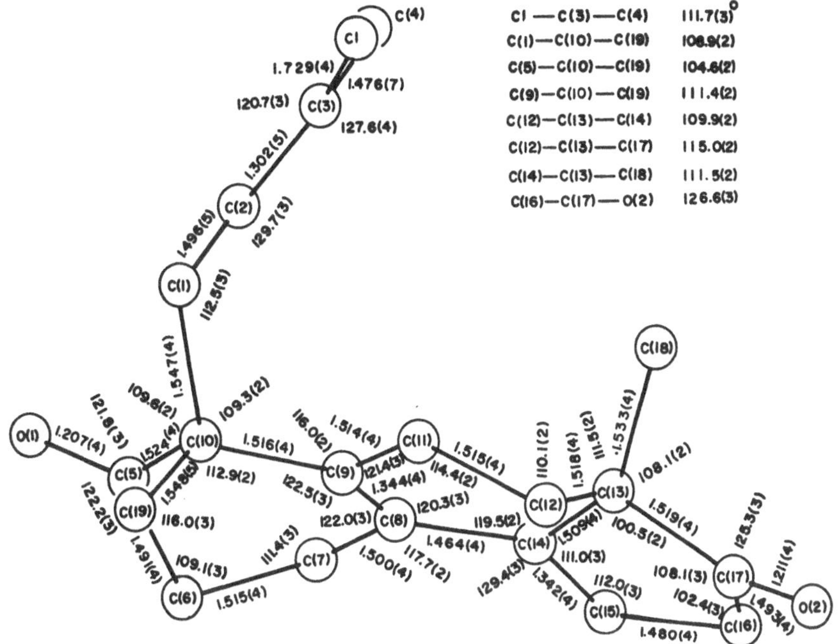

Fig. 1. Bond distances and angles in the benz[e]indene derivative $C_{19}H_{23}C\ell O_2$.

The two methyl groups in the molecule are established to be in trans configuration. The conjugated double bonds in the fused ring system confer an approximately coplanar arrangement on the nine atoms C(7)-C(11), C(13)-C(16) although significant deviations are apparent. Rings B and C are in extended boat conformations with ring D as a distorted envelope. Bond lengths and angles and the molecular conformation are shown in Fig. 1. Distances between molecules correspond to normal van der Waals interactions.

LEVOPIMARIC ACID

$C_{20}H_{30}O_2$

I. L. KARLE, 1972. Acta Cryst., B28, 2000-2007.

Orthorhombic, $P2_12_12_1$, a = 15.664, b = 19.387, c = 11.851 Å, Z = 8. Cu radiation, R = 0.060 for 3317 reflexions.

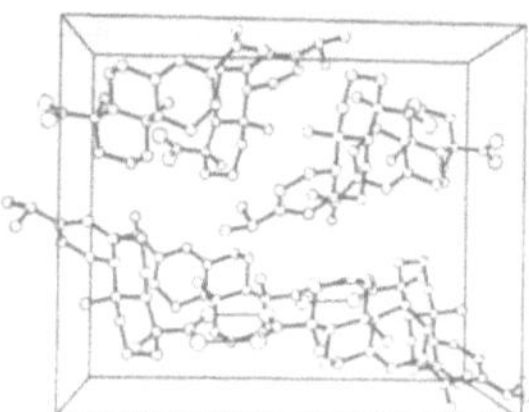

Fig. 1. View of the contents of a unit cell of levopimaric acid.

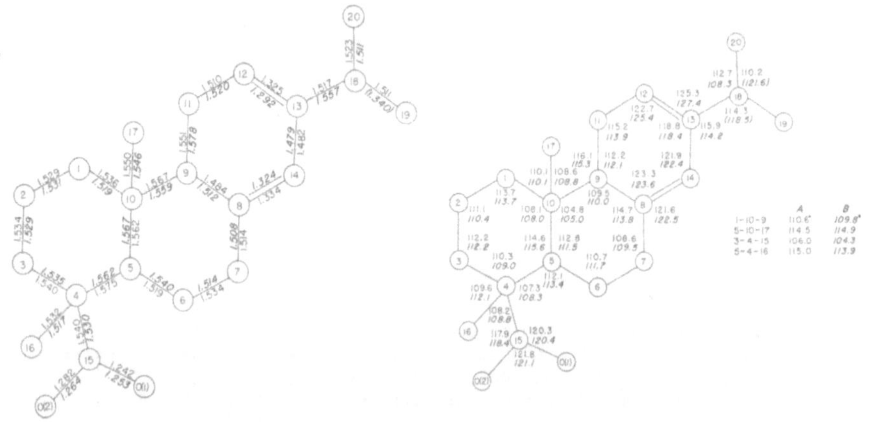

Fig. 2. Bond lengths and angles for molecule A (in regular type) and molecule B (in italics) for levopimaric acid. The standard deviations are 0.008 Å [except for C(18)-C(19) in molecule B (0.011 Å)] and 0.65°.

Fig. 3. Torsional angles for levopimaric acid. The values in brackets are deviations (in Å) from planarity in ring C.

Levopimaric acid is an optically active carboxylic acid which forms a dimer by hydrogen bonding between the carboxyl groups of the two independent molecules in the asymmetric unit (O-H···O = 2.644 and 2.628 Å) (Fig. 1). Bond lengths and angles are given in Fig. 2 and torsion angles in Fig. 3. The carboxyl groups, which are disordered with nearly equal C-O distances, assume an approximate symmetric conformation with torsion angles C(16)C(4)C(15)O(1) = 7.7° and C(16*)C(4*)C(15*)O(2*) = 20.9°. Ring A and B have the chair conformation while deviations from planarity in ring C are rather large (Fig. 3). There is a folded conformation between ring B and C and the diene system is twisted as postulated from the large negative Cotton effect. The dihedral angle between the least-squares planes through ring B and C is 43.7° for molecule A and 39.9° for molecule B.

12β-HYDROXYSANDARACOPIMARIC ACID

$C_{20}H_{30}O_3$

J. LAPASSET and J. FALGUEIRETTES, 1972. Acta Cryst., B28, 3321-3331.

Orthorhombic, $P2_12_12$, a = 22.83, b = 11.25, c = 7.069 Å, D_m = 1.10, Z = 4. Cu radiation, R = 0.062 for 1746 reflexions.

The molecule (Fig. 1), with O(23) oxygen in the 12β-position, has rings A and B in approximate chair conformations. Ring A is distorted by the steric interaction of the two methyl groups at C(18) and C(19) which have a separation of 3.38 Å. The double bond at C(14)-C(8) has little influence on ring B which has a normal chair conformation. Ring C which contains the double bond is severely distorted from the regular chair conformation; atoms C(7), C(8), C(9), C(13), and C(14) are in an approximate plane imposed by the trigonal hybridization of C(8) and C(14). The bond distances within the rings are normal. The molecules are held together by O-H···O hydrogen bonds (2.61, 2.81 Å) which link two carboxyl and two hydroxyl groups from four molecules about a twofold axis.

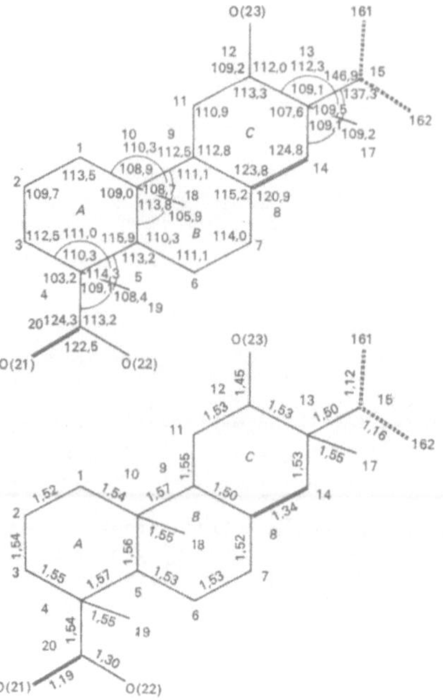

Fig. 1. Interatomic distances and angles for 12β-hydroxysandaracopimaric acid (σ = 0.011 Å and 0.7°). Atom C(16) is disordered.

5-(2'-BROMO-3',4'-DIHYDRONAPHTHYL)-ACENAPHTHENE

$C_{22}H_{17}Br$

J. BORDNER, R. L. GREENE, L. A. JONES and R. WATSON, 1972. Cryst. Struct. Comm., 1, 393-395.

Monoclinic, $P2_1$, a = 9.04, b = 7.15, c = 14.00 Å, β = 115.41, Z = 2. Cu radiation, R = 0.022 for 917 reflexions.

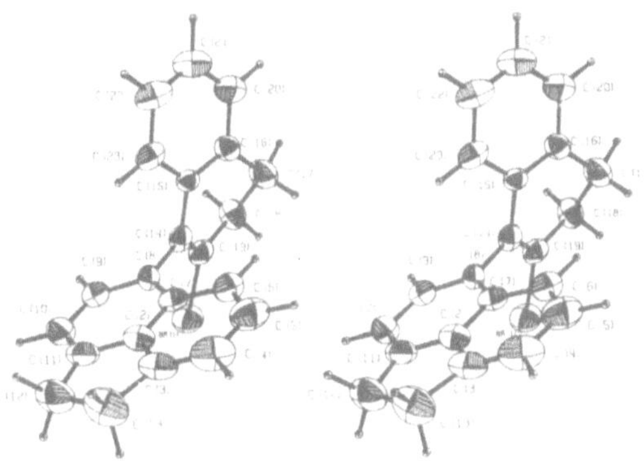

Fig. 1. The $C_{22}H_{17}Br$ acenaphthene derivative.

This compound resulted from a bromination reaction under allylic conditions. Subsequent reactions indicated that the allylic position had not been brominated and the X-ray analysis established that a vinylic not allylic substitution had resulted (Fig. 1). There are no unusual bonded distances.

PYRENE

(NEUTRON STUDY)

$C_{16}H_{10}$

A. C. HAZELL, F. K. LARSEN and M. S. LEHMANN, 1972. Acta Cryst., B28, 2977-2984.

Monoclinic, $P2_1/a$, a = 13.649, b = 9.256, c = 8.470 Å, β = 100.28°, D_m = 1.27, Z = 4. Neutron radiation, R = 0.034 for 1008 reflexions.

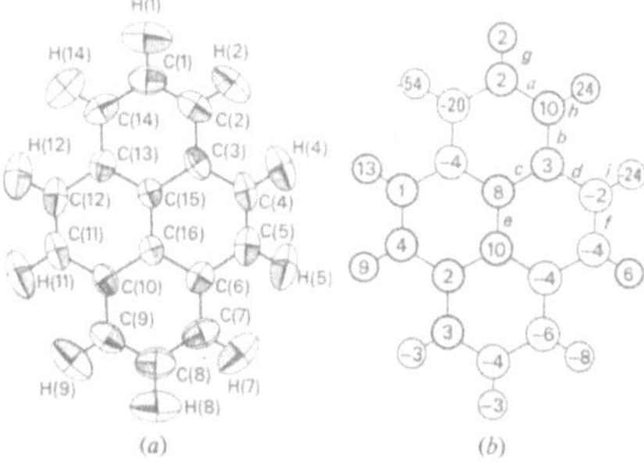

Fig. 1. Pyrene; numbers within atoms in (b) are deviations from the best plane (in Å x 10^3).

Pyrene has been extensively studied by X-ray methods (1). From this neutron study the molecule (Fig. 1) is shown to deviate significantly from mmm symmetry. Mean bond length values for bonds a-f (Fig. 1) are a = 1.395(2), b = 1.406(2), c = 1.425(2), d = 1.438(2), e = 1.430(2), f = 1.367(3) Å, in good agreement with those predicted by MO calculations.

1. Structure Reports, 35B, 131.

2,7-DI-t-BUTYLPYRENE

$C_{24}H_{26}$

A. C. HAZELL and J. G. LOMBORG, 1972. Acta Cryst., B28, 1059-1064.

Triclinic, P$\bar{1}$, a = 9.004, b = 6.138, c = 9.153 Å, α = 96.9, β = 109.1, γ = 108.3°, D_m = 1.18, Z = 1. Cu radiation, R = 0.068 for 1369 reflexions.

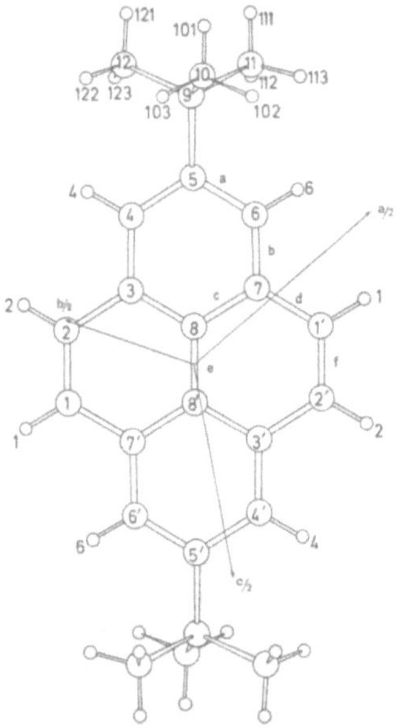

Fig. 1. The 2,7-di-t-butylpyrene molecule showing atom numbering and bond lettering.

The molecule is situated on a centre of symmetry and is shown in Fig. 1. The pyrene skeleton is planar to within experimental error with C(9) displaced 0.09 Å out of this plane. Intramolecular steric factors cause the butyl group to be rotated some degrees from the idealized configuration. The bond lengths in the pyrene skeleton corrected for thermal motion are: (a) = 1.394(3), (b) = 1.390(3), (c) = 1.417(3), (d) = 1.437(3), (e) = 1.418(4), and (f) = 1.346(5) Å. The bond lengths in the butyl group are 1.523(5), 1.535(6), 1.534(3), and 1.531(4) Å. The bond angles in the pyrene skeleton range from 118.2 to 122.8°, and in the butyl group from 107.3 to 109.9°.

5,6-DIHYDRODIBENZ[a,h]ANTHRACENE

$C_{22}H_{16}$

C. H. WEI and J. R. EINSTEIN, 1972. Acta Cryst., B28, 1478-1483.

Orthorhombic, $P2_12_12_1$, a = 8.465, b = 15.082, c = 11.616 Å, D_m = 1.23, Z = 4.
Cu radiation, R = 0.075 for 854 reflexions.

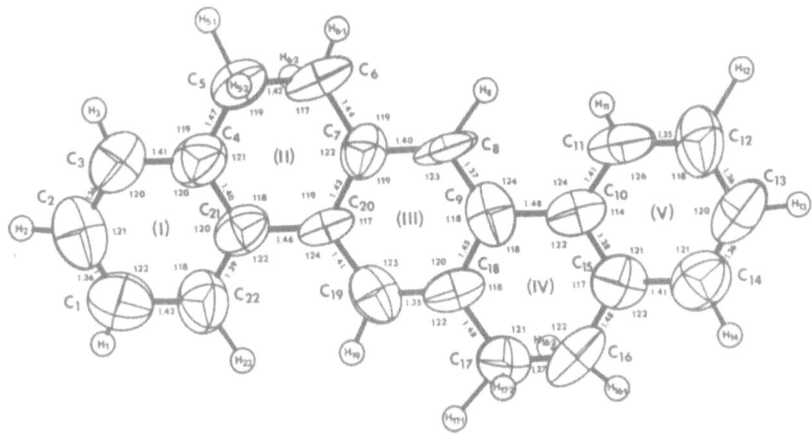

Fig. 1. Configuration of the 5,6-dihydrodibenz[a,h]anthracene molecular unit (the average molecule), consisting of two nearly superimposable alternative orientations with an approximate occupancy factor ratio of 1:1. Standard deviations for C-C bond distances are mostly 0.01 Å, except for the bonds between C(13) and C(17), where they are 0.02 Å. The standard deviation for each C-C-C angle is 1°.

The crystal structure of 5,6-dihydrodibenz[a,h]anthracene is composed of average molecules with configurations as depicted in Fig. 1, where C-C bond distances and C-C-C bond angles are also shown. Each asymmetric unit is considered to be made up of two fractional molecules of alternative configurations, roughly related to each other by a centre of symmetry located at the centre of the middle ring (III). Because of the disorder, the end aromatic rings (I) and (V) both appear to be twisted from the central ring (III) by 10.1 and 6.7° respectively, whereas the expected twist for the end ring adjacent to the ring containing the saturated carbon atoms is about 21°, and that expected for the ring at the opposite (aromatic) end is nearly zero.

5,6-DIHYDRODIBENZ[a,j]ANTHRACENE

$C_{22}H_{16}$

C. H. WEI, 1972. Acta Cryst., B28, 1466-1477.

Monoclinic, $P2_1/c$, a = 12.1434, b = 8.0864, c = 30.6369 Å, β = 101.130°, D_m = 1.23, Z = 8. Cu radiation, R = 0.082 for 3090 reflexions.

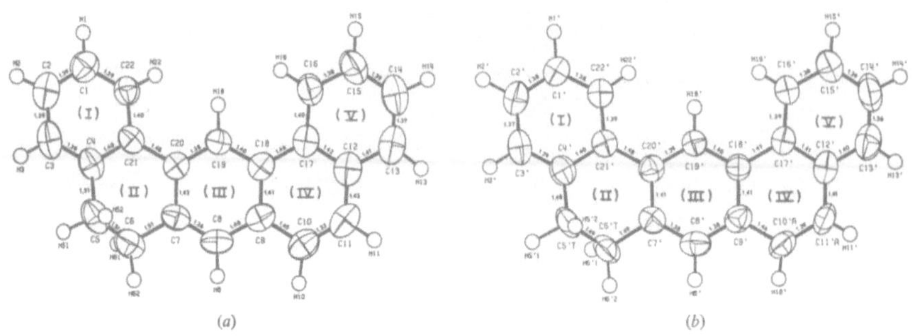

Fig. 1. Bond lengths (σ = 0.004 - 0.006 Å) and angles in 5,6-dihydrodibenz-[a,j]anthracene. (a) Molecule A, and (b) disordered molecule B.

One of the two independent molecules (A) (Fig. 1a) is normal, whereas the crystal-disordered model for the other molecule (B) (Fig. 1b), successfully utilized in the refinement, assumes a distribution of two partial molecules in two orientations approximately related to each other by a mirror plane through the C(8') and C(19') atoms of the middle ring. The ratio of occupancy factors for the two disordered orientations is 7:3. Each aromatic ring in both independent molecules is almost perfectly planar, and the phenanthrene moieties are approximately planar, but with an observable distortion of the rings, attributable to overcrowding of hydrogen atoms. The bond length in the electron-rich K region of the phenanthrene moiety is 1.334(6) Å for molecule A. The saturated carbon atoms are about 0.35 Å above and below the plane defined by the other carbon atoms of the ring, and the adjacent aromatic ring is twisted 21° from the plane of the central aromatic ring.

BIS-(2,2'-BIPHENYLENE)METHANE

$C_{25}H_{16}$

H. SCHENK, 1972. Acta Cryst., B28, 625-628.

Monoclinic, $P2_1/c$, a = 10.173, b = 10.241, c = 16.722 Å, β = 96.54°, D_m = 1.19, Z = 4. Cu radiation, R = 0.060 for 2500 reflexions.

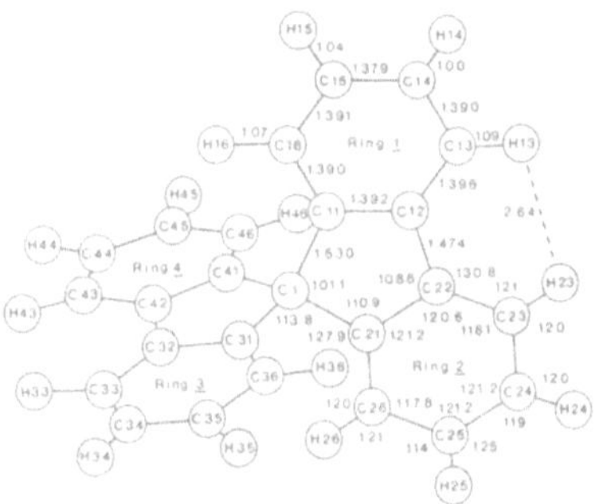

Fig. 1. Mean bond lengths (σ = 0.008 Å for C-C, 0.07 Å for C-H) and bond angles in bis-(2,2'-biphenylene)methane.

Mean molecular dimensions averaged over equivalent portions of the molecule are in Fig. 1. The individual aromatic rings are planar but both biphenyl groups of the molecule deviate from planarity. The angle between rings 1 and 2 is 1.2°; that between rings 3 and 4 is 4.2°. In the crystal short intermolecular C···C distances (3.22, 3.35 and 3.45 Å) are found approximately perpendicular to the planes of the benzene rings.

cis-PHOTODIMER OF 1-CYANOACENAPHTHYLENE

$C_{26}H_{14}N_2$

C. COURSEILLE, B. BUSETTA, M. HOSPITAL and A. CASTELLAN, 1972. Cryst. Struct. Comm., 1, 337-340.

Orthorhombic, I222, a = 10.794, b = 10.816, c = 16.051 Å, Z = 4. Cu radiation, R = 0.065 for 989 reflexions.

Fig. 1. cis-Photodimer of 1-cyanoacenaphthylene.

The analysis establishes the cis configuration of the photodimer (Fig. 1); the two cyano groups are disordered over the four positions. The angle between the acenaphthylene planes is 51° and the angle between an acenaphthylene plane and the cyclobutane ring is 116°.

(-)-2-BROMOHEXAHELICENE

$C_{26}H_{15}Br$

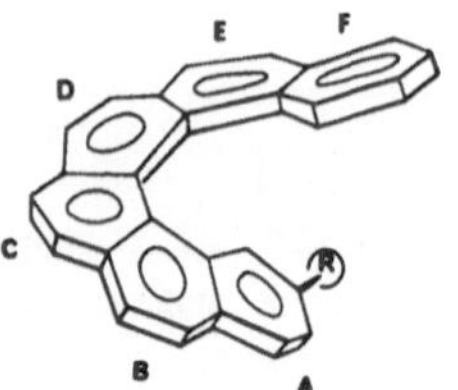

D. A. LIGHTNER, D. T. HEFELFINGER, T. W. POWERS, G. W. FRANK and K. N. TRUEBLOOD, 1972. J. Amer. Chem. Soc., 94, 3492-3497.

Orthorhombic, $P2_12_12_1$, a = 17.364, b = 13.344, c = 7.799 Å, D_m = 1.50, Z = 4. Cu radiation, R = 0.059 for 1462 reflexions.

The absolute configuration of (-)-2-bromohexahelicene is as shown above (R=Br) and as it can be converted to (-)-hexahelicene, it follows that (-)-hexahelicene also has the left-handed helical configuration (R=H). A view of the bromo-derivative is shown in Fig. 1. and details of the geometry in the overcrowded region of the molecule are in Fig. 2.

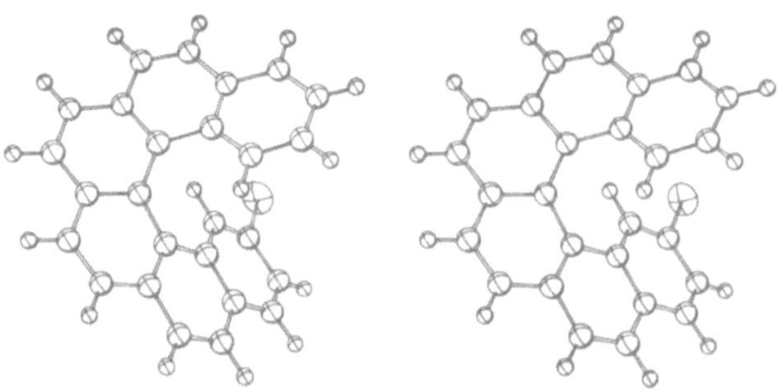

Fig. 1. View of (-)-2-bromohexahelicene.

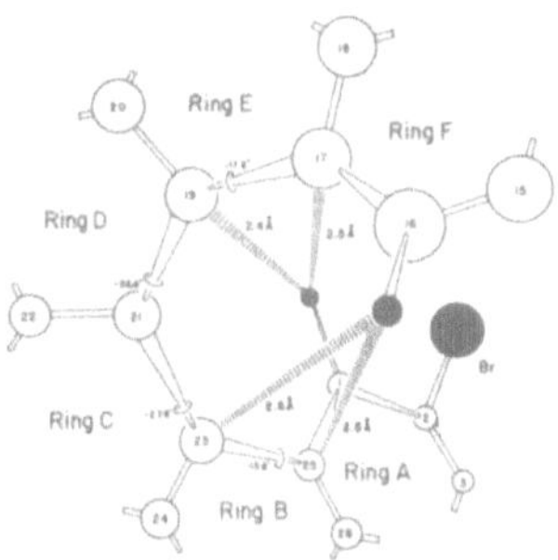

Fig. 2. Some interatomic distances and torsion angles in (-)-2-bromohexahelicene. Distances between calculated hydrogen positions and carbons in the inner core of the 2-bromohexahelicene molecule. Torsion angles about the inner core carbon-carbon bonds shown in degrees. Bromine atom and hydrogens on carbons 1 and 16 shown as solid black circles.

PYRENO-(1',2':1,2)-PYRENE

$C_{30}H_{16}$

K. W. MUIR and J. M. ROBERTSON, 1972. Acta Cryst., B28, 879-884.

Monoclinic, $P2_1/c$, a = 13.17, b = 7.71, c = 20.32 Å, β = 118.5°, D_m = 1.38, Z = 4. Mo radiation, R = 0.11 for 2284 reflexions.

As a result of intramolecular overcrowding the molecule is propeller shaped (Fig. 1), the minimum non-bonding C···C and H···H distances involving overcrowded atoms being 2.96 and 2.03 Å respectively. To a lesser extent deformation of bond angles (Fig. 2) plays a part in relieving steric strain. There are no apparent changes in bond lengths as a result of overcrowding; the observed lengths (Fig. 2) are in good agreement with theoretical predictions. In the crystals, the molecules are separated by normal van der Waals distances.

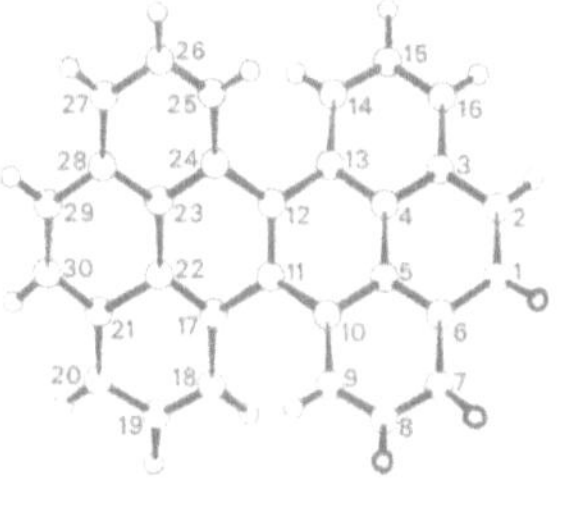

Fig. 1. Perspective view of pyreno-(1',2':1,2)-pyrene.

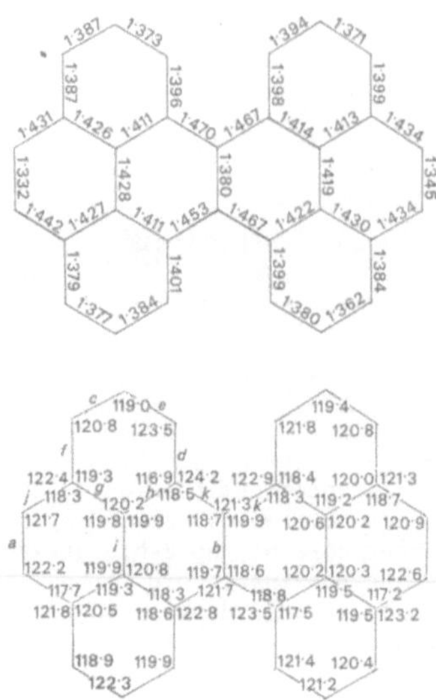

Fig. 2. Bond distances and angles in pyreno-(1',2':1,2)-pyrene. Standard deviations are 0.006 - 0.009 Å and 0.4 - 0.5°.

8,8-DICHLOROTRICYCLO[3,2,1,0]OCTANE

$C_8H_{10}Cl_2$

K. B. WIBERG, G. J. BURGMAIER, K.-W. SHEN, S. J. LA PLACA, W. C. HAMILTON and M. D. NEWTON, 1972. J. Amer. Chem. Soc., 94, 7402-7406.

Monoclinic, $P2_1/m$, a = 6.796, b = 8.721, c = 6.903 Å, β = 102.62°, Z = 2. Mo radiation, R = 0.067 for 242 reflexions.

The molecule has crystallographically imposed mirror symmetry (Fig. 1). The Cℓ-C(8)(bridge)-Cℓ angle is 108.7° and the C(8)-Cℓ bond distances average 1.761(10) Å. Each of the two mirror-related bridgehead carbons [C(1),C(5)] is displaced only 0.093 Å from the plane defined by its three nearly trigonal nonbridgehead neighbors. The bridgehead-bridgehead bond (1.572(15) Å) is considerably longer than the bridge-bridgehead bond (1.458(12) Å). The remaining crystallographically independent distances are C(1)-C(7) = 1.526(12), C(6)-C(7) = 1.512(20) Å and C(1)-C(2) = 1.483(12), C(2)-C(3) = 1.529(12) Å in the four- and five-membered rings, respectively.

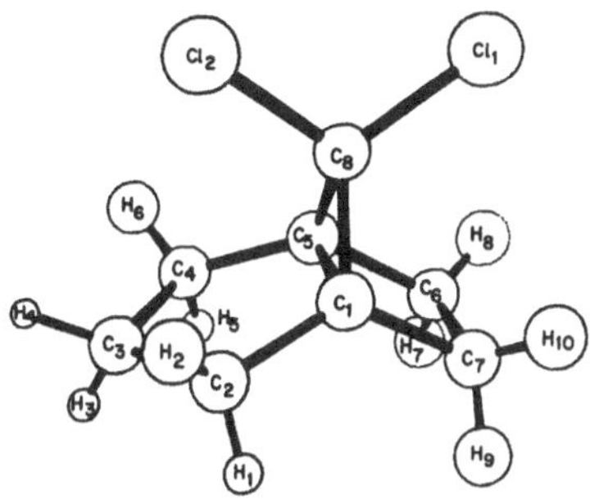

Fig. 1. 8,8-Dichlorotricyclo[3,2,1,0]octane (the crystallographic mirror plane contains the atoms Cℓ(1), Cℓ(2), C(8), C(3), H(4), and H(3)).

(-)-2-exo-AMINONORBORNANE-2-CARBOXYLIC ACID HYDROBROMIDE SALT

$C_8H_{14}BrNO_2$

P. A. APGAR and M. L. LUDWIG, 1972. J. Amer. Chem. Soc., 94, 964-967.

Orthorhombic, $P2_12_12_1$, a = 6.988, b = 22.673, c = 6.150 Å, Z = 4. Mo radiation, R = 0.040 for 923 reflexions.

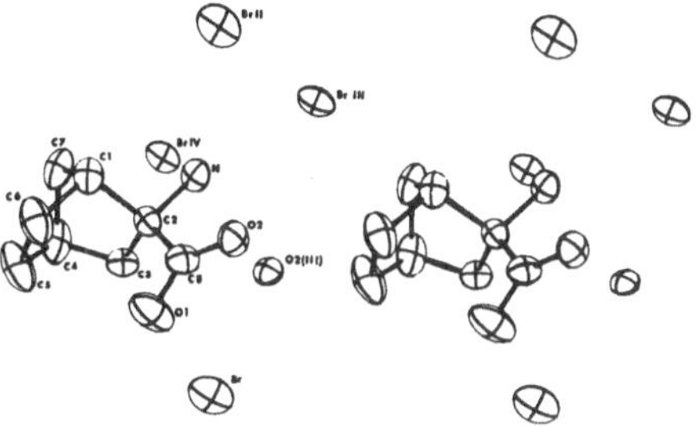

Fig. 1. Drawing showing the absolute stereochemistry (1R,2R,4S) as determined by Bijvoet's method. Hydrogen bonds connect O(1) with Br, and N with O(2)III, BrII and BrIV.

The absolute stereochemistry of this (-)-exo-isomer is depicted in Fig. 1. Bond distances and angles in the bicycloheptane ring are normal for this system but the location of the carboxyl group is counter to predictions from pK values. Each molecule makes four independent hydrogen bonds to adjacent moieties (NH···O = 2.862, NH···Br = 3.482 and 3.483, and OH···Br = 3.098 Å). In this way hydrogen bonded sheets are formed perpendicular to b. The sheets themselves are held together by van der Waals forces.

BICYCLO[2,2,1]HEPT-5-ENE-2,3-exo-DICARBOXYLIC ANHYDRIDE

$C_9H_8O_3$

G. FILIPPINI, C. M. GRAMACCIOLI, C. ROVERE and M. SIMONETTA, 1972. Acta Cryst., B28, 2869-2874.

Orthorhombic, $P2_12_12_1$, a = 7.9850, b = 7.6300, c = 12.7700 Å, Z = 4. Cu radiation, R = 0.086 for 704 reflexions.

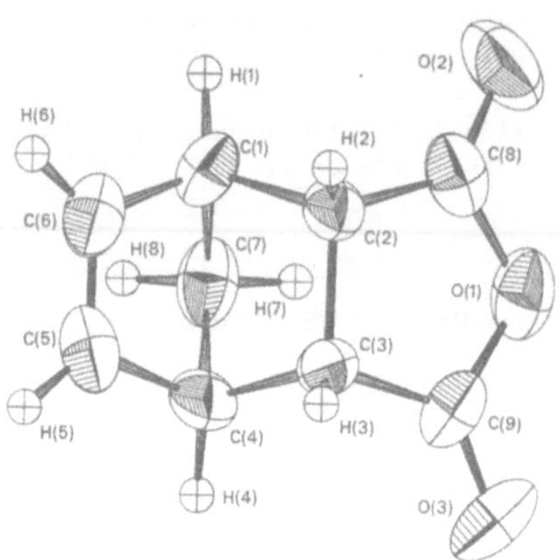

Fig. 1. The molecule of bicyclo[2,2,1]hept-5-ene-2,3-exo-dicarboxylic anhydride.

The molecule (Fig. 1) has pseudo m symmetry. The bond lengths and angles are in good agreement with those found for the endo isomer (1) and with other bicyclo[2,2,1]heptene skeletons (e.g. the bridge-head angle C(1)-C(7)-C(4) is 93.5°). There are no unusual intermolecular contacts.

1. R. DESTRO, G. FILIPPINI, C. M. GRAMACCIOLI and M. SIMONETTA, 1969. Acta Cryst., B25, 2467; Structure Reports, 34B.

TRICYCLO[4,4,0,0^{2,8}]DEC-3-ENE-7,10-DIONE

$C_{10}H_{10}O_2$

C. S. GIBBONS and J. TROTTER, 1972. J. Chem. Soc., Perkin II, 737-739.

Monoclinic, $P2_1/n$, a = 6.806, b = 15.171, c = 7.667 Å, β = 92.18°, D_m = 1.32, Z = 4. Cu radiation, R = 0.046 for 984 reflexions.

Fig. 1. Molecular diagram for tricyclo[4,4,0,$0^{2,8}$]dec-3-ene-7,10-dione; the crystallographic numbering scheme does not correspond to the nomenclature in the title. The double bond is C(8)-C(9) here.

The molecular configuration is shown in Fig. 1. The six-membered ring containing the carbonyl groups is in the boat conformation, with the C(3)···C(6) distance decreased significantly by the single carbon bridge [2.57 calculated for an ideal boat form decreased to 2.27 Å]. The angle at the apex of the bridge, C(3)-C(7)-C(6) is 94.4°. These features are common to several compounds of the norbornyl type. All other angles at C(7) are in the range 110 - 114°. The six-membered ring containing the double bond is in a half-chair conformation, having atoms C(7)-(10) in a plane, with C(1) and C(6) on opposite sides of the plane, displaced from it by 0.31 and 0.64 Å respectively. Bond lengths agree well with expected values and intermolecular contacts correspond to van der Waals distances.

7-syn-ACETOXY-6-endo-DEHYDROXYBICYCLO[2,2,1]HEPTANE-2-endo-CARBOXYLIC ACID α-LACTONE

$C_{10}H_{12}O_4$

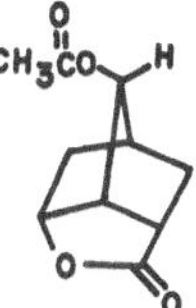

J. L. FLIPPEN, 1972. Acta Cryst., B28, 2046-2048.

Monoclinic, $P2_1/c$, a = 13.810, b = 6.206, c = 12.555 Å, β = 120.4, Z = 1. Cu radiation, R = 0.051 for 1418 reflexions.

In the acetoxylactone molecule (Fig. 1) the planar acetoxyl group is trans to the lactone moiety. The three five-membered rings in the bicyclo system are all in the envelope conformation while the six-membered ring C(2)C(3)C(7)-C(6)C(5)C(8) consists of two planes having atoms C(3) and C(5) in common with a dihedral angle of 72° between them. The C(3)-C(4)-C(5) angle, 96.2(2)°, and the C(2)-C(3)-C(7) angle, 97.8(2)°, are much less than tetrahedral as is not unusual in bridgehead situations. The molecules are held together by van der Waals forces and form interlocking chains such that the ester groups from neighbouring chains face each other. The lactone moieties from adjacent chains also face each other and the C=O bonds on the lactone ring are essentially parallel.

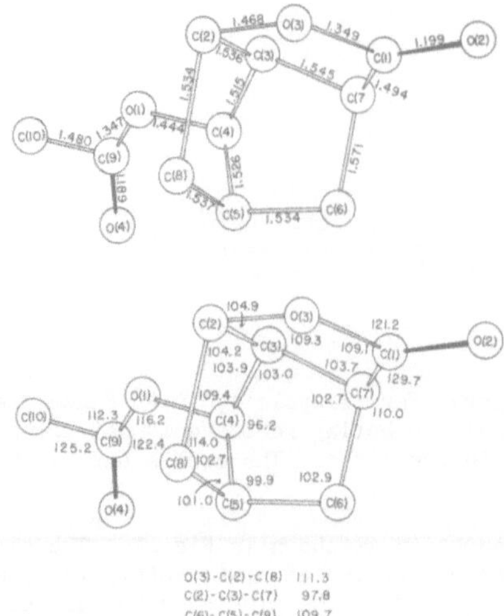

Fig. 1. Bond distances and angles for the acetoxylactone molecule (σ = 0.004 Å and 0.2°).

3,4,5,6,6,7-HEXACHLOROPENTACYCLO[6,4,$0^{2,10}$,$0^{3,7}$,$0^{5,9}$]DODEC-11-ENE

(PHOTOALDRIN)

$C_{12}H_8Cl_6$

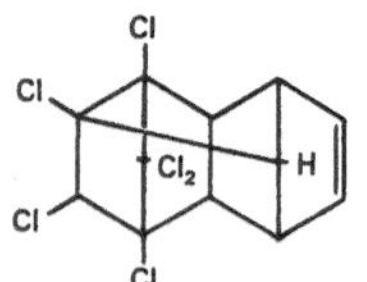

A. A. KHAN, W. H. BAUR and M. A. Q. KHAN, 1972. Acta Cryst., B28, 2060-2065.

Monoclinic, $P2_1/n$, a = 9.344, b = 15.756, c = 8.952 Å, β = 91.31°, D_m = 1.83, Z = 4. Ag radiation, R = 0.045 for 2787 reflexions.

The photoaldrin molecule (Fig. 1) is composed of two fused bicycloheptane rings bridged by one C-C bond, resulting in a cage structure consisting of six- and five-membered rings whose bond distances are given in Fig. 2. The mean C-Cl bond is 1.771(8) Å and the 13 single C-C bonds exhibit a wide variation (1.51 - 1.62 Å, mean 1.568 Å), but bond lengthening is not uncommon in cage molecules. An approximate correlation has been observed between the closest approach of the nonbonded atoms X-X and the lengthening of the C-C bond in an X-C-C-X arrangement as a result of repulsive interaction. The longest bond, C(3)-C(4), has, in addition to a C···C intramolecular contact of 2.24 Å, the closest intramolecular Cl···Cl contact (3.17 Å). All the C-C-Cl angles (mean 115.1°) are larger than tetrahedral; the constraints imposed by the polycyclic nature of the cage molecule result in smaller C-C-C angles (mean 102.6°).

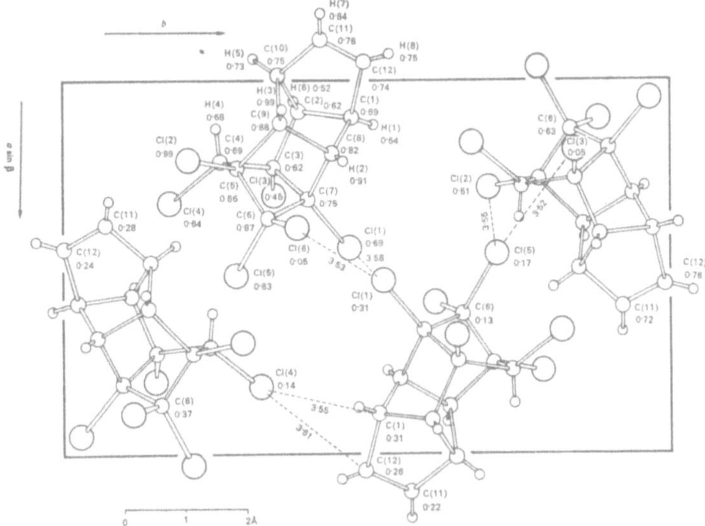

Fig. 1. The view of photoaldrin parallel to [001].

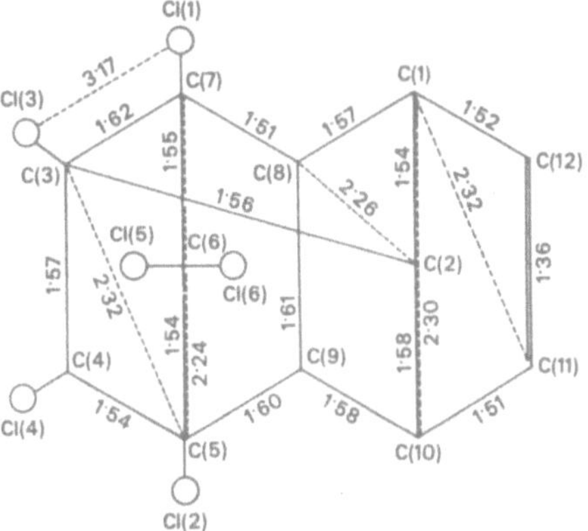

Fig. 2. Schematic diagram of the photoaldrin molecule showing the C-C bonds and some nonbonded intramolecular distances (σ = 0.006 Å and 0.4°).

5,8-DIMETHANONAPHTHALENE INSECTICIDES

T. P. DeLACY and C. H. L. KENNARD, 1972. J. Chem. Soc., Perkin II, 2153-2158.

ENDRIN
(1,2,3,4,10,10-HEXACHLORO-6,7-EPOXY-1,4,4a,5,6,7,8,8a-OCTAHYDRO-endo-1,4-endo-5,8-DIMETHANONAPHTHALENE)

$C_{12}H_8C\ell_6O$

Orthorhombic, $P2_12_12_1$, a = 15.304, b = 11.486, c = 8.202 Å, D_m = 1.75, Z = 4. Mo radiation, R = 0.044 for 753 reflexions.

ALDRIN
(1,2,3,4,10,10-HEXACHLORO-1,4,4a,5,8,8a-HEXA-HYDRO-endo-1,4-exo-5,8-DIMETHANONAPHTHALENE)

$C_{12}H_8Cl_6$

Monoclinic, $P2_1/n$, a = 10.806, b = 14.743, c = 8.992 Å, β = 93.03°, D_m = 1.73, Z = 4. Mo radiation, R = 0.048 for 1080 reflexions.

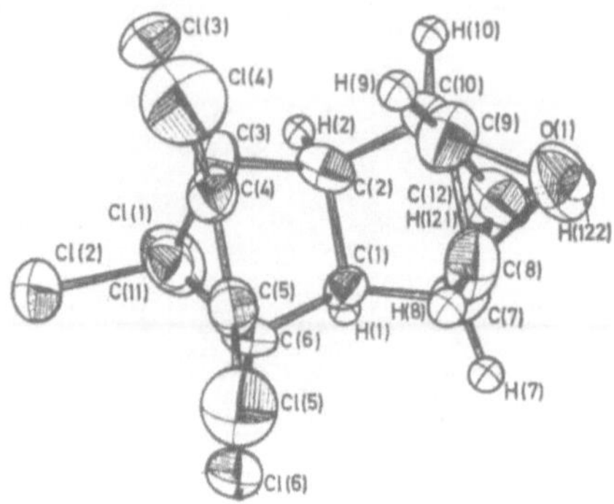

Fig. 1. The structure of endrin viewed perpendicular to the plane of the gross molecule.

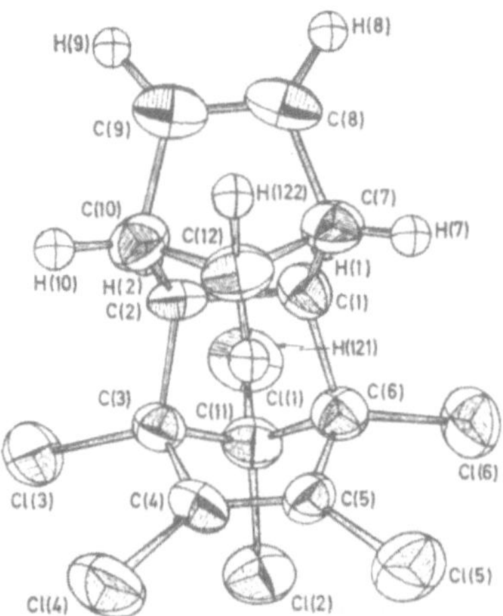

Fig. 2. The structure of aldrin viewed perpendicular to the plane of the gross molecule.

Endrin (Fig. 1) has six-membered rings in the endo-endo configuration with the epoxide exo to its adjoining ring. The molecule has approximately *m*-symmetry. The aldrin molecule is shown in Fig. 2; the six-membered rings have the endo-exo configuration and this molecule also has approximate *m*-symmetry.

The hexachloronorbornene moieties in both the endrin and aldrin molecules resemble each other closely. Most interatomic distances are in the expected range except that the olefinic Cℓ-C distances, Cℓ(4)-C(4) and Cℓ(5)-C(5) (1.694, 1.695, 1.706, and 1.703(5) Å) are less than the normal value (1.72 Å). The different geometrical configurations in endrin and aldrin and the effect of the epoxide oxygen in endrin produce small differences in bond lengths and angles in the unchlorinated section of the two molecules consistent with the presence of the epoxide ring. In both crystal structures the molecules are held together by van der Waals forces.

11-CHLORO-3,8-METHANO-[11]ANNULENONE

$C_{12}H_9C\ell O$

R. L. BEDDOES and O. S. MILLS, 1972. Israel J. Chem., 10, 485-492.

Monoclinic, $P2_1/c$, a = 6.854, b = 7.681, c = 19.324 Å, β = 108.54°, D_m = 1.41, Z = 4. Mo radiation, R = 0.045 for 1293 reflexions.

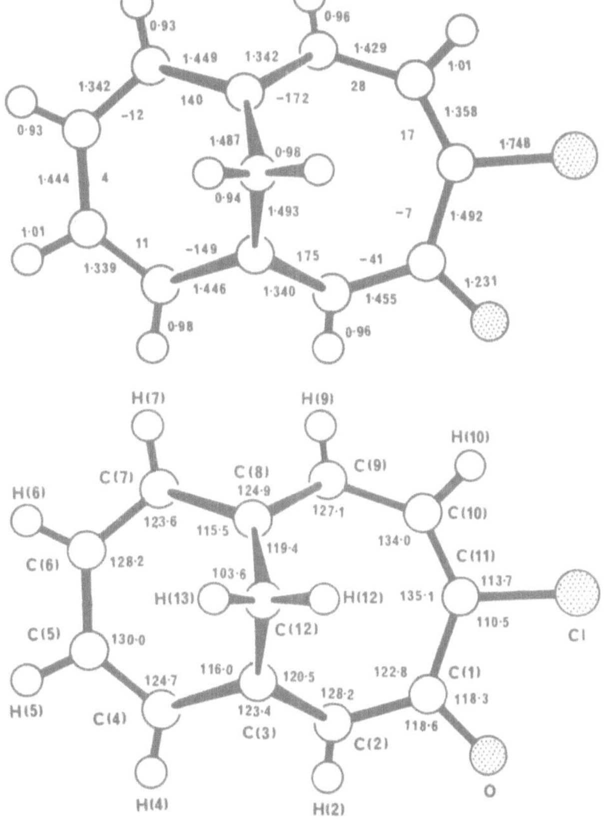

Fig. 1. Bond lengths (σ = 0.003 - 0.004 Å), torsion angles and bond angles (σ = 0.2 - 0.3°) in the methano-[11]annulenone derivative.

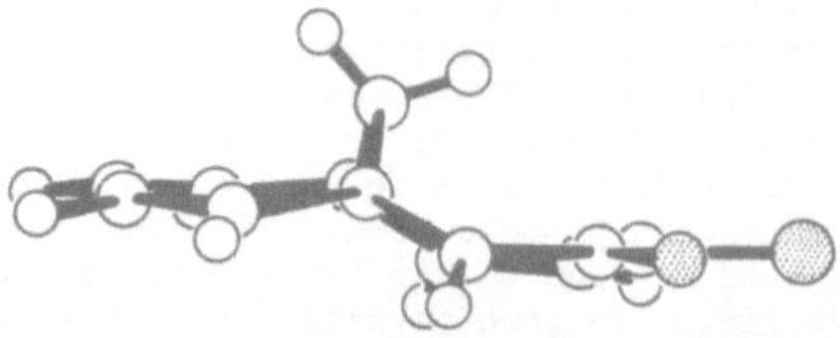

Fig. 2. Elevation of the 11-chloro-3,8-methano-[11]annulenone molecule showing the non-planarity of the 11-ring.

Details of the molecular geometry are given in Fig. 1. The molecule exhibits alternation of bond lengths but the data are consistent with some slight π-electron delocalisation. The non-planarity of the molecule is illustrated in Fig. 2. There are no abnormally short intermolecular contacts.

2,3,4-TRICHLOROPENTACYCLO[6,4,0,$0^{2,4}$,$0^{5,9}$]DODECA-6,11-DIENE

$C_{12}H_9C\ell_3$

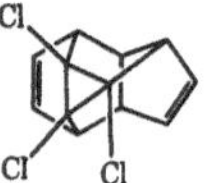

W. L. MOCK, C. M. SPRECHER, R. F. STEWART and M. G. NORTHOLT, 1972. J. Amer. Chem. Soc., 94, 2015-2020.

Orthorhombic, Pcab, a = 25.56, b = 12.148, c = 7.234 Å, D_m = 1.539, Z = 8. Mo radiation, R = 0.157 for 520 reflexions.

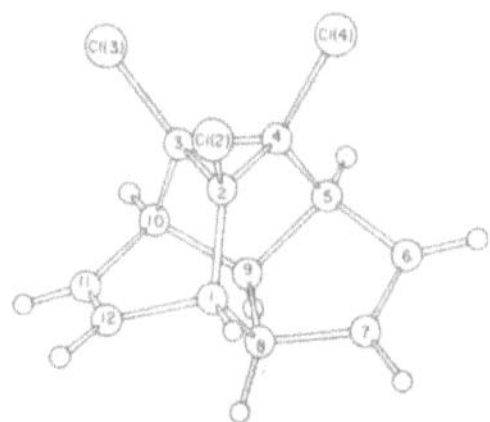

Fig. 1. Perspective drawing of the $C_{12}H_9C\ell_3$ molecule.

The molecular structure is shown in Fig. 1. The crystals suffered radiation damage during diffraction measurements (hence the relatively high R-factor) but the bond lengths are sufficient to establish the structure. There are no unusual intermolecular contacts.

12-METHYL-11,13-DIOXO-12-AZA-PENTACYCLO[$4,4,3,0,^{1,6}0^{2,10},0^{5,7}$]TRIDECA--3,8-DIENE

$C_{13}H_{11}O_2N$

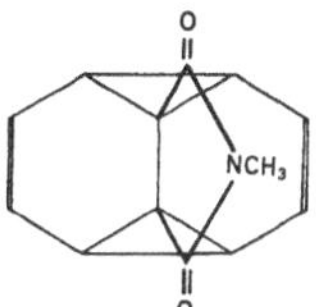

K. J. HWANG, J. DONOHUE and C.-C. TSAI, 1972. Acta Cryst., B28, 1727-1732.

Orthorhombic, Fdd2, a = 27.289, b = 13.108, c = 9.165 Å, Z = 8. Cu radiation, R = 0.046 for 377 reflexions.

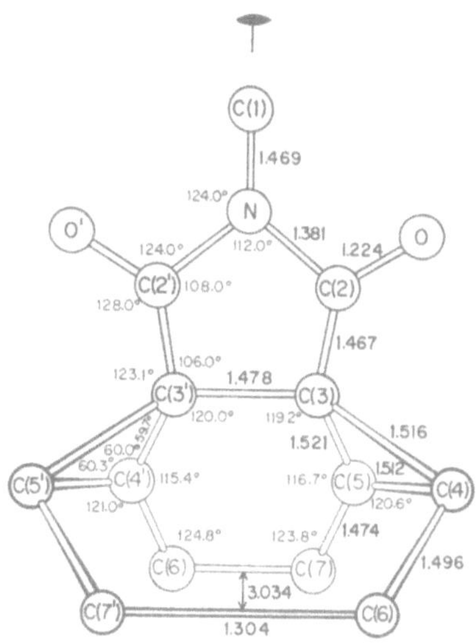

Fig. 1. Bond lengths (Å) and angles (°) in the pentacyclo[$4,4,3,0^{1,6}0^{2,10}0^{5,7}$]-trideca-3,8-diene system corrected for thermal motion. Average standard deviations are 0.008 Å and 0.5°.

The shape of the molecule, the first stable derivative of tetracyclo-[$4,4,0,0^{2,10},0^{5,7}$]deca-3,8-diene, is shown in Fig. 1. The shortening of the C-C bonds is explained by conjugation of the C=O and C=C bonds with the three-membered rings. With only 8 molecules in space group Fdd2, the molecules lie on twofold symmetry axes. The molecules are held together by van der Waals forces.

TETRACYCLO[7,2,1,$1^{5,8}0^{4,9}$]TRIDECAN-2,10-DIONE

$C_{13}H_{16}O_2$

M. PRZYBYLSKA, 1972. Acta Cryst., B28, 2814-2820.

Monoclinic, C2/c, a = 16.214, b = 6.194, c = 24.198 Å, β = 119.06°, D_m = 1.26, Z = 8. Cu radiation, R = 0.042 for 1563 reflexions.

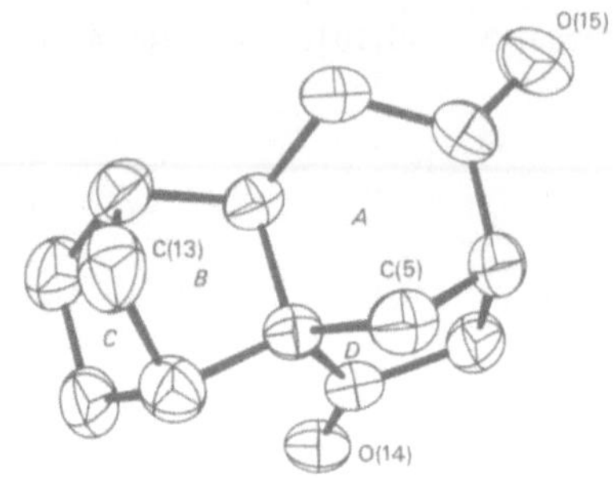

Fig. 1. Perspective view of the tetracyclic diketone $C_{13}H_{16}O_2$.

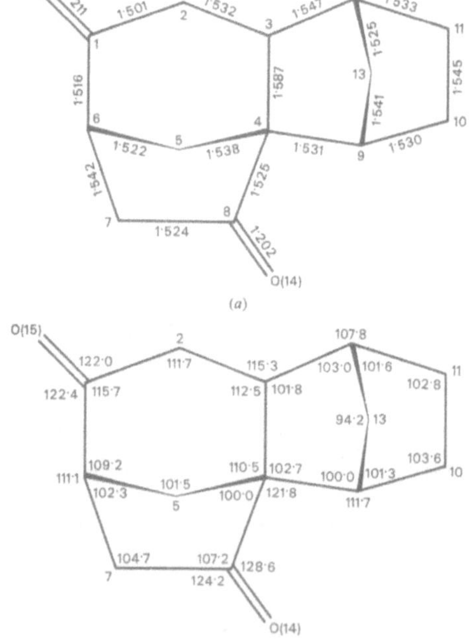

Fig. 2. Bond lengths (σ = 0.003 Å) and angles σ = 0.2° in the tetracyclic diketone $C_{13}H_{16}O_2$.

The molecular conformation is shown in Fig. 1. The three five-membered rings are envelope shaped and the torsion angles in the norbornane system show the presence of a synchro-twist which appears to be caused by substitution at C(3) and C(4) (for crystallographic numbering scheme see Fig. 2). Some of the bond lengths and angles (Fig. 2) are significantly different from usual values, e.g. C(5)-C(6) and C(12)-C(13) are shorter and C(3)-C(4) is longer than the normal C(sp^3)-C(sp^3) distance, 1.537(5) Å. The C(4)-C(8)-O(14) angle is considerably larger than the others and the increase (to 128.6°) appears due to repulsion by the hydrogen at C(10). The molecules are held together in the crystal by electrostatic attractions between carbonyl groups.

2-exo-NORBORNANOL p-TOLUENESULPHONATE

$C_{14}H_{18}O_3S$

C. ALTONA and M. SUNDARALINGAM, 1972. Acta Cryst., B28, 1806-1816.

Triclinic, $P\bar{1}$, a = 6.181, b = 11.092, c = 11.060 Å, α = 111.4, β = 103.6, γ = 94.8°, D_m = 1.309, Z = 2. Cu radiation, R = 0.041 for 2189 reflexions.

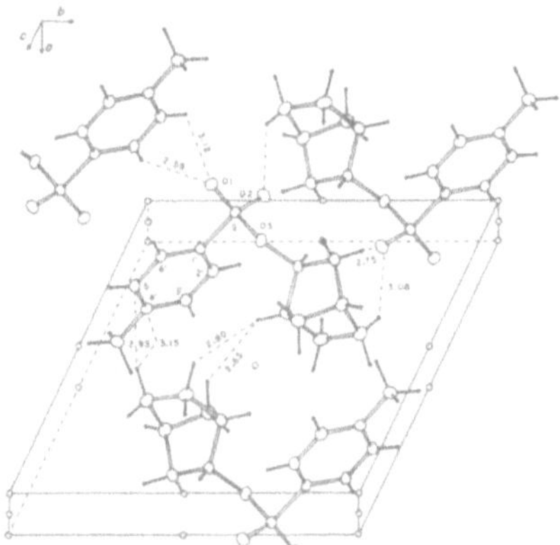

Fig. 1. View of the crystal structure of 2-exo-norbornanol p-toluenesulphonate showing intermolecular contacts.

There is partial disorder in this crystal structure. Atoms C(1), C(3), C(4), and C(6) of the norbornane group form a second enantiomeric cage, sharing the "non-disordered" atoms C(2), C(5), and C(7) with the original molecule. The crystal packing (Fig. 1) appears sufficiently loose to permit a statistical disruption of the regular arrangement of D and L molecules, about 24% of L being replaced by D and vice-versa. Bond distances are C-C(average cage) = 1.527(5), S=O(average) = 1.425 ± 2, S-O(3) = 1.573(2), S-C(1') = 1.758(2), and O(3)-C(2) = 1.473 Å.

[2.2]METACYCLOPHANE-1,9-DIENE

$C_{16}H_{12}$

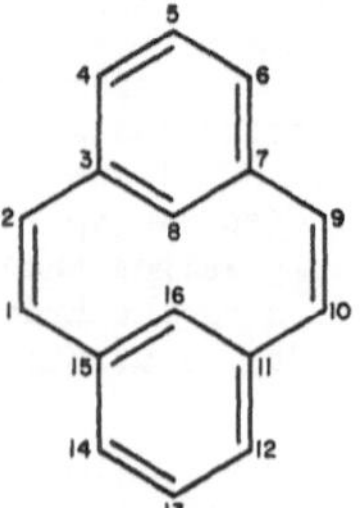

A. W. HANSON and M. RÖHRL, 1972. Acta Cryst., B28, 2032-2037.

Monoclinic, $P2_1/a$, a = 13.25, b = 5.640, c = 7.350 Å, β = 95.39°, D_m = 1.24, Z = 2. Cu radiation, R = 0.054 for 833 reflexions.

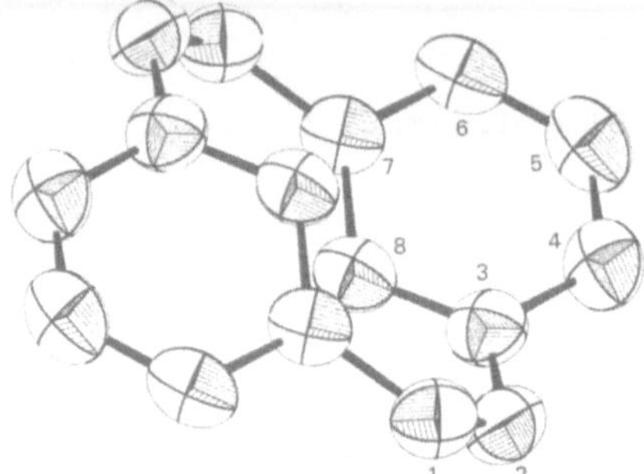

Fig. 1. The molecular configuration of [2.2]metacyclophane-1,9-diene.

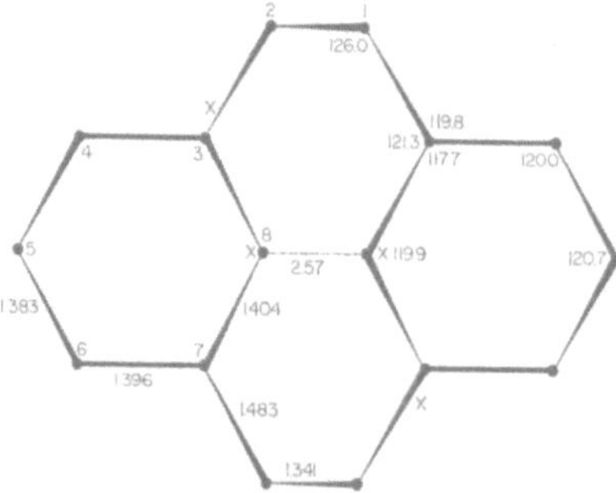

Fig. 2. Mean bond distances and angles for [2.2]metacyclophane-1,9-diene (σ = 0.003 Å and 0.2°).

The strained molecule (Fig. 1) has fairly precise 2/m symmetry although only an inversion centre is demanded by the space group, and all bond lengths (Fig. 2) are close to their normal values, with C(8) and C(8') constrained to lie only 2.57 Å apart. The mutual repulsion of these atoms causes severe boat deformation of the phenyl rings which are necessarily parallel but are displaced stepwise to avoid too close mutual contact. There is no indication of any elongation of the double bonds in the inter-phenyl bridges (C=C = 1.341(3) Å). Intermolecular contacts are normal.

PARACYCLOPHANE DERIVATIVES

H. HOPE, J. BERNSTEIN, and K. N. TRUEBLOOD, 1972. Acta Cryst., B28, 1733-1743.

[2.2]PARACYCLOPHANE

$C_{16}H_{16}$

Tetragonal, $P4_2/mnm$, a = 7.781, c = 9.290 Å, Z = 2. Cu radiation, R = 0.029 for 360 reflexions.

1,1,2,2,9,9,10,10-OCTAFLUORO-[2.2]PARACYCLOPHANE

$C_{16}H_8F_8$

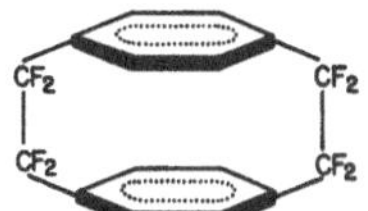

Monoclinic, $P2_1/n$, a = 7.994, b = 7.986, c = 10.855 Å, β = 97.84°, D_m = 1.704, Z = 2. Mo and Cu radiation, R = 0.037 for 956 reflexions.

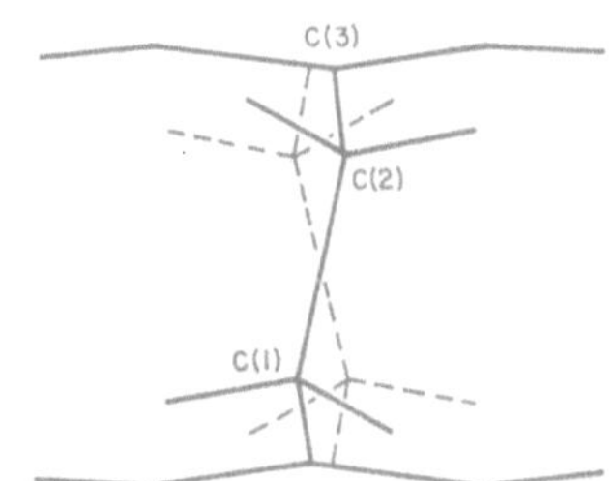

Fig. 1. End view of [2.2]paracyclophane illustrating the effect of disorder and the bending of the aromatic hydrogen atoms below the ring plane.

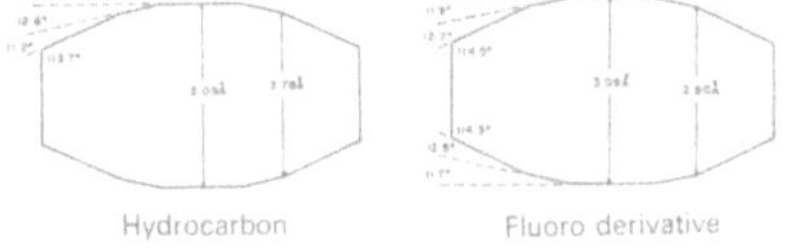

Fig. 2. Molecular profiles of the $C_{16}H_{16}$ and $C_{16}H_8F_8$ paracyclophane derivatives.

The results of the analyses reported here indicate clearly that the best description involves disordered methylene (Fig. 1) or difluoromethylene groups. The disorder is the same in each compound. Final refinement was carried out with each molecule statistically disordered to simulate a dynamic disorder that occurs by a twist of the aromatic rings in opposite directions about their common normal. The amplitude of this motion is about 3° for each molecule. In both compounds the bridge C-C distance is about 1.59 Å (1.57 - 1.58 Å after libration correction). The average C-F distance is 1.35 Å and the average F-C-F angle is 106°. The planes of the four unsubstituted C atoms of each ring are about 3.09 Å apart (Fig. 2). The para carbon atoms are bent 12° out of the plane of the other four atoms of the aromatic ring.

HEXACYCLO[10,3,1,$0^{2,10}$,$0^{3,7}$,$0^{6,15}$,$0^{9,14}$]HEXADECANE

$C_{16}H_{22}$

S. T. RAO and M. SUNDARALINGAM, 1972. Acta Cryst., B28, 694-699.

Monoclinic, $P2_1/c$, a = 9.151, b = 6.510, c = 19.733 Å, β = 98.62°, D_m = 1.220, Z = 4. Cu radiation, R = 0.044 for 1615 reflexions.

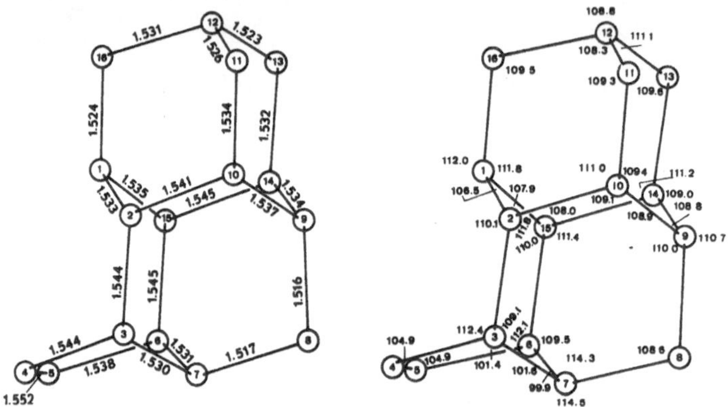

Fig. 1. Bond lengths (σ = 0.002 Å) and bond angles (σ = 0.13°) in the carbon skeleton of the hexacyclohexadecane $C_{16}H_{22}$.

The molecular geometry is summarised in Fig. 1. The cyclopentane ring is in an envelope form. The main carbon skeleton is not significantly twisted; the "twists" around the axes C(2)···C(15), C(10)···C(14), and C(3)···C(6) are less than 0.1°. There are no unusually short contacts in the crystal structure.

1,6:8,13-PROPANE-1,3-DIYLIDENE[14]ANNULENE

$C_{17}H_{14}$

A. GAVEZZOTTI, A. MUGNOLI, M. RAIMONDI and M. SIMONETTA, 1972. J. Chem. Soc., Perkin II, 425-430.

Orthorhombic, Fdd2, a = 19.4549, b = 36.358, c = 6.5716 Å, D_m = 1.248, Z = 16. Cu radiation, R = 0.062 for 826 reflexions.

Fig. 1. Structure and numbering scheme for 1,6:8,13-propane-1,3-diylidene[14]annulene.

The molecule (Fig. 1) has close to mm2 symmetry in the crystal structure. In spite of the distortion imposed by the propane bridge the annulene ring retains its aromatic character; the C-C bond distances along the annulene perimeter are in the range 1.381 - 1.417 Å with a mean value 1.401 Å. Other important mean distances (assuming mm2 symmetry) are C(1)-C(15) type = 1.517, C(15)-C(16) type = 1.518 Å.

[2.2.2](1,3,5)CYCLOPHANE-1,9,17-TRIENE

$C_{18}H_{12}$

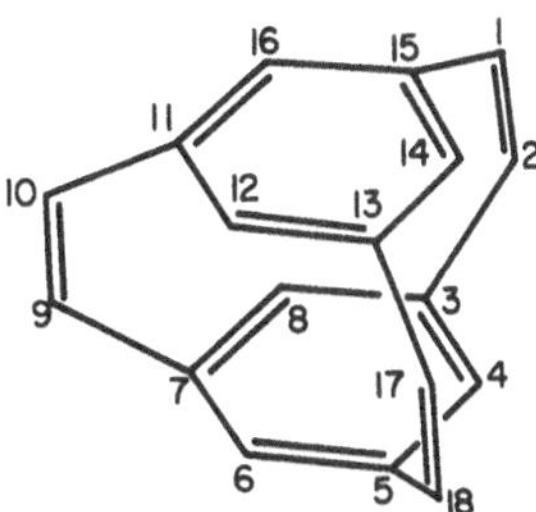

A. W. HANSON and M. RÖHRL, 1972. Acta Cryst., B28, 2287-2291.

Triclinic, $P\bar{1}$, a = 7.332, b = 11.663, c = 7.224 Å, α = 87.73, β = 100.24, γ = 105.52°, D_m = 1.29, Z = 2. Cu radiation, R = 0.043 for 1708 reflexions.

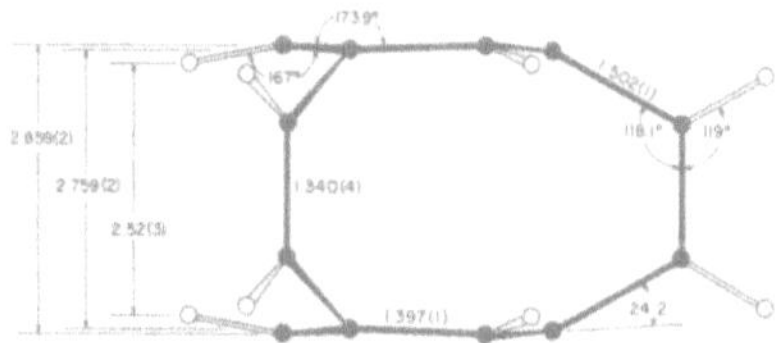

Fig. 1. An idealised molecule of [2.2.2](1,3,5)cyclophane-1,9,17-triene viewed normal to one of the mirror planes. Open circles are hydrogen atoms.

The molecule has fairly exact noncrystallographic $\bar{6}m2$ symmetry with the phenyl rings eclipsing each other with a mean interplanar spacing of 2.809 Å. The bond distances are normal (with mean values C=C = 1.340(4), C-C(exocyclic) = 1.502(1), and C-C(phenyl) = 1.397(1) Å) but some strain is evident in the angles (Fig. 1). The single bonds of the bridges are severely bent out of the ring planes; the average angle between a single bond and the plane of the three nearest phenyl-ring atoms is 24.2°. The phenyl rings themselves are chair-shaped, with successive atoms lying 0.024 Å above and below the plane of the remaining four.

1,6:8,13-BUTANE-1,4-DIYLIDENE[14]ANNULENE

$C_{18}H_{16}$

C. M. GRAMACCIOLI, A. MUGNOLI, T. PILATI, M. RAIMONDI and M. SIMONETTA, 1972. Acta Cryst., B28, 2365-2370.

Orthorhombic, Fmm2, a = 18.034, b = 11.399, c = 6.059 Å, D_m = 1.22, Z = 4. Cu radiation, R = 0.045 for 407 reflexions.

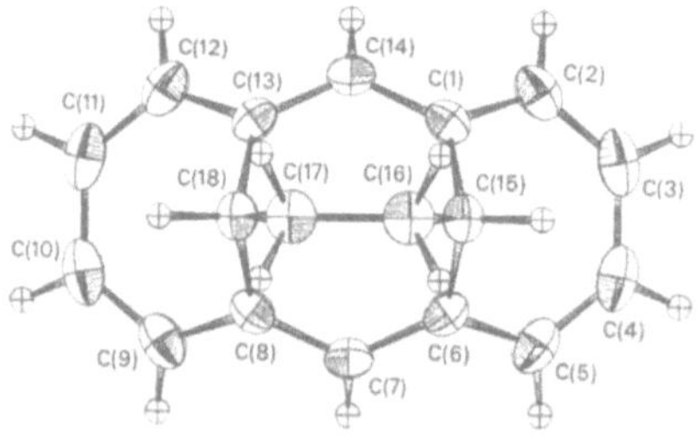

Fig. 1. The 1,6:8,13-butane-1,4-diylidene[14]annulene structure.

The molecule (Fig. 1) has crystallographic mm2 symmetry. The independent C-C distances around the annulene perimeter are C(1)-C(2) = 1.428, C(2)-C(3) = 1.368, C(3)-C(4) = 1.424, C(1)-C(4) = 1.402 Å, and are consistent with an assignment of "aromatic" character to the ring system. The molecular symmetry requires that the adjacent CH_2 groups be eclipsed. The molecules are separated by normal van der Waals distances.

exo-1α,7α,2β,6β-11,11-DIMETHYLTRICYCLO[5,4,0,0^{2,6}]UNDECAN-8-OL p-BROMOPHENYLURETHANE DERIVATIVE

$C_{20}H_{26}BrNO_2$

R. M. BOWMAN, C. CALVO, J. J. McCULLOUGH, P. W. RASMUSSEN and F. F. SNYDER, 1972. J. Org. Chem., 37, 2084-2091.

Monoclinic, $P2_1/c$, a = 13.00, b = 5.597, c = 27.20 Å, β = 109.46°, D_m = 1.40, Z = 4. Mo radiation, R = 0.10 for 610 reflexions.

The analysis was undertaken to establish the stereochemistry at the ring junctions; the five- and six-membered rings have the anti relationship and the two ring junctions are cis. Bond lengths have standard deviations in the range 0.03 - 0.06 Å and must be considered 'normal'.

(+)-2,5-DIMETHOXY-7-DIMETHYLAMINOTRIPTYCENE HYDROBROMIDE

$C_{24}H_{24}BrNO_2$ $[C_{24}H_{24}NO_2]^+Br^-$

N. SAKABE, K. SAKABE, K. OZEKI-MINAKATA and J. TANAKA, 1972. Acta Cryst., B28, 3441-3446.

Orthorhombic, $C222_1$, a = 8.733, b = 13.853, c = 34.728 Å, D_m = 1.39, Z = 8. Cu radiation, R = 0.044 for 1302 reflexions.

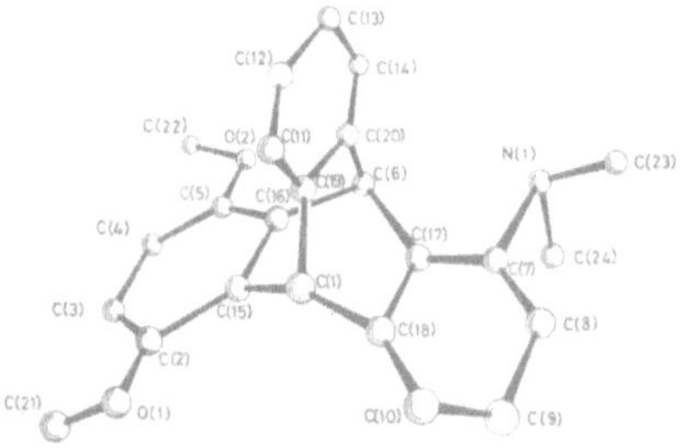

Fig. 1. The absolute configuration of (+)-2,5-dimethoxy-7-dimethylamino-triptycene hydrobromide.

The absolute configuration of (+)-2,5-dimethoxy-7-dimethylaminotriptycene has been determined as 1R, 6S. The structure (Fig. 1) has approximate three-fold symmetry. The angles subtended at the central carbon atoms C(1) and C(6), are all 4(1)° less than the tetrahedral angle, evidence of strain at these atoms. All other bond lengths and angles are normal. The nitrogen atom of the dimethylamino group is hydrogen bonded to the bromide (NH···Br = 3.16 Å); no other intermolecular contact is shorter than 3.5 Å.

SYMMETRICAL CEDRONE DIMETHYLFORMAMIDE SOLVATE

$C_{24}H_{34}N_2O_8$

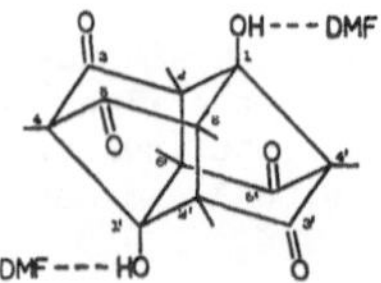

J.A. BEISLER and J.V. SILVERTON, 1972. Acta Cryst., B28, 298-304.

Monoclinic, $P2_1/c$, a = 9.322, b = 8.250, c = 15.804 Å, β = 101.86°, D_m = 1.34, Z = 2. Mo radiation, R = 0.058 for 1730 reflexions.

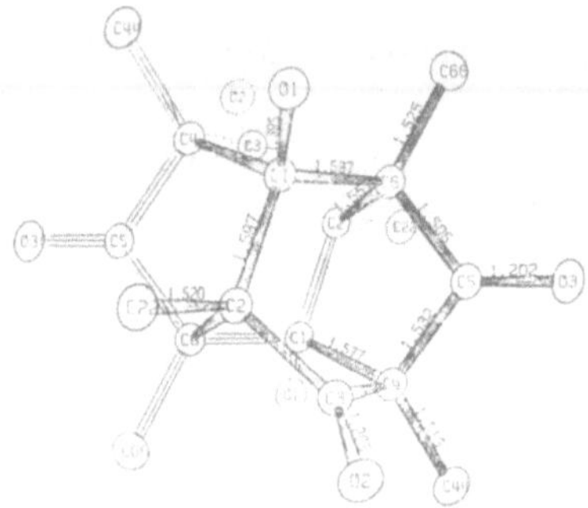

Fig. 1. Molecular skeleton and bond lengths (σ = 0.003 Å) of symmetrical cedrone.

Symmetrical cedrone, obtained by dimerization of trimethylphloroglucinol, has been shown to be a centrosymmetric pentacyclic structure in the crystal of its dimethylformamide solvate and has a considerable number of very long C-C bonds (Fig. 1). The molecule forms two strong hydrogen bonds (2.668(3) Å) to the dimethylformamide molecules and is possibly stabilized by these bonds. In the crystal structure discrete cedrone·2DMF units are separated by van der Waals distances.

DI-p-BROMOBENZOATE OF A NORBORNADIENE DIMER

$C_{28}H_{26}Br_2O_4$

N. ACTON, R.J. ROTH, T.J. KATZ, J.K. FRANK, C.A. MAIER and I.C. PAUL, 1972. J. Amer. Chem. Soc., 94, 5446-5456.

Monoclinic, $P2_1/c$, a = 14.991, b = 10.317, c = 19.236 Å, β = 123°45', D_m = 1.51, Z = 4. Cu radiation, R = 0.085 for 2759 reflexions.

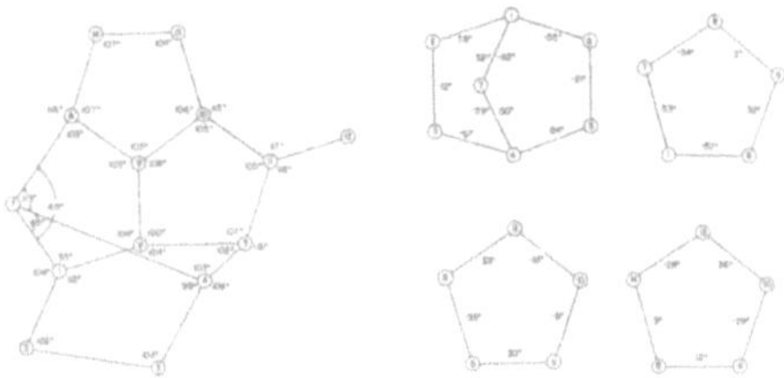

Fig. 1. (a) Bond angles in the fused ring system of the norbornadiene dimer derivative, σ = 0.8 - 1.2°. (b) Torsion angles in some of the constituent rings. The angles shown on the C(1)-C(2), C(3)-C(4), C(1)-C(6), and C(4)-C(5) bonds do not involve the bridge atom C(7).

The dimer from which the derivative was made is one of many complex molecules obtained by dimerisation of norbornadiene. The C-C-C bond angles in the fused ring system and the torsion angles around many of the bonds are shown in Fig. 1. These dimensions indicate the highly strained and unsymmetrical nature of the hydrocarbon moiety. The "norbornane" skeleton (C(1)-C(7)) adopts a synchro-twist, with quite large torsion angles about the C(2)-C(3) and C(5)-C(6) bonds. Two of the cyclopentane rings in the molecule [C(8), C(9), C(10), C(13), C(14) and C(5), C(6), C(9), C(10), C(11)] have distorted half-chair conformations, while the ring [C(1), C(6), C(9), C(8), C(7)] has an envelope conformation.

1-ACETYL-4-BROMOPYRAZOLE

$C_5H_5N_2OBr$

J. LAPASSET, A. ESCANDE and J. FALGUEIRETTES, 1972. Acta Cryst., B28, 3316-3321.

Monoclinic, $P2_1/c$, a = 7.727, b = 17.57, c = 7.230 Å, β = 137.6°, D_m = 1.92, Z = 4. Cu radiation, R = 0.067 for 1063 reflexions.

The X-ray results confirm that the molecule (Fig. 1) adopts the conformation predicted by an n.m.r. study. The molecule consists of two planes of atoms, the first comprising the pyrazole ring and the bromine atom and the second including the acetyl group; the angle between these planes is 3.3°. In the crystal structure one intermolecular contact distance Br···O(7) = 3.19 Å is shorter than the van der Waals distance.

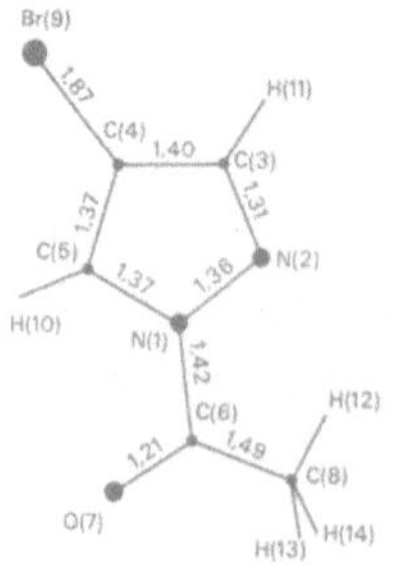

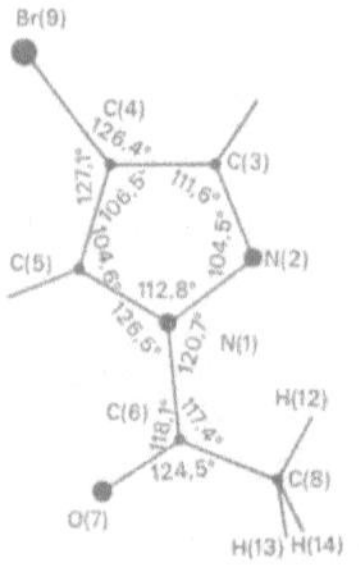

Fig. 1. Bond lengths and angles for 1-acetyl-4-bromopyrazole (σ = 0.024 Å and 1°).

1,3-DIMETHYLIMIDAZOLE-2(3H)-THIONE

$C_5H_8N_2S$

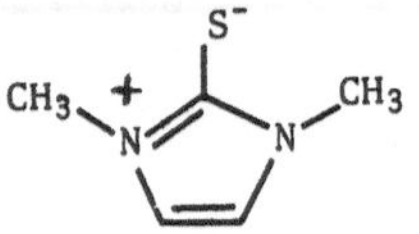

G.B. ANSELL, 1972. J. Chem. Soc., Perkin II , 841-843.

Orthorhombic, $Bm2_1b$, a = 8.475, b = 6.826, c = 11.206 Å, D_m = 1.29, Z = 4. Cu radiation, R = 0.027 for 195 reflexions. The space group may also be Bmmb and in this space group the structure refines to R = 0.028.

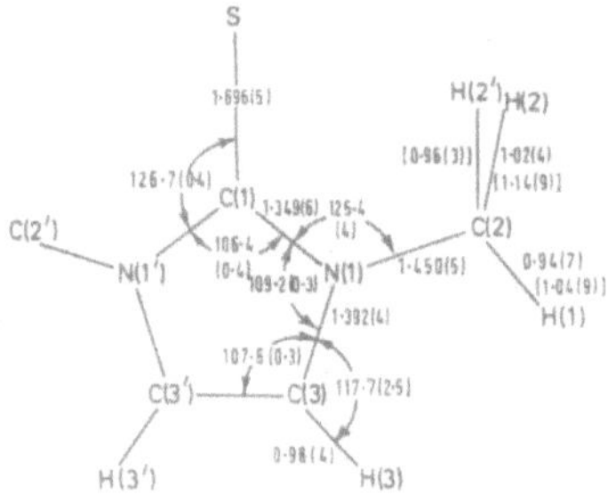

Fig. 1. Bond lengths and angles (with estimated errors) within the molecule for Bmmb refinement. For the $Bm2_1b$ refinement, bond lengths vary by amounts smaller than the estimated standard deviations for Bmmb except for those shown in square brackets. The C(3)-C(3') distance, 1.325 Å.

In space group Bmmb the molecule has crystallographic mm symmetry. The bond distances and angles (Fig. 1) indicate that the formulation shown above is a major contributor to the ground state structure. The planar molecules are stacked along the b-axis with a 3.41 Å perpendicular separation and no unusual intermolecular contacts. By contrast in space group $Bm2_1b$ the molecule is only required to have m symmetry and the mirror plane is perpendicular to the plane of the five-membered ring. A pseudo-mirror plane is still found passing through the eight heavy atoms of the structure. Bond lengths are almost identical to those found in the Bmmb refinement.

DIPROTONATED HISTAMINE TETRACHLOROCOBALTATE(II)

$C_5H_{11}C\ell_4CoN_3$ $C_5H_{11}N_3{}^{2+}\cdot CoC\ell_4{}^{2-}$

J.J. BONNET and Y. JEANNIN, 1972. Acta Cryst., B28, 1079-1085.

Orthorhombic, $Pna2_1$, a = 7.619, b = 21.401, c = 7.345 Å, D_m = 1.741, Z = 4. Mo radiation, R = 0.045 for 804 reflexions.

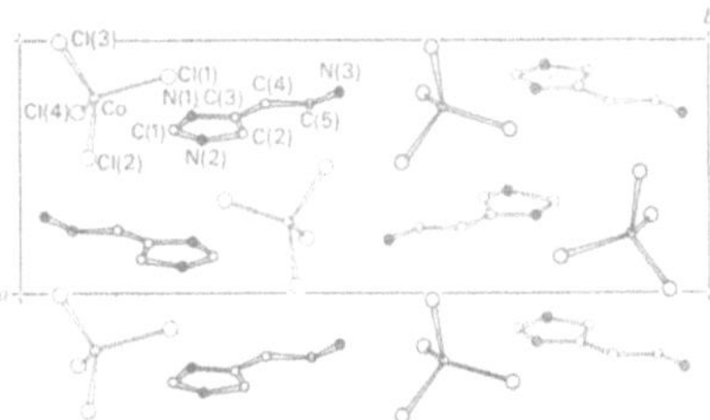

Fig. 1. Projection of the diprotonated histamine tetrachlorocobaltate(II) along c.

The crystal is formed from very distorted tetrahedral $CoC\ell_4{}^{2-}$ ions and almost planar diprotonated histamine cations (Fig. 1). The Co-Cℓ bond lengths vary from 2.233 to 2.295(3) Å, mean 2.273 Å; and the Cℓ-Co-Cℓ angles range from 106.2° to 115.3°, mean 109.5°. Two protons are attached to the histamine molecule through the nitrogen atoms of the cetimine group of the imidazole ring, and the primary amine group of the side chain. The atoms of the cation are localized in two planes, one containing the atoms of the imidazole ring, the other the atoms of the side chain. The dihedral angle between these two planes is 7°. Bond lengths in the cation are in accord with accepted values.

1,4-DINITRO-2-ISOPROPYLIMIDAZOLE

$C_6H_8N_4O_4$

R.S. GLASS, J.F. BLOUNT, D. BUTLER, A. PERROTTA and E.P. OLIVETO, 1972. Canad. J. Chem., 50, 3472-3477.

Monoclinic, $P2_1/c$, a = 15.488, b = 6.072, c = 10.030 Å, β = 99.35°, D_m = 1.34, Z = 4. Cu radiation, R = 0.044 for 1000 reflexions.

From the reaction of nitronium tetrafluoroborate with 1-trimethylsilyl-2-isopropylimidazole there was isolated a dinitro compound the structure of which was shown by X-ray crystallography to be 1,4-dinitro-2-isopropylimidazole. The imidazole ring is planar to within ±0.005 Å. The 1-nitro nitrogen (N(2)) is displaced 0.07 Å from the plane of the ring while N(4) and the isopropyl carbon bonded to the ring lie 0.04 and 0.06 Å, respectively, out of the plane on the opposite side from N(2). The nitro groups are both twisted slightly from the plane of the imidazole ring. The dihedral angles between the plane of the ring and the 1-nitro (N(2), O(1), O(2)) and 4-nitro (N(4), O(3), O(4)) groups are 7.3 and 10.0°, respectively.

1-(2-CHLOROETHYL)-3-(4-CARBAMOYLPYRAZOL-3-YL)-Δ^2-1,2,3-TRIAZOLIUM CHLORIDE

(NSC-82196)

$C_8H_{12}C\ell_2N_6O$

J.E. WHINNERY and W.H. WATSON, 1972. Acta Cryst., B28, 3635-3640.

Orthorhombic, Pbca, a = 14.151, b = 16.990, c = 10.344 Å, D_m = 1.489, Z = 8. Cu radiation, R = 0.050 for 1250 reflexions.

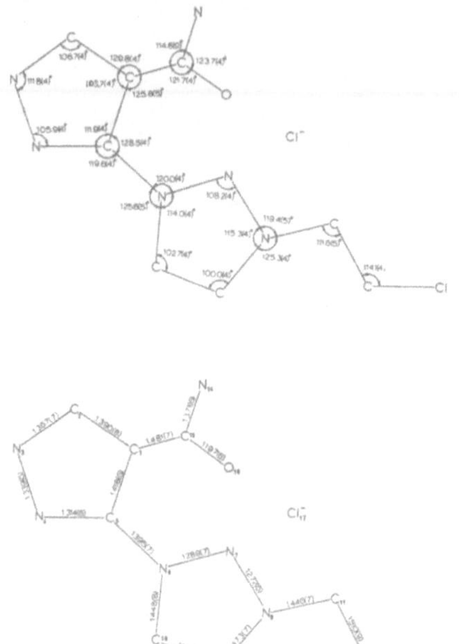

Fig. 1. Molecular geometry of 1-(2-chloroethyl)-3-(4-carbamoylpyrazol-3-yl)-Δ^2-1,2,3-triazolium chloride.

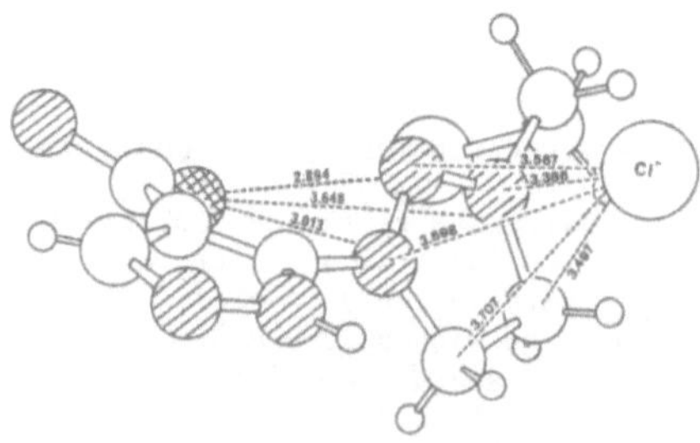

Fig. 2. Interactions between the triazolium ring, amide oxygen, and chloride ion of NSC-82196.

The bond lengths and angles are shown in Fig. 1. Both the triazolium and pyrazole rings are planar. The two planes are tilted at an angle of 50° and the plane of the amide group makes an angle of 22° with the pyrazole ring. A chloride ion is symmetrically placed above the triazolium ring (Fig. 2); the amide group interacts with the N-N-N group from the opposite side of the ring. Both the N(4)H and the N(3)H tautomers are present with the former in a four-fold excess. For the N(3)H tautomer the intermolecular H···Cℓ distance of 2.12 Å indicates a strong hydrogen bond. N(4)H is not involved in hydrogen bonding. Additionally, the amide nitrogen has a weak interaction with H(11) (N···H = 2.990 Å).

1-p-NITROPHENYL-3-METHYL-4-BROMOPYRAZOLE

$C_{10}H_8BrN_3O_2$

J. LAPASSET and J. FALGUEIRETTES, 1972. Acta Cryst., B28, 791-796.

Orthorhombic, $Pna2_1$, a = 12.63, b = 21.58, c = 3.88 Å, D_m = 1.78, Z = 4. Cu radiation, R = 0.044 for 1057 reflexions.

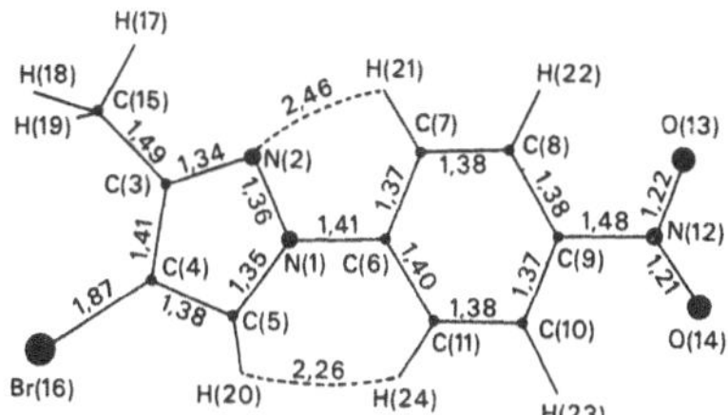

Fig. 1. Bond lengths (σ = 0.02 Å) in 1-p-nitrophenyl-3-methyl-4-bromopyrazole.

Bond lengths in this molecule (Fig. 1) are for the most part in accord with accepted values. The pyrazole ring and its substituent atoms are coplanar; the dihedral angle between the benzene ring plane and the pyrazole plane is 19.6° and between the benzene ring and the nitro group is 7.3°. In the crystal structure there is only one short intermolecular contact (3.14 Å between Br and a nitro oxygen atom).

ETHYL 4-ACETYL-3-ETHYL-5-METHYLPYRROLE-2-CARBOXYLATE

$C_{12}H_{17}NO_3$

R. BONNETT, M.B. HURSTHOUSE and S. NEIDLE, 1972. J. Chem. Soc., Perkin II, 902-906.

Monoclinic, $P2_1/n$, a = 6.772, b = 14.628, c = 12.187 Å, β = 94.20°, D_m = 1.24, Z = 4. Cu radiation, R = 0.062 for 1531 reflexions.

The bond lengths in the molecule are not at all similar to those found for pyrrole (1), but indicate the importance of mesomeric structures summarized in the formulation above. For example, the bond lengths in the pyrrole ring are N(1)-C(5) = 1.330(4), C(5)-C(4) = 1.404(4), C(4)-C(3) = 1.417(4), C(3)-C(2) = 1.382(4), C(2)-N(1) = 1.395(4) Å [numbering corresponding to chemical numbering scheme, not that adopted by the authors]. The pyrrole ring itself is planar but most of the substituent atoms deviate significantly from the ring plane. In the crystal structure molecules are linked into chains running approximately parallel to c by N-H···O (2.922 Å) hydrogen bonds.

1. L. NYGAARD, et al., 1969. J. Mol. Struct., 3, 491.

1-(2'-CARBOXYPHENYL)-2-ACETYL-5-METHYL-3-PYRAZOLONE

$C_{13}H_{12}N_2O_4$

G.D. ANDREETTI, G. BOCELLI, L. CAVALCA and P. SGARABOTTO, 1972. Gazz. Chim. Ital., 102, 106-116.

Monoclinic, $P2_1/c$, a = 10.04, b = 9.08, c = 14.67 Å, β = 111.5°, D_m = 1.38, Z = 4. Mo radiation, R = 0.059 for 2255 reflexions.

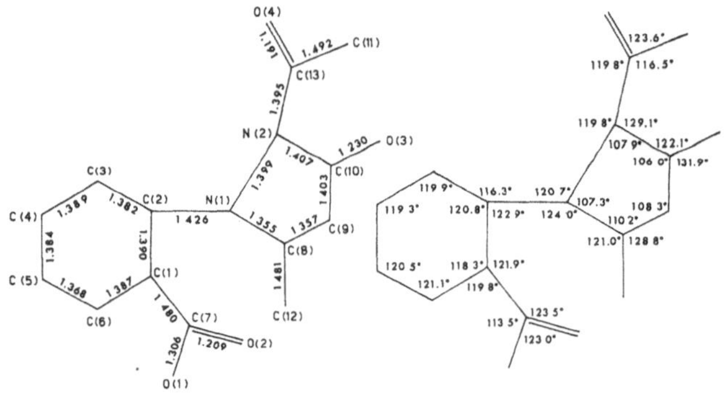

Fig. 1. Bond lengths (σ = 0.002 - 0.003 Å) and angles in 1-(2'-carboxyphenyl)-2-acetyl-5-methyl-3-pyrazolone.

The pyrazolone ring (Fig. 1) is slightly non-planar, the phenyl ring makes an angle of 73° with the pyrazolone plane, and the acetyl and carboxy groups are 20° and 27° respectively out of their ring planes. Bond lengths and angles (Fig. 1) indicate electron delocalization in the pyrazolone ring, but some non-planarity of bonds at the nitrogen atoms.

N-PHENYL-2-(p-BROMOPHENYL)-CYCLOPROPANE-1,3-DICARBOXIMIDE

$C_{17}H_{12}BrNO_2$

J.P. DECLERCQ, P. PIRET and M. van MEERSSCHE, 1972. Acta Cryst., B28, 328-330.

Monoclinic, $P2_1/a$, a = 16.062, b = 6.039, c = 17.484 Å, β = 118.63°, Z = 4. Mo radiation, R = 0.093 for 1191 reflexions.

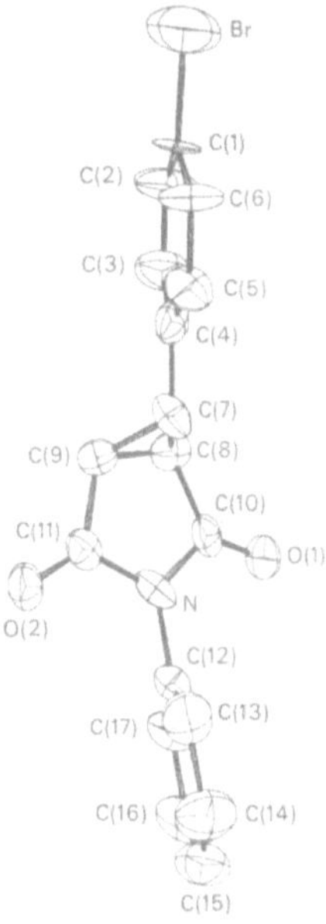

Fig. 1. Molecular configuration and numbering scheme in N-phenyl-2-(p-bromophenyl)-cyclopropane-1,3-dicarboximide.

The molecular structure is shown in Fig. 1. Bond lengths are normal with the exception of C(8)-C(9) = 1.44(2), C(10)-N = 1.40(2), and C(11)-N = 1.41(2) Å. The angle between the planes of the three- and five-membered rings is 74°.

1,3-BIS[4-(2-BROMOETHOXY)PHENYL]-5-TETRAZOLONE

$C_{17}H_{16}Br_2N_4O_3$

Y. IITAKA, S. UCHIYAMA and Z. TAMURA, 1972. Chem. Pharm. Bull. Japan, 20, 1181-1185.

Orthorhombic, $Pna2_1$, a = 32.8, b = 4.60, c = 12.3 Å, D_m = 1.731, Z = 4. Cu radiation, R = 0.16 for 111 $h0\ell$ reflexions.

The results of this two dimensional analysis establish that the material examined is correctly formulated as in the title.

1-(α-BENZOYLOXY-BENZYLIDENEAMINO)-4,5-DIMETHYL-1,2,3-TRIAZOLE

$C_{18}H_{16}N_4O_2$

H. BAUER, A.J. BOULTON, W. FEDELI, A.R. KATRITZKY, A. MAJID-HAMID, F. MAZZA and A. VACIAGO, 1972. J. Chem. Soc., Perkin II, 662-667.

Monoclinic, $P2_1/c$, a = 5.94, b = 10.76, c = 27.31 Å, β = 100.8°, D_m = 1.23, Z = 4. Cu radiation, R = 0.094 for 1576 reflexions.

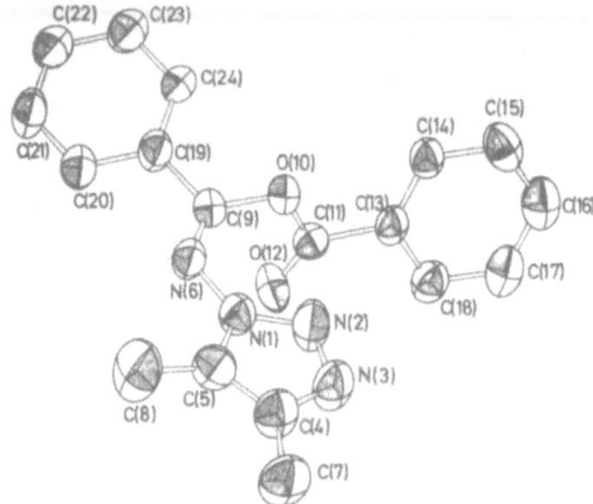

Fig. 1. The structure and conformation of the triazole derivative, $C_{18}H_{16}N_4O_2$.

The crystal structure analysis establishes the material examined to be the title compound. In the molecule (Fig. 1) the sequence N(6)-C(9)-O(10)-C(11) forms a helix-like chain. The triazole ring and benzene ring attached to C(9) lie in planes whose dihedral angle is 18.5°, while the angle between the triazole ring and the benzene ring attached to C(11) is 65.5°. Bond lengths and angles are normal but there are a number of short intramolecular contacts resulting in a very closely packed molecule (e.g. O(10)···N(2) = 2.71, O(10)···O(24) = 2.81, O(10)···C(14) = 2.77 Å).

trans-3-PHENYL-3-METHOXYCARBONYL-4-p-METHOXYPHENYL-1-PYRAZOLINE

$C_{18}H_{18}N_2O_3$

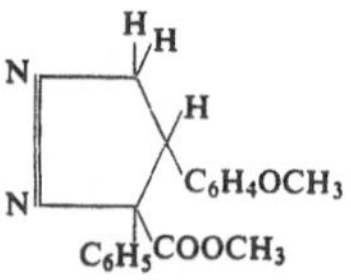

M.-P. ROUSSEAUX, J. MEUNIER-PIRET, J.-P. PUTZEYS, G. GERMAIN and M. van MEERSSCHE, 1972. Acta Cryst., B28, 1720-1724.

Monoclinic, $P2_1/c$, a = 14.183, b = 5.777, c = 22.681 Å, β = 120.20°, D_m not given, Z = 4. Cu radiation, R = 0.061 for 1630 reflexions.

Carbon atom C(4) (to which is bonded the p-methoxyphenyl group) is 0.49 Å out of the plane of the other atoms of the heterocyclic ring. Bond lengths in the five-membered ring are N=N = 1.240(3), N(2)-C(3) = 1.510(2), C(3)-C(4) = 1.548(3), C(4)-C(5) = 1.515(3), and N(1)-C(5) = 1.482(3) Å. The molecules are held in the crystal by van der Waals forces.

1,3-DIPHENYL-4-p-CHLOROBENZAL-5-PYRAZOLONE

$C_{22}H_{15}C\ell N_2O$

N=C–C_6H_5 ; C_6H_5–N ; =CHC_6H_4Cl ; O

B. BOVIO and S. LOCCHI, 1972. Cryst. Struct. Comm., 1, 253-256.

Triclinic, $P\bar{1}$, a = 13.96, b = 12.94, c = 7.29 Å, α = 93.0, β = 100.1, γ = 135.6°, Z = 2. Cu radiation, R = 0.098 for 3835 reflexions.

Bond lengths and angles are normal. The p-chlorophenyl group is coplanar with the αβ-unsaturated system. The other phenyl groups are rotated by 9° and 48° out of the pyrazolone plane. There is a short (2.98 Å) intramolecular contact between the carbonyl oxygen and a CH group of the p-chlorophenyl ring.

5,5'-DIETHOXYCARBONYL-3,3',4,4'-TETRAETHYLDIPYRROL-2-YLMETHANE

$C_{23}H_{34}N_2O_4$

R. BONNETT, M.B. HURSTHOUSE and S. NEIDLE, 1972. J. Chem. Soc., Perkin II, 1335-1340.

Monoclinic, $P2_1/c$, a = 8.655, b = 17.781, c = 15.465 Å, β = 102.70°, D_m = 1.16, Z = 4. Cu radiation, R = 0.071 for 2711 reflexions.

The molecular conformation is shown in Fig. 1, and bond lengths are in Fig. 2. The planar pyrrole rings are slightly distorted from C_{2v} symmetry; the interplanar angle is 71.9°. The ester functions on rings A and B take up different conformations in the crystal; in ring A the C=O group is syn with respect to the N-H vector whereas in ring B it is anti. The molecules are arranged in the crystal so as to form hydrogen bonded dimers related by a centre of symmetry with O(16) of one molecule associating with both imino hydrogens of the other molecule (N(10)H···O(16) = 2.883(3) and N(11)H···O(16) = 2.959(3) Å).

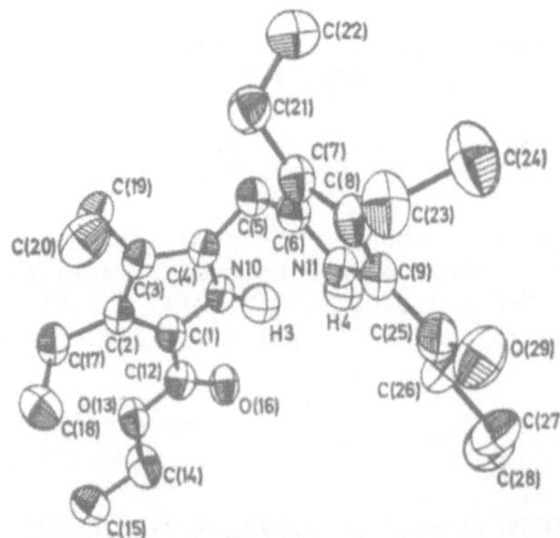

Fig. 1. The conformation of the dipyrrol-2-ylmethane derivative, $C_{23}H_{34}N_2O_4$.

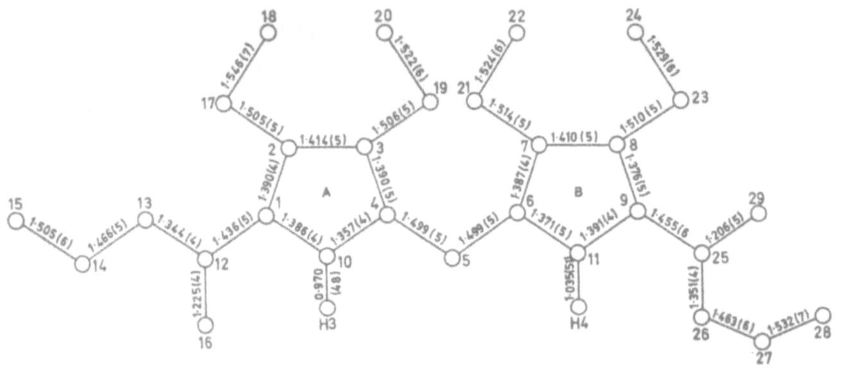

Fig. 2. Bond lengths in the dipyrrol-2-ylmethane derivative, $C_{23}H_{34}N_2O_4$.

1-p-IODOPHENYL-3-CARBOMETHOXY-4-PHENACYLIDEN-5-HYDROXY-5-PHENYLPYRAZOLINE

$C_{25}H_{19}IN_2O_4$

G.D. ANDREETTI, G. BOCELLI, L. CAVALCA and P. SGARABOTTO, 1972. Gazz. Chim. Ital., 102, 355-363.

Triclinic, $P\bar{1}$, a = 17.11, b = 13.13, c = 5.93 Å, α = 74.9, β = 83.9, γ = 115.8°, D_m not given, Z = 2. Mo radiation, R = 0.041 for 2766 reflexions.

The pyrazoline ring (Fig. 1) is slightly non-planar, with the tetrahedral carbon atom 0.03 Å out of plane; the molecule contains an intramolecular O-H···O hydrogen bond.

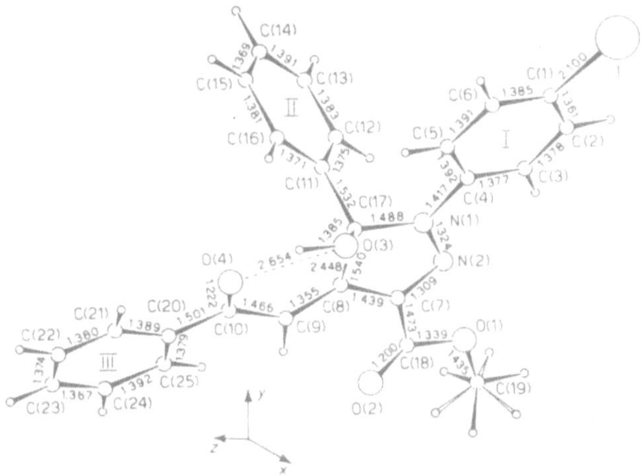

Fig. 1. Bond lengths in the $C_{25}H_{19}IN_2O_4$ molecule.

CYCLOTRIMETHYLENE-TRINITRAMINE

(RDX)

$C_3H_6N_6O_6$

C.S. CHOI and E. PRINCE, 1972. Acta Cryst., B28, 2857-2862.

Orthorhombic, Pbca, a = 13.182, b = 11.574, c = 10.709 Å, D_m = 1.816, Z = 8. Mo radiation, R = 0.039 for 836 reflexions.

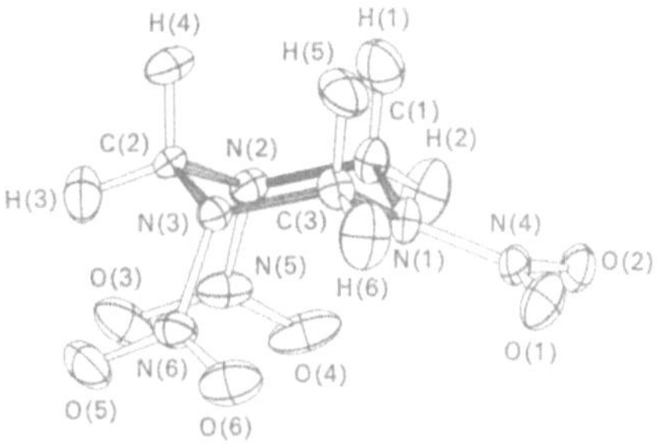

Fig. 1. View of the RDX molecule.

The molecule (Fig. 1) consists of alternate CH_2 and N-NO_2 groups in a puckered ring. The structure determination using three-dimensional photographic data has been reported previously (1). The molecule possesses an approximate mirror symmetry through atoms N(1)C(2)H(3) (Fig. 1). The mean plane of the N(1)-NO_2 group is inclined at 18° to the plane of the three carbon atoms whereas both the N(2)-NO_2 and N(3)-NO_2 groups are inclined at 62° to the carbon plane. The C-N distances are 1.440 - 1.468(4), N-N = 1.351 - 1.398(3), N-O = 1.201(5) - 1.233(4) Å. The methylene groups have bond angles in the range 109.5 ± 2°. One hydrogen on each CH_2 group is always located close to oxygen atoms of adjacent nitro groups on both sides with H···O = 2.22 - 2.30 Å.

1. P.M. HARRIS and P.T. REED, 1959. AFOSR-TR-59-165 Ohio State University Research Foundation, Columbus, Ohio.

2-AMINO-3-CHLOROPYRAZINE

$C_4H_4C\ell N_3$ $C_4H_2N_2C\ell NH_2$

J.C. MORROW and B.P. HUDDLE, 1972. Acta Cryst., B28, 1748-1753.

Orthorhombic, Pbca, a = 13.880, b = 10.685, c = 7.196 Å, D_m = 1.60, Z = 8. Cu radiation, R = 0.077 for 648 reflexions.

In the crystal structure (Fig. 1) pairs of planar molecules form centrosymmetric dimers through N-H···N hydrogen bonds (3.178(8) Å). The crystal consists of pleated sheets formed from dimers linked with zig-zag chains of other N-H···N bonds (3.275(7) Å). The pyrazine ring is essentially planar (mean deviation 0.009 Å) while the chlorine and amino nitrogen atoms are displaced +0.022(2) and +0.036(6) Å respectively from the ring plane. Bond distances and angles are in Fig. 2.

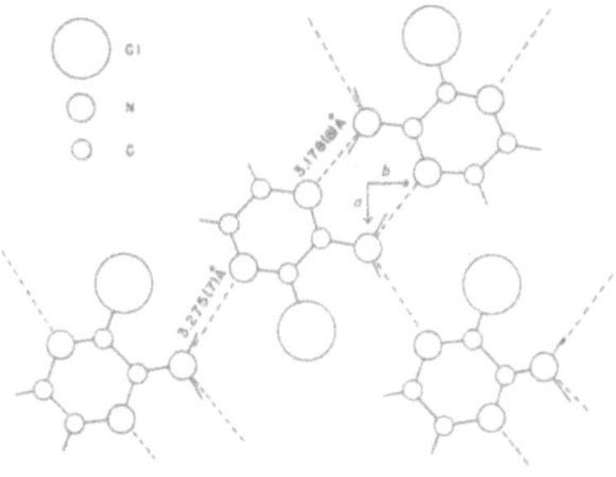

Fig. 1. Molecular packing and hydrogen bonding in 2-amino-3-chloropyrazine.

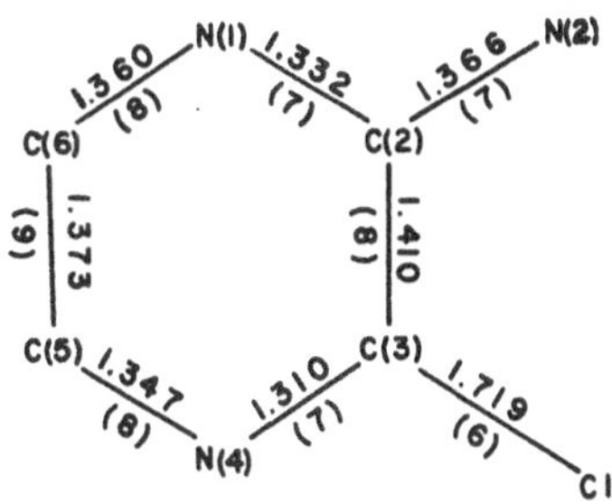

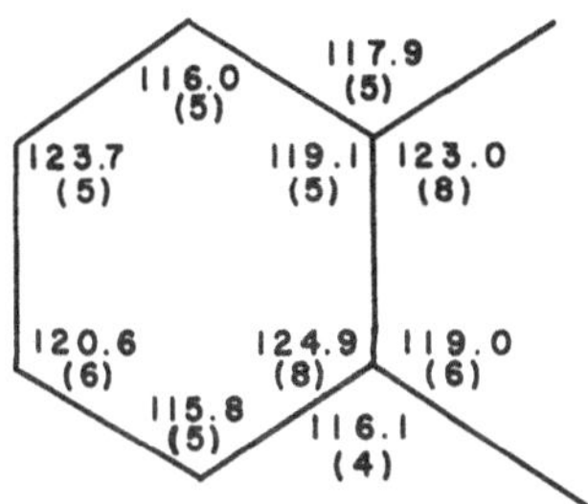

Fig. 2. Bond lengths (Å), bond angles and standard deviations in 2-amino-3-chloropyrazine.

5-CHLORO-2-PYRIDONE

$C_5H_4C\ell NO$

Å. KVICK and S.S. BOOLES, 1972. Acta Cryst., B28, 3405-3409.

Monoclinic, C2/c, a = 13.509, b = 8.887, c = 12.158 Å, β = 127.396°, Z = 8. Mo radiation, R = 0.049 for 694 reflexions.

Planar 5-chloro-2-pyridone molecules (Fig. 1) occur in the keto form and are linked to form dimers, via pairs of N-H···O hydrogen bonds, 2.796(3) Å, across a centre of symmetry. The deviations of the heavy atoms of the pyridine ring from the mean plane through the ring are all less than or equal to 0.006 Å; the chlorine and oxygen atoms are respectively 0.009 and 0.010 Å out of this plane. From the bond distances (Fig. 1) it is evident that bonds C(3)-C(4) and C(5)-C(6) have considerable double bond character.

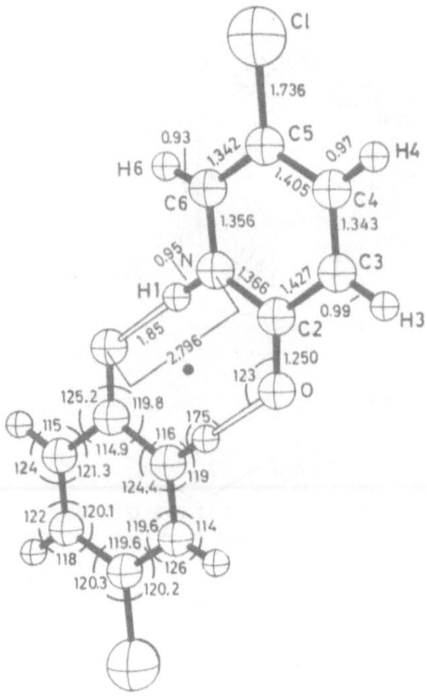

Fig. 1. Bond lengths (σ = 0.003 - 0.005 Å) and angles (σ = 0.2 - 0.3°) for 5-chloro-2-pyridone.

2,5-DICHLORO-3-METHOXYPYRAZINE

$C_5H_4Cl_2N_2O$

D.R. CARTER and F.P. BOER, 1972. J. Heterocyclic Chem., 9, 335-340.

Monoclinic, $P2_1/c$, a = 4.059, b = 15.773, c = 11.123 Å, β = 90.96°, Z = 4. Mo radiation, R = 0.057 for 1397 reflexions.

The molecule is planar except for the methyl group; the methyl carbon lies 0.072 Å off the best plane through the other atoms. This distortion is ascribed to crystal packing forces. The conformation of the methoxy substituent is *syn* with respect to N(4) (Fig. 1). Bond distances, angles and intermolecular contacts are normal.

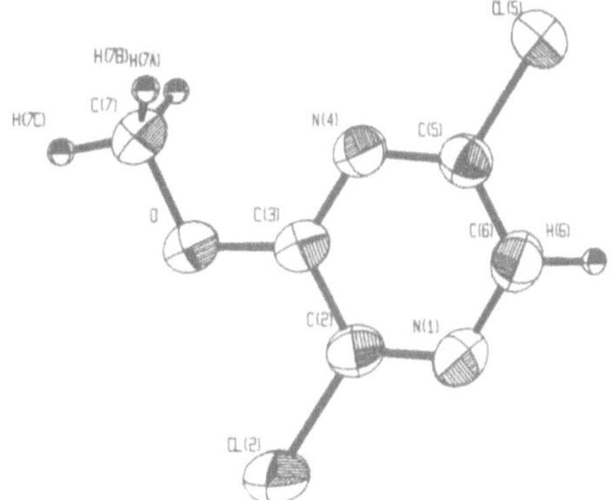

Fig. 1. A view of the 2,5-dichloro-3-methoxypyrazine molecule.

β-PYRAZINE-2-CARBOXAMIDE

$C_5H_5N_3O$

G. RØ and H. SØRUM, 1972. Acta Cryst., B28, 991-998.

Monoclinic, $P2_1/c$, a = 14.372, b = 3.711, c = 10.726 Å, β = 101.92°, D_m not given, Z = 4. Cu radiation, R = 0.055 for 813 reflexions.

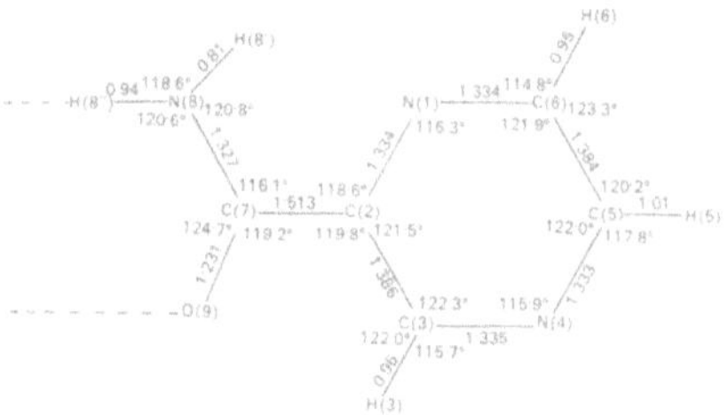

Fig. 1. Bond lengths (Å) and bond angles in β-pyrazine-2-carboxamide.

The bond lengths and angles with their estimated standard deviations for β-pyrazine-2-carboxamide are given in Fig. 1. All bond lengths except those containing O(9), N(8), or H were corrected for thermal motion. The average values of the bond lengths in the pyrazine ring are 1.334 and 1.385 Å for the C-N and C-C bond respectively. The ring is planar within experimental error, and the carboxamide group, which is also planar, deviates 3.2° from the plane of the ring. The molecules pack to form dimers across centres of symmetry, joined by N(8)-H(8")-O(9) hydrogen bonds of length 2.92 Å. There are several short intermolecular contacts less than the sum of van der Waals radii, N(4)···H(6) = 2.47 Å, N(1)···H(3) = 2.49 Å, and N(4)···H(5) = 2.58 Å. These contacts, however, are considerably longer than the distances expected for hydrogen bonds.

δ-PYRAZINE-2-CARBOXAMIDE

$C_5H_5N_3O$

G. RØ and H. SØRUM, 1972. Acta Cryst., B28, 1677-1684.

Triclinic, P$\bar{1}$, a = 5.728, b = 5.221, c = 9.945 Å, α = 96.81, β = 97.27, γ = 106.22°, D_m not given, Z = 2. Cu radiation, R = 0.061 for 595 reflexions.

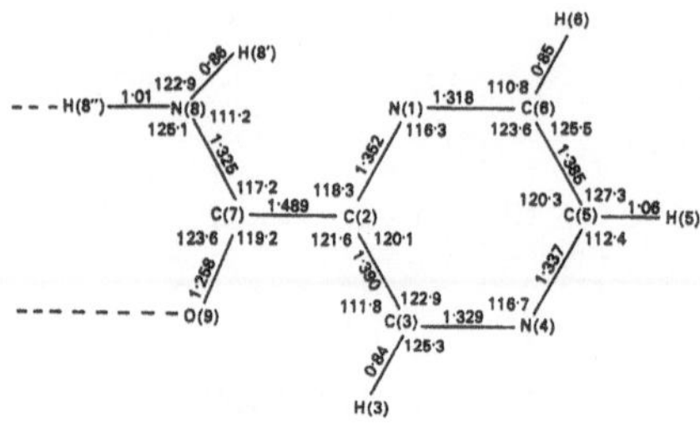

Fig. 1. Bond lengths (Å) and angles in δ-pyrazine-2-carboxamide. Mean standard deviations are 0.003 Å and 0.4°.

Pyrazincarboxamide is known to appear in four different crystalline modifications, α(1), β(2), γ(3), and δ. In the δ form the molecule is essentially planar; the carboxamide group is only slightly twisted (0.8°) around the exocyclic C(2)-C(7) bond out of the molecular plane. Bond lengths and angles are in Fig. 1. Pairs of molecules are connected by hydrogen bonds (N-H···O = 2.90 Å) across the symmetry centres to form dimers which are held in the crystal lattice by van der Waals forces.

1. Y. TAKAKI, Y. SASADA and T. WATANABE, 1960. Acta Cryst., 13, 693.
2. G. RØ and H. SØRUM, 1972. Acta Cryst., B28, 991; this volume preceding report.
3. C. TAMURA, H. KUWANO and Y. SASADA, 1961. Acta Cryst., 14, 693.

PYRIDINIUM HEXAIODOPENTAARGENTATE (AT -30°C)

$C_5H_6Ag_5I_6N$ $(C_5H_5NH)Ag_5I_6$

I. S. GELLER, 1972. Science, 176, 1016-1019.
II. S. GELLER and B.B. OWENS, 1972. J. Phys. Chem. Solids, 33, 1241-1250.

Hexagonal, P6/mcc, a = 11.97, c = 7.41 Å, D_m = 4.92, Z = 2. Mo radiation, R = 0.059 for 200 reflexions.

The crystal structure is unique in that face-sharing iodide octahedra as well as face-sharing tetrahedra and face-sharing between octahedra and tetrahedra provide paths for silver ion transport in this solid electrolyte (Fig. 1). Each iodide ion has 11 near neighbours, I···I = 4.119 - 4.866(9) Å. The tetrahedral Ag^+-I^- distance is 2.861(3). The pyridinium ion is centred at the origin of the lattice and is, of course, disordered.

In II it is reported that the structure was determined at six different temperatures between -30° and 125°C. Although a transition occurs at 50°C there is no obvious change in crystal structure at this temperature. The transition is described as one from a region of low disorder, in which the fraction of mobile Ag^+ ions increases fairly rapidly with increasing temperature, to one of high disorder in which essentially all the Ag^+ ions are mobile.

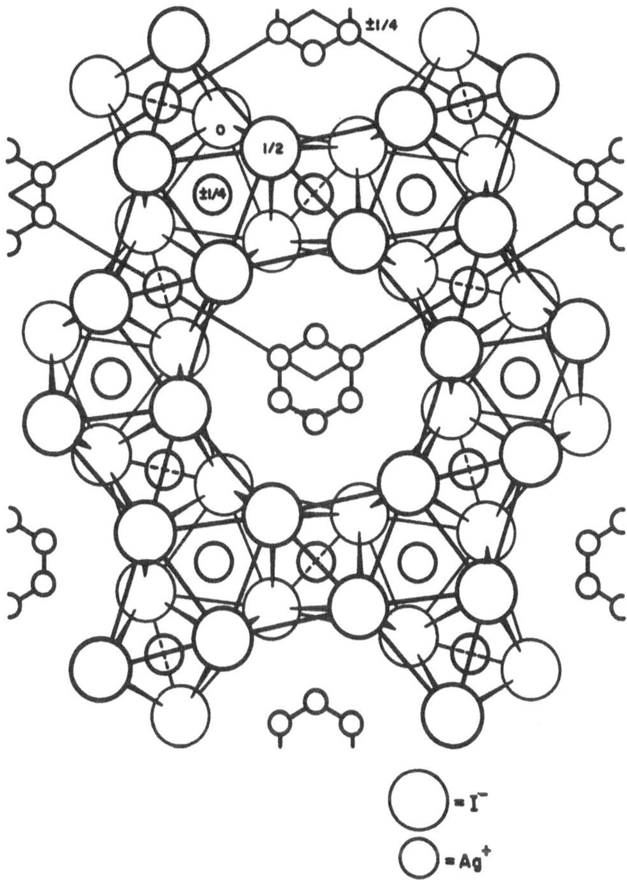

Fig. 1. The pyridinium hexaiodopentaargentate structure viewed along the c-axis.

2-THIOAMIDOPYRIDINE

$C_6H_6N_2S$ $C_5H_4N \cdot CS \cdot NH_2$

T.C. DOWNIE, W. HARRISON, E.S. RAPER and M.A. HEPWORTH, 1972. Acta Cryst., B28, 283-290.

Monoclinic, $P2_1/c$, a = 5.79, b = 7.42, c = 16.26 Å, β = 106.9°, D_m = 1.37, Z = 4. Cu radiation, R = 0.069 for 1004 reflexions.

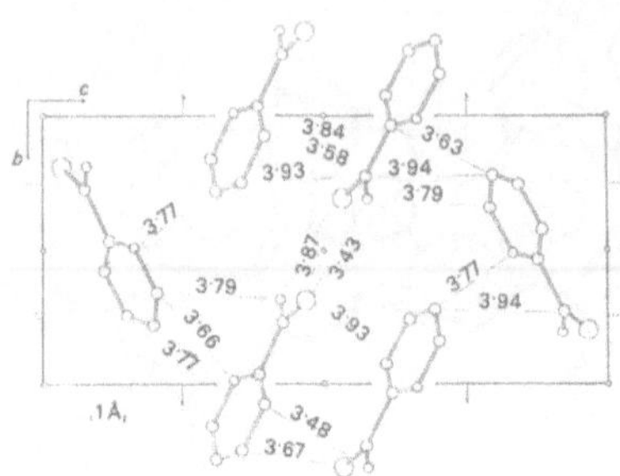

Fig. 1. 2-Thioamidopyridine packing diagram.

The mean bond lengths found in the analysis are C-C = 1.383, C-N = 1.336 Å in the pyridine ring and C-C = 1.505, C-N = 1.325, C-S = 1.657 (all ±0.006 Å) in the thioamido group. These data imply the ionic canonical form $-C\langle^{S^-}_{\overset{+}{N}H_2}$ as well as the amide form make contributions to the ground state structure. The molecule is not planar; the angle between the pyridine plane and that of the thioamido group is 10.5°. In the crystal structure (Fig. 1) pairs of N-H···S hydrogen bonds (3.43 Å) join molecules across centres of symmetry.

3,5-DICHLORO-2,6-DIMETHYL-4-PYRIDINOL

(CLOPIDOL)

$C_7H_7Cl_2NO$

F.P. BOER, 1972. Acta Cryst., B28, 3200-3206.

Monoclinic, $P2_1/m$, a = 6.795, b = 6.932, c = 8.746 Å, β = 103.56°, Z = 2. Mo radiation, R = 0.044 for 421 reflexions.

The molecule lies on a crystallographic mirror plane. Bond length and angle data (Fig. 1) for clopidol, and the location of the non-methyl hydrogen atom, eliminate formulation (I) and confirm that forms (II) and (III) make significant contributions to the ground state structure. In the crystal

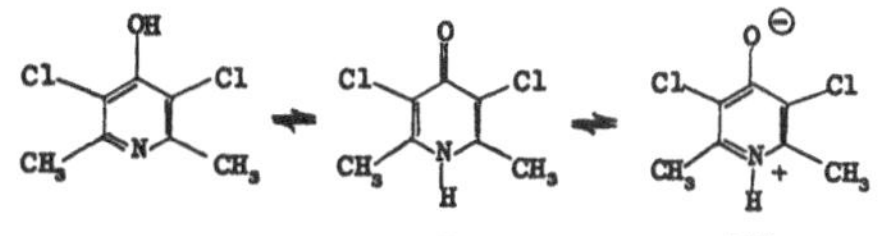

(I) (II) (III)

structure molecules are lined up head-to-tail and linked by N-H···O hydrogen bonds (2.755 Å) to yield a linear chain structure.

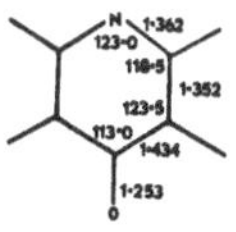

Fig. 1. Bond distances (σ = 0.007 - 0.009 Å) in the clopidol ring system (averaged values).

2-FORMYLPYRIDINE SELENOSEMICARBAZONE

$C_7H_8N_4Se$

A. CONDE, A. LÓPEZ-CASTRO and R. MARQUEZ, 1972. Cryst. Struct. Comm., 1, 155-158.

Monoclinic, $P2_1/c$, a = 9.320, b = 6.524, c = 16.275 Å, β = 90.53°, Z = 4. Cu radiation, R = 0.11 for 1180 reflexions.

The molecular conformation is similar to that found in 4-formylpyridine thiosemicarbazone (1); the Se-C-N-N-C side chain is fully extended. The N-N bond length is 1.41(2) Å, and the C-Se distance is 1.83(2) Å. The molecules are linked into dimers in the crystal by N-H···Se hydrogen bonds (3.52 Å); the dimer-like units are further linked by N-H···N hydrogen bonds (2.94 Å).

1. Structure Reports, 35B, 166.

trans-3-(6-METHYL-2-PYRIDYLTHIO)-PROPENIC ACID

$C_9H_9NO_2S$

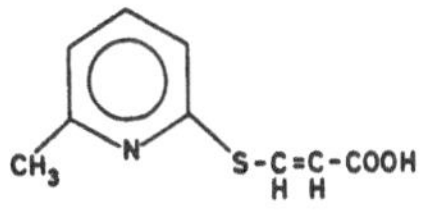

P. GROTH, K. DAVIDKOV and A. AASEN, 1972. Acta Chem. Scand., 26, 1141-1148.

Monoclinic, $P2_1/c$, a = 9.17, b = 5.86, c = 18.34 Å, β = 90.0°, D_m = 1.31, Z = 4. R = 0.081 for 896 reflexions.

In the crystal dimers are formed by hydrogen bonds, 2.620(8) Å, between molecules related by a centre of symmetry. The S-C(pyridine) and S-CH distances are 1.786(7) and 1.727(7) Å respectively, and the dihedral angle C=C-S-C is 174.9°. Aside from the hydrogen bonds, there are no short intermolecular contacts.

2,2,6,6-TETRAMETHYL-4-PIPERIDINONE-1-OXYL

$C_9H_{16}NO_2$

R.P. ŠIBAEVA, L.O. ATOVMJAN, M.G. NEIGAUZ, L.A. NOVAKOVSKAJA and S.L. GINZBURG, 1972. Ž. Strukt. Khim., 13, 887-892 [J. Struct. Chem., 13, 826-830].

Orthorhombic, $Pca2_1$, a = 7.94, b = 11.94, c = 10.54 Å, Z = 4. Cu radiation, R = 0.13 for 444 reflexions (film data).

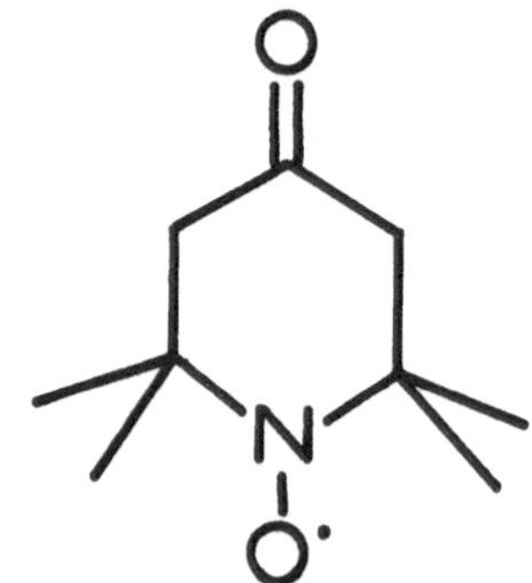

Fig. 1. 2,2,6,6-Tetramethyl-4-piperidinone-1-oxyl.

The six-membered ring (Fig. 1) has a twisted-boat conformation, and bond lengths and angles are close to normal values (N-O = 1.30 Å).

HEXAMETHYLMELAMINE

(2,4,6-TRIS(DIMETHYLAMINO)-1,3,5-TRIAZINE)

$C_9H_{18}N_6$

G.J. BULLEN, D.J. CORNEY and F.S. STEPHENS, 1972. J. Chem. Soc., Perkin II, 642-646.

Hexagonal, $P6_3/m$, a = 9.99, c = 7.11 Å, D_m = 1.10, Z = 2. Mo radiation, R = 0.073 for 255 reflexions.

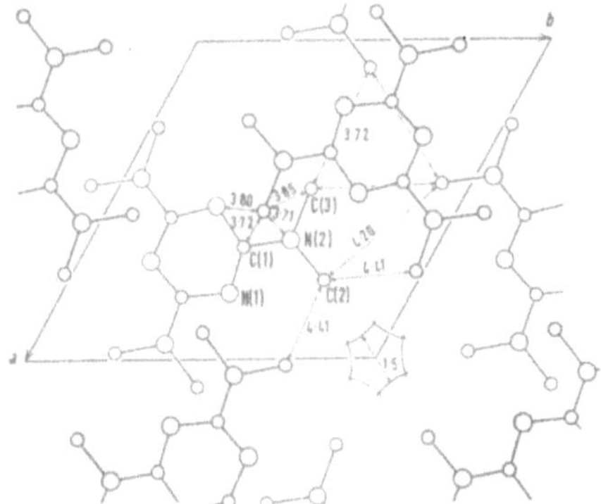

Fig. 1. The hexamethylmelamine structure viewed in projection down the c axis. Molecules at a height z = 1/4 are drawn in thin lines and those at a height z = 3/4 in thick lines. Intermolecular distances are marked in Å.

The molecule is planar (apart from the hydrogen atoms). It occupies a crystallographic site of symmetry 3/m but within experimental error the molecular symmetry is $\bar{6}2m$. Mean bond lengths corrected for molecular oscillations are: C-N(cyclic) = 1.345, C-N(exocyclic) = 1.366, and C(methyl)-N = 1.466 Å. The C-N-C and N-C-N angles in the ring are 112.7 and 127.3°. The van der Waals distances between the molecules are normal (3.7 - 3.8 Å) except in the region of the 6_3 screw axis where the molecules are further apart leaving a channel through the structure with a free diameter of 1.5 Å (Fig. 1).

PARAQUAT DIHALIDES

J.H. RUSSELL and S.C. WALLWORK, 1972. Acta Cryst., B28, 1527-1533.

N,N'-DIMETHYL-4,4'-BIPYRIDYLIUM DICHLORIDE

(PARAQUAT DICHLORIDE)

$C_{12}H_{14}C\ell_2N_2$

Orthorhombic, Pnma, a = 9.22, b = 10.76, c = 15.88 Å, D_m = 1.26, Z = 4. Mo radiation, R = 0.15 for 1131 reflexions.

The crystal structure (Fig. 1) consists of alternate layers of chloride and paraquat ions, at b/4 spacings perpendicular to the b-axis. The chloride ions lie on mirror planes and the cations across centres of symmetry. One of the two crystallographically independent chloride ions is approximately perpendicularly above or below the nitrogen atom of each paraquat plane at a distance of 3.359(9) Å. Bond lengths within the paraquat ion are normal.

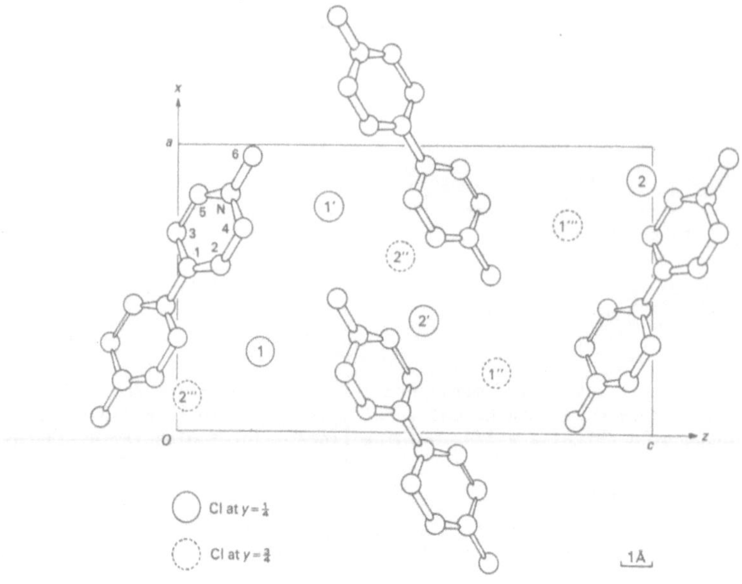

Fig. 1. Structure of paraquat dichloride viewed along b.

N,N'-DIMETHYL-4,4'-BIPYRIDYLIUM DIBROMIDE

(PARAQUAT DIBROMIDE)

$C_{12}H_{14}Br_2N_2$

Monoclinic, $P2_1/c$, a = 5.86, b = 8.226, c = 13.51 Å, β = 95.8°, D_m = 1.77, Z = 2. Mo radiation, R = 0.059 for 994 reflexions.

N,N'-DIMETHYL-4,4'-BIPYRIDYLIUM DIIODIDE

(PARAQUAT DIIODIDE)

$C_{12}H_{14}I_2N_2$

Monoclinic, $P2_1/c$, a = 6.04, b = 8.342, c = 14.27 Å, β = 96.2°, D_m = 2.03, Z = 2. Mo radiation, R = 0.070 for 1011 reflexions.

The diiodide salt is isomorphous with the dibromide derivative but not with the dichloride. In the two isomorphous structures, lines of paraquat ions along the a axis lie across centres of symmetry (Fig. 2) and between these are pleated pseudo-hexagonal layers of anions. Each anion lies almost perpendicularly above or below the nitrogen atom and an adjacent ring carbon atom of a paraquat ion with distances Br···N = 3.679(6), Br···C = 3.529(7), I···N = 3.841(8), I···C = 3.671(12) Å. These close approaches are consistent with spectroscopic evidence for charge transfer interaction in these compounds. Molecular dimensions of the paraquat ions are normal.

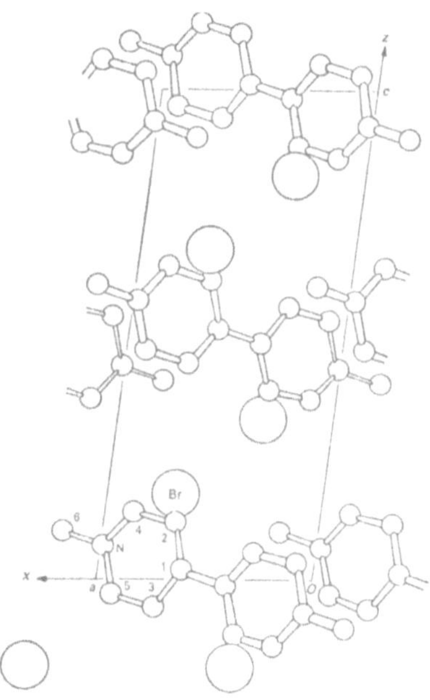

Fig. 2. Structure of the isomorphous paraquat dibromide and diiodide salts viewed along b.

SULPHONAMIDE DERIVATIVES

E. SHEFTER, Z.F. CHMIELEWICZ, J.F. BLOUNT, T.F. BRENNAN, B.F. SACKMAN and P. SACKMAN, 1972. J. Pharmaceutical Sciences, 61, 872-877.

	SULPHADIMETHOXINE	SULPHADOXINE	SULPHISOXAZOLE
	$C_{12}H_{14}N_4O_4S$	$C_{12}H_{14}N_4O_4S$	$C_{11}H_{13}N_3O_3S$
	Triclinic, P$\bar{1}$	Monoclinic, $P2_1/c$	Orthorhombic, Pcab
a(Å)	10.411	8.873	11.578
b	9.309	8.784	14.151
c	7.951	18.938	14.811
α(deg.)	114.11		
β	95.43	107.64(2)	
γ	93.38		
D_m	1.51	1.47	1.45
Z	2	4	8
Radiation	Cu	Cu	Cu
R	0.047	0.071	0.048
Reflexions	1577	1774	1379

Bond distances and angles for the three molecules are in Fig. 1. In each case the amido tautomer (H atom attached to N(1)) is the stable form. The molecules have similar conformations about their sulphonamide linkage (S-N(1)) but have markedly different orientations of their respective heterocyclic rings relative to the sulphanilamide portion of the molecule. The oxazole and phenyl rings of sulphisoxazole are folded closer toward each other than the pyrimidine and phenyl rings of the other two sulphonamides. The dihedral angle between the two rings is 68° for sulphisoxazole, 98° for sulphadoxine, and 106° for sulphadimethoxine. This folding effect can be seen clearly in Fig. 1. All three molecules form intermolecular hydrogen bonds in their crystal structures. For sulphadimethoxine, the dominant intermolecular binding consists of a pair of centrosymmetrically related hydrogen bonds between N(1)-H and O(2). In addition, there is another hydrogen bridge between N(2) and H(N4). Sulphadoxine also has cyclic centrosymmetric hydrogen bonds between N(1)-H and O(2). The ring nitrogen N(2), however, is not involved in hydrogen bonding. The O(1) atom of the sulphone moiety and the methoxy oxygen O(4) are hydrogen bonded to the two protons on the anilino nitrogen N(4) in this structure. The O(1) of sulphadimethoxine does not appear to be involved in any significant intermolecular interaction. Each sulphisoxazole molecule is involved in several short interactions, in which a carbon hydrogen appears to be the donor atom.

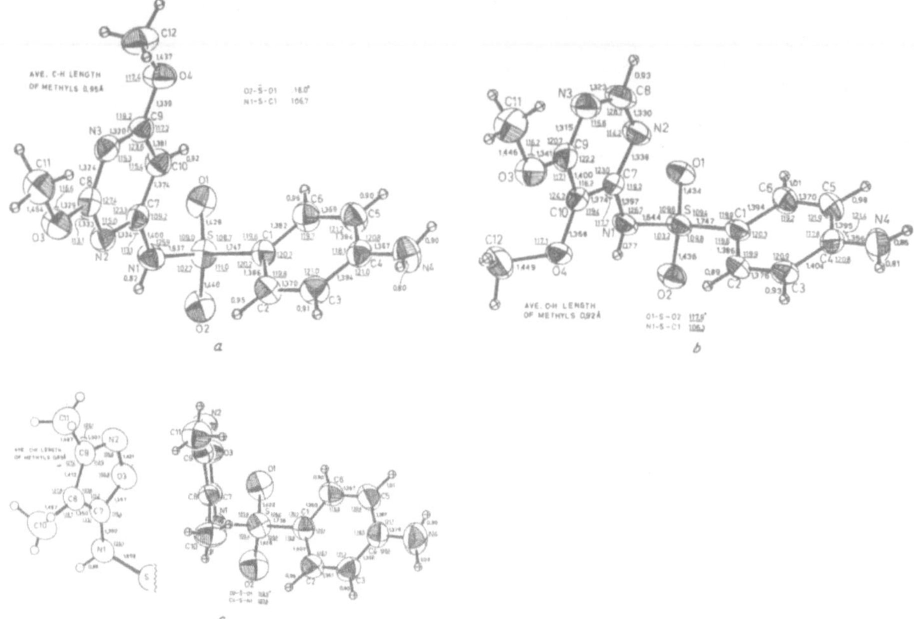

Fig. 1. Intramolecular bonding parameters for sulphadimethoxine (a), sulphadoxine (b), and sulphisoxazole (c). The molecules are seen with the bonds emanating from the sulphur atom oriented in the same perspective.

3,6-DIPHENYL-s-TETRAZINE

$C_{14}H_{10}N_4$ $C_6H_5C_2N_4C_6H_5$

N.A. AHMED and A.I. KITAIGORODSKY, 1972. Acta Cryst., B28, 739-742.

Monoclinic, $P2_1/c$, a = 5.415, b = 5.183, c = 20.603 Å, β = 101°30', D_m = 1.35, Z = 2. Cu radiation, R = 0.130 for 650 reflexions.

The molecule is centrosymmetrical and virtually planar; the angle between the phenyl ring and the tetrazine ring is 1°. The N-N distance is 1.31 Å, and mean C-N is 1.35 Å. From theoretical calculations of packing it is deduced that the planarity of the molecule is the result of intermolecular interactions rather π-electron energy.

2,3-DI(2-PYRIDYL)-2,3-BUTANEDIOL

(meso FORM)

$C_{14}H_{16}N_2O_2$

J.N. BROWN, R.M. JENEVEIN, J.H. STOCKER and L.M. TREFONAS, 1972. J. Org. Chem., 37, 3712-3718.

Monoclinic, $P2_1/c$, a = 9.293, b = 11.885, c = 13.615 Å, β = 123.98°, D_m = 1.29, Z = 4. Cu radiation, R = 0.052 for 1097 reflexions.

The configuration of the molecule was established as meso by the analysis. The conformation adopted (Fig. 1) involves the formation of hydrogen bridges between each pyridyl nitrogen and the more distant hydroxyl group (Fig. 2 gives details). Although not a space group requirement the molecule possesses a centre of symmetry well within the error limits of the analysis. The pyridyl rings are planar to within 0.005 Å. There are no unusual intermolecular distances.

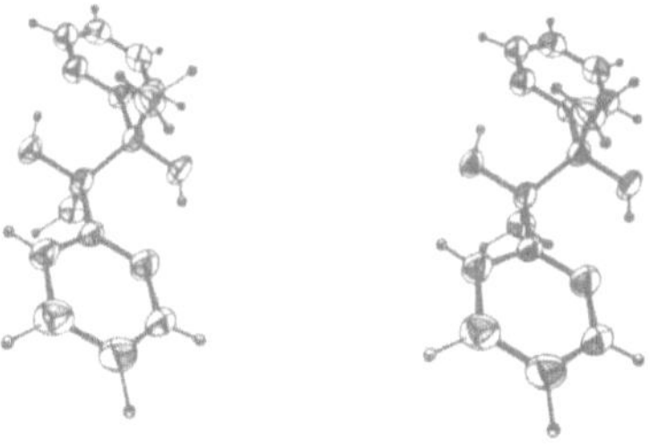

Fig. 1. View of meso-2,3-di(2-pyridyl)-2,3-butanediol.

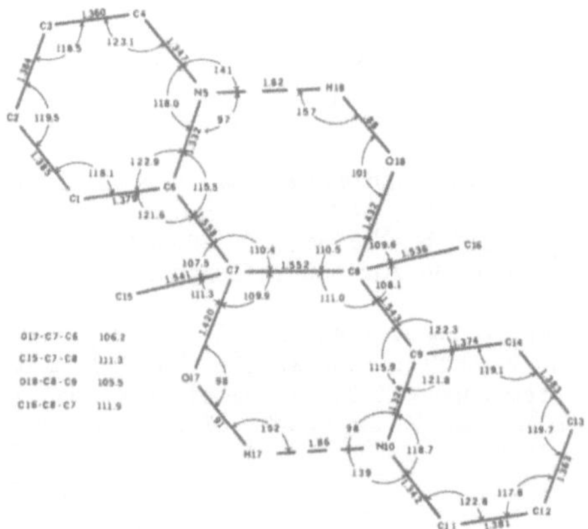

Fig. 2. 2,3-Di(2-pyridyl)-2,3-butanediol, showing bond distances (σ(C-C, C-N, C-O) = 0.008 Å), bond angles, and intramolecular hydrogen bonding scheme.

2-[(2)-DIMETHYLAMINOETHYL-2-THENYLAMINO]PYRIDINE HYDROCHLORIDE

$C_{14}H_{20}ClN_3S$

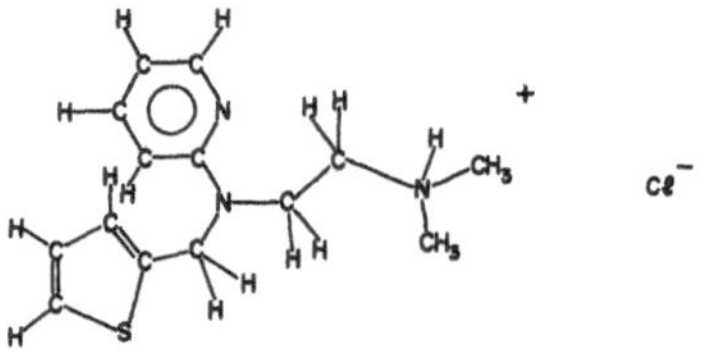

G.R. CLARK and G.J. PALENIK, 1972. J. Amer. Chem. Soc., <u>94</u>, 4005-4009.

Monoclinic, $P2_1/c$, a = 10.936, b = 10.417, c = 28.256 Å, β = 106.21°, Z = 8. Cu radiation, R = 0.055 for 2983 reflexions.

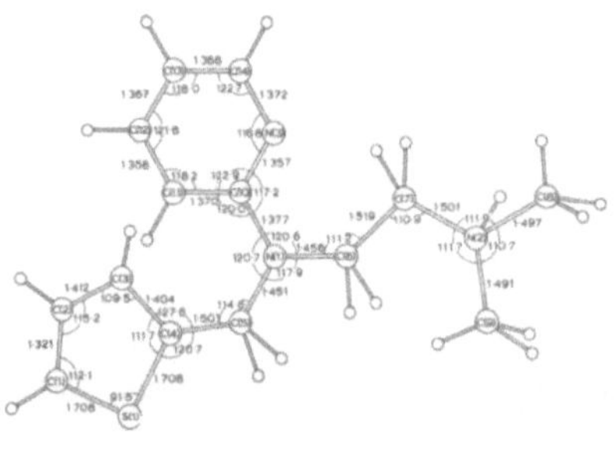

Fig. 1. The atomic numbering and average bond distances and angles in the 2-[(2)-dimethylaminoethyl-2-thenylamino]pyridinium ion. The values for the thiophene ring are the values found in the ordered ring (σ(C-S, C-N) = 0.006 Å, σ(C-C) = 0.006 - 0.009 Å).

There are two independent molecules in the asymmetric unit with different conformations and in addition the thiophene ring in one molecule is disordered. Equivalent bond lengths in the two molecules are equal (Fig. 1) except for those involving the disordered thiophene group. The nitrogen atom at the centre of the molecule is in the sp^2 hybridization state. Both molecules adopt the trans configuration about the carbon-carbon bond of the ethylamine side chain (as was found for histamine itself), thus adding support for the competitive site-occupancy theory of antihistaminic activity.

(±)-α-1,2a,5e-TRIMETHYL-4e-PHENYLPIPERIDIN-4a-OL

[(±)-α-PROMEDOL ALCOHOL]

$C_{14}H_{21}NO$

F.R. AHMED and W.H. de CAMP, 1972. Acta Cryst., B28, 3489-3494.

Monoclinic, $P2_1/c$, a = 8.311, b = 15.970, c = 10.434 Å, β = 109.62°, D_m = 1.109, Z = 4. Cu radiation, R = 0.049 for 1227 reflexions.

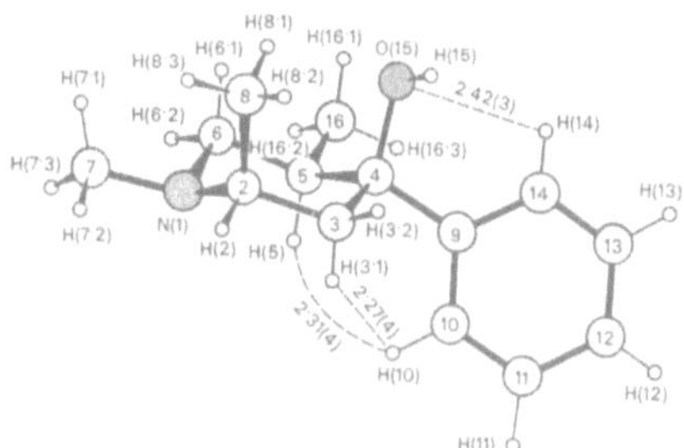

Fig. 1. Molecular structure of (±)-α-1,2a,5e-trimethyl-4e-phenylpiperidin-4a-ol. The dotted lines are short intramolecular contacts (Å).

In the molecular structure (Fig. 1) the piperidine ring has a slightly distorted chair form; its mean plane is at a dihedral angle of 86.6° to the equatorially substituted phenyl ring. The bond distances and angles are normal. The molecules are joined by OH···N hydrogen bonds (2.16 Å) involving O(15) and N(1), which link molecules of opposite chirality into infinite chains.

(±)-β-1,2,5-TRIMETHYL-4-PHENYLPIPERIDIN-4-OL

((±)-β-PROMEDOL ALCOHOL)

$C_{14}H_{21}NO$

I. W.H. de CAMP and F.R. AHMED, 1972. Acta Cryst., B28, 1796-1800 (monoclinic form).

II. Idem, 1972. Ibid., B28, 3484-3489 (rhombohedral form).

Monoclinic, $P2_1/n$, a = 13.298, b = 7.721, c = 12.776 Å, β = 90.09°, Z = 4. Cu radiation, R = 0.054 for 1585 reflexions.

Rhombohedral, $R\bar{3}$, hexagonal cell has a = 29.754, c = 7.713 Å, D_m = 1.110, Z = 18. Cu radiation, R = 0.054 for 1410 reflexions.

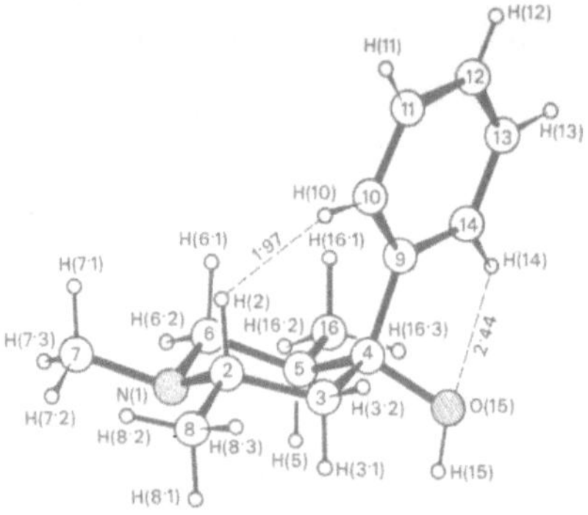

Fig. 1. Molecular structure of β-promedol alcohol (monoclinic form), showing the short intramolecular contacts.

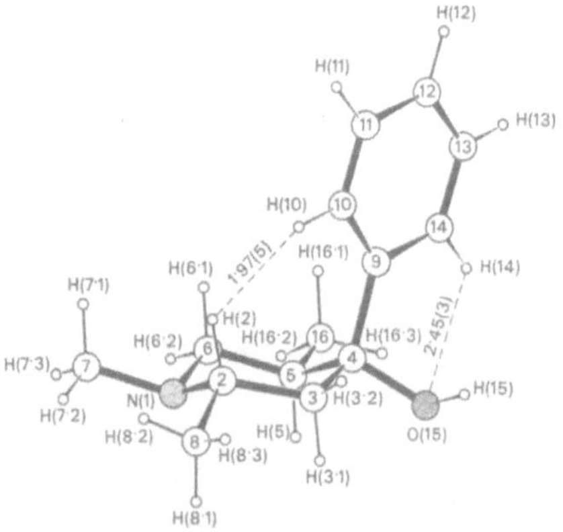

Fig. 2. Molecular structure of β-promedol alcohol (rhombohedral form)

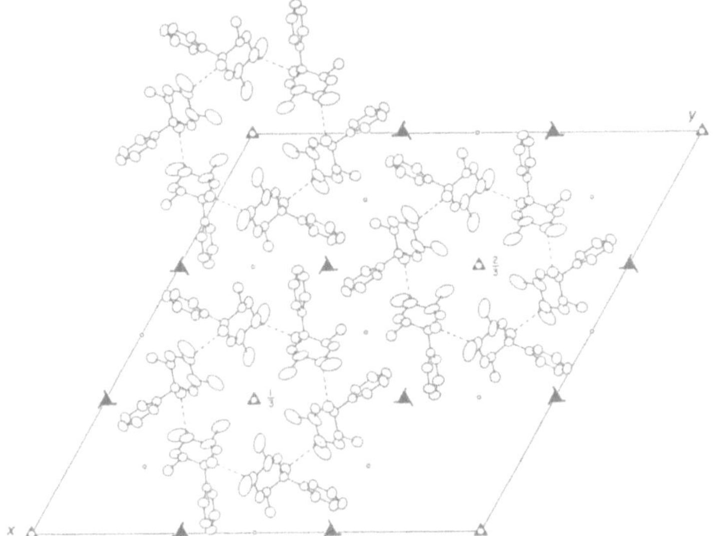

Fig. 3. The rhombohedral β-promedol alcohol structure viewed down the c axis.

The piperidine rings in both crystalline forms have slightly skewed chair conformations with the phenyl rings axial and other substituents equatorial (Figs. 1 and 2). The angle between the phenyl ring plane and that through atoms C(7), N(1), C(4), C(9), O(15) is 40.9° for the monoclinic form and 37.6° for the rhombohedral form. The major difference in the two structures is in the direction of the O-H bond (see Figs.1 and 2). In the monoclinic form, intermolecular hydrogen bonds (O-H···N' = 2.811(3) Å) link molecules to form enantiomeric chains along b. In the rhombohedral form the O-H···N' hydrogen bonds (2.953 Å) link molecules of identical chirality to form hexameric rings (Fig. 3).

(±)-γ-1,2,5-TRIMETHYL-4-PHENYLPIPERIDIN-4-OL

((±)-γ-PROMEDOL ALCOHOL)

$C_{14}H_{21}NO$

W.H. de CAMP and F.R. AHMED, 1972. Acta Cryst., B28, 1791-1796.

Monoclinic, $P2_1/c$, a = 10.806, b = 11.569, c = 10.460 Å, β = 98.75°, D_m = 1.116, Z = 4. Cu radiation, R = 0.048 for 1565 reflexions.

Fig. 1. Molecular structure of γ-promedol alcohol, showing the short intramolecular contacts (Å).

In γ-promedol alcohol (Fig. 1), the piperidine ring has a chair conformation with the hydroxyl group axial and the other substituents equatorial. The phenyl ring is oriented at 73.7° from the mean plane of the piperidine ring, and its ortho hydrogen atoms make rather short intramolecular contacts H···H = 2.07 and H···O = 2.39 Å. Intermolecular hydrogen bonds (O-H···N = 2.939(2) Å) link molecules to form infinite chains along c.

2,6-BIS(BROMOMETHYL)-1,4-DIPHENYLPIPERAZINE

$C_{18}H_{20}Br_2N_2$

B. MOROSIN and J. HOWATSON, 1972. J. Chem. Soc., Perkin II, 1087-1089.

Monoclinic, $P2_1/n$, a = 10.717, b = 17.955, c = 9.124 Å, β = 91.28°, D_m = 1.58, Z = 4. Mo radiation, R = 0.074 for 1562 reflexions.

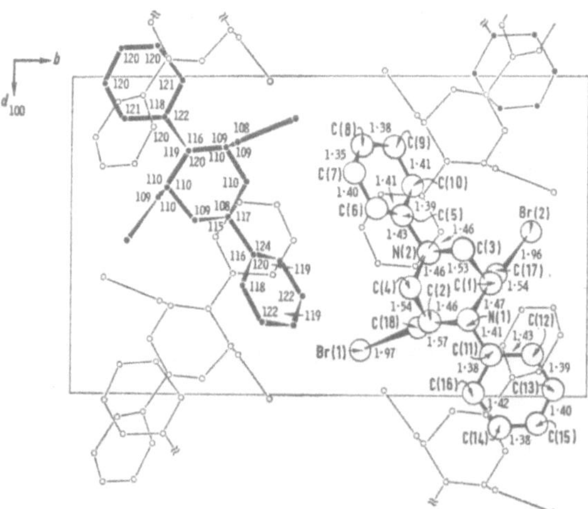

Fig. 1. The structure of 2,6-bis(bromomethyl)-1,4-diphenylpiperazine viewed along the c axis.

The crystal and molecular structure, bond lengths and angles are in Fig. 1. The piperazine ring is in a distorted chair conformation. The interior angles of the heterocyclic ring are approximately tetrahedral with the exception of C(1)-N(1)-C(2) which is 120°. Other bond angles and bond lengths have normal values.

α-CYCLOHEXYL-α-PHENYL-1-PIPERIDINEPROPANOL

(TRIHEXYPHENIDYL)

$C_{20}H_{31}NO$

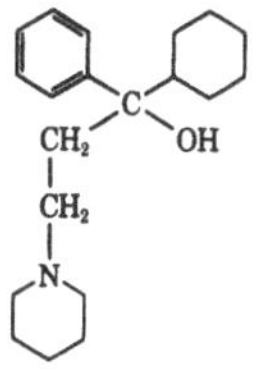

N. CAMERMAN and A. CAMERMAN, 1972. J. Amer. Chem. Soc., 94, 8553-8556.

Monoclinic, C2/c, a = 31.059, b = 5.713, c = 21.889 Å, β = 112.67°, D_m = 1.113, Z = 8. Cu radiation, R = 0.051 for 1665 reflexions.

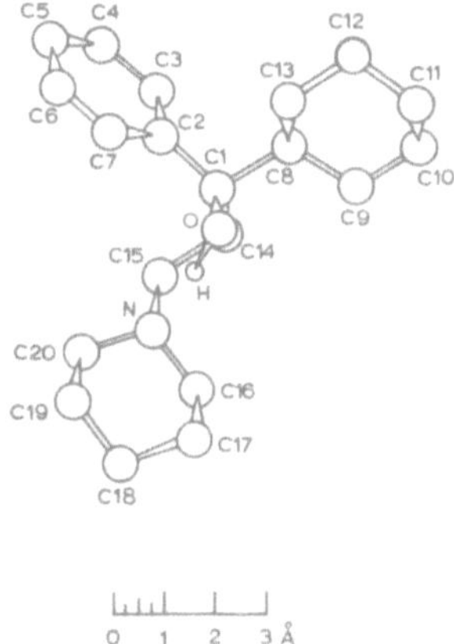

Fig. 1. Perspective drawing of one molecule of trihexyphenidyl, viewed along [010].

The solid-state conformation of this anticonvulsant is shown in Fig. 1. The cyclohexyl ring is in the chair conformation, with the ring atoms lying at distances of 0.22 - 0.25 Å alternately above and below the "best plane" through all six cyclohexyl carbons as one progresses around the ring. The angle between normals to this "best plane" and that of the phenyl group is 96°. The nitrogen heterocycle is also in the chair conformation, with ring atoms lying alternately ±0.22 - 0.26 Å above and below a mean plane through all six atoms. The normal to this mean plane makes obtuse angles of 152 and 106° with normals to the plane of the phenyl ring and to the mean plane through the cyclohexyl group, respectively. The bond lengths and angles in trihexyphenidyl are all near normal. There is an intramolecular hydrogen bond between the hydroxyl group and the heterocyclic nitrogen in trihexyphenidyl; the hydrogen-nitrogen separation is 1.85 Å, and O-H···N angle is 150° (a value closer to linearity is prevented by the geometry of the molecule and the intramolecular nature of the contact). The intermolecular distances correspond to normal van der Waals interactions.

BUTYROPHENONE DERIVATIVES

M.H.J. KOCH and G. GERMAIN, 1972. Acta Cryst., B28, 121-125.

4'-FLUORO-4-{1-[4-HYDROXY-4-(4'-FLUORO)-PHENYLPIPERIDINO]}-BUTYROPHENONE

$C_{21}H_{23}F_2NO_2$

Monoclinic, $P2_1/c$, a = 7.855, b = 8.924, c = 28.060 Å, β = 105.45°, D_m = 1.266, Z = 4. Mo radiation, R = 0.10 for 1229 reflexions.

4'-FLUORO-4-{1-[4-HYDROXY-4-(4'-FLUORO)-PHENYLPIPERIDINO]}-BUTYROPHENONE HYDROCHLORIDE

$C_{21}H_{24}C\ell F_2NO_2$

Monoclinic, $P2_1/a$, a = 16.611, b = 7.056, c = 17.458 Å, β = 102.42°, D_m = 1.319, Z = 4. Mo radiation, R = 0.07 for 1467 reflexions.

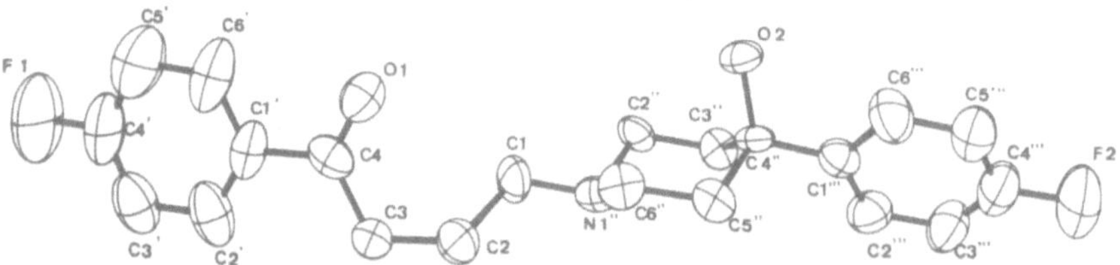

Fig. 1. Conformation of the free base butyrophenone derivative, $C_{21}H_{23}F_2NO_2$.

Bond distances and angles in both molecules are in accord with accepted values. The conformation of the molecule in both crystals is quite different; that of the free base is shown in Fig. 1. The differences in conformation are described in terms of the conformational angles as follows (with the values for the base given first and those for the hydrochloride in parentheses): C(4)-C(3) = 3(354), C(3)-C(2) = 112(18), C(2)-C(1) = 13(355), C(1)-N = 1(257), C(4'')-C(1''') = 74(90)°. In the base crystal one O-H···N hydrogen bond (2.85 Å) is found; in the hydrochloride salt one O-H···O hydrogen bond (2.89 Å) occurs.

4,5,6-TRIS-[p-METHOXYPHENYL]-1,2,3-TRIAZINE

$C_{24}H_{21}N_3O_3$ $(CH_3OC_6H_4)_3C_3N_3$

E. OESER and L. SCHIELE, 1972. Chem. Ber., 105, 3704-3708.

Monoclinic, $P2_1/c$, a = 9.49, b = 10.58, c = 20.91 Å, β = 90.1°, Z = 4. Cu radiation, R = 0.078 for 2537 reflexions.

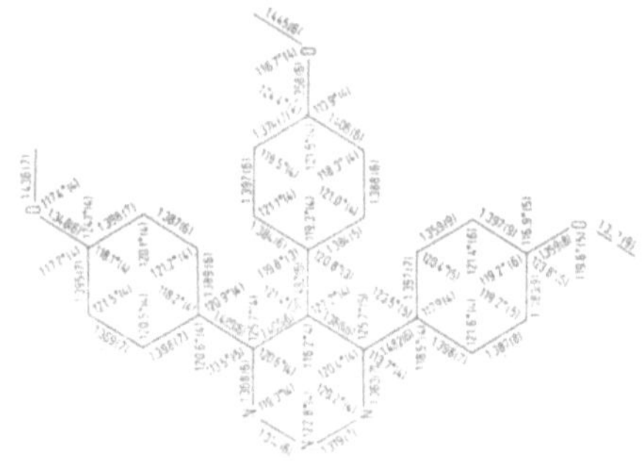

Fig. 1. Bond lengths and angles in the 1,2,3-triazine derivative.

Details of the molecular geometry are in Fig. 1. The planes of the phenyl rings at the 4, 5 and 6 positions make angles of 37°, 68° and 37° respectively with that of the triazine ring.

1-TOSYL-1,2-DIAZEPINE

$C_{12}H_{12}N_2O_2S$

R. ALLMANN, A. FRANKOWSKI and J. STREITH, 1972. Tetrahedron, 28, 581-595.

Orthorhombic, $P2_12_12_1$, a = 12.179, b = 16.547, c = 5.914 Å, D_m = 1.38, Z = 4. Cu radiation, R = 0.037 for 1253 reflexions.

The crystal structure is isomorphous with that of 1-(p-bromobenzene-sulphonyl)-azepine (1). The molecular conformation and details of geometry are in Fig. 1. The 1,2-diazepine ring exists in a boat conformation with the double bonds localised at N(2)-C(3), C(4)-C(5) and C(6)-C(7).

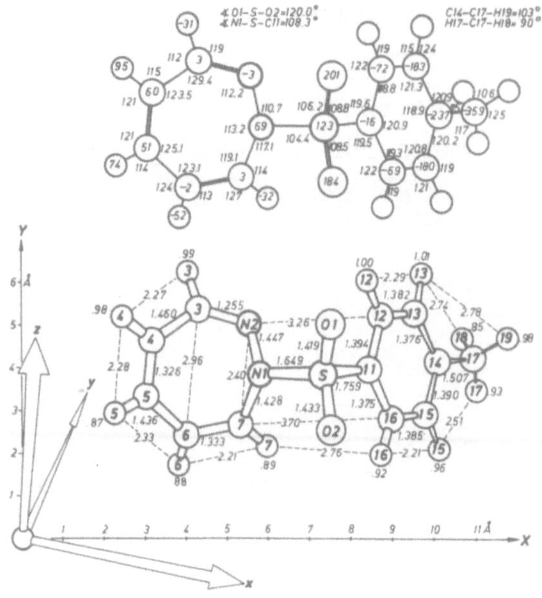

Fig. 1. Bond lengths (in Å, lower part) and bond angles (upper part) in 1-tosyl-1,2-diazepine (uncorrected for thermal vibration). The molecule is projected onto the best plane through N(2), C(3), C(6) and C(7) of the diazepine ring. The numbers within the circles of the lower part correspond to the numbering of the atoms, those in the upper part give the deviation from the projection plane in 10^{-2} Å. The standard deviations of the bond lengths are about 0.005 Å (0.04 Å if hydrogen atoms are involved), and those of the bond angles are about 0.3° (2.5° involving H).

1. I.C. PAUL, S.M. JOHNSON, L.A. PAQUETTE, J.H. BERRETT and R.J. HALUSKA, 1968. J. Amer. Chem. Soc., 90, 5023.

N-(PHENOXYCARBONYL)AZEPINE

$C_{13}H_{11}NO_2$

H.J. LINDNER and B. von GROSS, 1972. Chem. Ber., 105, 434-440.

Monoclinic, $P2_1$, a = 8.64, b = 6.15, c = 10.64 Å, β = 97.61°, Z = 2. Cu radiation, R = 0.058 for 873 reflexions.

The azepine ring in this derivative exists in a boat conformation with localized double bonds (Fig. 1). The nitrogen atom is 0.51 Å and carbon atoms C(3), C(4) 0.45 Å from the best plane through atoms C(1)C(2)C(5)C(6). The phenyl ring plane is inclined at an angle of 60.8° to that of N(7)C(8)O(9)O(10).

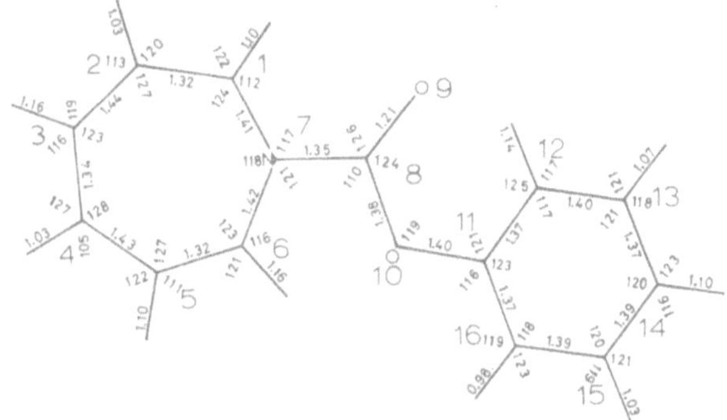

Fig. 1. Bond lengths and angles in N-(phenoxycarbonyl)azepine. For bonds not involving hydrogen σ = 0.009 Å, σ(C-H) = 0.07 Å; for angles σ = 0.6°, except those involving hydrogen (3°).

syn-N-METHYLTHIOALKYLLACTAMS

J.L. FLIPPEN, 1972. Acta Cryst., B28, 3618-3624.

syn-N-METHYLTHIOLAURYLLACTAM

$C_{13}H_{25}NS$

Orthorhombic, $P2_12_12_1$, a = 12.266, b = 17.193, c = 6.435 Å, D_m = 1.34, Z = 4. Cu radiation, R = 0.053 for 1309 reflexions.

syn-N-METHYLTHIOCAPRYLLACTAM

$C_9H_{17}NS$

Orthorhombic, $Pna2_1$, a = 16.621, b = 8.729, c = 6.908 Å, D_m = 1.32, Z = 4. Cu radiation, R = 0.043 for 872 reflexions.

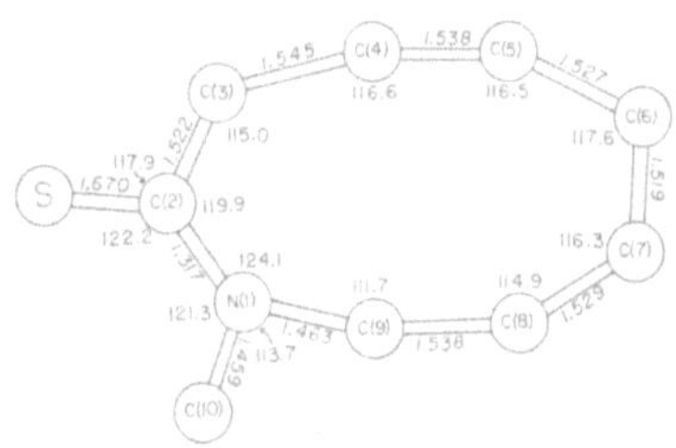

Fig. 1. Bond distances and angles of syn-N-methylthiocapryllactam. (σ = 0.010 Å for bonds and 0.6° for angles.)

Fig. 2. Bond distances and angles for syn-N-methylthiolauryllactam. (σ = 0.007 Å for bonds and 0.4° for angles.)

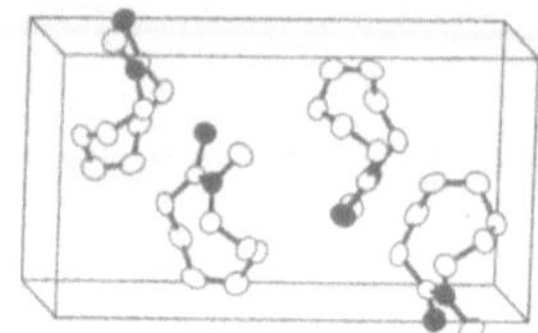

Fig. 3. Crystal packing and molecular conformation of syn-N-methylthio-capryllactam.

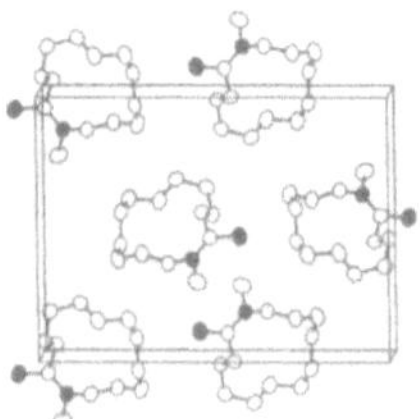

Fig. 4. Crystal packing and molecular conformation of syn-N-methylthio-lauryllactam.

The bond distances and angles in the two compounds are shown in Fig. 1 and Fig. 2. The average C-C distance in both structures is 1.530(10) Å; the average C-C-C angle in the capryllactam is 116.2(6)° and 113.4(4)° in the lauryllactam. The smaller average bond angle in the lauryllactam is consistent with less Baeyer strain for the 13-membered ring compared to the 9-membered capryllactam ring. The CH_3-N-C=S group is planar and has syn conformation in both compounds. The molecular conformations and crystal packing are shown in Fig. 3 and Fig. 4. There are no abnormal intermolecular contacts.

3H-AZEPINE DERIVATIVES

E. CARSTENSEN-OESER, 1972. Chem. Ber., 105, 982-990.

3,5-DIPHENYL-2,4,6,7-TETRAKIS-METHOXYCARBONYL-3H-AZEPINE

$C_{26}H_{23}NO_8$

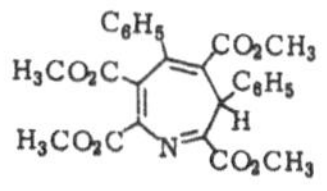

B1

Monoclinic, $P2_1/a$, a = 17.39, b = 11.46, c = 12.40 Å, β = 98.4°, Z = 4. Cu radiation, R = 0.062 for 2900 reflexions.

4,6-DIPHENYL-2,3,5,7-TETRAKIS-METHOXYCARBONYL-3H-AZEPINE

$C_{26}H_{23}NO_8$

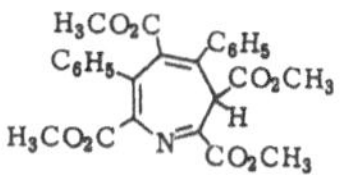

C1

Monoclinic, $P2_1/n$, a = 17.83, b = 11.93, c = 11.97 Å, β = 101.6°, Z = 4. Cu radiation, R = 0.065 for some 3000 reflexions.

Crystal structure analyses have determined the constitution of the two related azepine derivatives B1 and C1. Details of their molecular geometries are in Figs. 1 and 2 respectively. Both azepine rings adopt boat conformations with the lone hydrogen atom on the seven-membered ring in the equatorial position. Both molecules have the same values for the angles α(57°) and β(32°) in Fig. 1.

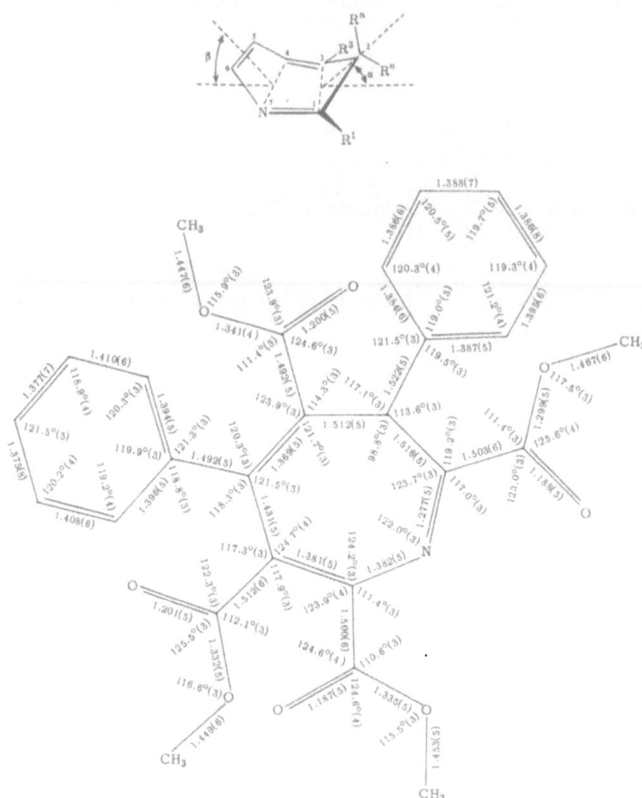

Fig. 1. Bond lengths and angles for azepine B1.

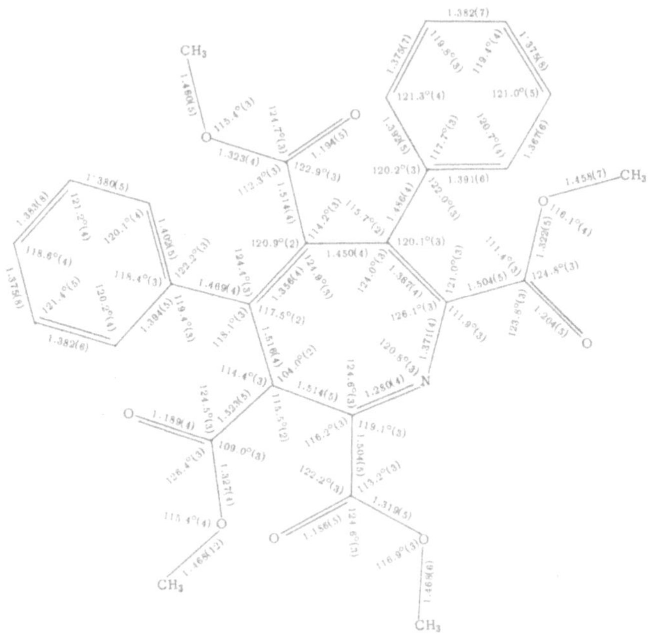

Fig. 2. Bond lengths and angles for azepine C1.

3-METHOXY-1H-PYRAZOLO[4,3-e]as-TRIAZINE

$C_5H_5N_5O$

H. J. LINDNER and G. SCHADEN, 1972. Chem. Ber., 105, 1949-1955.

Monoclinic, $P2_1$, a = 12.24, b = 6.78, c = 3.88 Å, β = 91.10, D_m = 1.57, Z = 2. Cu radiation, R = 0.055 for 568 reflexions.

The yellow pigment normethylpseudoiodinine was identified as the title compound by X-ray analysis. Bond lengths and angles (Fig. 1) within the molecule are normal and the pyrazolo-triazine ring system is planar. In the crystal, molecules are linked by N-H···N (2.90 Å) hydrogen bonds.

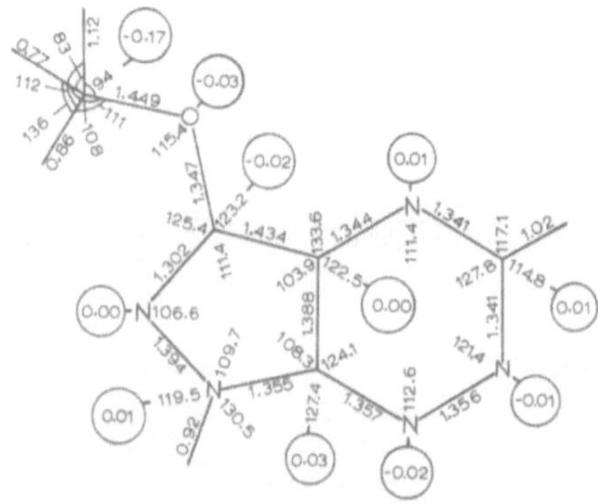

Fig. 1. Bond lengths (σ = 0.009), angles (σ = 0.7) and deviations from the molecular plane in the triazine derivative, $C_5H_5N_5O$.

PHTHALIMIDE

(KLADNOITE)

$C_8H_5NO_2$ $C_6H_4(CO)_2NH$

E. MATZAT, 1972. Acta Cryst., B28, 415-418.

Monoclinic, $P2_1/n$, a = 22.83, b = 7.651, c = 3.810 Å, β = 91.36°, D_m = 1.47, Z = 4. Cu radiation, R = 0.094 for 1058 reflexions.

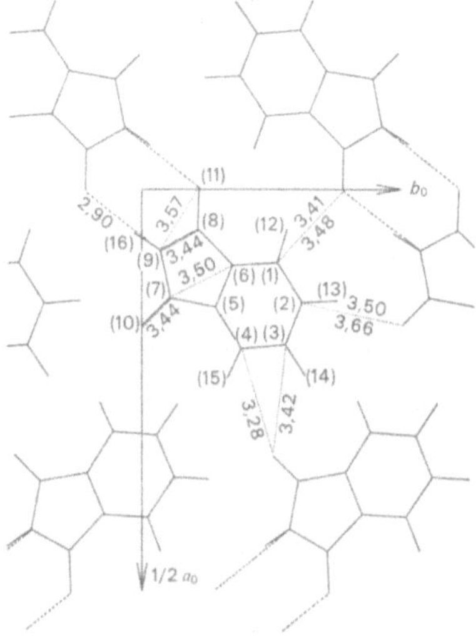

Fig. 1. Projection of the structure of phthalimide on (001). Shorter intermolecular contacts are shown.

In this structure pairs of planar molecules are associated by N-H···O hydrogen bonds (2.895(4) Å) across centres of symmetry (Fig. 1). Bond distances are normal.

2,6-DIMETHYL-4,8-DICHLORO-2H,6H-PYRIDAZINO[4,5-d]PYRIDAZIN-1,5-DIONE

$C_8H_6C\ell_2N_4O_2$

C. SABELLI and P.F. ZANAZZI, 1972. Acta Cryst., B28, 1173-1177.

Monoclinic, $P2_1/n$, a = 9.189, b = 5.546, c = 10.230 Å, β = 109.82°, D_m = 1.77, Z = 2. Cu radiation, R = 0.075 for 814 reflexions.

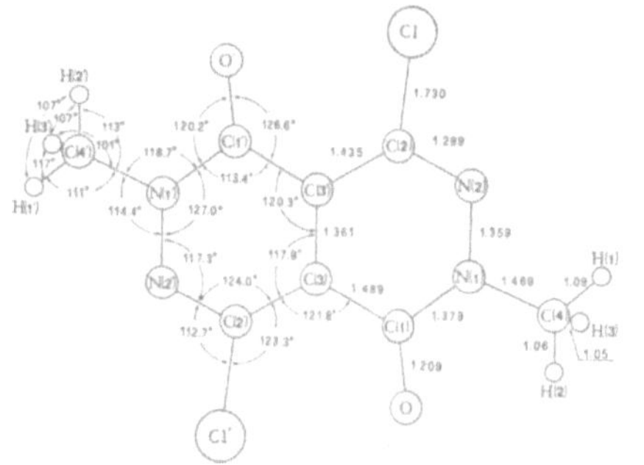

Fig. 1. Bond lengths and angles in 2,6-dimethyl-4,8-dichloro-2H,6H-pyridazino[4,5-d]pyridazin-1,5-dione.

The heterocyclic ring is approximately planar. The maximum displacement for the ring atoms is 0.027 Å. The methyl group is out of the plane by 0.089 Å; the adjacent oxygen and chlorine atoms are displaced in opposite directions by 0.096 and 0.149 Å respectively. This fact as well as the splaying apart of C-Cℓ and C-O bonds by 3 and 7° respectively (Fig. 1) is due to steric hindrance. From consideration of bond distances and angles, a mesomeric form of the type

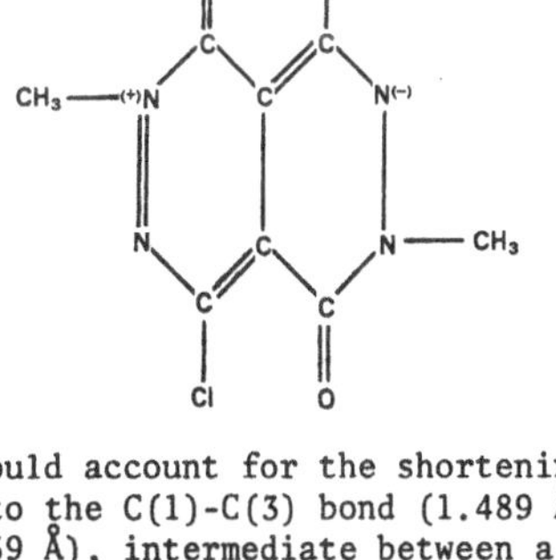

seems favoured. This could account for the shortening of the C(2)-C(3') bond (1.435 Å) with respect to the C(1)-C(3) bond (1.489 Å), and for the value of the N(1)-N(2) bond (1.359 Å), intermediate between a single and a double bond. In addition it may explain N(1) being closer to sp^2 than sp^3 hybridization.

2,3-DIAZANAPHTHALENE

$C_8H_6N_2$

C. HUISZOON, B.W. van de WAAL, A.B. van EGMOND and S. HARKEMA, 1972. Acta Cryst., B28, 3415-3419.

Orthorhombic, Pbca, a = 13.695, b = 10.557, c = 9.285 Å, D_m = 1.279, Z = 8. Mo radiation, R = 0.039 for 703 reflexions.

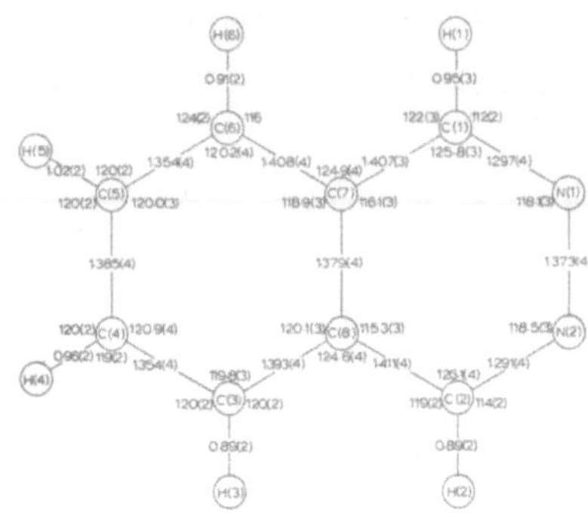

Fig. 1. Bond lengths and angles of 2,3-diazanaphthalene.

The molecule (Fig. 1) is not completely planar, in agreement with the observation that non-planarity can be expected when the lone pairs of electrons on the nitrogen atoms are close to each other. The separation of the nitrogen atoms, measured perpendicularly to the best plane through the heavy atoms of the molecule is about 0.01 Å. Bond lengths are normal and the smallest bond angles are found at the nitrogen atoms, 118.1(3) and 118.5(3)°. The molecules are grouped in parallel pairs in the unit cell; the separation between the molecular planes of a parallel pair is 3.53 Å.

1,8-NAPHTHYRIDINE

$C_8H_6N_2$

A. CLEARFIELD, M.J. SIMS and P. SINGH, 1972. Acta Cryst., B28, 350-355.

Monoclinic, $P2_1/c$, a = 6.170, b = 10.485, c = 11.454 Å, β = 117.66°, D_m = 1.317, Z = 4. Cu radiation, R = 0.048 for 831 reflexions.

The molecule is non-planar (Fig. 1); the non-planarity is ascribed to repulsion of lone-pair electrons on the nitrogen atoms. Bond lengths are in Fig. 2. There are no unusual intermolecular contacts.

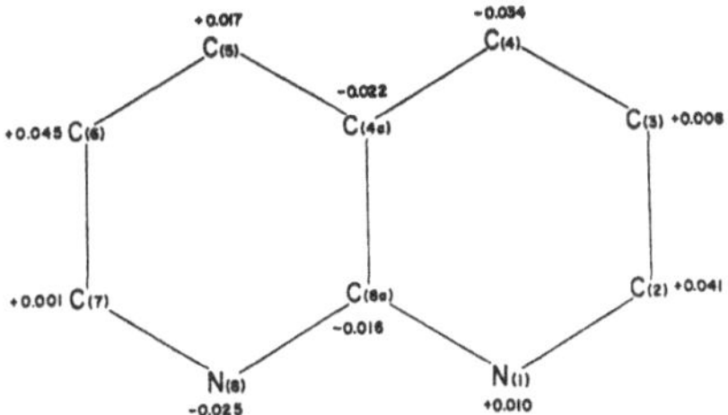

Fig. 1. 1,8-Naphthyridine: displacements of the ring atoms from their mean plane (σ = 0.002 - 0.0026 Å).

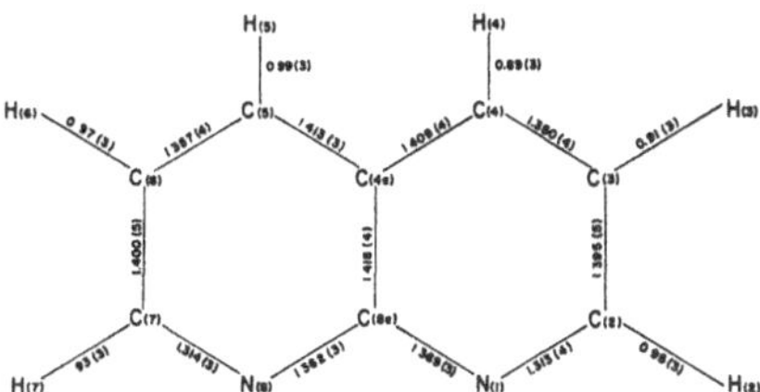

Fig. 2. Bond distances (with standard deviations in parentheses) in 1,8-naphthyridine.

1,4,5,8-TETRAMETHOXYPYRIDAZINO[4,5-d]PYRIDAZINE

$C_{10}H_{12}N_4O_4$

L. FANFANI, P.F. ZANAZZI and C. SABELLI, 1972. Acta Cryst., B28, 1178-1182.

Monoclinic, $P2_1/n$, a = 4.312, b = 16.826, c = 7.727 Å, β = 101.01°, D_m = 1.46, Z = 2. Cu radiation, R = 0.057 for 637 reflexions.

The bond lengths and angles are given in Fig. 1, and have been corrected for thermal motion. Agreement between observed and calculated U tensors supports the rigid-body assumption. Dimensions of the heterocyclic ring are very similar to those occurring in 2,6-dimethyl-4,8-dichloro-2H,6H-pyridazino-[4,5-d]pyridazin-1,5-dione (1). They may be explained by taking into account the contribution of a 'naphthalene-like' mesomeric formula (1). Bond lengths between oxygen atoms of the methoxyl groups and the ring carbons have a mean value of 1.353 Å. The methyl-oxygen distances are also regular; mean 1.442 Å. The shorter intermolecular distances correspond to van der Waals interactions.

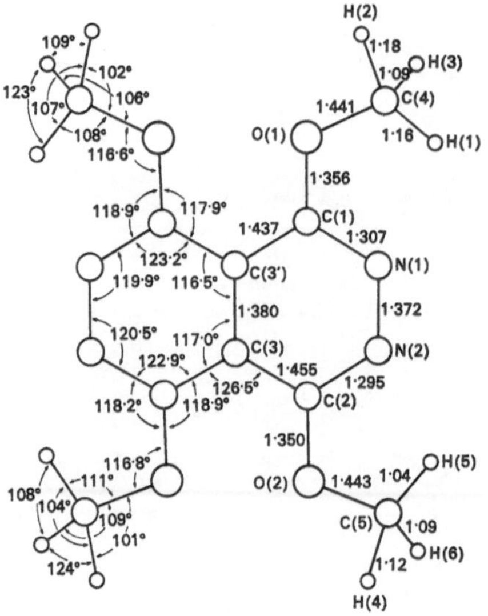

Fig. 1. Bond distances and angles in 1,4,5,8-tetramethoxypyridazino[4,5-d]-pyridazine.

1. This volume, p. 283.

2-METHOXY-4-HYDROXY-5-OXO-BENZ[f]AZEPINE

$C_{11}H_9NO_3$

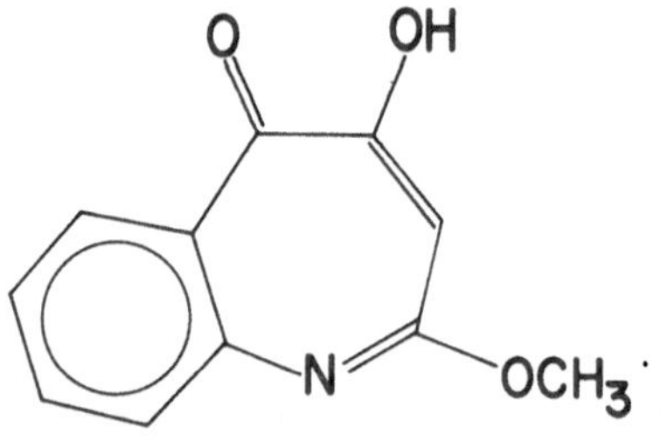

W.A. DENNE and M.F. MACKAY, 1972. Tetrahedron, 28, 1795-1802.

Orthorhombic, $P2_12_12_1$, a = 14.107, b = 16.924, c = 3.982 Å, Z = 4. Cu radiation, R = 0.033 for 876 reflexions.

The molecular skeleton is planar to within ±0.04 Å and the only atoms lying out of the plane are two hydrogens of the methyl group. The bond lengths (Fig. 1) are indicative of little π-electron delocalization in the seven-membered ring. In the crystal structure the hydroxyl group is involved in a bifurcated hydrogen bond; the hydroxyl hydrogen is only 2.007 Å from the adjacent keto-oxygen atom in the same molecule and, in addition, it is only 1.950 Å from the keto-oxygen of an adjacent molecule symmetry related by a twofold screw axis. The OH···O distances are 2.524 Å (intramolecular) and 2.827 Å (intermolecular). All other intermolecular contacts are of the van der Waals variety.

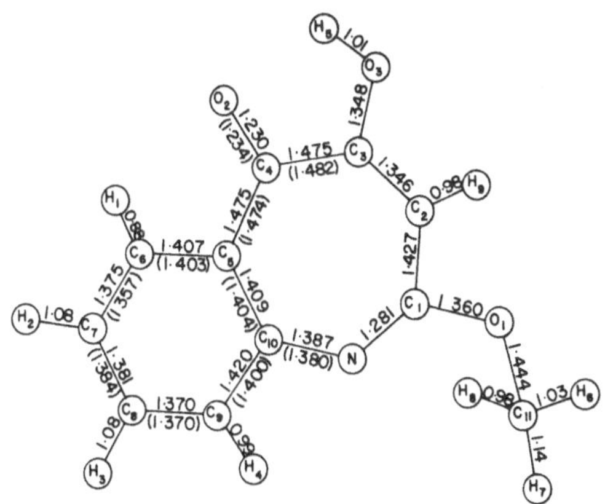

Fig. 1. Bond lengths in 2-methoxy-4-hydroxy-5-oxo-benz[f]azepine. The standard deviations for lengths involving H were 0.04 Å and for all others 0.006 Å. Some bond lengths for 5H-10,11-dioxodihydrodibenzo-[b,f]-azepine have been included in parenthesis (1).

1. W.A. DENNE and M.F. MACKAY, 1970. Tetrahedron, 26, 4435.

3-CARBETHOXY-4-OXO-6-METHYLHOMOPYRIMIDAZOLE

$C_{12}H_{12}N_2O_3$

N COOC₂H₅ CH₃ O

K. SASVÁRI, J.C. HORVAI and K. SIMON, 1972. Acta Cryst., B28, 2405-2416.

Monoclinic, $P2_1/a$, a = 8.085, b = 21.62, c = 12.91 Å, β = 98.8°, D_m = 1.38, Z = 8. Cu radiation, R = 0.102 for 1356 reflexions.

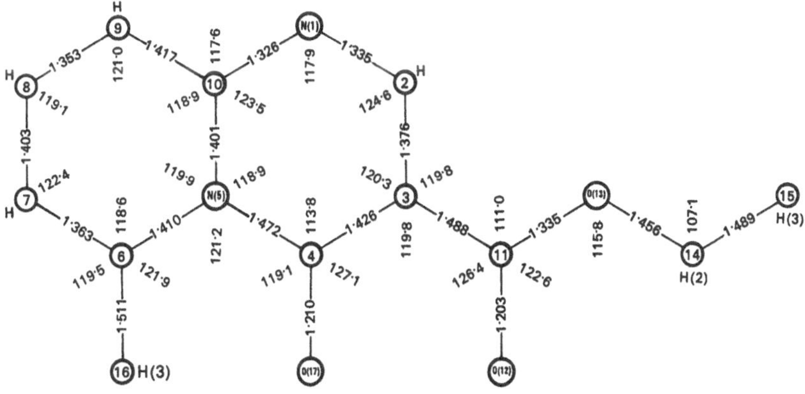

Fig. 1. Bond lengths (σ = 0.007 - 0.010 Å) and angles (σ = 0.4 - 0.6°) in 3-carboethoxy-4-oxo-6-methylhomopyrimidazole.

The averaged bond lengths and angles of the two crystallographically independent molecules in the unit cell are shown in Fig. 1. The bicyclic ring system is planar in both molecules, provided C(4) is excluded. The ring containing C(4) is thus in a flattened half-chair conformation. The molecules appear to be separated by van der Waals contacts only.

DILUMAZINE TRIHYDRATE

(DI-PTERIDINE-2,4-DIONE TRIHYDRATE)

$C_{12}H_{14}N_8O_7$ $2C_6H_4N_4O_2 \cdot 3H_2O$

R. NORRESTAM, B. STENSLAND and E. SÖDERBERG, 1972. Acta Cryst., B28, 659-666.

Monoclinic, $P2_1/c$, a = 14.865, b = 16.691, c = 6.829 Å, β = 109.80°, Z = 4. Cu radiation, R = 0.043 for 1543 reflexions.

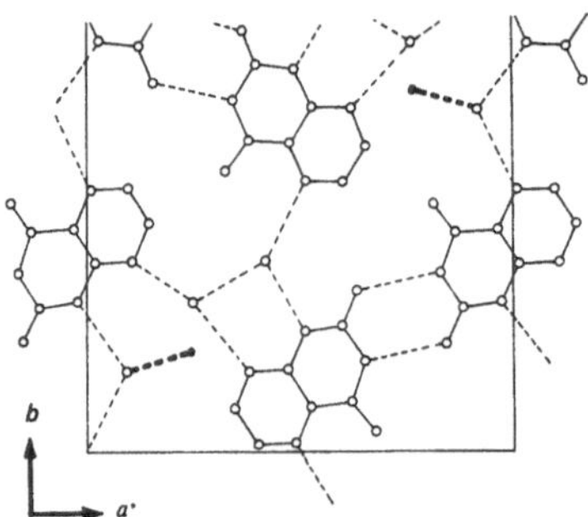

Fig. 1. Schematic drawing of the hydrogen bonds in dilumazine trihydrate between molecules at z ∿ 0.25. The thick dashed lines indicate hydrogen bonds in the c direction between two of the water molecules.

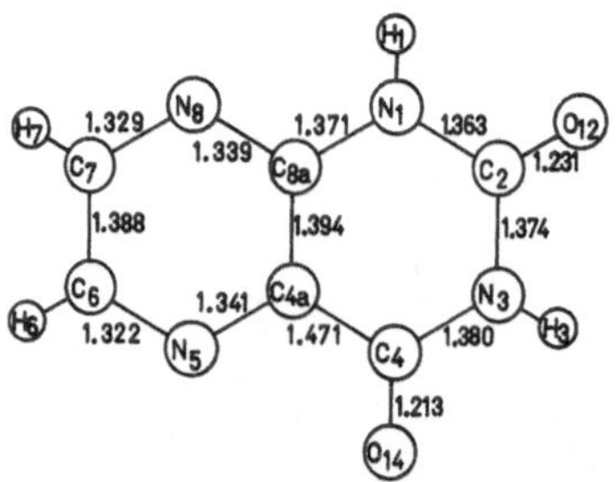

Fig. 2. Bond lengths (σ = 0.003 - 0.004 Å) in lumazine (values averaged over the two molecules in the asymmetric unit).

As shown in Fig. 1 the crystal structure is built up of an almost planar arrangement of hydrogen bonded dimers of lumazine. The lumazine molecules are almost planar and are held together in one plane by hydrogen bonds involving the water molecules. In both independent lumazine molecules in the asymmetric unit, all the nitrogens, but only one of the two keto-oxygens, are involved in the hydrogen bonding scheme. The other keto oxygen is involved in a short (3.16 Å) intermolecular contact with C(7). The π-bond scheme consistent with the observed molecular dimensions (Fig. 2) classifies the electrons of the π-pyrazinoid ring as delocalised.

5-HYDROXY-(N,N)-DIMETHYLTRYPTAMINE

(BUFOTENINE)

$C_{12}H_{16}N_2O$

G. FALKENBERG, 1972. Acta Cryst., B28, 3219-3228.

Monoclinic, $P2_1/a$, a = 17.95, b = 11.52, c = 14.24 Å, β = 131.29°, D_m = 1.205, Z = 8. Mo radiation, R = 0.054 for 1748 reflexions.

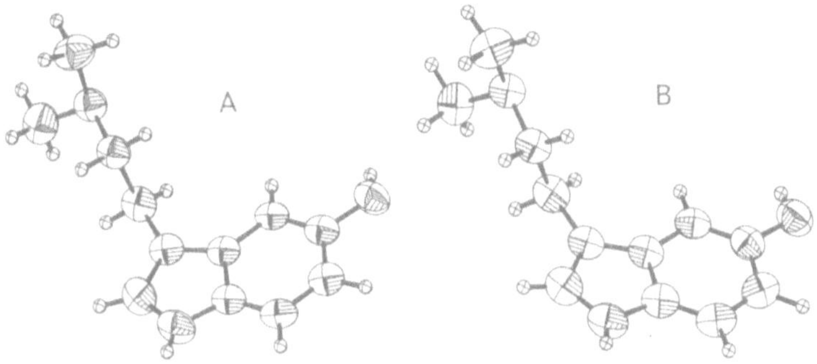

Fig. 1. Molecular conformation of molecules A and B of bufotenine.

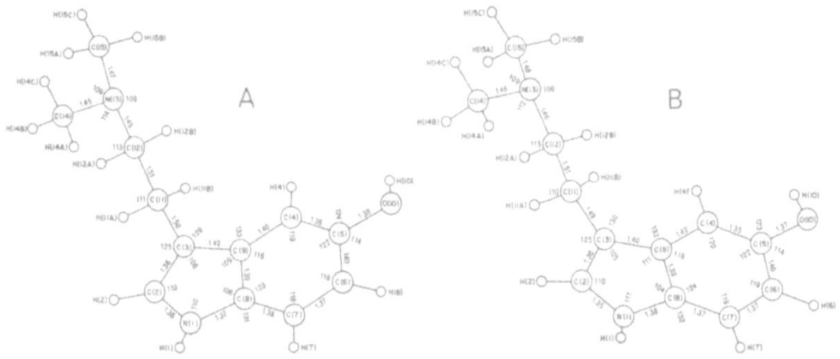

Fig. 2. Bond lengths (σ = 0.007 Å) and angles (σ = 0.5°) for molecules A and B of bufotenine.

The conformations of the two molecules in the asymmetric unit are almost identical as seen in Fig. 1. The indole nucleus of the A-molecule is planar; in the B-molecule there is a small angle of 1.5° between the pyrrole and benzene portions of the ring system. The ethylamine chain is fully extended, and lies approximately in a plane which is nearly perpendicular to the indole nucleus. This conformation is defined by the torsion angles C(2)C(3)C(11)C(12) = 87° and 72° for the A- and B-molecules, respectively. There are normal O-H···N hydrogen bonds linking the A-molecules (2.68 Å) and the B-molecules (2.72 Å) separately in two similarly arranged hydrogen bonding systems. The bond lengths and angles (Fig. 2) are consistent with values found for similar molecules.

1,6-DIMETHYL-3-CARBETHOXY-4-OXO-1,6,7,8,9,10-HEXAHYDRO-HOMOPYRIMIDAZOLE

$C_{13}H_{20}N_2O_3$

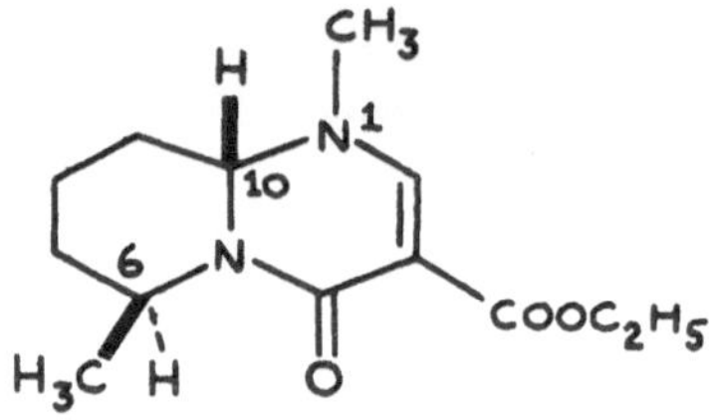

K. SIMON and K. SASVÁRI, 1972. Cryst. Struct. Comm., 1, 419-422.

Monoclinic, $P2_1/n$, a = 11.407, b = 9.441, c = 13.123 Å, β = 106.9°, D_m = 1.255, Z = 4. Cu radiation, R = 0.129 for 2143 reflexions.

The saturated six-membered ring in this molecule is in a chair conformation with the 6-methyl and 10-hydrogen axial. The N(1)-C(2) bond is very short (1.326(6) Å) compared with N(5)-C(4) (1.372(6) Å).

3-PHENYL-2,4-(1H,3H)-QUINAZOLINEDIONE

$C_{14}H_{10}N_2O_2$

Y. KITANO, M. KASHIWAGI and Y. KINOSHITA, 1972. Acta Cryst., B28, 1223-1231.

Monoclinic, $P2_1/c$, a = 5.799, b = 8.339, c = 22.962 Å, β = 94.16°, D_m = 1.369, Z = 4. Cu radiation, R = 0.104 for 2255 reflexions.

The bond lengths and angles in the benzene and quinazoline rings are normal. The mean C-N bond distance in the heterocyclic ring is 1.389 Å. The C-O distances are 1.201 and 1.213 Å (Fig. 1). The molecules pack in pairs around a centre of symmetry and are connected by two CO···HN hydrogen bonds of length 2.821 Å. The hydrogen bonding involves only O(1). The average deviation of the ten atoms of the quinazoline ring from the least-squares plane is 0.03 Å, with the maximum deviation of 0.07 Å for C(8). O(1) and O(2) are displaced from this plane by 0.173 and -0.220 Å respectively. The benzene ring is rotated by 65° from the quinazoline ring.

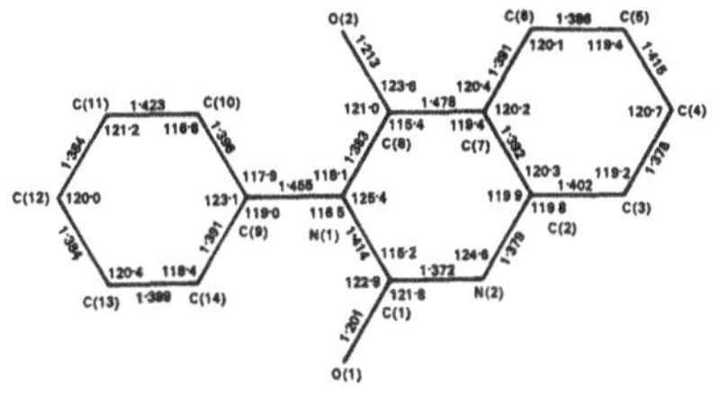

Fig. 1. Bond lengths and angles in 3-phenyl-2,4-(1H,3H)-quinazolinedione (σ = 0.005 Å and 0.4°).

7-CHLORO-1,3-DIHYDRO-1-METHYL-5-PHENYL-2H-1,4-BENZODIAZEPIN-2-ONE

(DIAZEPAM)

$C_{16}H_{13}C\ell N_2O$

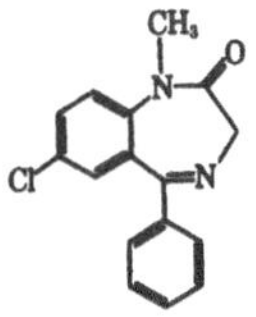

A. CAMERMAN and N. CAMERMAN, 1972. J. Amer. Chem. Soc., 94, 268-272.

Monoclinic, $P2_1/a$, a = 12.928, b = 13.354, c = 7.976 Å, β = 90.01°, D_m = 1.39, Z = 4. Mo radiation, R = 0.039 for 2438 reflexions.

The molecular structure is composed of a non-planar seven-atom heterocyclic ketone fused to a planar chlorophenyl ring and substituted with another planar phenyl group. The obtuse angle between the normals to the best planes through the two phenyl rings is 125.3°. Bond lengths are in Fig. 1. Both N atoms appear to be essentially trigonal in configuration and the shortening of the C(5)-N(8) bond from the single bond length of 1.47 Å suggests that there is some electron delocalisation between the chlorophenyl ring and the adjacent amide group. The conformation of the seven-membered ring is boat shaped with a fold between C(6) and C(12).

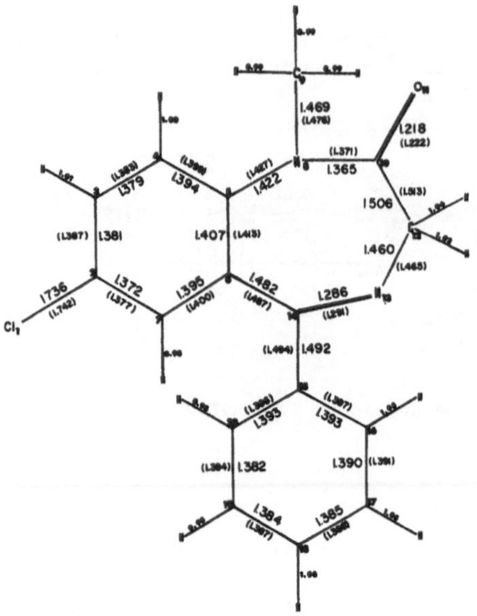

Fig. 1. Bond lengths in diazepam before correction for thermal motion of the atoms, and after correction (in parentheses). Standard deviations are 0.002 Å for bonds between non-hydrogen atoms and 0.02 Å for bonds involving hydrogen.

1-METHYL-3(N-METHYL-4'-CHLOROANILINO)-5-CHLOROINDOLE

$C_{16}H_{14}Cl_2N_2$

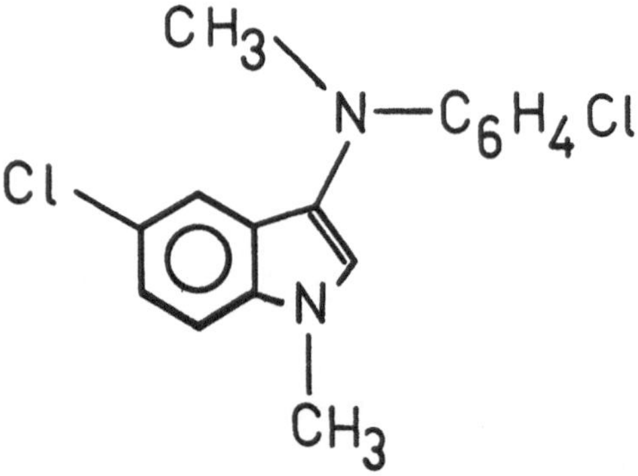

M. ZOCCHI, G. TIEGHI and A. ALBINATI, 1972. Cryst. Struct. Comm., 1, 139-142.

Monoclinic, $P2_1/c$, a = 13.96, b = 13.34, c = 8.96 Å, β = 114°33', Z = 4. Mo radiation, R = 0.086 for 1408 reflexions.

The interatomic distances and bond angles are generally in good agreement with already published values. The average C-C bond length is 1.392 Å with a r.m.s. deviation of 0.022 Å (average calculated on the three aromatic rings). The planes defined by the indole rings and by the anilinic moiety are almost perpendicular.

INDOLE DERIVATIVES

N.W. GILMAN, J.F. BLOUNT and L.H. STERNBACH, 1972. J. Org. Chem., 37, 3201-3206.

5-CHLORO-3-CYANO-2-DIMETHYLAMINO-3-PHENYL-3H-INDOLE

$C_{17}H_{14}C\ell N_3$

Triclinic, P$\bar{1}$, a = 10.758, b = 11.899, c = 12.423 Å, α = 88.91, β = 72.41, γ = 83.08°, D_m = 1.30, Z = 4. Cu radiation, R = 0.041 for 3920 reflexions.

5-CHLORO-2-DIMETHYLAMINO-3-METHOXYIMINOMETHYL-3-PHENYL-3H-INDOLE

$C_{18}H_{18}C\ell N_3O$

Orthorhombic, $P2_12_12_1$, a = 10.698, b = 10.252, c = 15.259 Å, D_m = 1.30, Z = 4. Cu radiation, R = 0.048 for 724 reflexions.

The structures were determined during the course of a study of quinazoline and 1,4-benzodiazepine reactions. For both molecules, bond lengths and angles are normal. The stereochemistry of $C_{17}H_{14}C\ell N_3$ is shown in Fig. 1 and that for $C_{18}H_{18}C\ell N_3O$ is in Fig. 2.

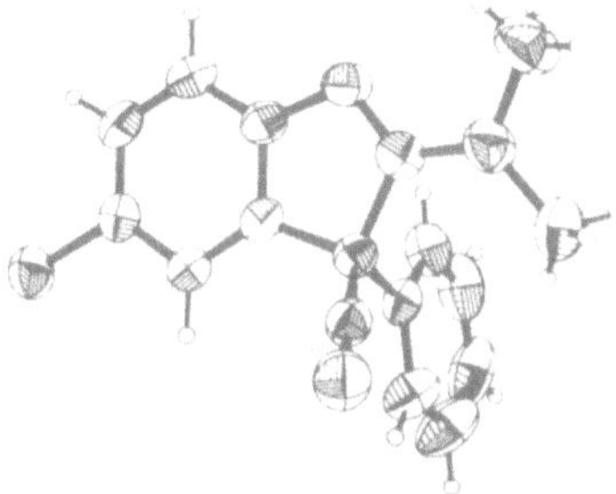

Fig. 1. The stereochemistry of the indole, $C_{17}H_{14}C\ell N_3$.

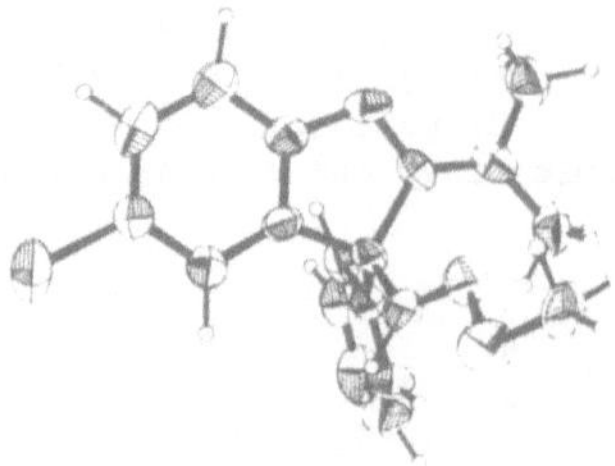

Fig. 2. The stereochemistry of the indole, $C_{18}H_{18}C\ell N_3O$.

1-(p-CHLOROBENZOYL)-5-METHOXY-2-METHYLINDOLE-3-ACETIC ACID (INDOMETHACIN)

$C_{19}H_{16}C\ell NO_4$

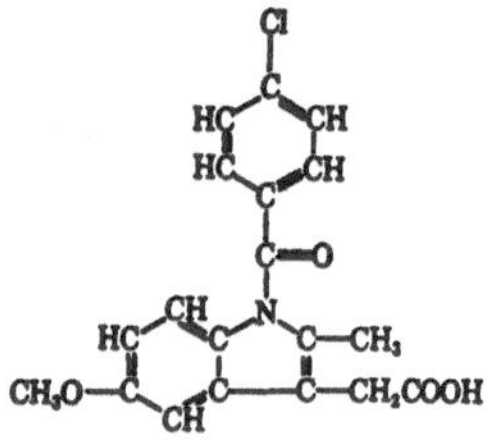

T.J. KISTENMACHER and R.E. MARSH, 1972. J. Amer. Chem. Soc., 94, 1340-1345.

Triclinic, $P\bar{1}$, a = 9.295, b = 10.969, c = 9.742 Å, α = 69.38, β = 110.79, γ = 92.78°, D_m = 1.38, Z = 2. Cu radiation, R = 0.059 for 3678 reflexions.

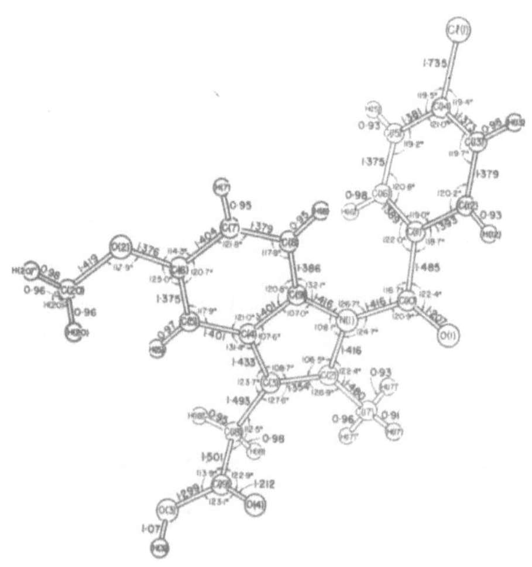

Fig. 1. The molecular geometry and numbering scheme in indomethacin.

The molecular dimensions are shown in Fig. 1. The methyl substituent at C(2) prevents the carbonyl group C(10)-O(1) from being coplanar with the indole ring system, thus reducing the double bond character of the N(1)-C(10) bond; the torsion angle O(1)-C(10)-N(1)-C(2) is -25.5(3)°. Bond angles at N(1) are also distorted. Even with these distortions the O(1)···C(17) intramolecular distance is only 2.853 Å. Steric requirements also preclude the benzene ring from being coplanar with the carbonyl group. The indole, p-chlorophenyl and carboxylic acid groups are each nearly planar. The relative orientation of the p-chlorophenyl and indole groups is unusual and may play an important role in the chemistry of indomethacin. In the crystal, pairs of molecules are linked by O-H···O hydrogen bonds (2.667 Å) about centres of symmetry. There is also overlap of the indole ring with the acetic acid group of another molecule.

2-CYANO-2-[3-CYANO-4-DIETHYLAMINO-1H-QUINOLYLIDENE-2]-N,N-DIETHYLACETAMIDE

$C_{21}H_{25}N_5O$

J.A. LERBSCHER and J. TROTTER, 1972. J. Cryst. Mol. Struct., 2, 67-77.

Monoclinic, $P2_1/c$, a = 12.680, b = 11.844, c = 13.550 Å, β = 104.13°, D_m = 1.24, Z = 4. Cu radiation, R = 0.076 for 2340 reflexions.

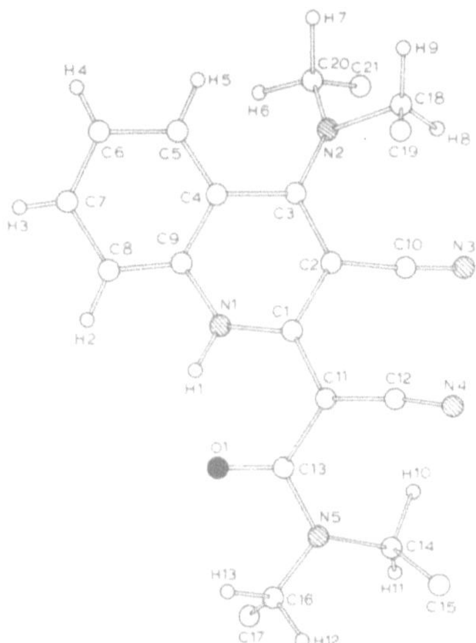

Fig. 1. A view of the 2-cyano-2-[3-cyano-4-diethylamino-1H-quinolylidene-2]-N,N-diethylacetamide molecule.

Neither the six-membered homocyclic ring nor the six-membered heterocyclic ring is exactly planar. The hydrogen atom which is bonded to the ring nitrogen atom forms an intramolecular hydrogen bond with the carbonyl oxygen atom, O···N = 2.540 Å (Fig. 1). Bond lengths are: C-C = 1.364 - 1.522, C-N = 1.151 - 1.480, C-O = 1.256 Å, σ(C-C, C-N, C-O) = 0.004 - 0.007 Å. The molecular geometry and the detailed bond distances suggest that electron delocalization into the side chains is stabilized by the intramolecular hydrogen bond.

DIMETHYL 2,3-DIHYDRO-2-INDOL-3-YLBENZ[b]AZEPINE-3,4-DICARBOXYLATE

$C_{22}H_{20}N_2O_4$

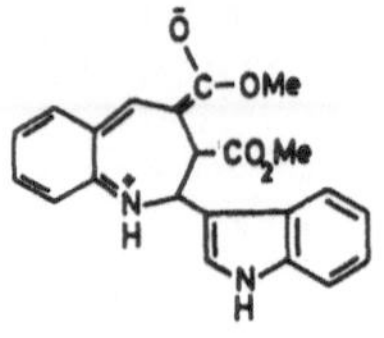

R.M. ACHESON, J.N. BRIDSON and T.S. CAMERON, 1972. J. Chem. Soc., Perkin I, 968-975.

Triclinic, P$\bar{1}$, a = 10.46, b = 11.50, c = 8.85 Å, α = 91.5, β = 114.6, γ = 106.2°, D_m = 1.31, Z = 2. Cu radiation, R = 0.114 for 2989 reflexions.

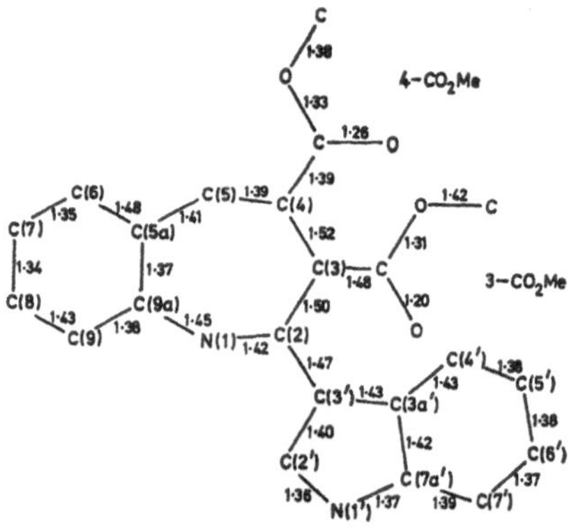

Fig. 1. Bond lengths in the $C_{22}H_{20}N_2O_4$ molecule.

The analysis confirms the structure deduced from chemical data. Bond lengths and numbering scheme are in Fig. 1. The angles in the seven-membered ring are all larger than expected for a strain-free conformation, the angle at the nitrogen atom (131°) being particularly large. A feature is that the N(1), C(4), C(5), C(5a), and C(9a) and the 4-carbonyl group with the associated oxygen atom, are coplanar. Resonance interactions between this carbonyl group and the lone pair on the nitrogen atom can therefore occur, and the small deviations observed from the accepted mean bond lengths between carbon, nitrogen and oxygen atoms are consistent with a contribution from the canonical form shown above.

DIMETHYL 1,2-BIS-(4-BROMOPHENYL)-3,5-DIMETHYL-4-OXO-6,7-DIAZABICYCLO[3,2,0]HEPT-2-ENE-6,7-DICARBOXYLATE

$C_{23}H_{20}Br_2N_2O_5$

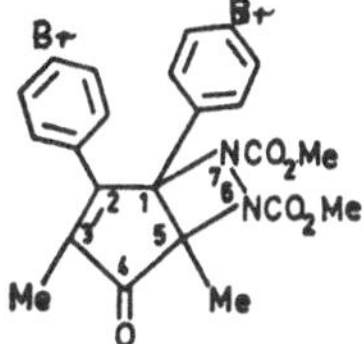

P.C. CHIEH, D. MACKAY and L. WONG, 1972. J. Chem. Soc., Perkin II, 2094-2097.

Monoclinic, $P2_1/n$, a = 12.395, b = 15.527, c = 11.773 Å, β = 92.8°, D_m = 1.64, Z = 4. Cu radiation, R = 0.071 for 1959 reflexions.

In the [3,2,0]-bicyclic system the five-membered ring is planar but the four-membered ring is slightly buckled with its component atoms ±0.057 Å from their best plane. The angle between the four-atom ring plane and that of the five-membered ring is 112°. Bonded distances in the four-membered ring are C-C = 1.582(17), C-N = 1.550 and 1.529(15), and N-N = 1.467(14) Å.

5,5',6,6'-TETRACHLORO-1,1',3,3'-TETRAETHYLBENZIMIDAZOLO-CARBOCYANINE IODIDE (SOLVATES WITH METHANOL AND ACETONITRILE)

$C_{27}H_{35}Cl_4IN_4O_2$ $C_{25}H_{27}Cl_4IN_4 \cdot 2CH_3OH$

$C_{27}H_{30}Cl_4IN_5$ $C_{25}H_{27}Cl_4IN_4 \cdot CH_3CN$

D.L. SMITH and H.R. LUSS, 1972. Acta Cryst., B28, 2793-2806.

	Methanol solvate (DYEM)	Acetonitrile solvate (DYEA)
Crystal system	Monoclinic	Triclinic
Space group	$P2_1/a$	P1
a(Å)	22.547	10.392
b	11.036	8.242
c	13.375	9.284
α (deg.)		93.92
β	107.48	107.81
γ		77.71
D_m	1.49	1.51
Z	4	1

Mo radiation, R = 0.033 and 0.040 for 2971 and 3137 reflexions.

The distances and angles (Fig. 1) in the cations of the two structures agree very well with each other and with accepted values. In spite of extended conjugation the cations are only approximately planar. In both structures the cations pack plane-to-plane and edge-to-edge in sheets parallel to (100), separated by sheets containing anions and solvent molecules. The $C_{25}H_{27}Cl_4IN_4$

compound is an efficient spectral sensitizer of silver halide photographic systems and there is a remarkable similarity of the (100) projections of the cation sheets in the two structures to each other and to the arrangement of ions in (111) faces of AgBr.

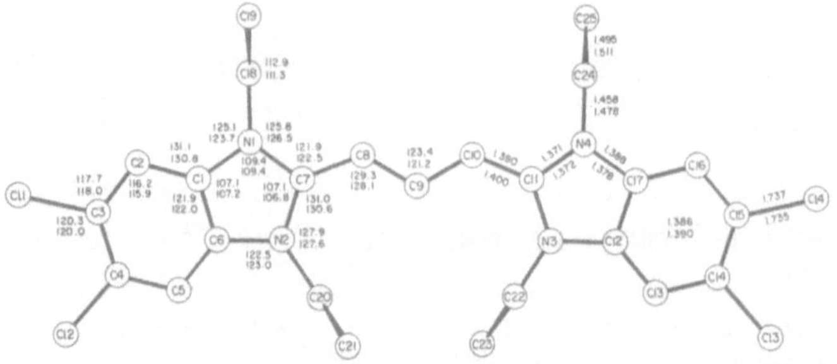

Fig. 1. Mean bond lengths and angles for the cations in DYEM and DYEA.

1,1-BIS(5-METHYL-2-PHENYL-1-BENZOYL-3-INDOLIZINYL)ETHYLENE

$C_{46}H_{34}N_2O_2$

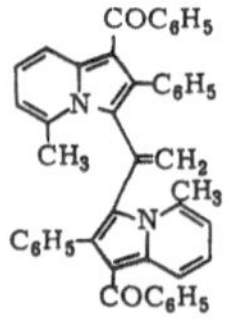

E. OESER, 1972. Chem. Ber., 105, 2351-2357.

Monoclinic, $P2_1/c$, a = 16.83, b = 12.17, c = 19.31 Å, β = 117.7°, Z = 4. Cu radiation, R = 0.073 for 4217 reflexions.

The material $C_{46}H_{34}N_2O_2$ obtained from 5-methyl-2-phenyl-1-benzoylindolizine and acetic anhydride has been shown to be the title compound by X-ray analysis. Details of the molecular geometry are in Fig. 1. The molecule almost has two-fold (non-crystallographic) symmetry, as can be seen in Fig. 2.

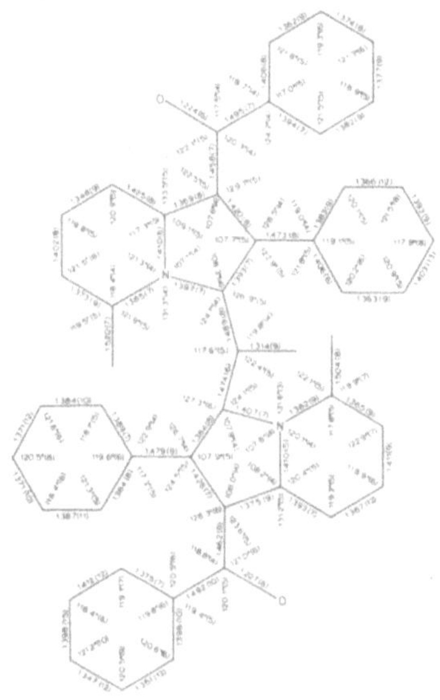

Fig. 1. Bond lengths and angles (with standard deviations in parentheses) in $C_{46}H_{34}N_2O_2$.

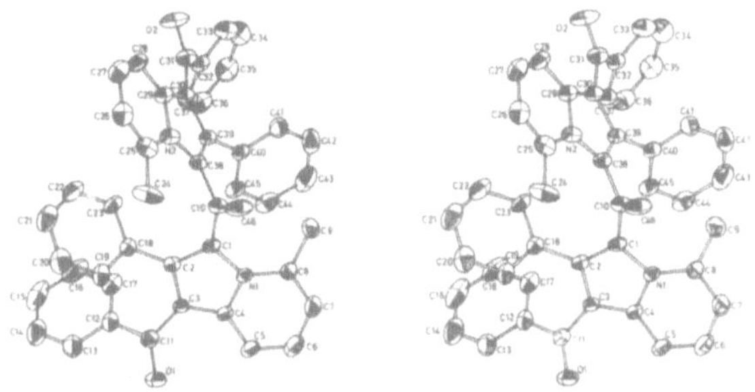

Fig. 2. View of the $C_{46}H_{34}N_2O_2$ molecule.

5-KETO-1,5-DIHYDROBENZ[cd]INDOLE

$C_{11}H_7NO$

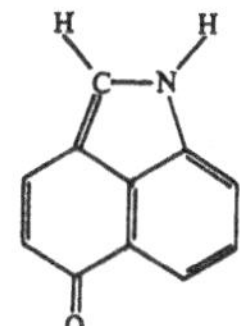

M.B. LAING, R.A. SPARKS and K.N. TRUEBLOOD, 1972. Acta Cryst., B28, 1920-1926.

Monoclinic, $P2_1/c$, a = 3.893, b = 13.39, c = 15.37 Å, β = 90°40', D_m = 1.414, Z = 4. Cu radiation, R = 0.071 for 1207 reflexions.

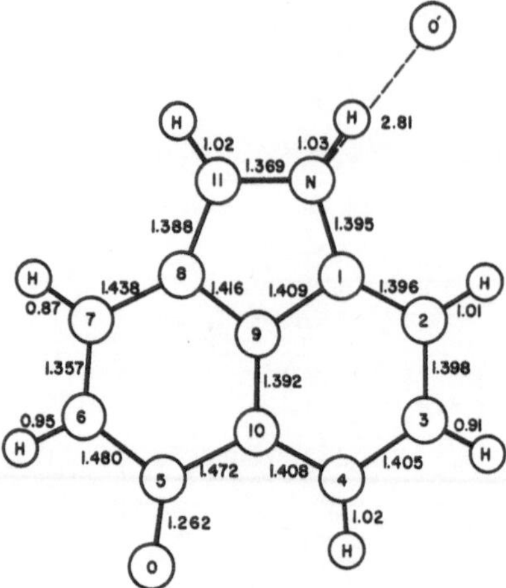

Fig. 1. Bond distances in 5-keto-1,5-dihydrobenz[c,d]indole. The standard deviations are about 0.004 Å for C-C and C-N bonds, 0.01 Å for C-O and N···O distances, and 0.04 Å for C-H and N-H distances.

The molecule exists predominantly in the quinonoid form shown above as evidenced by the bond lengths (Fig. 1). The strain introduced by the exocyclic bridge buckles the molecule, the angles between the normals to the planes of the fused six-membered rings being 1.4°. The molecules pack in a herringbone fashion and are linked in infinite chains by hydrogen bonds (N-H···O = 2.81 Å) parallel to c.

9,10-DIAZAPHENANTHRENE

$C_{12}H_8N_2$

H. van der MEER, 1972. Acta Cryst., B28, 367-370.

Monoclinic, $P2_1/c$, a = 6.945, b = 11.938, c = 11.802 Å, β = 110.84°, D_m = 1.33, Z = 4. Cu radiation, R = 0.05 for 1317 reflexions.

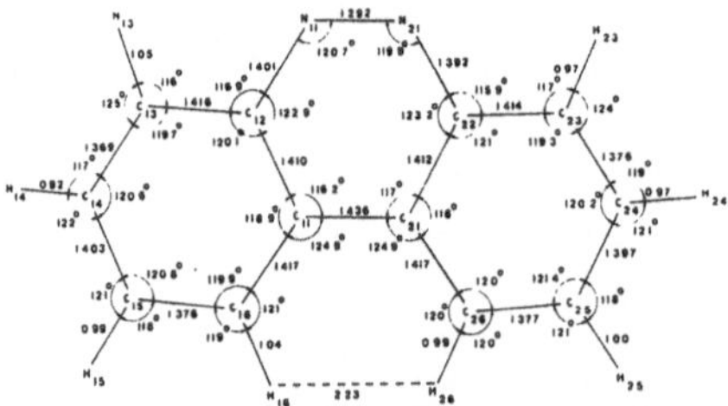

Fig. 1. Bond distances (σ = 0.003 - 0.005 Å; 0.03 - 0.04 Å for bonds involving H atoms) and angles in 9,10-diazaphenanthrene.

Intramolecular distances and angles in the molecule are in Fig. 1. The molecule is not planar and displacements from the mean molecular plane range from -0.030 Å (C(15)) to +0.039 Å (C(26)). In the crystal, molecules are arranged in pairs 3.49 Å apart, and the crystal structure can be considered as an assembly of isolated pairs of molecules.

DIPYRIDO[1,2-a:2',1'-c]PYRAZINIUM DIBROMIDE MONOHYDRATE

$C_{12}H_{12}Br_2N_2O$ $(C_{12}H_{10}N_2)^{2+}\cdot 2Br^-\cdot H_2O$

J.E. DERRY and T.A. HAMOR, 1972. Acta Cryst., B28, 1244-1247.

Orthorhombic, Pbca, a = 17.06, b = 22.50, c = 6.89 Å, Z = 8. Cu radiation, R = 0.094 for 1897 reflexions.

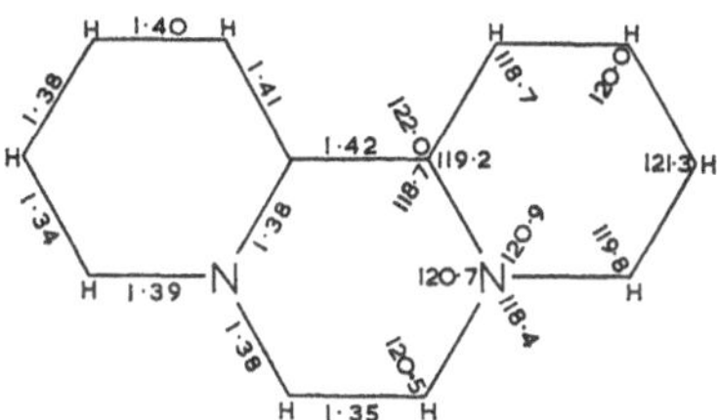

Fig. 1. Mean dimensions for the dipyrido[1,2-a:2',1'-c]pyrazinium ion. For bond lengths σ = 0.011 - 0.018 Å.

The crystal structure consists of near-planar organic cations interleaved with bromide ions approximately halfway between successive cations. A second crystallographically independent bromide ion and the water of crystallization are linked by hydrogen bonds in an infinite chain parallel to the cation (Br···HO = 3.19 and 3.32 Å). The organic moiety possesses a pseudo mirror plane and there are no significant differences between pairs of chemically equivalent lengths or angles (mean values are in Fig. 1). Valency angles are generally normal, but there appear to be some anomalies in the bond lengths. The three rings are each planar within experimental error; however, the two outer rings are displaced slightly in opposite directions from the plane of the central ring. This effect is probably a result of intramolecular overcrowding.

N,N-DIMETHYLTRYPTAMINE

$C_{12}H_{16}N_2$

G. FALKENBERG, 1972. Acta Cryst., B28, 3075-3083.

Monoclinic, $P2_1/c$, a = 12.99, b = 12.08, c = 18.38 Å, β = 127.85°, D_m = 1.080, Z = 8. Mo radiation, R = 0.048 for 2080 reflexions.

Fig. 1. Hydrogen bonding in dimethyltryptamine. Nitrogen atoms are represented by dotted circles; hydrogen bonds are dashed.

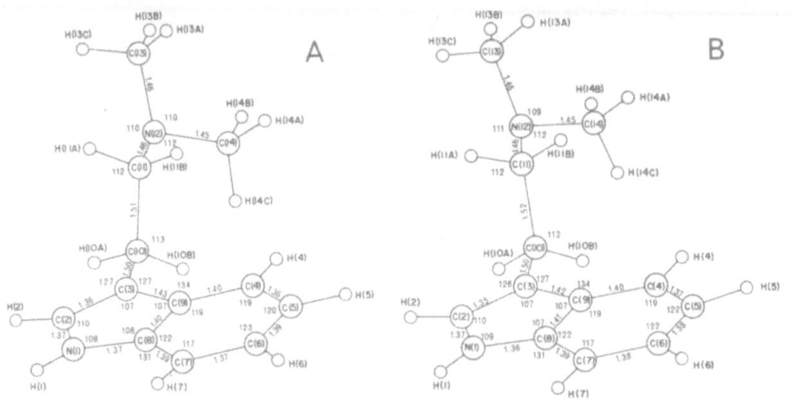

Fig. 2. Bond lengths (σ = 0.003 - 0.006 Å) and bond angles (σ = 0.3 - 0.5°) in dimethyltryptamine.

The two independent molecules in the asymmetric unit differ only slightly in conformation. The indole nuclei are planar and the ethylamine side chains are almost fully extended (torsion angles 176 and 188° for the two molecules). Each ethylamine side chain is approximately perpendicular to its indole nucleus. N-H···N hydrogen bonds link the molecules in two independent hydrogen bonding systems around screw axes (Fig. 1). Details of bond lengths and angles in the two molecules are in Fig. 2.

METHYL OROTATE trans-syn-PHOTODIMER DIHYDRATE

$C_{12}H_{16}N_4O_{10}$

G.I. BIRNBAUM, 1972. Acta Cryst., B28, 1248-1254.

Monoclinic, C2/c, a = 14.024, b = 6.722, c = 17.006 Å, β = 105.11°, D_m = 1.63, Z = 4. Cu radiation, R = 0.042 for 1230 reflexions.

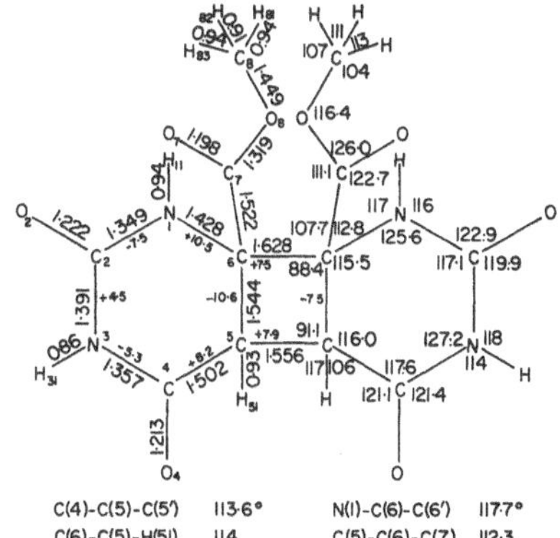

Fig. 1. Bond lengths, angles, and torsion angles in methyl orotate trans-syn-photodimer. For bonds not involving hydrogen all σ are 0.002 Å except C(8)-O(8) (0.003 Å).

The structure is that of a trans-syn cyclobutane dimer (Fig. 1). The molecule is situated on a C_2 axis bisecting the two interpyrimidine bonds. The pyrimidine ring is not planar but has the conformation of a very flat chair. The two pyrimidine rings are fused to a slightly puckered (dihedral angle, 169.5°) cyclobutane ring. The pyrimidine rings are twisted by 7° with respect to one another. The bond lengths are normal except for the long C(6)-C(6') bond (1.628 Å) due to the full substitution and the resulting steric repulsion. There is an extensive network of intermolecular hydrogen bonds. Each molecule is connected to two others by pairs of N(1)···O(4) bonds. The same molecules are also joined via water molecules, each of which is hydrogen-bonded to two pyrimidine rings. Each water molecule also joins a pair of pyrimidine rings to a third molecule of the dimer.

1,10-DICHLORO-3,8-DIMETHYL-4,7-PHENANTHROLINE

$C_{14}H_{10}C\ell_2N_2$

F.H. HERBSTEIN, M. KAPON and D. RABINOVICH, 1972. Israel J. Chem., 10, 537-558.

Triclinic, P$\bar{1}$, a = 7.651, b = 7.792, c = 12.301 Å, α = 108.48, β = 82.74, γ = 116.53°, D_m = 1.479, Z = 2. Mo radiation, R = 0.067 for 2000 reflexions.

The molecule has approximate twofold symmetry and is appreciably distorted from planarity (Fig. 1). The chlorine atoms are displaced +1.12 and -1.15 Å from the mean molecular plane. Bond lengths are close to normal values but there are appreciable distortions in bond angles and torsional angles (Fig. 2). There is hexagonal close packing of stacks of molecules, with van der Waals binding between stacks. Antiparallelism of adjacent molecules in a stack suggests that dipole-dipole forces are important in the intrastack binding.

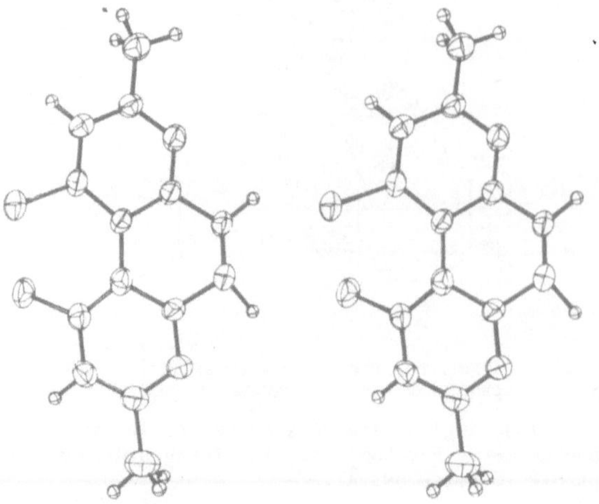

Fig. 1. View of 1,10-dichloro-3,8-dimethyl-4,7-phenanthroline showing the non-planarity of the molecule.

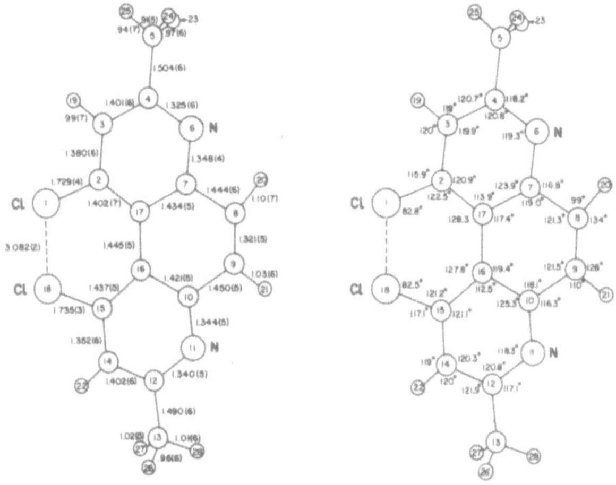

Fig. 2. Bond lengths and angles in 1,10-dichloro-3,8-dimethyl-4,7-phenanthroline.

3,7,8,10-TETRAMETHYLISOALLOXAZINE

(3-METHYL-LUMIFLAVIN)

$C_{14}H_{14}N_4O_2$

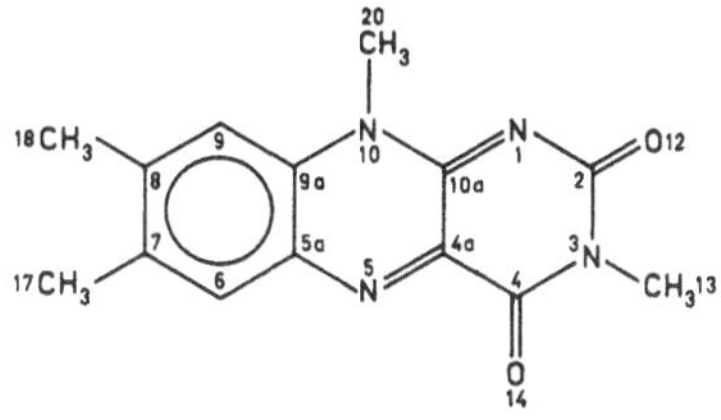

R. NORRESTAM and B. STENSLAND, 1972. Acta Cryst., B28, 440-447.

Monoclinic, $P2_1/c$, a = 15.509, b = 9.624, c = 8.865 Å, β = 107.88°, D_m not given, Z = 4. Cu radiation, R = 0.047 for 1538 reflexions.

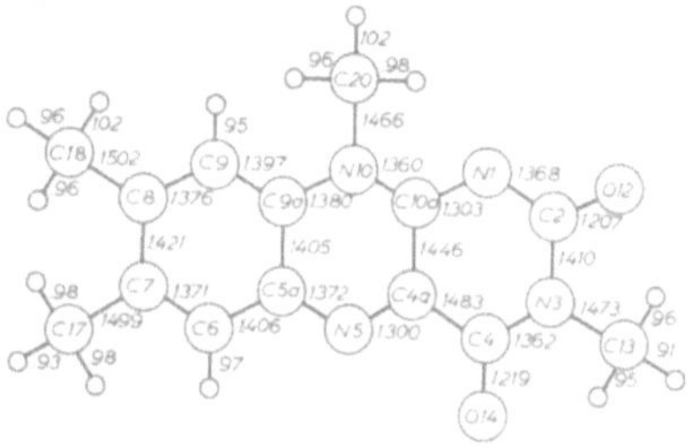

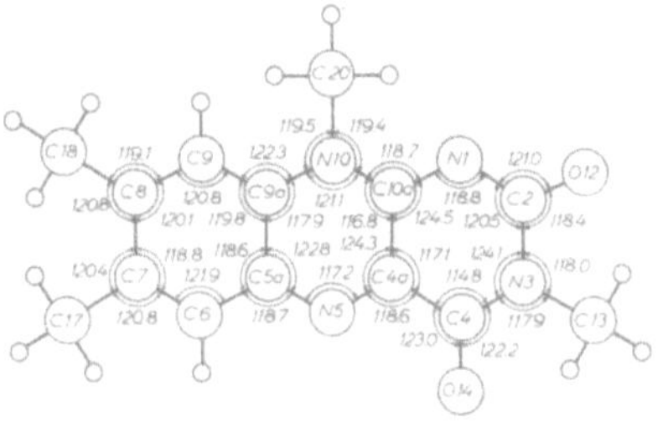

Fig. 1. Bond lengths (σ = 0.003 Å) and angles (σ = 0.2°) in 3-methyl-lumiflavin.

The isoalloxazine ring system of the molecule is essentially planar, as was suggested by interpretation of spectroscopic data. Bond distances and angles are in Fig. 1. Most intermolecular distances correspond to van der Waals contacts except between keto oxygen O(12) of one molecule and methyl carbon C(20) of another (3.114 Å), and also between O(12) and methyl carbon C(13) of an adjacent molecule (3.287 Å).

9-BROMO-3,7,8,10-TETRAMETHYLISOALLOXAZINE MONOHYDRATE

$C_{14}H_{15}BrN_4O_3$

M. von GLEHN and R. NORRESTAM, 1972. Acta Chem. Scand., 26, 1490-1502.

Monoclinic, $P2_1/c$, a = 13.047, b = 7.053, c = 17.957 Å, β = 120.40°, Z = 4. Mo radiation, R = 0.051 for 1283 reflexions.

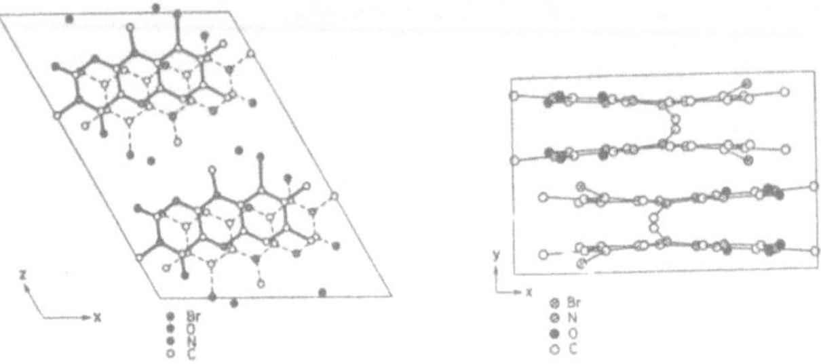

Fig. 1. Molecular packing diagrams for 9-bromo-3,7,8,10-tetramethyliso-alloxazine monohydrate.

The crystal structure is built up of an infinite stacking in the [010] direction of almost planar molecules with parallel molecular orientation. The intermolecular separation in the [010] direction is about 3.5 Å, a normal van der Waals separation (Fig. 1). The water molecules, located between the stacks, seem to be involved in a number of weak hydrogen bonds in the range 3.1 - 3.4 Å. The isoalloxazine ring system is almost planar, the major deviations from planarity occurring with C(20) and Br to relieve intramolecular overcrowding effects. The bond lengths are in accordance with the commonly accepted scheme for an oxidized flavin derivative as shown above.

9-BROMO-1,3,7,8,10-PENTAMETHYL-1,5-DIHYDROISOALLOXAZINE

$C_{15}H_{17}BrN_4O_2$

R. NORRESTAM and M. von GLEHN, 1972. Acta Cryst., B28, 434-440.

Monoclinic, C2/c, a = 15.680, b = 15.440, c = 12.845 Å, β = 103.03°, D_m not given, Z = 8. Cu radiation, R = 0.042 for 1892 reflexions.

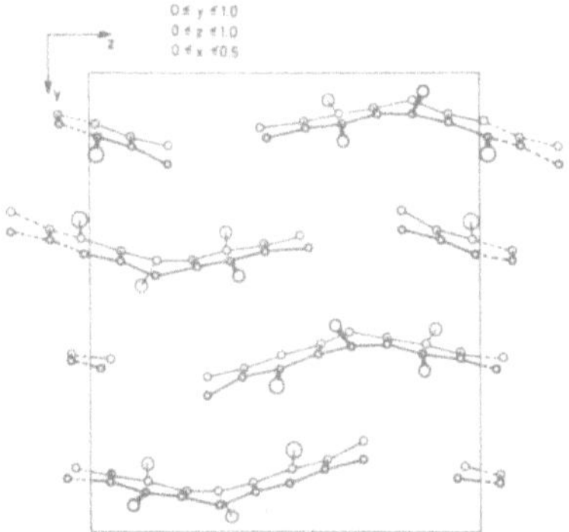

Fig. 1. Molecular packing diagram of the dihydroisoalloxazine derivative, $C_{15}H_{17}BrN_4O_2$, viewed along the a-axis. Only half of the cell content ($0<x<\frac{1}{2}$) is shown.

Fig. 2. Bond lengths (σ = 0.007 - 0.008 Å) and π-bond orders in the dihydroisoalloxazine derivative $C_{15}H_{17}BrN_4O_2$.

The atoms forming the alloxazine ring system of the molecule are close to two planes, one through the benzenoid ring and the other through the pyrimidinoid ring intersecting at an angle of 149°; this is shown clearly in Fig. 1. The molecules are held by van der Waals forces in the [010] and [001] directions but in the [100] directions molecules are connected in pairs by N-H···O(14) hydrogen bonds (H···O = 2.23 Å). Interatomic distances and estimated π-bond orders are in Fig. 2.

1,3,7,8,10-PENTAMETHYLISOALLOXAZINIUM IODIDE MONOHYDRATE

$C_{15}H_{19}IN_4O_3$ $C_{15}H_{17}N_4O_2^+ \cdot I^- \cdot H_2O$

R. NORRESTAM, L. TORBJÖRNSSON and F. MÜLLER, 1972. Acta Chem. Scand., 26, 2441-2454.

Orthorhombic, $P2_12_12_1$, a = 12.857, b = 18.221, c = 7.049 Å, D_m = 1.73, Z = 4. Cu radiation, R = 0.047 for 1365 reflexions.

In the crystal structure the iodine atoms, situated between two alloxazinium ions, are related to each other by one unit translation along the c axis, in the stacking sequence ...alloxazinium-iodide-alloxazinium-iodide... along z. Fairly short contact distances, 3.57 and 3.58 Å, occur between iodine and the two carbon atoms C(4a) and C(10a) of the alloxazinium ring. The molecular stacking, the short contact distances, and the deep red colour of the crystals suggest charge transfer interactions between the iodide and the alloxazinium ions to yield an axial charge transfer complex. The water molecule is hydrogen bonded to the iodide ion with O-H···O = 3.57 Å. The ring skeleton of the cation (Fig. 1) is nearly planar except for a slight (7.6°) departure of the pyrimidine ring from the plane. The bond lengths are consistent with the bonding pattern indicated in Fig. 1 but all the bonds in the pyrazinoid and pyrimidinoid rings have some double-bond character.

Fig. 1. Schematic drawing of the 1,3,7,8,10-pentamethylisoalloxazinium ion.

NAPHTHO[2,1-c]CINNOLINE

$C_{16}H_{10}N_2$

A.K. BHUIYA, E. STANLEY and J. CHAN, 1972. Z. Kristallogr., 135, 408-415.

Orthorhombic, Fdd2, a = 40.8, b = 10.8, c = 10.8 Å, D_m = 1.27, Z = 16. Cu radiation, R = 0.12 for 140 hk0 relfexions, and 0.10 for 75 h0ℓ reflexions.

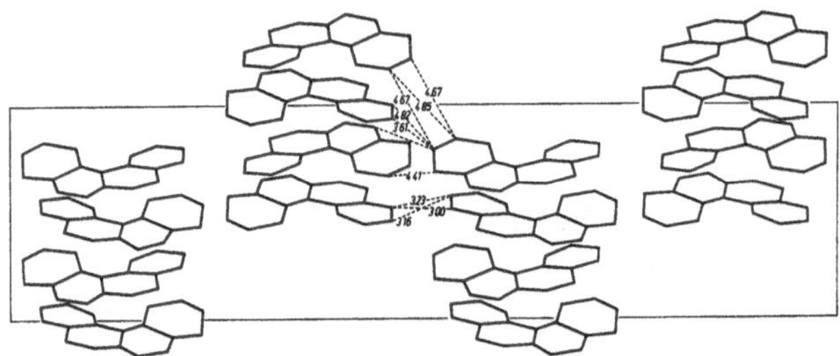

Fig. 1. Packing of naphtho[2,1-c]cinnoline seen in the [001] projection.

The molecule, an analogue of 3,4-benzophenanthrene, was only studied in the [001] and [010] projections. The positions of the nitrogen atoms could not be distinguished from carbon atoms. The molecular conformation and packing are shown in Fig. 1. The separation of the most overcrowded atoms C(1) and C(12) is 2.83(4) Å.

10-CHLORO-3,4-DIHYDRO-4,4,7,8-TETRAMETHYL-2-METHOXYPYRIMIDO[5,4-b]QUINOLINE

$C_{16}H_{18}C\ell N_3O$

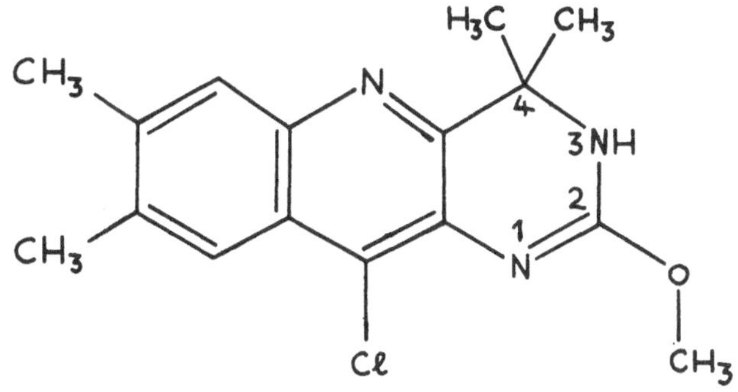

E. SHEFTER, E.M. LEVINE, P. SACKMAN and T.J. BARDOS, 1972. Cryst. Struct. Comm., 1, 283-286.

Monoclinic, $P2_1/c$, a = 10.154, b = 5.571, c = 26.603 Å, β = 97.22°, Z = 4. Cu radiation, R = 0.062 for 1545 reflexions.

The molecular parameters obtained for the quinoline portion of this molecule are quite similar to those found in the benzopyrazine portion of various isoalloxazine molecules. The resonance state in the pyrimidine part differs substantially from that found in other isoalloxazine structures whose structures have been reported. (Pyrimidine bond lengths are N(1)-C(2) = 1.265, C(2)-N(3) = 1.327, N(3)-C(4) = 1.456, C(4)-C(4a) = 1.527, C(4a)-C(1a) = 1.424, C(1a)-N(1) = 1.392 Å.) The difference is undoubtedly influenced by the presence of the C(4) gem-dimethyls. The N(3) proton is involved in an intermolecular hydrogen bond to the methoxyl oxygen (N-H···O = 3.197 Å).

5-ETHYL-3,7,8,10-TETRAMETHYLISOALLOXAZINIUM PERCHLORATE

$C_{16}H_{19}C\ell N_4O_6$

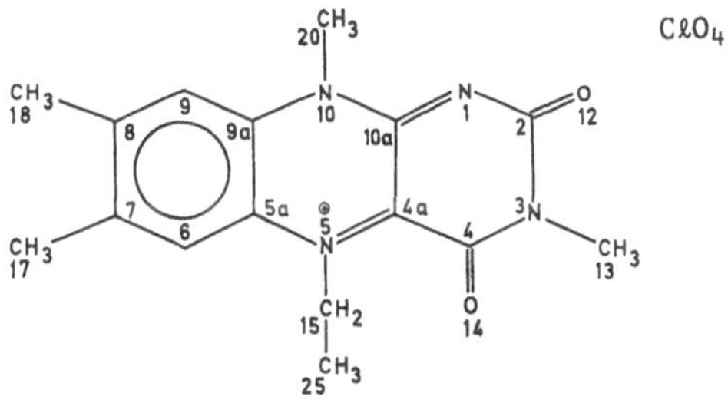

R. NORRESTAM and O. TILLBERG, 1972. Acta Cryst., B28, 1704-1712.

Orthorhombic, $P2_12_12_1$, a = 7.768, b = 13.255, c = 17.647 Å, D_m not given, Z = 4. Cu radiation, R = 0.042 for 1342 reflexions.

The isoalloxazine ring system is almost planar. The largest deviation from the least-squares plane is 0.08 Å. The deviations from planarity are consistent with a slight twist of the molecule along its elongation, possibly caused by repulsive forces between the methylene group at C(15) and the keto oxygen atom

O(14). From the molecular geometry it would appear that the π-electrons are more delocalised than shown above and the positive charge is probably in the C(10a) region. This is also supported by very short (2.8 - 3.0 Å) intermolecular contacts between N(10), C(10a), and C(4a) and various oxygen atoms of adjacent ions, of the same order as those found in crystal structures of strong charge-transfer complexes. Another interesting feature is the long (1.529(6) Å) bond N(5)-C(15).

2-[2-(6-CHLORO-2-METHOXY-9-ACRIDINYLAMINO)ETHYLAMINO]ETHANOL

$C_{18}H_{20}C\ell N_3O_2$

$NH\text{-}CH_2\text{-}CH_2\text{-}NH\text{-}CH_2\text{-}CH_2\text{-}OH$

OCH_3

Cl N

J.P. GLUSKER, J.A. MINKIN and W. OREHOWSKY, 1972. Acta Cryst., B28, 1-8.

Monoclinic, $P2_1/c$, a = 23.027, b = 7.676, c = 9.923 Å, β = 100.12°, D_m = 1.30, Z = 4. Cu radiation, R = 0.066 for 2148 reflexions.

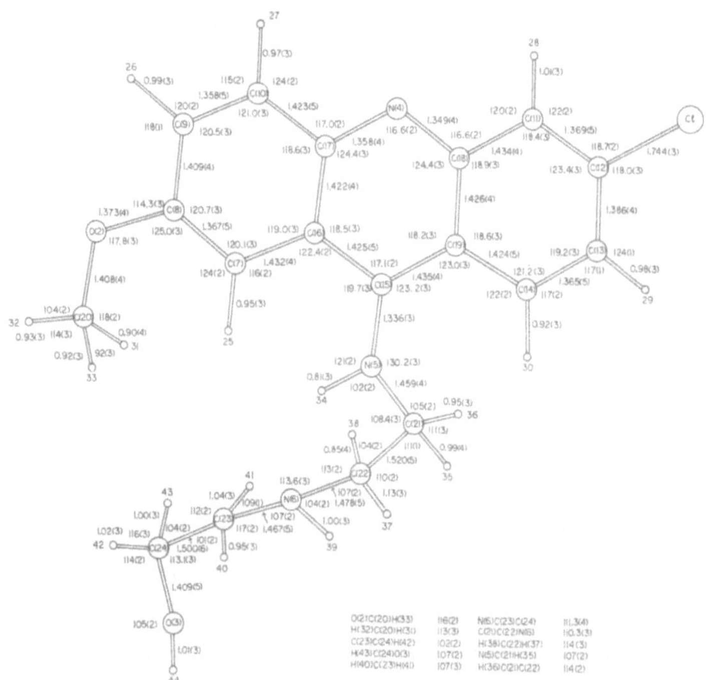

Fig. 1. Interatomic distances and interbond angles in ICR-171-OH.

The acridine portion of the molecule is not planar, the buckling being not simply a folding of the molecule but corresponding to an angle of 7.5° between the planes of the two outer rings which are themselves planar. Interatomic distances and angles are in Fig. 1. There is an intramolecular N-H···N hydrogen bond (2.72 Å) between nitrogen atoms of the side chain. The molecules are held in the crystal by O-H···N (2.80 Å) and N-H···O (2.94 Å) hydrogen bonds. The compound has been given the code number ICR-171-OH and is the precursor of a compound which is a mutagen for Salmonella but not an active antitumor agent.

4a-ALLYL-3,5,7,8,10-PENTAMETHYL-4a,5-DIHYDROISOALLOXAZINE

$C_{18}H_{22}N_4O_2$

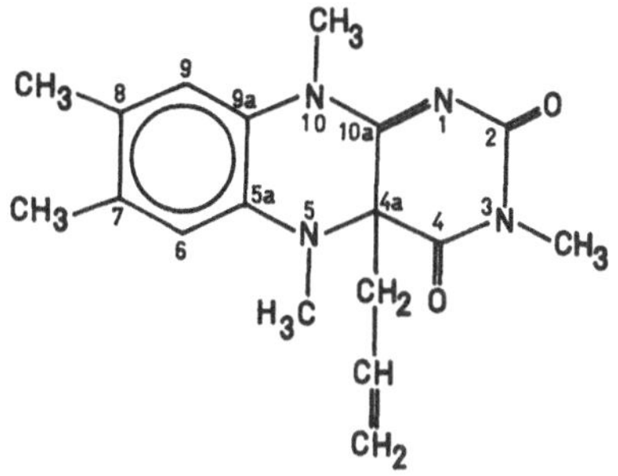

R. NORRESTAM, 1972. Acta Cryst., B28, 1713-1720.

Orthorhombic, $P2_12_12_1$, a = 17.422, b = 12.777, c = 7.564 Å, D_m not given, Z = 4. Cu radiation, R = 0.035 for 1608 reflexions.

The isoalloxazine ring system is not planar. The largest deviations from the least-squares plane occur for atoms C(4), C(4a), and N(5) (-0.23, -0.46 and +0.25 Å respectively). This distortion is due to the C(sp^3) character of the C(4a) atom. The observed bond lengths imply some delocalisation of π-electrons over the region N(10)-C(10a)-N(1)-C(2)-O(12). The bond C(4a)-C(14a) between the isoalloxazine ring and the allyl group is long, 1.571(3) Å. Intermolecular contacts suggest that the crystal structure is held together by van der Waals forces.

2-[3-(6-CHLORO-2-METHOXY-9-ACRIDINYLAMINO)PROPYLAMINO]ETHANOL

(ICR-191-OH)

$C_{19}H_{22}ClN_3O_2$

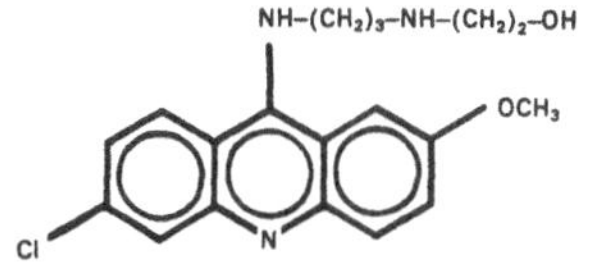

H.L. CARRELL, 1972. Acta Cryst., B28, 1754-1759.

Monoclinic, A2/a, a = 23.957, b = 10.086, c = 15.519 Å, β = 110.92°, D_m = 1.36, Z = 8. Cu radiation, R = 0.055 for 2520 reflexions.

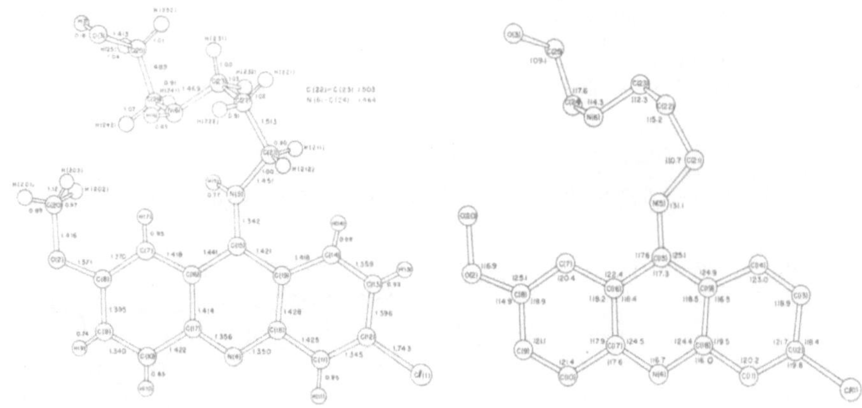

Fig. 1. Bond lengths (Å) and angles (°) in ICR-191-OH. Standard deviations are 0.003 Å except for those involving hydrogen (0.03 Å).

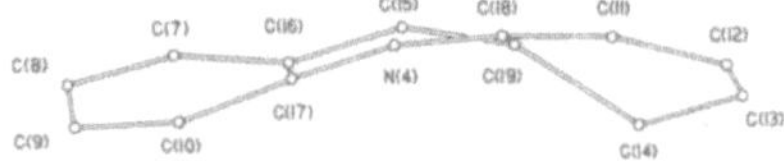

Fig. 2. A distorted view of the acridine portion of the ICR-191-OH molecule (vertical scale has been increased eightfold).

Bond lengths and angles in the molecule are shown in Fig. 1. The acridine portion of the molecule is not planar (Fig. 2); the fused six-membered rings both fold and twist about the C(15)···N(4) line. The angle between the best least-squares planes of the outer rings of the acridine system is 4.9°. The side chain is curled rather than in a zig-zag pattern and there is an intramolecular (N(5)-H···N(6) = 2.88 Å) hydrogen bond. In the crystal structure chains of molecules are linked by hydrogen bonds, O(3)-H···N(4) = 2.75 Å and N(6)-H···O(3) = 3.11 Å.

CONDENSATION PRODUCTS OF SUCCINALDEHYDE AND p-BROMOPHENYL HYDRAZINE

I.K. LARSEN, 1972. Acta Cryst., B28, 1136-1151.

(4aRS,4bSR,13bRS)-12-BROMO-1-(p-BROMOPHENYL)-1,4a,4b,5,6,13b-HEXAHYDRO-4H-DIPYRIDAZINO-[1,6-a:4,3-c]QUINOLINE

$C_{20}H_{18}Br_2N_4$

Racemic form

Monoclinic, $P2_1/c$, a = 12.20, b = 10.12, c = 16.37 Å, β = 110.4°, D_m = 1.66,

Z = 4. Mo radiation, R = 0.101 for 1402 reflexions.

Optically active form

Orthorhombic, $P2_12_12_1$, a = 13.11, b = 14.96, c = 9.412 Å, D_m = 1.70, Z = 4. Mo radiation, R = 0.108 for 660 reflexions.

(4aRS,4bRS,13bRS)-12-BROMO-1-(p-BROMOPHENYL)-1,4a,4b,5,6,13b-HEXAHYDRO-4H-DIPYRIDAZINO-[1,6-a:4,3-c]QUINOLINE

$C_{20}H_{18}Br_2N_4$

Racemic form

Monoclinic, $P2_1/c$, a = 7.429, b = 15.444, c = 15.991 Å, β = 93.47°, D_m = 1.70, Z = 4. Mo radiation, R = 0.086 for 1327 reflexions.

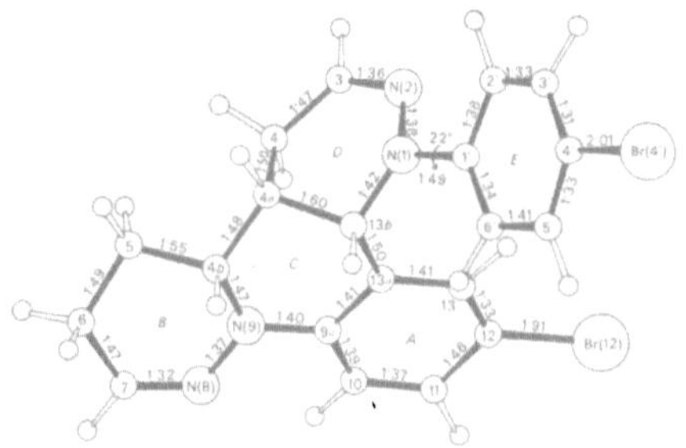

Fig. 1. The molecular structure of the racemic form (A) of $C_{20}H_{18}Br_2N_4$.

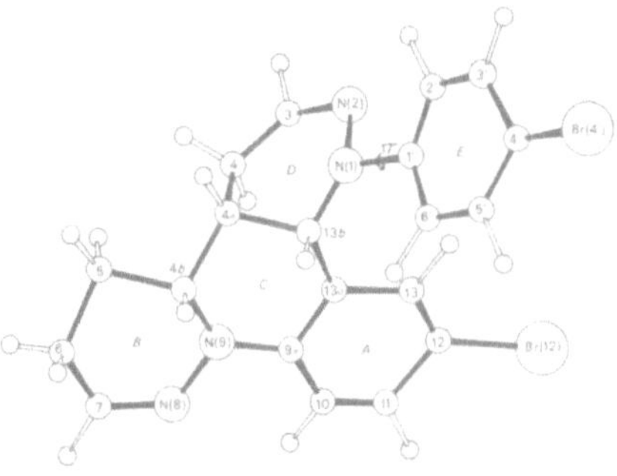

Fig. 2. The molecular structure of the optically active form (B) of $C_{20}H_{18}Br_2N_4$.

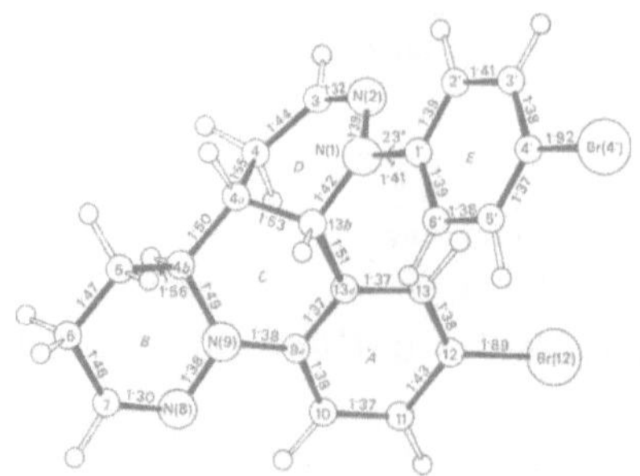

Fig. 3. The molecular structure of the racemic form (C) of $C_{20}H_{18}Br_2N_4$.

The geometry of the molecules is shown in Figs. 1 to 3; these will be referred to as (A), (B), and (C) respectively. Each of the molecules has three asymmetric carbon atoms, i.e., C(4b), C(4a), and C(13b). The racemates (A) and (C) differ only in the chirality at C(4b), while (B) is one of the enantiomers of (A) (the absolute configuration was not, however, determined).

Ring A of the condensed ring system of each compound and the rings E are aromatic and planar within experimental error. The B rings of (A) and (B) have conformations between sofa and half-chair with C(5) as the top atoms, while the B ring of (C) has nearly a sofa conformation with C(4b) as the top atom. The C rings of (A) and (B) are sofas with C(4a) as top atoms, while the C ring of (C) is nearly a boat form. The conformation of the D ring is nearly a half-chair conformation in all three molecules. The different conformations in the ring system of (C) are ascribed to the opposite chirality at C(4b). The dihedral angle about the N(1)-C(1') bond is close to 20° in each molecule. The bond lengths in the three structures are all close to normal values, but there are some deformations in (A) shown by the bond angles: N(1)-C(1')-C(6') = 125°; N(1)-C(1')-C(2') = 113°; Br(4')-C(4')-C(3') = 124°; and Br(4')-C(4')-C(5') = 111°. These distortions are ascribed to intermolecular steric effects. There are no unusual bond angles in (B) and (C).

2-METHOXY-6-CHLORO-9-[3-(ETHYL-2-HYDROXYETHYL)AMINOPROPYLAMINO]ACRIDINE

$C_{21}H_{26}ClN_3O_2$

$NH\text{-}CH_2\text{-}CH_2\text{-}CH_2\text{-}N(CH_2\text{-}CH_2\text{-}OH)(C_2H_5)$, OCH_3, Cl, N

(ICR-170-OH)

H.M. BERMAN and J.P. GLUSKER, 1972. Acta Cryst., B28, 590-596.

Monoclinic, $P2_1/c$, a = 13.644, b = 11.587, c = 12.613 Å, β = 100.13°, D_m = 1.31, Z = 4. Cu radiation, R = 0.068 for 1966 reflexions.

Each fused six-membered ring in the acridine moiety is slightly puckered, the outer ring with the methoxyl group showing the least deviation from planarity. The entire ring system both folds and twists about N(4)···C(15). The angle between the best planes through the outer rings is 10.7°. Bond distances and angles are shown in Fig. 1. The side chain has an intramolecular hydrogen bond NH···N between N(5) and N(6) of 2.979 Å. Molecules are associated in pairs by OH···N hydrogen bonds 2.755 Å (Fig. 2) and two of the six-membered rings of molecules that are related by an inversion centre partially overlap. The closest ring-ring distance is 3.50 Å. The compound has been given the code number ICR-170-OH and is an antitumor agent and a mutagen for Neurospora.

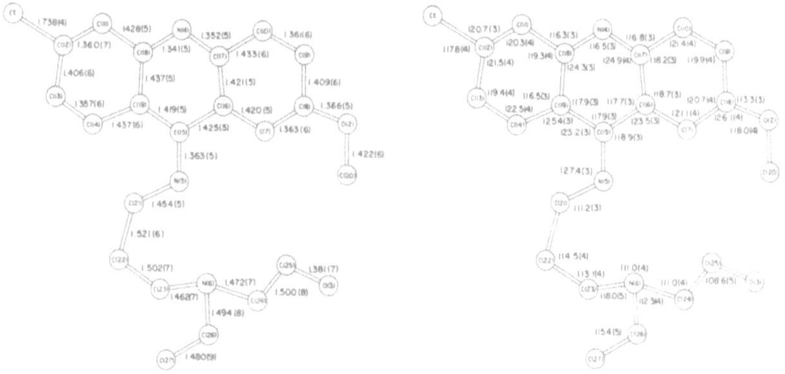

Fig. 1. Interatomic distances and interbond angles in ICR-170-OH.

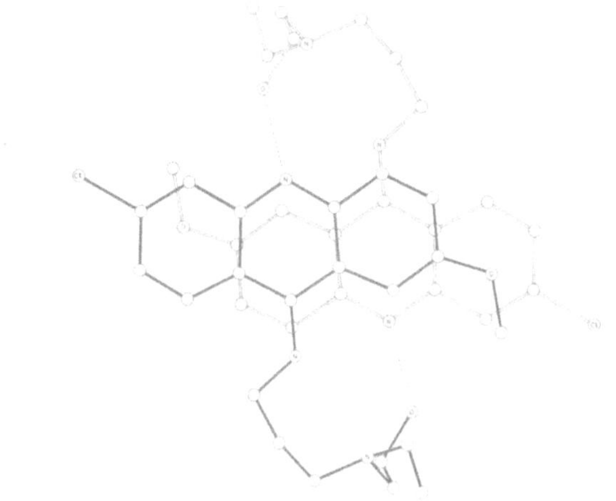

Fig. 2. Overlap of the ring systems of ICR-170-OH forming a hydrogen bonded dimer, viewed normal to the plane of the ring systems.

cis-5-ACETYL-5,5a,6,7,8,10,11,11a-OCTAHYDRO-9H-CYCLOOCT[b]INDOL-9-ONE p-BROMOPHENYLHYDRAZONE

$C_{22}H_{24}BrN_3O$

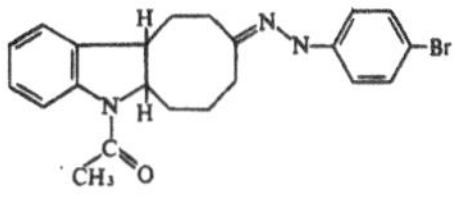

D.J. DUCHAMP and C.G. CHIDESTER, 1972. Acta Cryst., B28, 1092-1099.

Triclinic, P$\bar{1}$, a = 10.880, b = 7.684, c = 12.531 Å, α = 95.235, β = 106.09, γ = 100.08°, Z = 2. Cu radiation, R = 0.057 for 3974 reflexions.

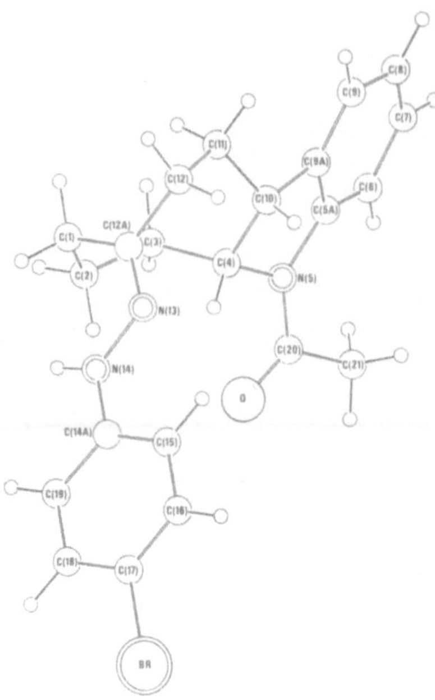

Fig. 1. Conformation and numbering scheme of the cis-5-acetyl-5,5a,6,7,8,-10,11,11a-octahydro-9H-cyclooct[b]indol-9-one p-bromophenylhydrazone molecule.

The configuration at the ring junction of the 5- and 8-membered rings is <u>cis</u> (see Fig. 1). The C(11)C(10)C(4)C(3) torsion angle is -32.4°. The heterocyclic ring has an envelope conformation with C(4) at the flap. C(3) is axial, to place it as far removed as possible from the acetyl group. The cyclooctane ring has the "boat-chair" form distorted by the ring fusion. The p-bromophenylhydrazone moiety is approximately planar; the deviations are probably due to crystal packing. There are intermolecular hydrogen bonds involving the amide oxygen and the hydrogen of N(14). The bond lengths in the molecule are all close to normal values. Several bond angles show the effects of steric repulsion. The C(1)-C(12a)-N(13) angle has increased to 126.2° to accommodate the closeness of C(1) and N(14). There is also a large (11.5°) difference between the C(5a)N(5)C(20) and C(4)N(5)C(20) angles also due to steric factors.

BIS(10-METHYL-9-ACRIDINE)MONOAZAMONOMETHINECYANINE PERCHLORATE

$C_{28}H_{22}ClN_3O_4$ $C_{28}H_{22}N_3^+ \cdot ClO_4^-$

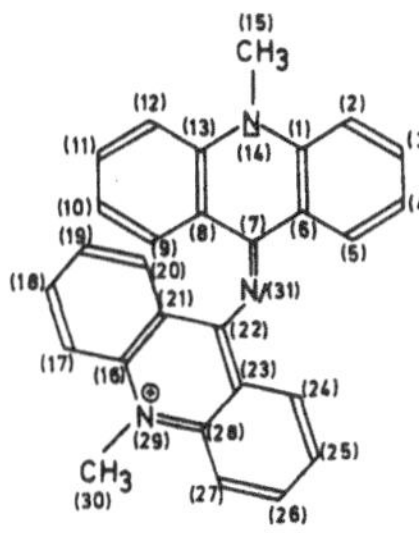

J. PREUSS, A. GIEREN, K. ZECHMEISTER, E. DALTROZZO, W. HOPPE and V. ZANKER, 1972. Chem. Ber., 105, 203-216.

Orthorhombic, $P2_12_12_1$, a = 19.76, b = 13.94, c = 8.43 Å, D_m = 1.42, Z = 4. Cu radiation, R = 0.083 for 2364 reflexions.

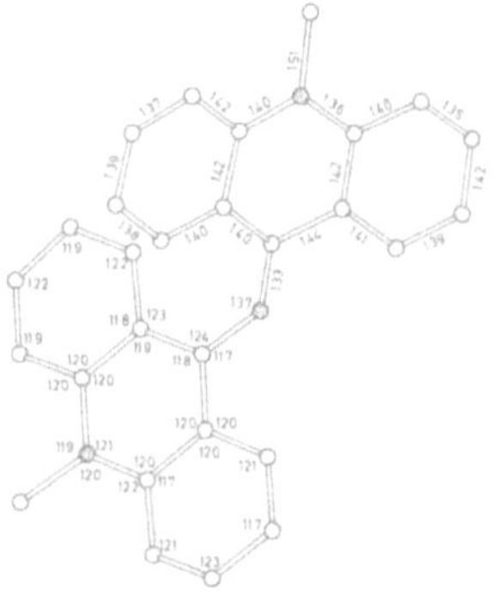

Fig. 1. Projection of the $C_{28}H_{22}N_3^+$ cation on the (001) plane with molecular geometry details averaged over the two acridine moieties (σ = 0.01 - 0.02 Å for bonds and 1.1 - 1.5° for angles).

Steric hindrance in the diacridine cation leads to a compromise between rotation of the ring systems (65.5°) and an increase in bond angles (for example, that at the central nitrogen is 137°). Details of molecular geometry averaged over the two independent acridine moieties are given in Fig. 1. In the crystal structure the cations are stacked in columns in the c-direction with perchlorate ions filling gaps between columns. It is interesting to note that in space group $P2_12_12_1$ only one of the two enantiomeric forms of the material is present, whereas the solution is racemic.

4,4-DICHLORO-2a-AZA-A-HOMO-COPROSTANE-3-ONE:ACETONE (2:1 ADDUCT)

$C_{57}H_{96}Cl_4N_2O_3$ $2[C_{27}H_{45}NOCl_2]\cdot(CH_3)_2CO$

H. ALTENBURG, D. MOOTZ and B. BERKING, 1972. Acta Cryst., B28, 567-574.

Orthorhombic, $P2_12_12_1$, a = 7.944, b = 44.904, c = 15.937 Å, D_m not given, Z = 4. Cu radiation, R = 0.059 for 2615 reflexions.

The conformations of the two independent homo-coprostane-3-one molecules are very similar. The ε-lactam ring adopts a half-chair conformation; the five-membered ring conformation is midway between a half-chair and an envelope (Fig. 1). Bond lengths and angles (Fig. 2) are normal. In the crystal structure hydrogen bonds N-H···O (2.93 and 2.99 Å) link molecules around screw axes parallel to a into two independent kinds of infinite chains (Fig. 3). The acetone molecules are held by van der Waals forces.

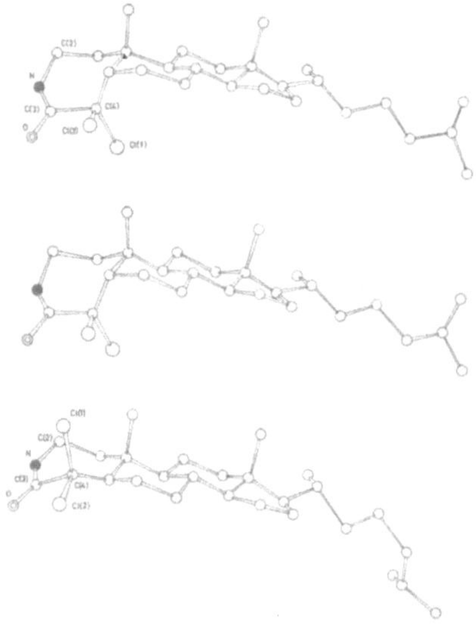

Fig. 1. The conformation of the two independent molecules (upper two diagrams) and that of the 5-β-derivative (1) of 4,4-dichloro-2a-aza-A-homocoprostane-3-one.

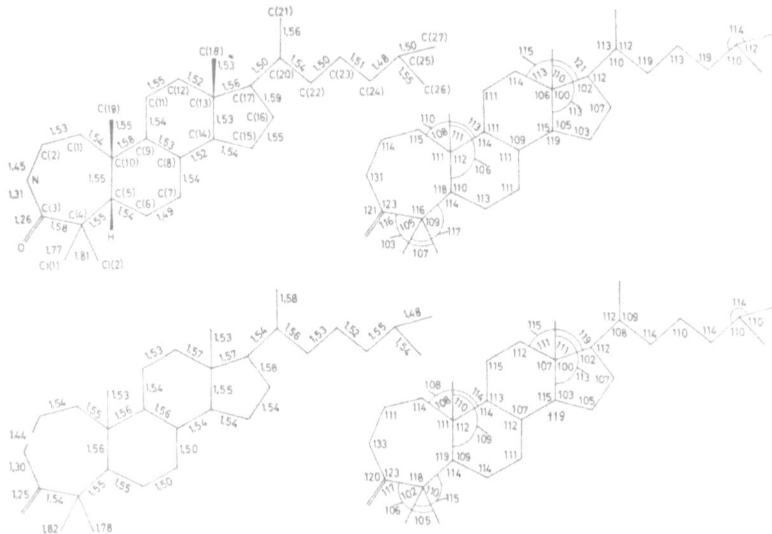

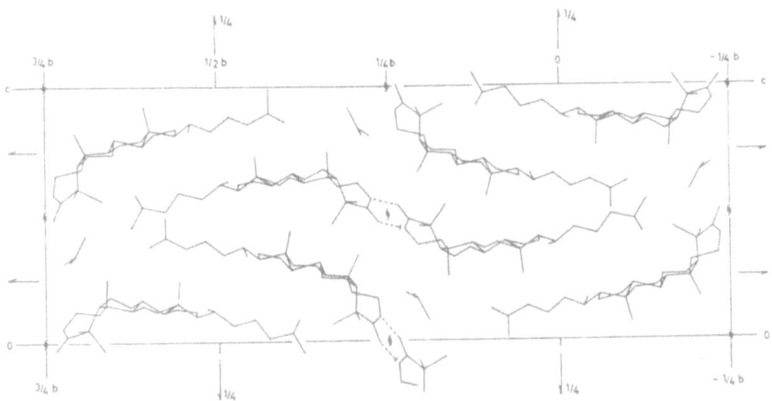

Fig. 2. Bond lengths (σ = 0.01 Å) and bond angles in 4,4-dichloro-2a-aza-A-homocoprostane-3-one.

Fig. 3. The crystal structure of the 2:1 adduct $2[C_{27}H_{45}NOCl_2]\cdot(CH_3)_2CO$ viewed along a.

1. Structure Reports, 35B, 342.

NORTROPANE DERIVATIVES

H. SCHENK and P. BENCI, 1972. Acta Cryst., B28, 538-543.

1,3-DINITRO-6-HYDROXYMETHYLENE-3,4-METHANONORTROPANE

$C_9H_{13}N_3O_5$

Monoclinic, $P2_1/c$, a = 6.423, b = 6.937, c = 24.416 Å, β = 99.88°, Z = 4. Cu radiation, R = 0.048 for 1380 reflexions.

1,3-DINITRO-6-HYDROXYMETHYLENE-N-HYDROXY-3,4-METHANONORTROPANE

$C_9H_{13}N_3O_6$

Monoclinic, $P2_1/c$, a = 13.304, b = 6.350, c = 13.364 Å, β = 105.30°, D_m not given, Z = 4. Cu radiation, R = 0.046 for 1833 reflexions.

Fig. 1. Mean values of corresponding bond lengths in the nortropane structures, (σ = 0.003 - 0.005 Å).

The molecular dimensions of the nortropane derivatives are equal within the limits of accuracy and are shown in Fig. 1. The only unusual bond length is that of C(6)-N(2) which, at 1.53 Å, is longer than the normal value (1.48 Å). The piperidine ring is distorted into an envelope conformation. The nitro group at the cyclopropane ring junction adopts a conformation with an oxygen atom pointing towards the cyclopropane group, in agreement with extended Hückel calculations of the rotation barrier in nitrocyclopropane. The N-hydroxy derivative has an intramolecular O(1)H···N(1) bond (2.798 Å) and successive molecules are linked by O(2)H···O(1) (2.659 Å) hydrogen bonds. In the $C_9H_{13}N_3O_5$ structure there is no intramolecular hydrogen bond and the intermolecular hydrogen bonds are longer (and weaker) (O(2)H···O(6) = 3.016, N(1)H···O(1) = 3.131 Å).

(R)-(-)-3-ACETOXYQUINUCLIDINE METHIODIDE

$C_{10}H_{18}INO_2$ $C_{10}H_{18}NO_2^{+} \cdot I^{-}$

R.W. BAKER and P.J. PAULING, 1972. J. Chem. Soc., Perkin II, 2340-2343.

Triclinic, P1, a = 7.610, b = 7.410, c = 6.909 Å, α = 69.75, β = 111.88, γ = 112.38°, D_m = 1.59, Z = 1. Mo radiation, R = 0.045 for 1637 reflexions.

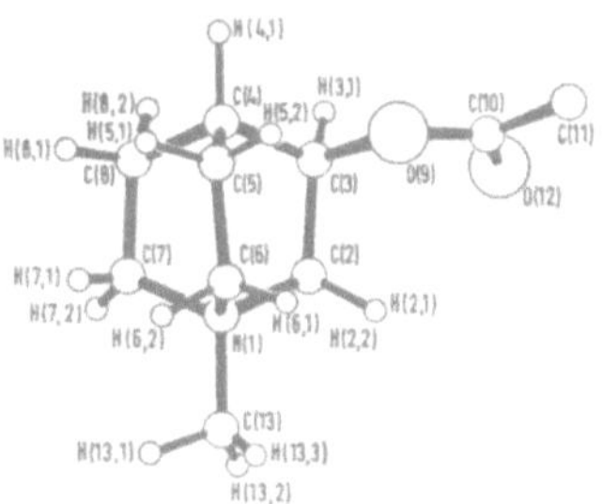

Fig. 1. Conformation of 3-acetoxyquinuclidine as observed in crystals of the methiodide.

The structure and absolute configuration of the cation are shown in Fig. 1. The acetoxy group joined to the quinuclidine cage is planar and the quinuclidine cage slightly twisted (e.g. torsion angles N(1)-C(2)-C(3)-C(4) = -9°, N(1)-C(6)-C(5)-C(4) = -12°, N(1)-C(7)-C(8)-C(4) = -14°). Bond lengths have expected values. There are no abnormal intermolecular contacts.

1-AZABICYCLO[3,3,3]UNDECANE HYDROCHLORIDE

(MANXINE HYDROCHLORIDE)

$C_{10}H_{20}C\ell N$ $C_{10}H_{20}N^{+} \cdot C\ell^{-}$

A.H.-J. WANG, R.J. MISSAVAGE, S.R. BYRN and I.C. PAUL, 1972. J. Amer. Chem. Soc., 94, 7100-7104.

Triclinic, $P\bar{1}$, a = 10.830, b = 7.122, c = 7.014 Å, α = 97°58', β = 90°24', γ = 96°17', D_m = 1.24, Z = 2. Cu radiation, R = 0.047 for 1644 reflexions.

The cation (Fig. 1) has effective (non-crystallographic) C_3 symmetry about the N···C(5) axis. Each of the three constituent eight-membered rings is in the boat-chair conformation and there is no evidence from this study to indicate that inversion of conformation of the cation takes place in the crystal. The conformation found results from a compromise between torsional, bond angle, and transannular H···H nonbonded effects. The bond angles C-C-C (115 - 120°), C-N-C (115 - 116°), and N-C-C (117°) are all greater than tetrahedral and can be correlated with spectral features of the molecule. In the crystal structure the chloride ions are associated with the cations via N^{+}-H···$C\ell^{-}$ hydrogen bonding (2.978(3) Å).

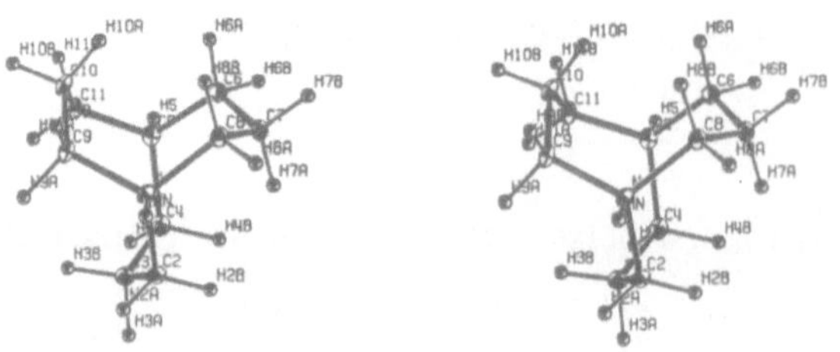

Fig. 1. View of the manxine hydrochloride cation.

2,9-DIACETYL-9-AZABICYCLO[4,2,1]NON-2,3-ENE

$C_{12}H_{17}NO_2$

C.S. HUBER, 1972. Acta Cryst., B28, 2577-2582.

Tetragonal, $P4_12_12$, a = 7.335, c = 42.294 Å, D_m = 1.19, Z = 8. Cu radiation, R = 0.0483 for 1089 reflexions.

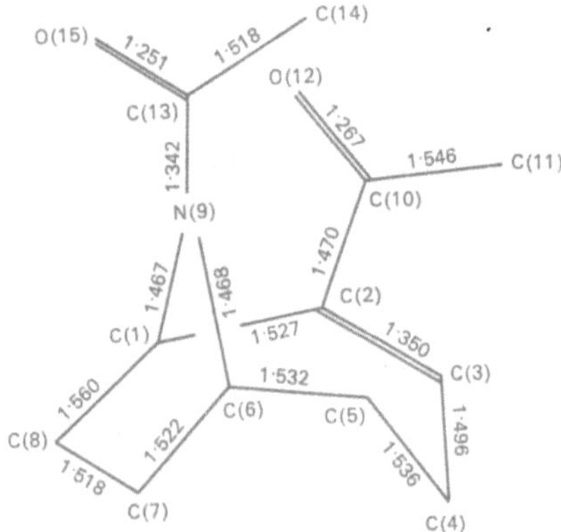

Fig. 1. Bond lengths (corrected for libration) in $C_{12}H_{17}NO_2$.

The absolute configuration (corresponding to 1R,6R) and details of molecular geometry are shown in Fig. 1. For the most part bond lengths are close to normal values, but bonds C(1)-C(8) and C(1)-C(2) are longer than expected; the latter bond is between more-substituted atoms. The hydrogen atom on C(1) is only 2.18 Å from O(12) and is suggested to be hydrogen bonded to it. The C(13)-N(9) bond has some double bond character and atoms C(13), C(14), O(15), C(1), C(6), N(9) are almost coplanar. The pyrrolidine ring is in envelope conformation with N(9) displaced 0.49 Å from the C(1)C(6)C(7)C(8) plane. The azacycloheptene ring has a distorted chair form. Intermolecular contacts correspond to van der Waals distances.

N-p-BROMOBENZOYL-exo-2,3-AZIRIDINOBICYCLO[2,2,1]HEPTANE

$C_{14}H_{14}BrNO$ $BrC_6H_4CONC_7H_{10}$

E.M. GOPALAKRISHNA, 1972. Acta Cryst., B28, 2754-2759.

Monoclinic, $P2_1/a$, a = 12.43, b = 17.13, c = 6.08 Å, β = 104.2°, D_m = 1.552, Z = 4. Cu radiation, R = 0.052 for 1924 reflexions.

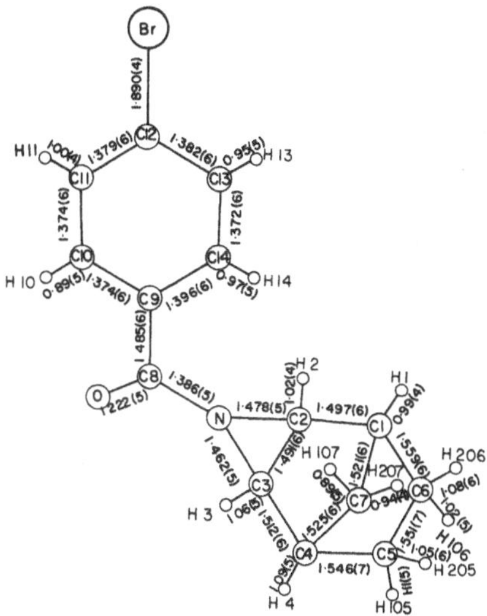

Fig. 1. Bond lengths in $BrC_6H_4CONC_7H_{10}$.

Bond lengths in the molecule are shown in Fig. 1. The C-C bonds in the bicyclo[2,2,1]heptane system which are remote from the aziridine ring have normal values whereas the ones which are close to it are shortened slightly. Molecular strain is shown in the bond angles, and, as is usual in these systems, the angle C(1)-C(7)-C(4) (95.4°) is much reduced from tetrahedral. The aziridine ring plane is inclined at 110.7° to the plane through C(1)C(2)C(3)C(4), and at 124.7° to the C(8)C(9)NO plane. There are no unusual intermolecular contacts.

LUPANINE-N-OXIDE PERCHLORATE

$C_{15}H_{25}C\ell N_2O_6$

Z. KALUSKI, A.I. GUSIEV, Ju.T. STRUCHKOV, J. SKOLIK, P. BARANOWSKI and M. WIEWIOROWSKI, 1972. Bull. Acad. Polon. Sci., 20, 1-14.

Orthorhombic, $P2_12_12_1$, a = 11.450, b = 13.510, c = 10.676 Å, D_m = 1.495, Z = 4. Cu radiation, R = 0.152 for 830 reflexions.

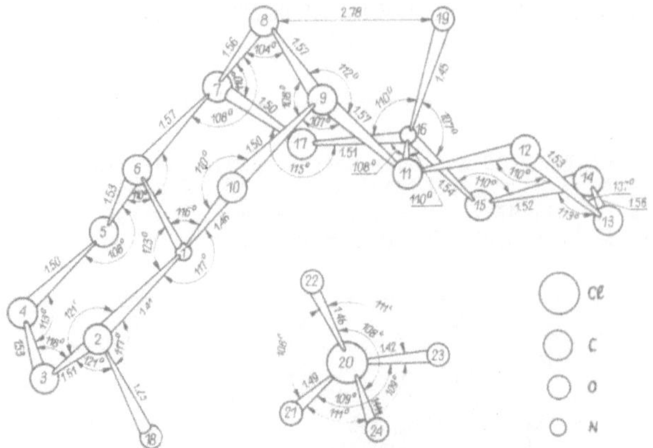

Fig. 1. Diagram of the molecule of lupanine-N-oxide perchlorate showing the bond lengths and angles (σ = 0.03 - 0.04 Å and 1° - 2°).

In the molecule (Fig. 1) rings A and B have chair conformations distorted by trigonal hybridisation of nitrogen and carbon atoms in the lactam group. Ring C has a boat and ring D a chair conformation. The crystal structure contains chains of molecules linked by N-O-H···O=C hydrogen bonds (2.61 Å).

OXALATE COMPLEX OF SYNTHETIC DELPHININE INTERMEDIATE

$C_{24}H_{31}NO_9$ $[C_{22}H_{30}O_5N]^+[C_2HO_4]^-$

K.B. BIRNBAUM, 1972. Acta Cryst., B28, 1551-1560.

Monoclinic, C2/c, a = 23.972, b = 10.346, c = 18.656 Å, β = 93.16°, D_m = 1.372, Z = 8. Cu radiation, R = 0.043 for 3115 reflexions.

This study was undertaken as part of a projected total synthesis of the alkaloid delphinine. The free base of the crystals is shown to have structure (I) and necessitates the structure of delphinine to be revised to (III). Bond

(I: R_1=OCH_3; R_2=H)
(II R_1=H, R_2=OCH_3)

(III)

lengths and angles of the protonated form of the free base and of the oxalate ion are shown in Figs. 1 and 2 respectively. There are several bicyclic systems fused to each other, resulting in considerable strain in the cation (Fig. 3). The acid oxalate anion is close to being planar; the angle of twist about the central C-C bond is only 3.0°. Pairs of cations are hydrogen bonded to pairs of oxalate anions, a centre of symmetry being located between each pair. Each cation is linked to two anions by one normal and two bifurcated hydrogen bonds.

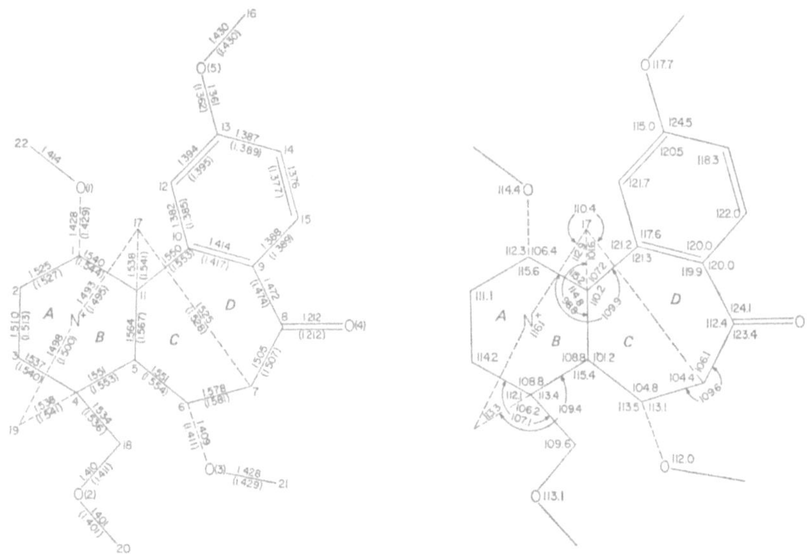

Fig. 1. Bond lengths (with corrected values in parentheses) and bond angles in the cation $[C_{22}H_{30}O_5N]^+$.

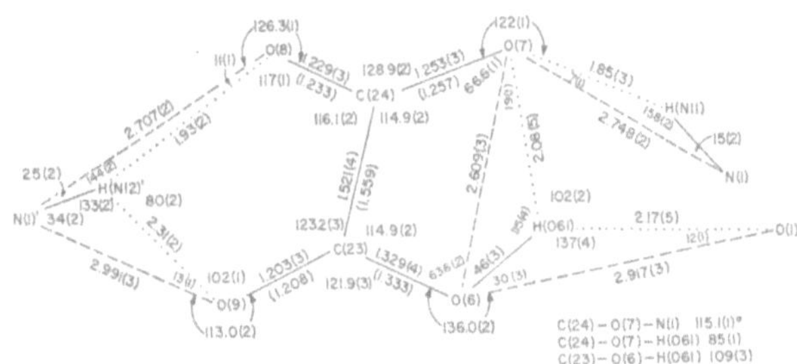

Fig. 2. Geometry of the acid oxalate ion and the hydrogen bonds. The values in parentheses are corrected values.

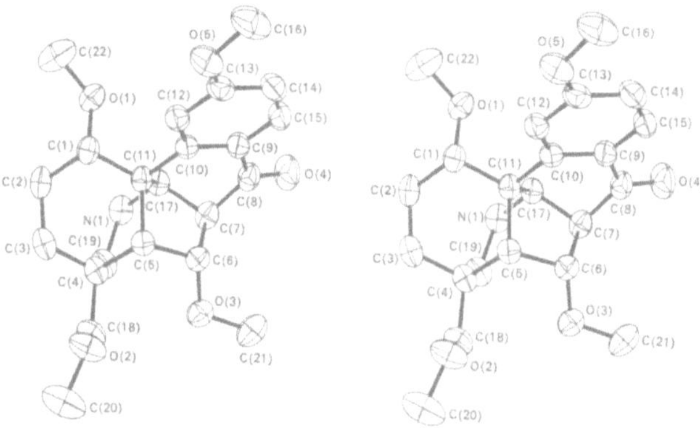

Fig. 3. View of the cation $[C_{22}H_{30}O_5N]^+$.

PYRIMIDINE TETRAMER TRIHYDRATE

$C_{24}H_{28}N_8O_6 \cdot 3H_2O$

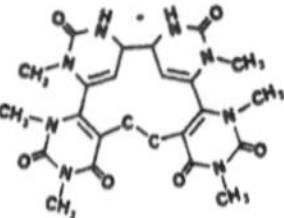

J.L. FLIPPEN, R.D. GILARDI and I.L. KARLE, 1972. Acta Cryst., B28, 360-367.

Monoclinic, $P2_1/c$, a = 7.117, b = 21.348, c = 18.237 Å, β = 98.62°, D_m not given, Z = 4. Cu radiation, R = 0.085 for 3035 reflexions.

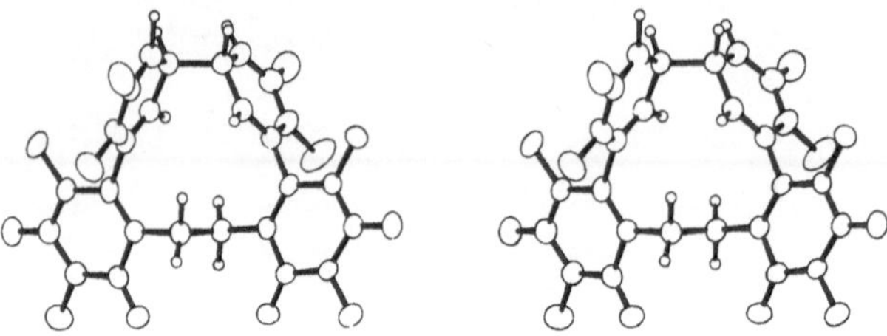

Fig. 1. Diagram of the tetramer molecule.

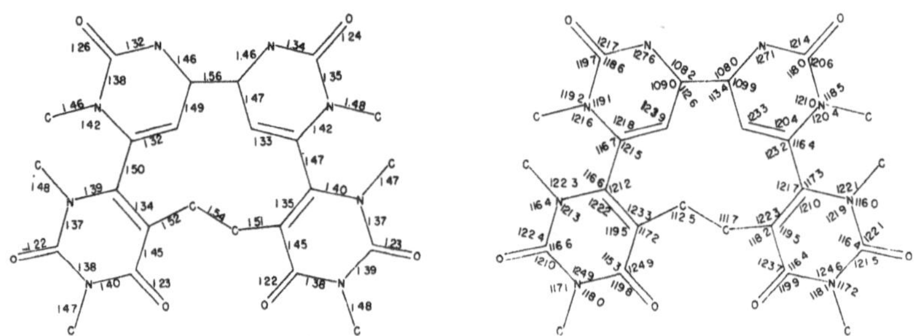

Fig. 2. Bond distances and angles (σ = 0.008 Å (C-C), 0.007 Å (C-N or C-O), 0.5° for angles) in pyrimidine tetramer trihydrate.

The molecule was produced by irradiating an aqueous solution of a thiamine-uracil adduct, 6,4'-(pyrimidin-2'-one)-thymine, with UV light (310 - 360 nm). The molecular conformation is shown in Fig. 1. Bond lengths and angles are in Fig. 2. There are eight independent hydrogen bonds involving the tetramer and the three water molecules (O-H···O = 2.83 - 3.12 Å, N-H···O = 2.86 - 2.92 Å).

FURAN (AT 152°K AND 123°K)

C_4H_4O

R. FOURME, 1972. Acta Cryst., B28, 2984-2991.

152°K, Phase I

Orthorhombic, Cmca, a = 8.65, b = 6.70, c = 6.75 Å, Z = 4. Mo radiation, R = 0.078.

123°K, Phase II

Tetragonal, $P4_12_12$, a = 5.69, c = 11.92 Å, Z = 4. Mo radiation, R = 0.065.

Phase I is disordered; each molecule randomly occupies four coplanar positions related by space group symmetry. Phase II has ordered, planar molecules with dimensions C-O = 1.368(6), C=C = 1.322(6), C-C = 1.428(7) Å, C-O-C = 106.2(6)°, O-C=C = 110.1(4)°, C=C-C = 106.8(2)°. All intermolecular contacts correspond to or are greater than van der Waals distances.

β-CHLOROGLUTARIC ANHYDRIDE

$C_5H_5ClO_3$

F.J. KOER, A.J. de KOK and C. ROMERS, 1972. Rec. Trav. Chim. Pays-Bas, 91, 691-699.

Monoclinic, C2/c, a = 23.63, b = 6.07, c = 20.83 Å, β = 122.1°, D_m = 1.58, Z = 16. Cu radiation, R = 0.122 for 1284 reflexions.

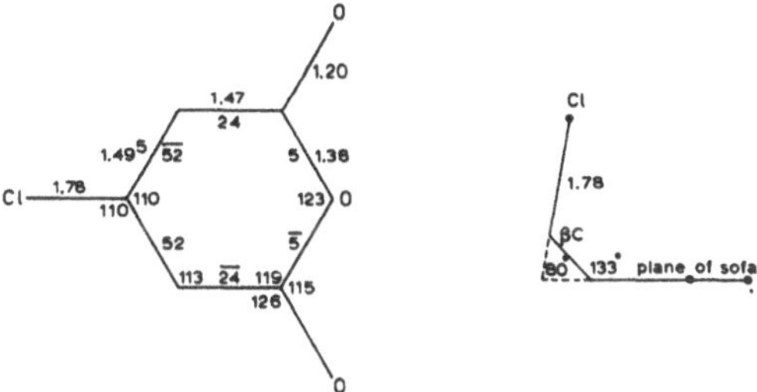

Fig. 1. Molecular geometry of β-chloroglutaric anhydride. Mean molecular dimensions (including torsion angles) are given at left and a side view of the average structure at right.

Details of the molecular structure are given in Fig. 1. There is a slight difference in conformation between the two crystallographically independent molecules in the asymmetric unit. The six-membered ring has a sofa conformation, the β carbon atom being 0.63 Å out of the plane of the other ring atoms. The CO-O-CO moiety is coplanar with the ring atoms, the C-Cl bond has axial orientation.

7-OXABICYCLO[2,2,1]HEPT-5-ENE-2,3-exo-DICARBOXYLIC ANHYDRIDE

$C_8H_6O_4$

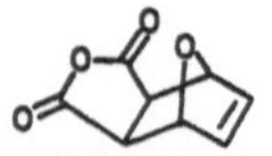

S. BAGGIO, A. BARRIOLA and P.K. de PERAZZO, 1972. J. Chem. Soc., Perkin II, 934-937.

Orthorhombic, $P2_12_12_1$, a = 7.00, b = 18.93, c = 5.38 Å, D_m = 1.58, Z = 4. Cu radiation, R = 0.087 for 750 reflexions.

The molecule has the structure indicated above with non-crystallographic m-symmetry. The incorporation of an oxygen atom at position 7 of the norbornene skeleton has resulted in a C-O-C angle of 96.2(7)° and C-O lengths of 1.44(1) Å, whereas in the anhydride ring C-O-C is 108.6(8)° and the mean C-O length is 1.38(1) Å. The C=C bond is long, 1.392(16) Å.

cis-1,4,5,8-TETRAOXADECALIN

$C_6H_{10}O_4$

B. FUCHS, I. GOLDBERG and U. SHMUELI, 1972. J. Chem. Soc., Perkin II, 357-364.

Monoclinic, C2/c, a = 16.003, b = 7.077, c = 11.654 Å, β = 92.81°, D_m = 2.0, Z = 8. Cu radiation, R = 0.07 for 658 reflexions.

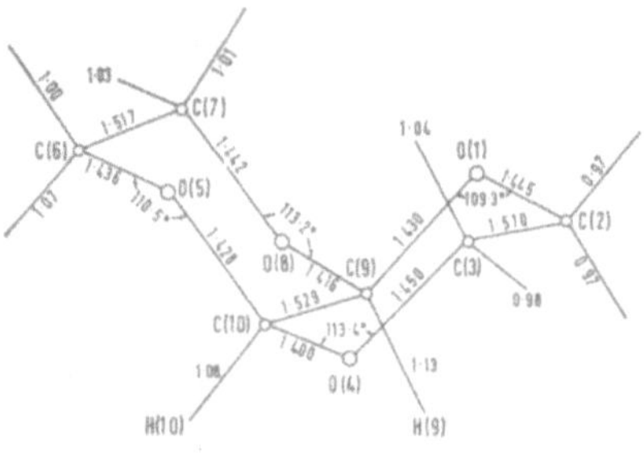

Fig. 1. A projection of the molecular framework of cis-1,4,5,8-tetraoxadecalin. The bond lengths and angles are corrected for libration (excluding the C-H bonds).

The chemically equivalent but crystallographically different bond lengths and angles (Fig. 1) imply that the molecule has a twofold axis of symmetry within the error limits of the analysis. The molecule has the cis-configuration and double chair conformation. The ideal chair conformation is somewhat distorted since the C(3)-O(4)-C(10) and C(7)-O(8)-C(9) angles (113.4 and 113.2°) are significantly larger than those in the other pair of related C-O-C angles. This distortion is probably primarly due to intramolecular non-bonded interactions, e.g. C(3)···O(5) (2.87 Å) and C(7)···O(1) (2.84 Å).

3-CHLOROMETHYL-1,5,7-TRIMETHYL-2,4,6,8-TETRAOXATRICYCLO-[3.3.1.0^{3,7}]NONANE

$C_9H_{13}C\ell O_4$

R.P. DODGE, 1972. Cryst. Struct. Comm., 1, 173-176.

Monoclinic, Ia, a = 10.88, b = 10.01, c = 9.47 Å, β = 93.5°, Z = 4. Mo radiation, R = 0.114 for 812 reflexions.

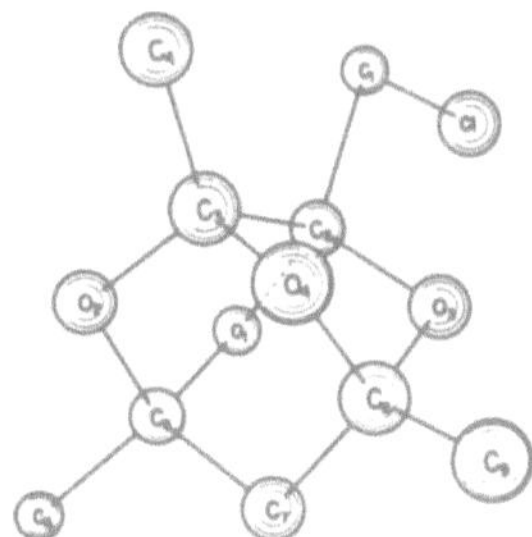

Fig. 1. Structure of the condensation product of acetylacetone and chlorobiacetyl.

The structure of the condensation product of acetylacetone and chlorobiacetyl is confirmed to be the title compound (Fig. 1). Carbon atoms C(1), C(2), C(3), and C(4) lie on a plane which makes an angle of 90° with the plane through atoms C(5), C(6), C(7), C(8), and C(9).

1,3,8,10-TETRAOXACYCLOTETRADECANE

$C_{10}H_{20}O_4$

I.W. BASSI, R. SCORDAMAGLIA and L. FIORE, 1972. J. Chem. Soc., Perkin II, 1726-1729.

Monoclinic, $P2_1/n$, a = 8.107, b = 14.721, c = 4.724 Å, β = 100.25°, D_m = 1.20, Z = 2. Mo radiation, R = 0.058 for 775 reflexions.

A view of the centrosymmetric molecule is given in Fig. 1. The C-O bond lengths are in the range 1.395 - 1.431(3) Å, whereas the C-C bonds are in the range 1.506 - 1.523(4) Å. The overall shape and conformation of the fourteen-membered ring is similar to that of the cyclotetradecane ring which may be derived from the diamond structure. The 2/m symmetry of ideal cyclotetradecane is lost. The carbon and oxygen atoms of the asymmetric unit in the structure approximately define two nearly perpendicular planes (83°), one of them defined by C(1), O(1), C(2), C(3), C(4), the five atoms of the lateral chain (root-mean-square distance 0.158 Å) the other containing C(4), C(5), O(2), C(1), the four atoms of the fold (root-mean-square distance 0.091 Å). The deviations of the ring skeleton of the molecule from the ideal structure may be tentatively accounted for on the basis of intramolecular van der Waals repulsions between hydrogen atoms five bonds apart, and of the presence of heteroatoms in the ring. There are no unusual intermolecular contacts.

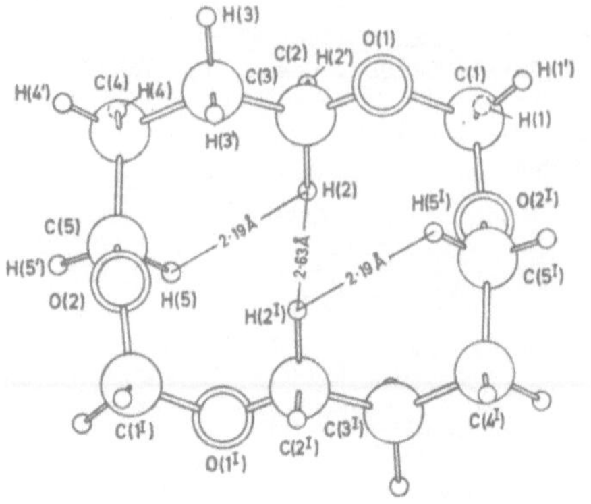

Fig. 1. The 1,3,8,10-tetraoxacyclotetradecane molecule, viewed in its c-axis projection.

2,3,7,8-TETRACHLORODIBENZO-p-DIOXIN

$C_{12}H_4Cl_4O_2$

F.P. BOER, F.P. van REMOORTERE, P.P. NORTH and M.A. NEUMAN, 1972. Acta Cryst., B28, 1023-1029.

Triclinic, P$\bar{1}$, a = 3.783, b = 9.975, c = 15.639 Å, α = 94.14, β = 95.20, γ = 92.77°, Z = 2. Mo radiation, R = 0.036 for 2381 reflexions.

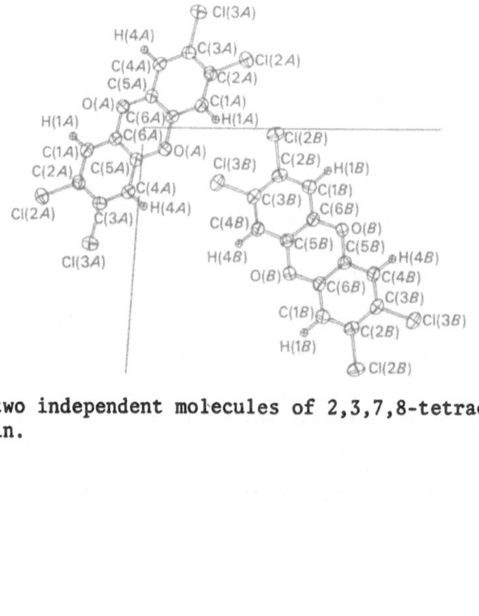

Fig. 1. The two independent molecules of 2,3,7,8-tetrachlorodibenzo-p-dioxin.

The structure consists of two independent molecules situated on inversion centres at (0,0,0) and (½,½,½) and stacked along [100]. The molecules are very nearly planar and approximate mmm (D_{2h}) symmetry. No non-hydrogen atom deviates more than 0.018 Å from its least-squares molecular plane. The four unique C-Cℓ distances range from 1.726 to 1.730, mean 1.728(3) Å, and the four C-O distances from 1.377 to 1.379, mean 1.378(3) Å. The mean values of the four chemically distinct types of C-C bonds (Fig. 1) are: C(1)-C(2) = 1.383 Å, C(2)-C(3) = 1.385 Å, C(4)-C(5) = 1.378 Å, and C(5)-C(6) = 1.387(3) Å. The C-O-C angles in the central ring are 115.6 and 115.8°; as a result the C-C-O angles in the heterocycle average 122.2°. The internal angles in the benzene rings are all close to 120.0°. Repulsion between adjacent Cℓ atoms may account for the opening of the C(2)-C(3)-Cℓ(3) and C(3)-C(2)-Cℓ(2) angles to 121.0°.

2,7-DICHLORODIBENZO-p-DIOXIN

$C_{12}H_6C\ell_2O_2$

F.P. BOER and P.P. NORTH, 1972. Acta Cryst., B28, 1613-1618.

Triclinic, P$\bar{1}$, a = 3.878, b = 6.755, c = 10.265 Å, α = 99.46, β = 100.63, γ = 99.73°, Z = 1. Mo radiation, R = 0.057 for 1030 reflexions.

The molecules lie on crystallographic inversion centres, are planar, and have approximate molecular symmetry C_{2h}. The C-Cℓ bond distance is 1.742(4) Å, the C-O distances are 1.380 and 1.382(4) Å, and the six independent C-C(aromatic) distances vary between 1.370 and 1.397(5) Å. The C-O-C angle in the central heterocyclic ring is 116.3(2)°. The crystal contains only one unusually short intermolecular interaction: a Cℓ···Cℓ contact of 3.402 Å.

2,8-DICHLORODIBENZO-p-DIOXIN

$C_{12}H_6C\ell_2O_2$

F.P. BOER, M.A. NEUMAN and O. ANILINE, 1972. Acta Cryst., B28, 2878-2880.

Orthorhombic, Pnam, a = 5.983, b = 7.114, c = 24.64 Å, Z = 4. Mo radiation, R not quoted for 294 reflexions.

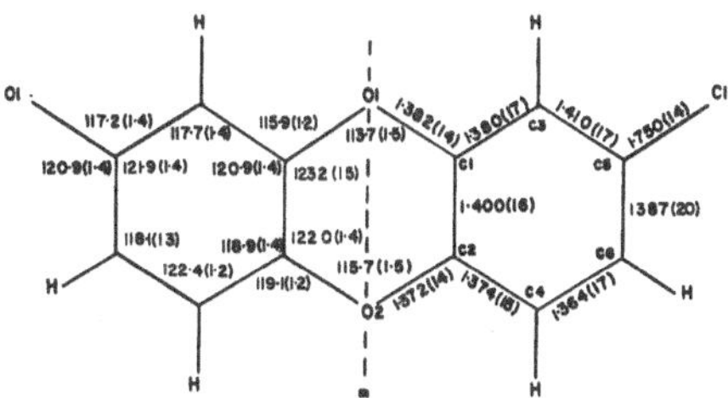

Fig. 1. Bond lengths and angles for 2,8-dichlorodibenzo-p-dioxin.

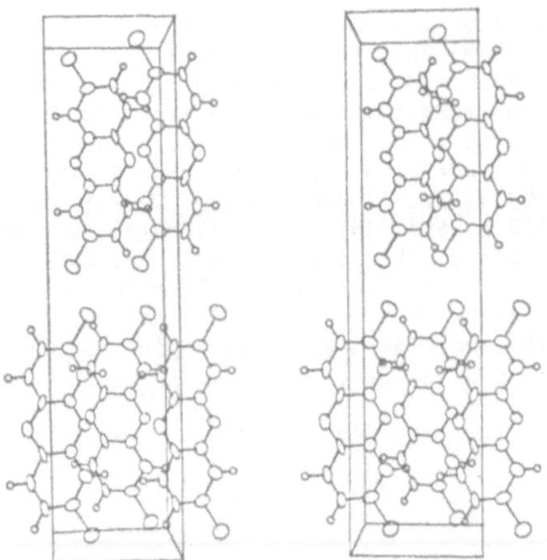

Fig. 2. View of the molecular packing of 2,8-dichlorodibenzo-p-dioxin.

The molecule (Fig. 1) is slightly non-planar; the dihedral angle between the two halves of the molecule related by the transverse mirror plane is 175.2°. Bond lengths and angles (also shown in Fig. 1) are all normal. In the crystal structure (Fig. 2) there are two distinct layers of molecules centred on mirror planes at $\underline{c}/4$ and $3\underline{c}/4$. Within each layer herringbone packing occurs.

DIBENZOFURAN

$C_{12}H_8O$

O. DIDEBERG, L. DUPONT and J.M. ANDRÉ, 1972. Acta Cryst., B28, 1002-1007.

Orthorhombic, Pnam, a = 7.702, b = 5.825, c = 19.185 Å, D_m = 1.29, Z = 4. Cu radiation, R = 0.073 for 700 reflexions.

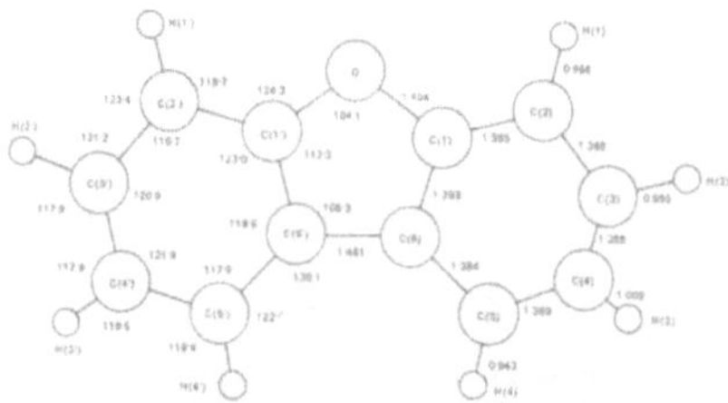

Fig. 1. Bond lengths and angles of dibenzofuran (values are corrected for thermal motion).

The bond lengths and angles corrected for thermal motion are shown in Fig. 1. The molecule is situated on a crystallographic mirror plane passing through the oxygen atom. The individual five- and six-membered rings are strictly planar, but the molecule as a whole shows a small, but significant, deviation from planarity. The dihedral angles between the five-membered ring and the six-membered rings are 1.12°.

XANTHOTOXIN

$C_{12}H_8O_4$

N.R. SEMPLE and W.H. WATSCN, 1972. Acta Cryst., B28, 2485-2489.

Orthorhombic, $Pna2_1$, a = 12.911, b = 15.804, c = 4.882 Å, D_m = 1.440, Z = 4. Cu radiation, R = 0.046 for 731 reflexions.

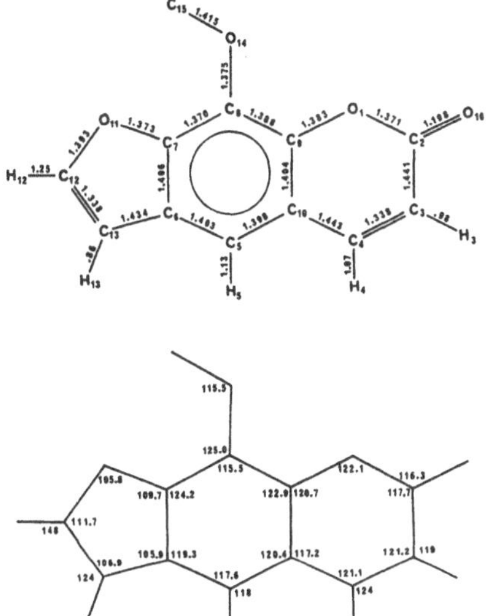

Fig. 1. Bond lengths (σ = 0.008 Å) and angles (σ = 0.5°) in xanthotoxin.

The molecular structure and details of geometry are shown in Fig. 1. The ring system is almost planar but the methoxyl methyl group is forced out of the plane (0.94 Å) due to steric interactions. The molecules are only weakly held together and the packing arrangement facilitates cleavage along (100) and (010).

1-(p-BROMOPHENYL)-1,2-EPOXYCYCLOHEXANE

$C_{12}H_{13}BrO$

S. MERLINO, G. LAMI, B. MACCHIA, F. MACCHIA and L. MONTI, 1972. J. Org. Chem., 37, 703-706.

Monoclinic, $P2_1/c$, a = 9.91, b = 5.72, c = 19.41 Å, β = 102°20', Z = 4. Cu radiation, R = 0.072 for 753 reflexions.

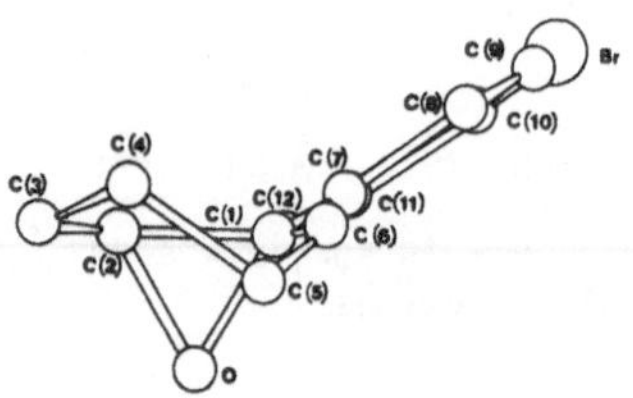

Fig. 1. The molecular structure of 1-(p-bromophenyl)-1,2-epoxycyclohexane, viewed along the direction normal to the epoxydic ring.

The cyclohexane ring has half-chair conformation (Fig. 1). The plane through atoms C(1),C(2),C(3),C(6) makes an angle of 100° with that of the epoxy group. Bond lengths are normal (e.g. C-C(epoxy) = 1.48(2), C-O(epoxy) = 1.47(2) Å). The phenyl ring plane makes an angle of 83° with the epoxy ring plane, thus pseudoconjugative interaction between the phenyl π-system and the three-membered ring is facilitated.

OCTACHLORODIBENZO-p-DIOXIN

$C_{12}Cl_8O_2$

M.A. NEUMAN, P.P. NORTH and F.P. BOER, 1972. Acta Cryst., B28, 2313-2317.

Monoclinic, $P2_1/c$, a = 12.009, b = 3.828, c = 16.297 Å, β = 101.14°, Z = 2. Mo radiation, R = 0.058 for 1459 reflexions.

Bond distances and angles of this centrosymmetric molecule are in Fig. 1. Small deviations from planarity occur whereby the Cl atoms are displaced from the plane of the carbon-oxygen skeleton (Cl(1) -0.016, Cl(2) -0.025, Cl(3) 0.022, and Cl(4) 0.053 Å). The crystal structure is based on a herringbone packing between stacks of molecules centred on (000) and ($0\frac{1}{2}\frac{1}{2}$). There are two short Cl···Cl intermolecular contacts (3.487 and 3.555 Å), two short C···Cl contacts (3.481 and 3.521 Å), and one short C···C contact (C(1)···C(2')) = 3.181 Å).

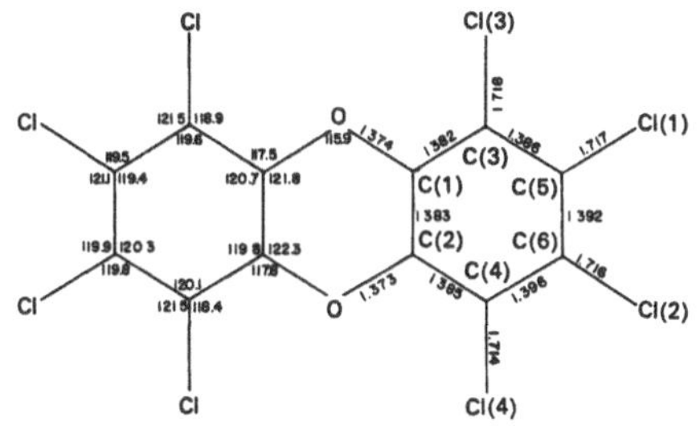

Fig. 1. Bond distances and angles in octachlorodibenzo-p-dioxin. Standard deviations are C-Cℓ, C-O 0.004 Å, C-C 0.005 Å, angles 0.3 - 0.4°.

DL-4-METHOXY-6-STYRYL-5,6-DIHYDRO-α-PYRONE

(KAVAIN)

$C_{14}H_{14}O_3$

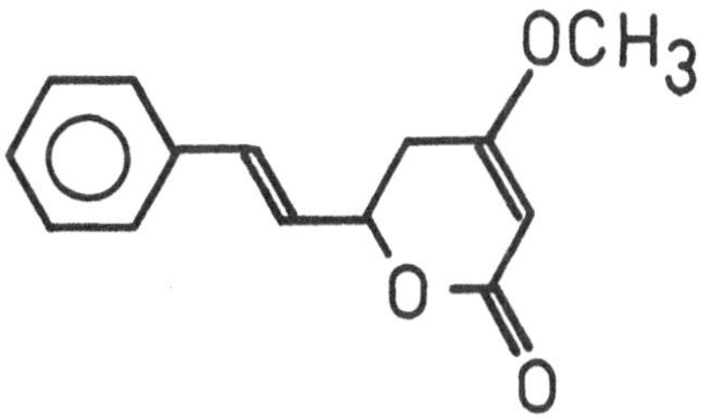

A. YOSHINO and W. NOWACKI, 1972. Z. Kristallogr., <u>136</u>, 66-80.

Monoclinic, $P2_1/a$, a = 23.032, b = 9.430, c = 5.738 Å, β = 104°14', D_m = 1.27, Z = 4. Cu radiation, R = 0.076 for 1384 reflexions.

The bond lengths and angles have close to expected values for the bonding scheme outlined above. There appears to be little conjugation between the phenyl ring and the ethylenic group; the angle between their respective planes is 16.7°. The conjugated portion of the pyrone ring is planar; the ring oxygen lies 0.09 Å off this plane. There are no abnormal intermolecular contacts.

4-METHOXY-6-(1',2'-DIHYDROSTYRYL)-5,6-DIHYDRO-α-PYRONE

(DIHYDROKAVAIN)

$C_{14}H_{16}O_3$

P. ENGEL and W. NOWACKI, 1972. Z. Kristallogr., <u>136</u>, 453-467.

Monoclinic, $P2_1/c$, a = 10.921, b = 7.605, c = 17.014 Å, β = 119.5°, Z = 4. Cu radiation, R = 0.09 for 2352 reflexions.

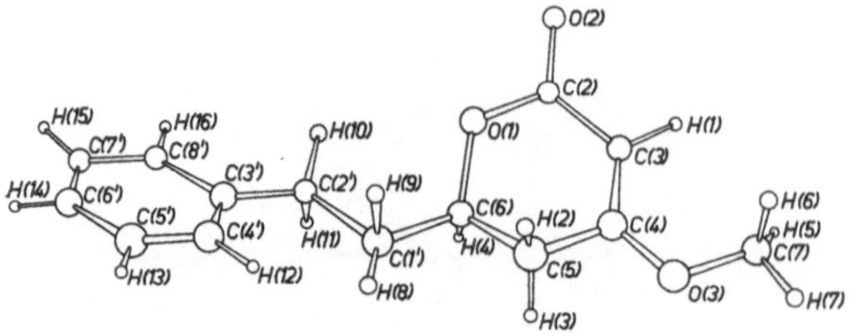

Fig. 1. The dihydrokavain molecular conformation.

In this molecule (Fig. 1) the dihydropyrone ring is not planar; the carbon atom (C(6)') bonded to the dihydrostyryl moiety is 0.581 Å from the plane of the other pyrone atoms. The dihydrostyryl group is planar and is inclined at an angle of 34.5° to the pyrone plane. Bond lengths and angles have expected values. The molecules are held together in the crystal lattice by van der Waals forces.

PENTAMETHYLHYDROCOUMARIN

$C_{14}H_{18}O_2$

J.M. KARLE and I.L. KARLE, 1972. J. Amer. Chem. Soc., 94, 9182-9189.

Monoclinic, $P2_1$, a = 10.148, b = 8.766, c = 7.802 Å, β = 115.45°, Z = 2. Cu radiation, R = 0.083 for 1086 reflexions.

The conformation is shown in Fig. 1. Bond lengths and angles are in Fig. 2. The values for the bond lengths are normal, but the angles deviate significantly from idealized values of 120° in order to accommodate the CH_3 groups in positions 10, 11, and 12, as well as those at 13 and 14. The increase in angle C(5)-C(4)-C(12) to 124.0° and in angle C(4)-C(5)-C(7) to 126.7° results in a distance of 3.03 Å between C(11) and C(12) which are separated by three atoms as compared to 2.99 Å between C(13) and C(14) which are separated by two atoms. The C(10)···C(12) distance is 3.80 Å.

Despite the large angular distortions, the six atoms of the phenyl group are nearly coplanar with a maximum deviation of 0.023 Å. The atoms C(13) and C(14) deviate by ±0.05 Å from the least-squares plane of the phenyl group and C(12) deviates by +0.125 Å. In addition, O(1) and C(7) of the lactone moiety, which may be expected to be coplanar with the phenyl group, also show significant deviations of +0.133 and -0.068 Å, respectively. The lactone ring which contains two saturated carbon atoms is markedly puckered, while the C(8)-C(9)-O(1)-O(2) moiety is planar to within ±0.003 Å.

In the crystal the molecules are disordered. [For another example of a molecule whose conformation is also restricted by a trialkyl lock see pentamethyl-o-hydroxyhydrocinnamyl alcohol (1).]

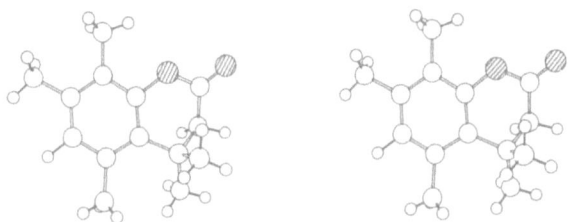

Fig. 1. Diagram showing the conformation of pentamethylhydrocoumarin.

Fig. 2. Bond lengths (σ = 0.008 Å) and angles (σ = 0.5°) in pentamethylhydrocoumarin.

1. This volume, p. 148.

2-(4-BROMOPHENYL)-r-2,4,4,c-6-TETRAMETHYL-1,3-DIOXAN

$C_{14}H_{19}BrO_2$

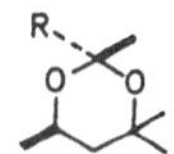

R = 4-bromophenyl

G.M. KELLIE, P. MURRAY-RUST and F.G. RIDDELL, 1972. J. Chem. Soc., Perkin II, 2384-2387.

Orthorhombic, Pbca, a = 27.65, b = 7.95, c = 13.16 Å, D_m = 1.385, Z = 8. Mo radiation, R = 0.12 for 635 reflexions.

The 1,3-dioxan ring exists in a deformed chair conformation with the 2- and 6-methyl groups in cis-equatorial positions and the bromophenyl group axial (Fig. 1). The 1,3-diaxial repulsion between the phenyl group and the axial 4-methyl group is not severe enough to force the ring into a twist conformation. Both the phenyl ring and the axial methyl group are displaced outwards so that the axial bonds (parallel in a perfect chair) intersect at an angle of 30° (the phenyl C(11)···C(42) methyl distance is 3.20 Å). In addition the C(2) end of the molecule is considerably flattened.

Fig. 1. A view of the dioxan derivative $C_{14}H_{19}BrO_2$.

AQUO-(2,3-BENZO-1,4,7,10,13-PENTAOXACYCLOPENTADEC-2-ENE)-SODIUM IODIDE (AQUO(BENZO-15-CROWN-5)-SODIUM IODIDE)

$C_{14}H_{22}INaO_6$

M.A. BUSH and M.R. TRUTER, 1972. J. Chem. Soc., Perkin II, 341-344.

Orthorhombic, $P2_12_12_1$, a = 12.271, b = 9.550, c = 15.719 Å, D_m = 1.58, Z = 4. Mo radiation, R = 0.048 for 887 reflexions.

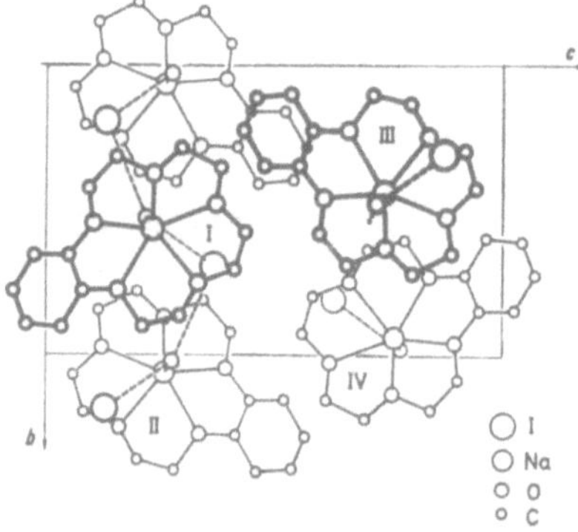

Fig. 1. The aquo(benzo-15-crown-5)-sodium iodide structure projected down the a-axis. Broken lines indicate the chain of interactions tentatively assigned to I···O(6) hydrogen bonds. Distances are O(6)···I, 3.51 and I···O(6^I), 3.47 Å.

The sodium ion is surrounded by a pentagonal pyramid of oxygen atoms. The base is formed by those of the cyclic ether with Na-O = 2.35 - 2.43(1) Å and the sodium ion is 0.75 Å from this plane towards the apical water molecule. Na-O = 2.29(1) Å. The iodide ion does not interact with the sodium ion but may form hydrogen bonds to water molecules [O···I 3.47 and 3.51(1) Å], linking the complexes in chains along the b-axis (Fig. 1). Bond lengths in the cyclic ether are similar to those in other such complexes. The absolute configuration of the complex was also determined.

4-METHOXY-6-(5',6'-DIOXYMETHYLENE-STYRYL)-5,6-DIHYDRO-α-PYRONE

(METHYSTICIN)

$C_{15}H_{14}O_5$

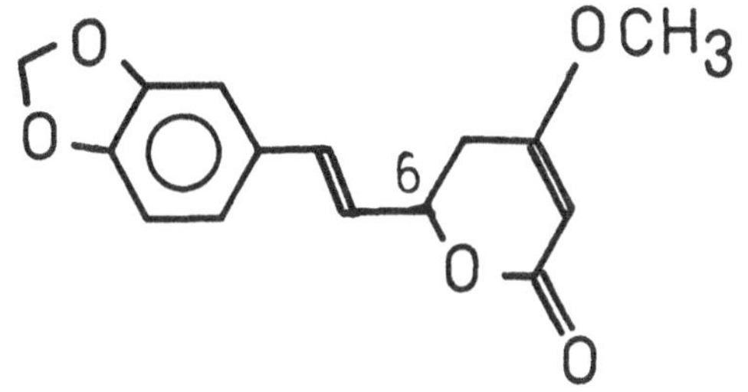

P. ENGEL and W. NOWACKI, 1972. Z. Kristallogr., 136, 437-452.

Orthorhombic, $P2_12_12_1$, a = 28.916, b = 8.504, c = 5.348 Å, Z = 4. Cu radiation, R = 0.070 for 1501 reflexions.

In the pyrone ring the carbon atom, C(6), bonded to the ethylenic carbon is 0.62 Å from the best plane through the remaining five pyrone ring atoms. The dioxymethylene-styryl group is planar and is inclined at an angle of 73.2° to the pyrone ring plane. Bond lengths are in accord with values expected for the molecular skeleton shown above. The molecules are separated in the crystal by normal van der Waals distances.

2-BROMOBENZO[b]INDENO[1,2-e]PYRAN

$C_{16}H_9BrO$

M.F.C. LADD and D.C. POVEY, 1972. J. Cryst. Mol. Struct., 2, 243-250.

Monoclinic, $P2_1/c$, a = 7.508, b = 5.959, c = 26.172 Å, β = 92.55°, D_m = 1.68, Z = 4. Cu radiation, R = 0.072 for 1027 reflexions.

The molecule is non-planar; the maximum deviations from the best plane occur at both ends of the length of the molecule, which results in distortion of the central portion from the expected geometry. The closest approach of two molecules, those related by a 2_1 axis, is 3.5 Å. Fig. 1 shows the bond distances and angles in the molecule with standard deviations in parentheses, σ(C-H) = 0.08 Å, σ(C-C-H) = 5°.

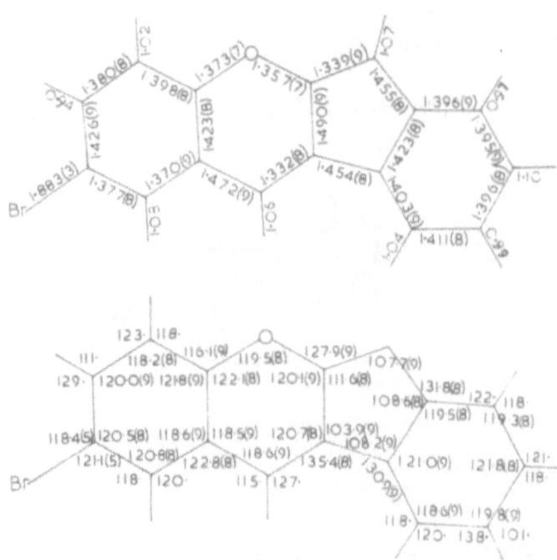

Fig. 1. Bond lengths (Å) and angles (degrees) in 2-bromobenzo[b]indeno-[1,2-e]pyran.

6,7-DIHYDRO-6,6-DIMETHYL-3-PHENYLBENZOFURAN-2,4(3H,5H)-DIONE-2-OXIME

$C_{16}H_{17}NO_3$

N.D. JONES and M.O. CHANEY, 1972. Acta Cryst., B28, 3190-3196.

Monoclinic, $P2_1/n$, a = 13.158, b = 8.657, c = 12.205 Å, β = 98.84°, D_m = 1.317, Z = 4. Cu radiation, R = 0.044 for 1534 reflexions.

The oxime is exclusively anti with respect to the oxygen of the lactone ring and, in the crystalline state, is hydrogen-bonded to the carboxyl oxygen of an adjacent molecule (Fig. 1). The aromatic rings from adjacent molecules overlap to form a stack which is coxaxial with a twofold screw-axis. Bond lengths (Fig. 2) and angles are normal. The lactone ring is nearly planar and the phenyl ring is turned at an angle of 74.2° to it. The cyclohexenone ring assumes a half-chair conformation and shows a near perfect staggering of the hydrogen substituents on C(5) and C(7).

Fig. 1. Packing diagram of the lactone oxime structure as viewed down the c* axis.

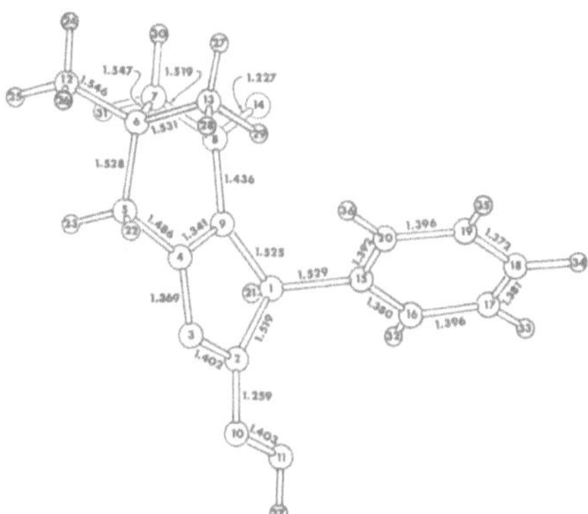

Fig. 2. Bond lengths (σ = 0.005 Å) for the lactone oxime, $C_{16}H_{17}NO_3$.

3,3'-METHYLENE(BIS-6-BROMO-4-HYDROXYCOUMARIN)

(DIBROMODICOUMAROL)

$C_{19}H_{10}Br_2O_6$

N.W. ALCOCK and E. HOUGH, 1972. Acta Cryst., B28, 1957-1960.

Tetragonal, $P4_12_12$, a = 5.496, c = 54.49 Å, D_m = 1.979, Z = 4. Cu radiation, R = 0.054 for 922 reflexions.

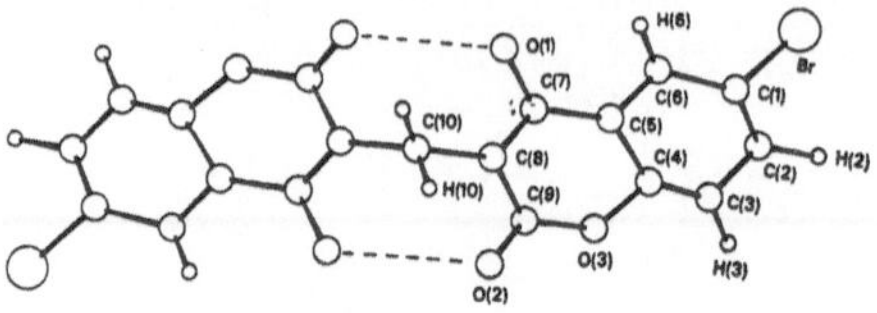

Fig. 1. The dibromodicoumarol molecule.

The molecule (Fig. 1) consists of two approximately planar halves related by a twofold (crystallographic) axis through C(10). Each half is itself composed of two closely planar portions hinged on the line O(1)-C(7)-O(3) with a dihedral angle 4.8°. Molecular dimensions correspond to standard values except for C(5)-C(7) = 1.44, C(7)-C(8) = 1.38, and C(8)-C(9) = 1.42 Å. Tautomerism is not found in the molecule which is entirely in the C(7)-O(1)-H and C(9)=O(2) form. The hydrogen bond distance O(1)-H···O(2) is 2.69 Å. The unusual unit cell is attributed to packing of the molecules in layers, the 4_1 axis arising as a result of good fit between layers after a rotation of 90°.

3',5,5',6-TETRAMETHOXYFLAVONE

$C_{19}H_{18}O_6$

H.-Y. TING, W.H. WATSON and X.A. DOMÍNGUEZ, 1972. Acta Cryst., B28, 1046-1051.

Monoclinic, $P2_1/c$, a = 7.335, b = 11.304, c = 20.249 Å, β = 104.5°, D_m = 1.42, Z = 4. Cu radiation, R = 0.086 for 1249 reflexions.

A schematic diagram of the molecule is shown in Fig. 1. The heterocyclic ring, which is fused to the benzene ring to form the γ-benzopyrone portion of the molecule, is slightly puckered with atom C(3) 0.1 Å out of the plane. The dihedral angle between the plane of the benzene ring and that of the heterocyclic ring is 4°. The phenyl ring is also planar and makes an angle of 28° with the γ-benzopyrone moiety. Three of the methoxy groups make angles of 2, 7, and 8° with the benzene rings while the fourth [C(13)O(12)] makes an angle of 84° because of steric interactions. The bond lengths and angles in the molecule are normal and are shown in Fig. 1.

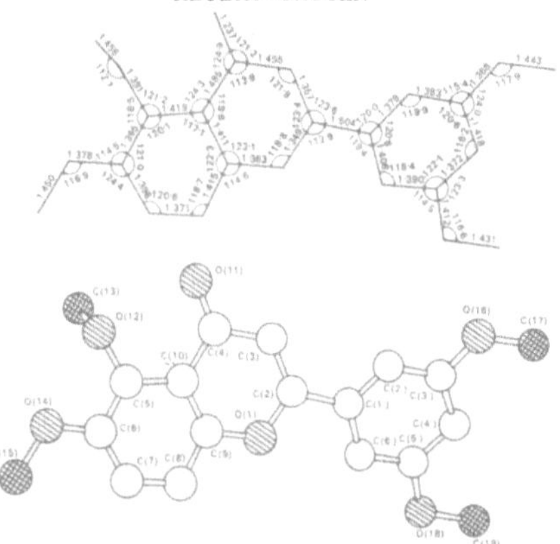

Fig. 1. Molecular structure of 3',5,5',6-tetramethoxyflavone (σ = 0.007 - 0.010 Å).

AVERUFIN

$C_{20}H_{16}O_7$

Y. KATSUBE, T. TSUKIHARA, N. TANAKA, K. ANDO, T. HAMASAKI and Y. HATSUDA, 1972. Bull. Chem. Soc. Japan., 45, 2091-2096.

Monoclinic, $P2_1/a$, a = 23.70, b = 7.24, c = 9.48 Å, β = 105.5°, D_m = 1.562, Z = 4. Mo radiation, R = 0.083 for 2909 reflexions.

The molecular structure (Fig. 1) is as previously proposed from chemical and n.m.r. studies (1,2). Bond lengths are close to expected values, the molecules are joined into centrosymmetrical dimers by O-H···O hydrogen bonds, and there are two intramolecular hydrogen bonds.

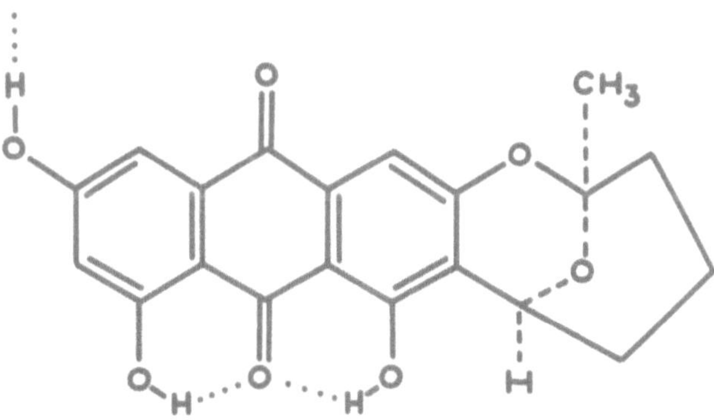

Fig. 1. One enantiomorph of averufin.

1. P. ROFFEY and M.V. SARGENT, 1966. Chem. Comm., 913.
2. J. GRANDJEAN, 1972. Ibid., 1060.

POTASSIUM IODIDE - 2,3,5,6,8,9,11,12-OCTAHYDRO-1,4,7,10,13-BENZOPENTAOXACYCLOPENTADECIN (BENZO-15-CROWN-5) (1:2 COMPLEX)

$C_{28}H_{40}IKO_{10}$

P.R. MALLINSON and M.R. TRUTER, 1972. J. Chem. Soc., Perkin II, 1818-1823.

Tetragonal, P4/n, a = 17.84, c = 9.750 Å, D_m = 1.52, Z = 4. Mo radiation, R = 0.09 for 1603 reflexions.

The complex cation (Fig. 1) has crystallographic symmetry $\bar{1}$, with the potassium sandwiched between two centrosymmetrically related ligand molecules. In each of these the five ether-oxygen atoms are approximately coplanar, the cation lying 1.67 Å away from this plane, so that the ten-coordination is an irregular pentagonal antiprism. The K-O distances range from 2.777 to 2.955(8) Å (Fig. 1). Iodide ions occupy two sets of positions with 4 or $\bar{4}$ symmetry and the arrangement of complex cations and anions resembles the caesium chloride structure; there is no interaction between the anions and the metal (Fig. 2).

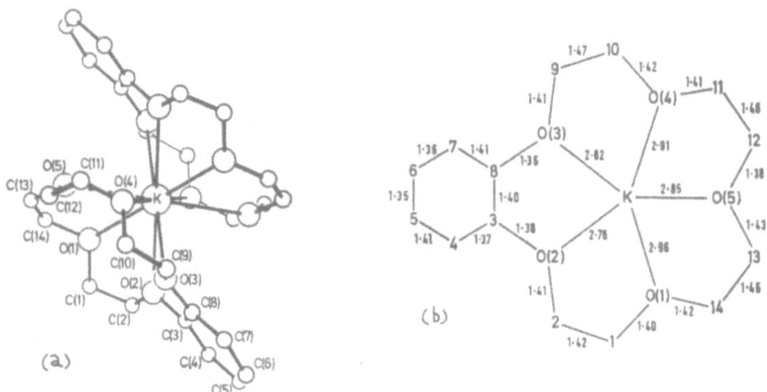

Fig. 1. The K(benzo-15-crown-5)$_2$ complex: (a) structure and (b) bond lengths.

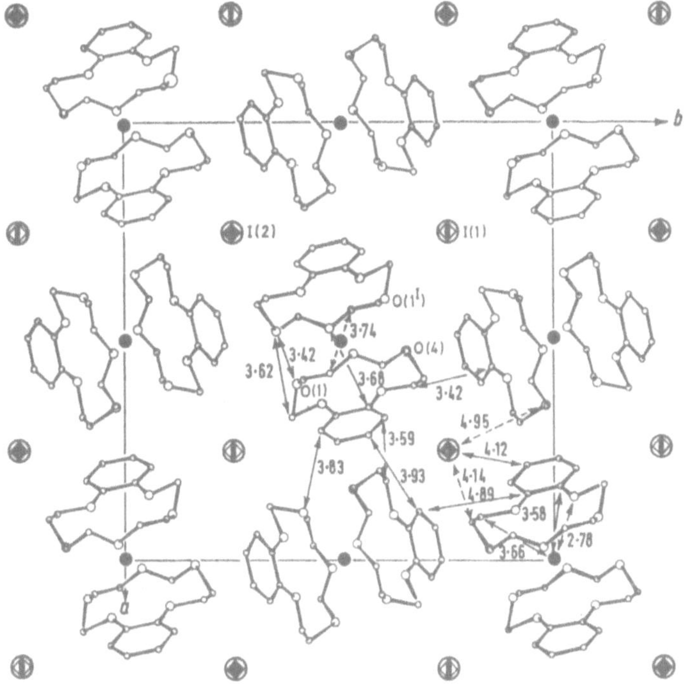

Fig. 2. Projection of the $[K(benzo\text{-}15\text{-}crown\text{-}5)_2]^+I^-$ structure down [001]. The designations of a few atoms are shown. Distances (Å) between atoms are shown along arrows connecting the atoms. Broken lines indicate translation ±1 along c. Filled circles represent potassium ions on centres of symmetry.

DIBENZO-30-CROWN-10 AND ITS POTASSIUM IODIDE COMPLEX

M.A. BUSH and M.R. TRUTER, 1972. J. Chem. Soc., Perkin II, 345-350.

DIBENZO-30-CROWN-10

(2,3:17,18-DIBENZO-1,4,7,10,13,16,19,22,25,28-DECAOXACYCLOTRIACONTA-2,17-DIENE)

$C_{28}H_{40}O_{10}$

Monoclinic, $P2_1/c$, a = 16.960, b = 8.920, c = 9.096 Å, β = 90.03°, D_m = 1.31, Z = 2. Mo radiation, R = 0.056 for 1414 reflexions.

(DIBENZO-30-CROWN-10)-POTASSIUM IODIDE

$C_{28}H_{40}IKO_{10}$

Orthorhombic, Pnna, a = 19.576, b = 12.405, c = 12.965 Å, D_m = 1.51, Z = 4. Mo radiation, R = 0.048 for 1102 reflexions.

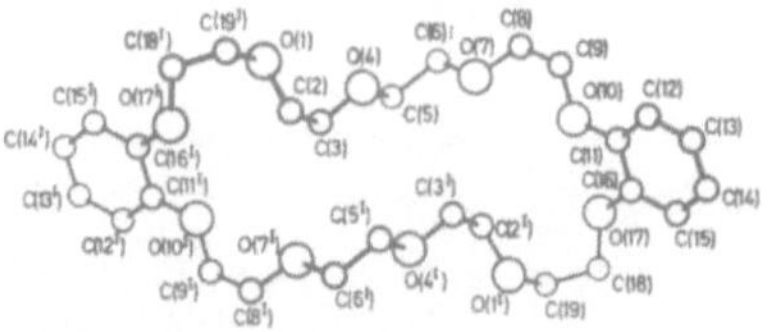

Fig. 1. Dibenzo-30-crown-10 showing the designations of the atoms. The Roman superscript I denotes relation by a crystallographic element of symmetry.

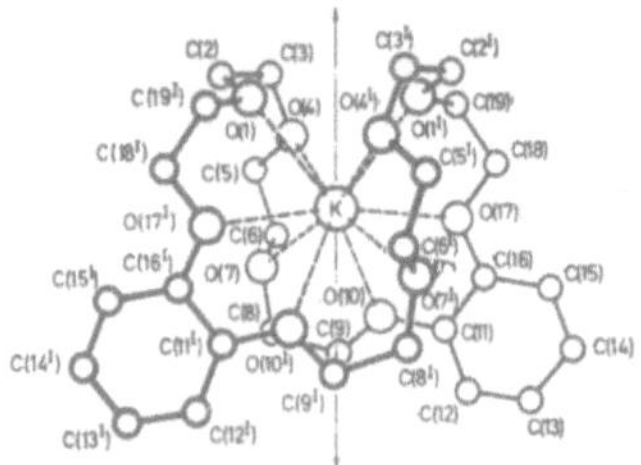

Fig. 2. One complex cation (dibenzo-30-crown-10)-potassium; the crystallographic twofold axis shown by the arrows relates the halves of the ligand. The ligand is wrapped round the cation "like the seam of a tennis ball".

The molecules of the cyclic ether are centrosymmetric and form loops (Fig. 1) approximately parallel to the a-axis. Bond lengths and angles are normal and there are no unusual features about the packing.

In the complex (Fig. 2) each potassium ion lies on a twofold axis, and is enclosed by a ligand molecule wrapped round to give ten K-O distances in the range 2.850 - 2.931(6) Å. The iodide ions also lie on twofold axes and are not in contact with potassium. The packing of the complex cations and the iodide ions is a distorted sodium chloride structure.

Bond lengths and angles in the complexed and free molecules are the same; the conformations are different. Four of the C-C-O-C angles change from trans in the free molecule to gauche in the complex; the remaining torsion angles are of the same type in both forms but corresponding ones differ by as much as 30° and the sign may be reversed.

1,3-DITHIOLANE-2-THIONE

$C_3H_4S_3$

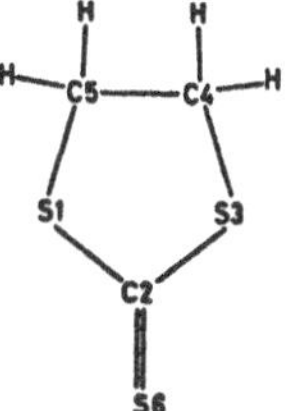

B. KLEWE and H.M. SEIP, 1972. Acta Chem. Scand., 26, 1860-1868.

Monoclinic, $P2_1/c$, a = 8.63, b = 7.18, c = 9.23 Å, β = 100.0°, Z = 4. Mo radiation, R = 0.048 for 983 reflexions.

The crystal structure contains well separated molecules; the shortest intermolecular contacts are S···H = 2.84 Å and S···S = 3.58 Å. The molecule is puckered with nearly C_2 symmetry; the torsional angle about the C-C bond is 44°. Bond lengths, corrected for thermal motion, are C-C = 1.519(4), S-C(sp^3) = 1.818(3), S(1)-C(2) = 1.748(3), S(3)-C(2) = 1.727(3), S(6)-C(2) = 1.652(2) Å. These data correspond to about 86% double bond character in the exocyclic C-S bond.

3,4-EPOXYSULPHOLANE

$C_4H_6O_3S$

D.E. SANDS, 1972. Acta Cryst., B28, 2463-2468.

Orthorhombic, $P2_12_12_1$, a = 8.475, b = 10.742, c = 5.899 Å, D_m = 1.66, Z = 4. Mo radiation, R = 0.055 for 1282 reflexions.

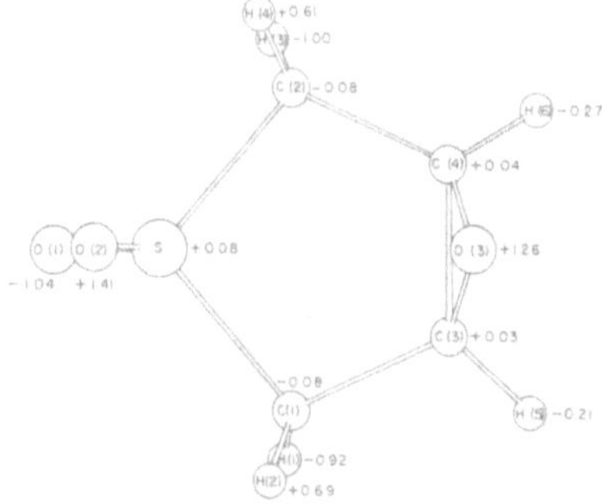

Fig. 1. The 3,4-epoxysulpholane molecule projected onto the mean plane through the S and C atoms. Numbers beside the atoms are distances (Å) from the plane.

The molecule is folded (Fig. 1) so that the plane through the epoxide atoms O(3)C(3)C(4) makes an angle of 105.1° with the C(1)C(2)C(3)C(4) plane. Principal bond lengths (corrected for libration) are S-O = 1.455, S-C = 1.807, C-O(epoxide) = 1.452, C(3)-C(4) = 1.460, C(1)-C(3) = 1.510 Å. The least librational oscillation is about the normal to the sulphur-carbon mean plane.

1-THIACYCLOBUTANE-3-CARBOXYLIC ACID-1-OXIDE

$C_4H_6O_3S$

CH_2
$O{=}S$ $CHCO_2H$
CH_2

S. ABRAHAMSSON and G. REHNBERG, 1972. Acta Chem. Scand., 26, 494-500.

Orthorhombic, $Pna2_1$, a = 9.89, b = 5.41, c = 10.81 Å, Z = 4. Cu radiation, R = 0.097 for 532 reflexions.

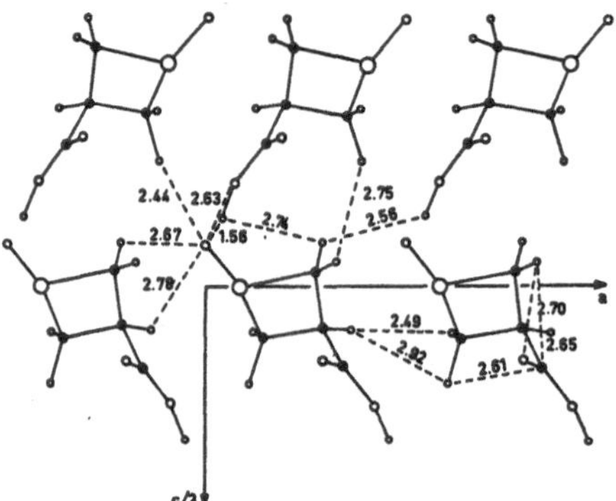

Fig. 1. Molecular packing of $C_4H_6O_3S$ showing short intra- and intermolecular contacts.

This study shows that the high-melting isomer of $C_4H_6O_3S$ has the trans-configuration with a puckered four-membered ring (dihedral angle 153°). Interatomic distances and angles are normal (e.g. C-S = 1.84(1), S-O = 1.53(1) Å). The molecules are held together by hydrogen bonds between the carboxyl and sulphoxide groups (Fig. 1).

3-ACETOXYTHIETANE 1,1-DIOXIDE

$C_5H_8O_4S$

G.D. ANDREETTI, G. BOCELLI and P. SGARABOTTO, 1972. Cryst. Struct. Comm., 1, 423-425.

Monoclinic, $P2_1/m$, a = 10.13, b = 6.51, c = 5.37 Å, β = 94.7°, Z = 2. Cu radiation, R = 0.058 for 672 reflexions.

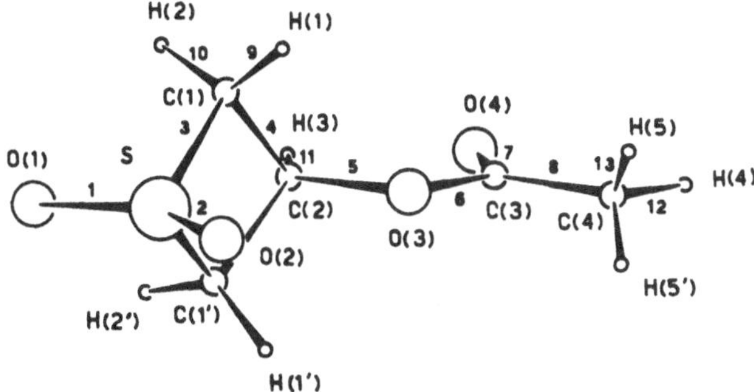

Fig. 1. The 3-acetoxythietane 1,1-dioxide molecule.

The molecule (Fig. 1) has m symmetry. The four-membered ring is not planar; the dihedral angle between the C(1)C(2)C(1') and C(1)SC(1') planes is 162°, and the acetoxy group is axial. The C-S distance is 1.789(4) Å and the S-O distances are 1.426 and 1.419(4) Å.

1,3,5,7,9-PENTATHIO-CYCLODECANE

(MONOCLINIC FORM)

$C_5H_{10}S_5$

G. VALLE, A. PIAZZESI and A. DEL PRA, 1972. Cryst. Struct. Comm., 1, 289-292.

Monoclinic, $P2_1/c$, a = 14.929, b = 4.844, c = 14.817 Å, β = 116.4°, Z = 4. Mo radiation, R = 0.08 for 633 reflexions.

The molecular conformation is shown in Fig. 1; the molecule has a pseudo twofold axis of symmetry through C(4) and S(9). C-S bond lengths are in the range 1.79 - 1.83(2) Å. Valence angles at the sulphur atoms lie between 99.7 and 102.5(10)°, while at the carbon atoms the S-C-S angles are in the range 112.0 - 116.5(10)°.

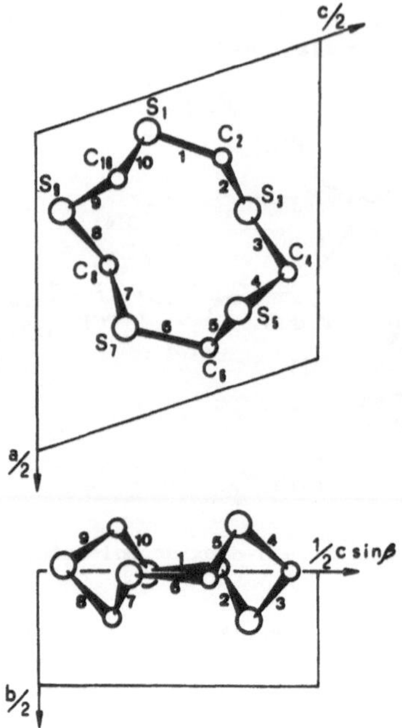

Fig. 1. Views of the pentathiocyclodecane molecule (monoclinic form).

1,3,5,7,9-PENTATHIO-CYCLODECANE

(ORTHORHOMBIC FORM)

$C_5H_{10}S_5$

G. VALLE, A. PIAZZESI and V. BUSETTI, 1972. Cryst. Struct. Comm., 1, 293-296.

Orthorhombic, $P2_12_12_1$, a = 13.376, b = 4.944, c = 14.862 Å, Z = 4. Cu radiation, R = 0.068 for 595 reflexions.

The molecular parameters are almost the same as found in the monoclinic modification (1), with C-S in the range 1.79 - 1.84(2) Å, C-S-C = 101.0 - 104.3(6)°, and S-C-S = 112.3 - 115.7(8)°. The similarity of the boat-like conformation in the two ring skeletons is also confirmed by comparison of the internal rotation angles.

1. Preceding report.

ACETYLTHIOPHENECARBOXYLIC ACIDS

M. GRIFFE, F. DURANT and A.F. PIERET, 1972. Bull. Soc. Chim. Belges, 81, 319-332.

2-CARBOXY-3-ACETYLTHIOPHENE

$C_7H_6O_3S$

T_{23}

Monoclinic, $P2_1/m$, a = 8.68, b = 6.61, c = 7.97 Å, β = 124.6°, D_m = 1.519, Z = 2. Cu radiation, R = 0.08 for 620 reflexions.

3-CARBOXY-2-ACETYLTHIOPHENE

$C_7H_6O_3S$

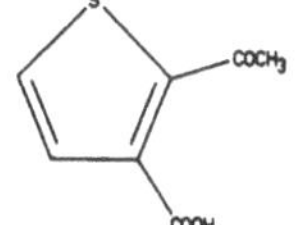

T_{32}

Orthorhombic, Pnam, a = 13.28, b = 8.56, c = 6.58 Å, D_m = 1.510, Z = 4. Cu radiation, R = 0.08 for 632 reflexions.

3-CARBOXY-4-ACETYLTHIOPHENE

$C_7H_6O_3S$

T_{34}

Orthorhombic, Pbcm, a = 7.53, b = 14.89, c = 6.64 Å, D_m = 1.512, Z = 4. Cu radiation, R = 0.10 for 274 reflexions.

The three compounds (coded T_{23}, T_{32} and T_{34}) are required to be planar as all occupy special positions (m) in their space groups. The molecules do not form intermolecular O-H···O hydrogen bonds via the carboxyl groups but rather the keto and carboxyl groups form intramolecular hydrogen bonds. In both T_{23} and T_{34} the "carboxylic" proton was found closer to the keto oxygen O(3) than carboxyl oxygen O(1) (O(3)-H = 1.06 Å in T_{23} and 1.26 Å in T_{32}), which suggests that enol forms play an important role in the ground state structure. Bond lengths are close to expected values, e.g. C-S = 1.69 - 1.72, C-C(exocyclic) = 1.48 - 1.50, C-C(ring) = 1.34 - 1.45. The molecules are separated by van der Waals distances in the crystals.

4,5-DIHYDROTHIEPIN 1,1-DIOXIDE

$C_6H_8O_2S$

H.L. AMMON, M.R. SMITH and E. KELSO, 1972. Acta Cryst., B28, 246-252.

Monoclinic, $P2_1/n$, a = 11.415, b = 6.476, c = 9.597 Å, β = 108.02°, D_m = 1.45, Z = 4. Mo radiation, R = 0.049 for 1247 reflexions.

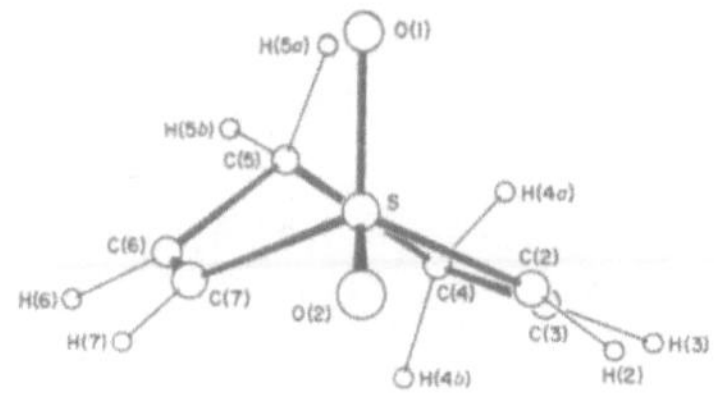

Fig. 1. Conformation of 4,5-dihydrothiepin 1,1-dioxide.

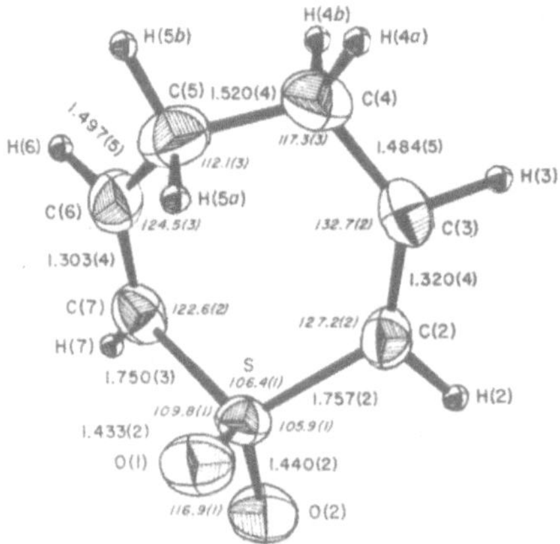

Fig. 2. Bond lengths and angles (angles not shown are Q(1)-S-C(2) = 109.9(1) and O(2)-S-C(7) = 107.6(1)°) in 4,5-dihydrothiepin 1,1-dioxide.

The molecular conformation (Fig. 1) is far from symmetrical; the planes of atoms S-C(2)=C(3)-C(4)-C(5) and S-C(7)=C(6)-C(5) intersect at an angle of 57°. Bond lengths and angles are in Fig. 2; the C=C distances (C(2)=C(3), C(6)=C(7)) are significantly less than normal for ethylenic bonds. There are no unusual intermolecular distances; only normal van der Waals contacts are found between molecules.

1,2,3,4-TETRATHIADECALIN

$C_6H_{10}S_4$

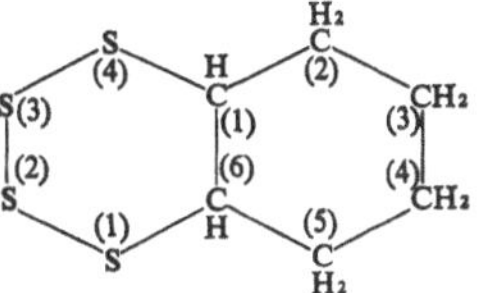

F. FEHÉR, A. KLAEREN and K.-H. LINKE, 1972. Acta Cryst., B28, 534-537.

Triclinic, P$\bar{1}$, a = 9.287, b = 8.606, c = 6.309 Å, α = 107.7, β = 102.4, γ = 98.6°, D_m = 1.5, Z = 2. Mo radiation, R = 0.129 for 962 reflexions.

The molecule exists in the chair-chair conformation; the sulphur atoms bound directly to the cyclohexane ring are trans-equatorial with respect to it. Mean bond lengths are S-S = 2.057(7), S-C = 1.80(2), C-C = 1.53(3) Å. The dihedral angles at the S-S bonds lie between 66 and 69°.

1,4-DITHIONABICYCLO[2,2,2]OCTANE TETRACHLOROZINCATE

$C_6H_{12}Cl_4S_2Zn$ $C_6H_{12}S_2^{2+}\cdot ZnCl_4^{2-}$

E. DEUTSCH, 1972. J. Org. Chem., 37, 3481-3486.

Orthorhombic, Pnma, a = 13.18, b = 8.38, c = 11.55 Å, D_m = 1.83, Z = 4. Mo radiation, R = 0.011 for 908 reflexions.

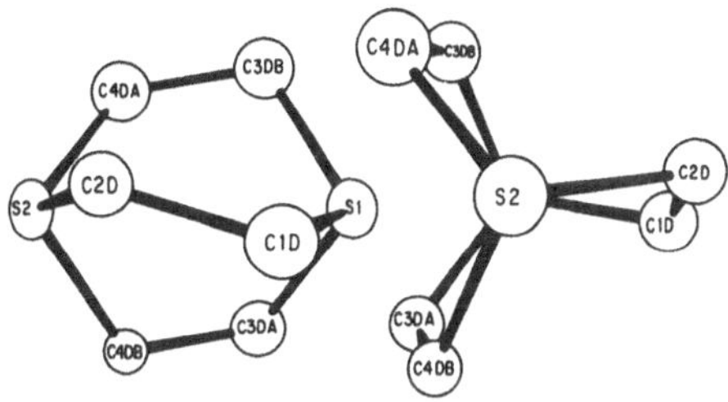

Fig. 1. Two views of the 1,4-dithionabicyclo[2,2,2]octane cation.

Both $C_6H_{12}S_2^{2+}$ and $ZnCl_4^{2-}$ ions lie on planes of symmetry but the cation is disordered. Views of the cation, which has approximate D_3 symmetry, are in Fig. 1. The S···S intramolecular distance is 3.130(4) Å, and C-S-C angles are in the range 98.2 - 101.5(7)°. The average angle of twist of the skeleton from D_{3h} symmetry about the S···S axis is 7.8°. The $ZnCl_4^{2-}$ ion is a distorted tetrahedron with Zn-Cl = 2.259 - 2.282(3) Å and angles Cl-Zn-Cl = 107.2 - 112.1(1)°.

2-FQRMYLPYRIDINE SELENOSEMICARBAZONE

$C_7H_8N_4Se$

A. CONDE, A. LÓPEZ-CASTRO and R. MÁRQUEZ, 1972. Acta Cryst., B28, 3464-3469.

Monoclinic, $P2_1/c$, a = 9.320, b = 6.524, c = 16.275 Å, β = 90.53°, D_m = 1.48, Z = 4. Cu radiation, R = 0.11 for 1180 reflexions.

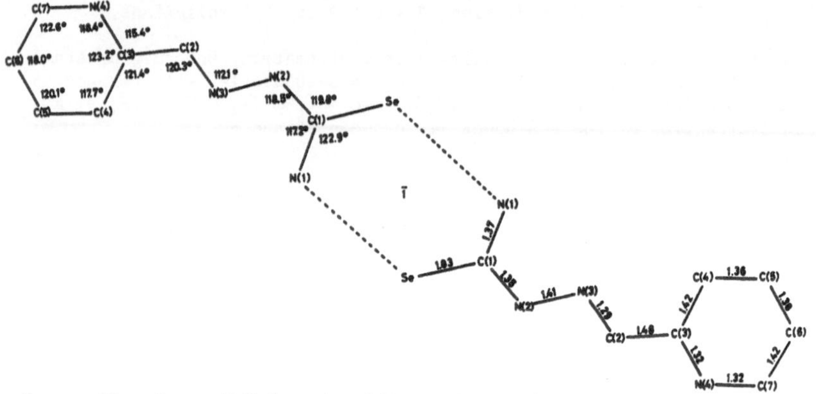

Fig. 1. Structure of 2-formylpyridine selenosemicarbazone.

The structure (Fig. 1) has normal bond lengths with the exception of the Se-C distance of 1.83(2) Å which is indicative of partial double bond character. The molecules are linked by rather weak NH···Se (3.52 Å) hydrogen bonds to form dimers about centres of symmetry.

4,8,9,10-TETRACHLORO-2,6-DITHIAADAMANTANE

$C_8H_8Cl_4S_2$

T. HIGUCHI and W. NOWACKI, 1972. Z. Kristallogr., 135, 56-72.

Monoclinic, $P2_1/c$, a = 7.59, b = 13.31, c = 12.77 Å, β = 118.2°, D_m = 1.82, Z = 4. Cu radiation, R = 0.082 for 1269 reflexions.

The molecule has an adamantane cage structure and (non-crystallographic) $\bar{4}$ symmetry along the line of the two S atoms. Fig. 1 shows the molecular structure and gives details of bond lengths and angles. The average bond angle at the S atoms is 96.4° whereas those at the carbon atoms are almost tetrahedral. The intermolecular distances correspond to van der Waals separations.

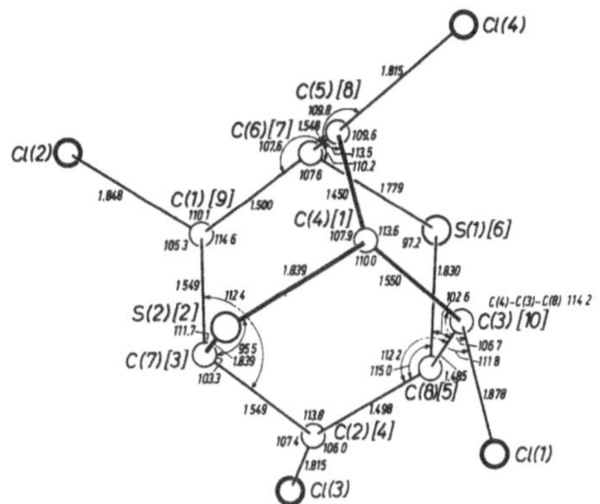

Fig. 1. Projection of a molecule of 4,8,9,10-tetrachloro-2,6-dithiaadamantane along the b axis. Numbers in square brackets indicate the ordinal numbers of the atoms in the adamantane molecule (chemical nomenclature). Mean standard deviations in the C-S, C-Cℓ, and C-C bond lengths are 0.011, 0.014, and 0.011 Å respectively.

DL-6-THIOCTIC ACID

$C_8H_{14}O_2S_2$

R.M. STROUD and C.H. CARLISLE, 1972. Acta Cryst., B28, 304-307.

Monoclinic, $P2_1/c$, a = 11.744, b = 9.895, c = 9.246 Å, β = 109.6°, D_m = 1.343, Z = 4. Cu radiation, R = 0.12 for 1319 reflexions.

For details of this structure see Structure Reports, 32B, 186, where this paper was dealt with in conjunction with that of Karle, Estlin and Britts (Acta Cryst., 1967, 22, 567).

2,5-BIS-(DICYANOMETHYLENE)-2,5-DIHYDROTHIOPHENE

$C_{10}H_2N_4S$

B. AURIVILLIUS, 1972. Acta Chem. Scand., 26, 3612-3618.

Orthorhombic, Pbnm, a = 10.969, b = 14.074, c = 6.473 Å, D_m = 1.32, Z = 4. Cu radiation, R = 0.068 for 355 reflexions.

The structure of the compound $C_{10}H_2N_4S$, a derivative of an unexpected product of the reaction of 2,3,5-tribromothiophene with tetracyanoethylene oxide, has been established by this analysis. The molecule lies on a crystallographic mirror plane at z = 1/4. Bond lengths and angles are in accord with the values expected for this structure; the C-S distances are 1.766(12) and 1.745(13) Å. Intermolecular CH···N contacts of 3.33 - 3.45(1) Å are also found.

3,5-BIS(N,N-DIETHYLIMONIUM)-1,2,4-TRITHIOLANE TETRAIODODI-μ-IODODIMERCURATE(II)

$C_{10}H_{20}Hg_2I_6N_2S_3$ $(S_3C_2N_2(C_2H_5)_4)Hg_2I_6$

P.T. BEURSKENS, W.P.J.H. BOSMAN and J.A. CRAS, 1972. J. Cryst. Mol. Struct., 2, 183-188.

Monoclinic, $P2_1/c$, a = 12.574, b = 15.777, c = 14.560 Å, β = 90.83°, Z = 4. Cu radiation, R = 0.05 for 1380 reflexions.

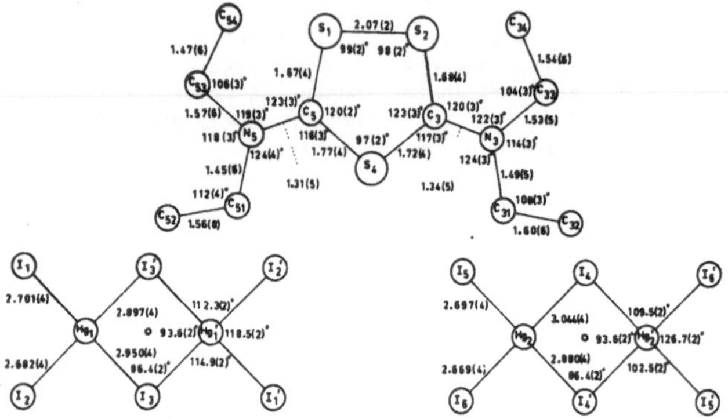

Fig. 1. Bond distances (Å) and bond angles (degrees), with standard deviations, of the $(S_3C_2N_2(C_2H_5))^{2+}$ ion and of the two independent centrosymmetric $Hg_2I_6^{2-}$ ions.

The structure consists of (3,5-bis(N,N-diethylimonium)-1,2,4-trithiolane)$^{2+}$ and $Hg_2I_6^{2-}$ ions. The cation contains a five-membered ring which can be formed by oxidation of dithiocarbamato and thiuramdisulphide complexes. Bond distances and angles are shown in Fig. 1.

TETRAMETHYLENESELENIUM DIMEDONYLIDE

$C_{12}H_{18}O_2Se$

V.V. SOAT-SAZOV, R.A. KJANDŽETSIAN, S.I. KUZNECOV, N.N. MAGDESIEVA and T.L. KHOCJANOVA, 1972. Dokl. Akad. Nauk SSSR, 206, 1130-1132.

Monoclinic, $P2_1/a$, a = 12.40, b = 20.73, c = 5.59 Å, β = 123°, D_m = 1.49, Z = 4. Cu radiation, R = 0.115 for 403 reflexions.

The five-membered ring is non-planar and the Se-C in the ring is 2.00(2) Å: the Se atom occupies the vertex of an irregular trigonal pyramid with C-Se-C angles 90, 104, and 105(2)°. The Se-dimedonyl ring distance is 1.88(2) Å and the six-membered ring is in a half-chair conformation. The bond-length pattern in the six-membered ring is explained by the formulation above.

7-(5-t-BUTYL-1,2-DITHIOLE-3-YLIDENE)-4,5,6,7-TETRAHYDRO-1,2-BENZODITHIOLE-3-THIONE

$C_{14}H_{16}S_5$

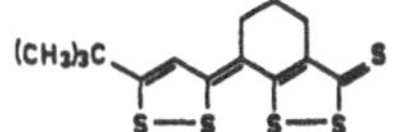

J. SLETTEN, 1972. Acta Chem. Scand., 26, 873-888.

Monoclinic, C2/c, a = 22.264, b = 9.122, c = 15.216 Å, β = 91.683°, D_m = 1.49, Z = 8. Mo radiation, R = 0.077 for 1848 reflexions.

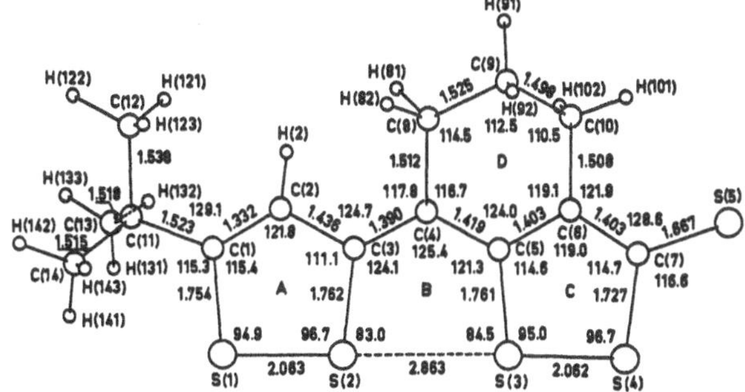

Fig. 1. Bond distances and angles in $C_{14}H_{16}S_5$. Mean standard deviations in S-S, S-C, C-C, and C-H bond lengths are 0.003, 0.008, 0.011, and 0.09 Å respectively.

In the molecule (Fig. 1) the sequence S(1), S(2), S(3) is approximately linear, while the fourth atom deviates from this line (S(1)-S(3)-S(4) = 163°). Rings A and C are individually planar with a dihedral angle of 17° between them. Ring B is non-planar and ring D is described as a slightly twisted boat with one stem (C(4), C(5), C(6)) flattened. The shortest intermolecular S···S contact is 3.374(3) Å, indicating that approximately 3.4 Å is a reasonable value for the S···S van der Waals distance.

2S,3S-1-CYANO-2-HYDROXY-3,4-EPITHIOBUTANE-α-NAPHTHYLURETHANE

$C_{16}H_{14}N_2O_2S$

R.B. BATES, R.A. GRADY and T.C. SNEATH, 1972. J. Org. Chem., 37, 2145-2147.

Orthorhombic, $P2_12_12_1$, a = 4.824, b = 10.803, c = 28.679 Å, Z = 4. Cu radiation, R = 0.048 for 935 reflexions.

Bond distances in the episulphide ring are in Fig. 1; this is the first X-ray study of such a ring system. The naphthalene ring and the urethane group form a dihedral angle of 55.1°. In the crystal, the molecules are linked by NH...O (2.89(8) Å) hydrogen bonds parallel to the a axis.

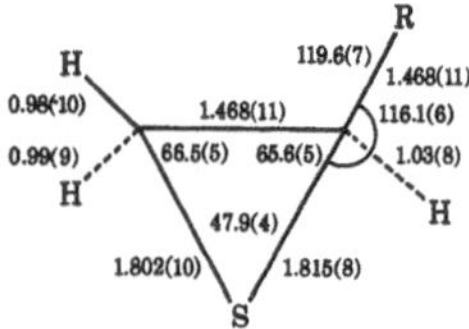

Fig. 1. Bond lengths and angles in the episulphide ring.

2-N,N-DIETHYLAMINO-3-CARBOMETHOXYNAPHTHO-[2,3-b]THIOPHENE-4,9-DIONE

$C_{18}H_{17}NO_4S$

C.S. GIBBONS, J.A. LERBSCHER and J. TROTTER, 1972. J. Cryst. Mol. Struct., 2, 235-242.

Monoclinic, $P2_1/c$, a = 8.273, b = 15.115, c = 13.465 Å, β = 98.40°, D_m = 1.369, Z = 4. Cu radiation, R = 0.059 for 1484 reflexions.

The ring system as a whole is planar (Fig. 1). The carboxyl group is rotated 76° from the plane of the amino group. The grouping at the nitrogen atom is planar due to probable electron delocalization involving the thiophene and amino groups. Bond lengths are as follows: C-C(ar) = 1.392, $C(sp^2)-C(sp^2)$ = 1.474, $C(sp^3)-C(sp^3)$ = 1.514, C-S = 1.732, C-H = 1.02 (preceding are mean values); C-O = 1.197 - 1.335, C-N = 1.365 - 1.471 Å, σ(S-C) = 0.005, σ(C-O, C-N, C-C) = 0.006 - 0.012 Å, σ(C-H) = 0.05 - 0.13 Å.

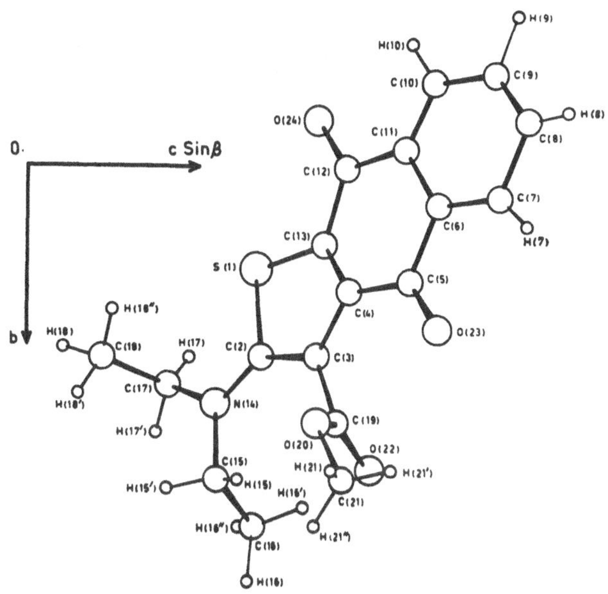

Fig. 1. The $C_{18}H_{17}NO_4S$ molecule viewed along a.

2,11,20-TRITHIA[3.3.3](1,3,5)CYCLOPHANE

$C_{18}H_{18}S_3$

A.W. HANSON and E.W. MACAULAY, 1972. Acta Cryst., B28, 1255-1260.

Monoclinic, $P2_1/c$, a = 13.19, b = 33.35, c = 7.06 Å, β = 93.18°, Dm = 1.39, Z = 8. Cu radiation, R = 0.064 for 837 reflexions.

The structure is disordered (Fig. 1) and the average molecule appears as two half-molecules related by twofold rotation. The essential features of the molecule are still observed however. The phenyl rings are planar, deviate from parallelism by less than 1°, and are separated by 3.192 Å. The C-C bonds of the bridges are inclined by 6° to the ring planes. The molecule possesses a non-crystallographic mirror plane mid-way between the phenyl rings. The structure is described in terms of a unit cell for which b is halved, and for which the space group is C2/c. The disorder in the crystal lattice is probably a reflection of a true conformational disorder of the sulphur bridge. The bond lengths have mean values C-C(aromatic) = 1.38, $C(sp^2)-C(sp^3)$ = 1.50, C-S = 1.82 Å.

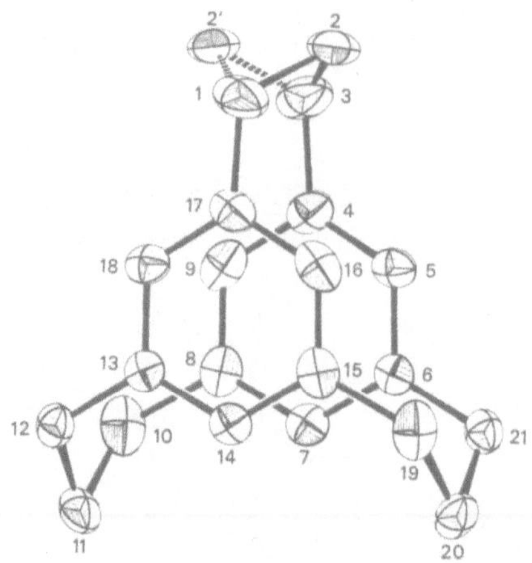

Fig. 1. The 2,11,20-trithia[3.3.3](1,3,5) cyclophane molecule.

2-(p-DIMETHYLANILINO)-4-PHENYL-6a-THIATHIOPHTHENE

$C_{19}H_{17}NS_3$

A. HORDVIK and L.J. SAETHRE, 1972. Acta Chem. Scand., 26, 3114-3130.

Monoclinic, $P2_1/c$, a = 7.158, b = 7.413, c = 33.237 Å, β = 90.14°, Dm = 1.332, Z = 4. Cu radiation, R = 0.044 for 2378 reflexions.

Equal S-S distances occur in the linear three sulphur sequence of the molecule (Fig. 1), and the p-dimethylanilino group has a quinoid structure, bonds C(2)-C(6), C(7)-C(8), C(10)-C(11), and C(9)-N possessing considerable double bond character. Both the 6a-thiathiophthene system and the p-dimethylanilino group are nearly planar. The phenyl group is twisted 81.8° about the C(4)-C(14) bond and the p-dimethylanilino group is twisted 12.7° about the C(2)-C(6) bond. There are no unusual intermolecular contacts.

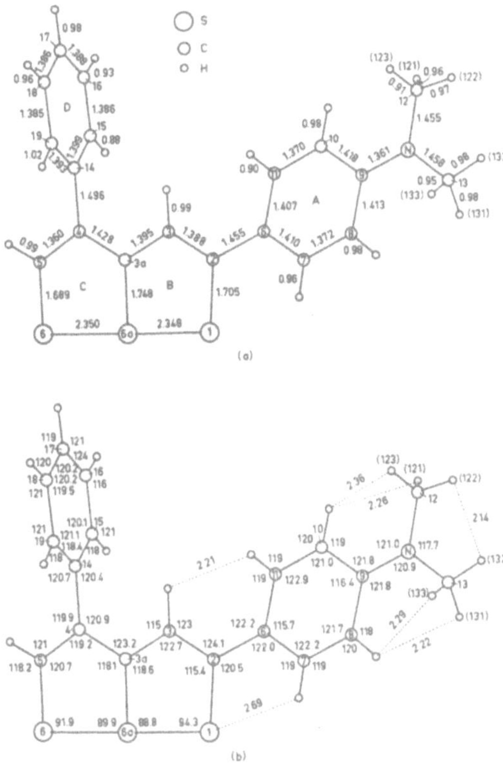

Fig. 1. (a) Bond lengths (Å) in the 2-(p-dimethylanilino)-4-phenyl-6a-thiathiophthene molecule. (b) Bond angles and intramolecular non-bonding distances (Å). Standard deviations are S-S = 0.001 Å, C-S = 0.003 - 0.004 Å, mean C-C, C-N = 0.004 Å, C-H = 0.03 - 0.05 Å, S-S-C = 0.1°, for other non-hydrogen angles 0.3 - 0.4°.

N-(p-BROMOPHENYLCARBAMOYL)THIAMINE ANHYDRIDE (N,S-cis-TYPE)

$C_{19}H_{20}BrN_5O_2S$

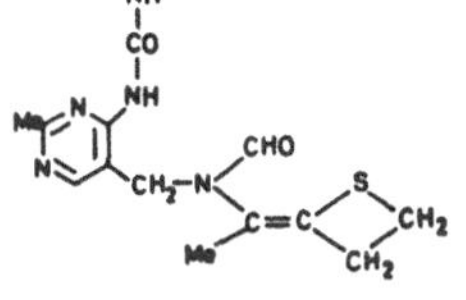

H. NAKAI and H. KOYAMA, 1972. J. Chem. Soc., Perkin II, 248-252.

Monoclinic, $P2_1/c$, a = 8.736, b = 11.668, c = 20.451 Å, β = 99°21', D_m = 1.463,

Z = 4. Mo radiation, R = 0.065 for 2583 reflexions.

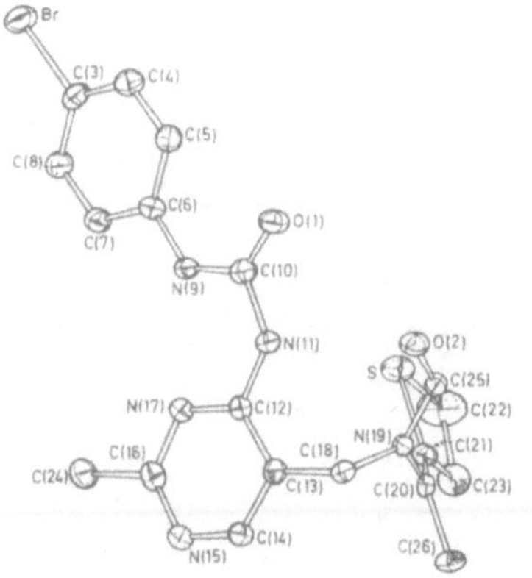

Fig. 1. The configuration of N(p-bromophenylcarbamoyl)thiamine anhydride (N,S-cis-type).

The molecular configuration is shown in Fig. 1 (1). The pyrimidine and thiacyclobutane rings are slightly distorted from planarity; the dihedral angle in the thiacyclobutane ring is 175°. The conformation of the molecule is determined by two intramolecular hydrogen bonds, N(11)H...O(2) = 3.040(8) and N(9)H...N(17) = 2.639(8) Å. Bonded distances and angles are not significantly different from expected values.

1. For details of the corresponding trans-N,S compound see Structure Reports, 37B, 212.

9β-3-METHOXY-17-ACETOXY-7-THIAOESTRA-1,3,5(10),8(14)-TETRAENE

$C_{20}H_{24}O_3S$

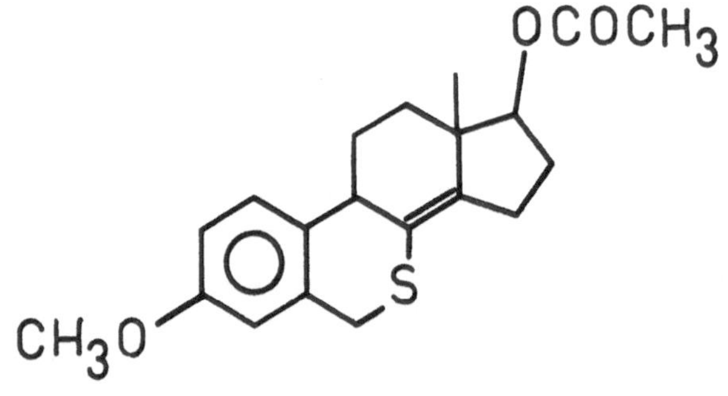

OCOCH3 S CH3O

C.F.W. van de VEN and H. SCHENK, 1972. Cryst. Struct. Comm., 1, 121-124.

Orthorhombic, $P2_12_12_1$, a = 9.006, b = 26.148, c = 7.529 Å, Z = 4. Cu radiation, R = 0.039.

Ring A is planar within the accuracy of the analysis; ring B is in a boat conformation while rings C and D are distorted half-chairs. Bond lengths are normal, e.g., S-C(sp^3) = 1.824(4), S-C(sp^2) = 1.773(4) Å.

METHIXENE HYDROCHLORIDE MONOHYDRATE

$C_{20}H_{26}ClNOS$ $C_{20}H_{23}NS.HCl.H_2O$

S.S.C. CHU, 1972. Acta Cryst., B28, 3625-3632.

Monoclinic, $P2_1/a$, a = 15.320, b = 9.118, c = 13.862 Å, β = 94.75°, D_m = 1.235, Z = 4. Mo radiation, R = 0.068 for 2228 reflexions.

In the molecule (I) the outer benzenoid rings have normal geometry; their mean

I

planes make a dihedral angle of 137.9° (Fig. 1). The meso atoms C(9) and S are

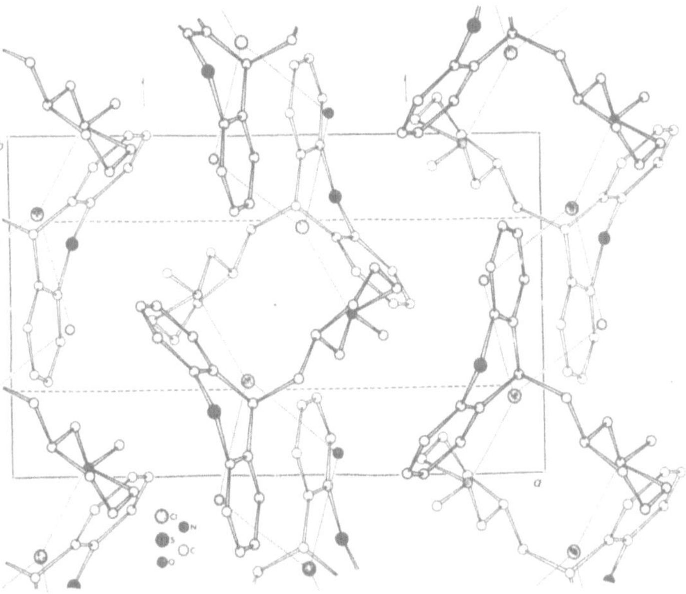

Fig. 1. Molecular conformation and packing in methixene hydrochloride monohydrate. Dashed lines are hydrogen bonds.

displaced from the appropriate benzene planes by -0.074 and -0.099 Å (for S), and 0.054 and 0.102 Å (for C(9)). The piperidyl ring is in a chair conformation with its substituent groups equatorial. All interatomic distances are normal. Each $C\ell^-$ ion is associated with three hydrogen bonds (Fig. 1), one to the quaternary nitrogen (NH...Cℓ = 3.051 Å) and the other two to two different water molecules (OH...Cℓ = 3.203, 3.280 Å)

1H,4H-NAPHTHO[1,8]DISELENEPINE DIMER

$C_{24}H_{20}Se_4$

S. ALEBY, 1972. Acta Cryst., B28, 1509-1518.

Monoclinic, C2/c, a = 22.907, b = 5.1459, c = 18.091 Å, β = 97.97°, Z = 4. Cu radiation, R = 0.029 for 1388 reflexions.

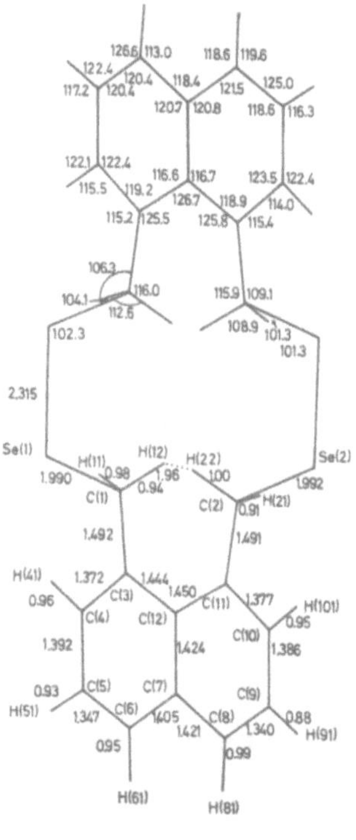

Fig. 1. Bond lengths (standard deviations are Se-Se = 0.002, Se-C = 0.015, C-C = 0.021, C-H = 0.12 Å) and bond angles in 1H,4H-naphtho[1,8] diselenepine. The dihedral angle C-Se-Se-C is 88.1°.

The molecule has twofold crystallographic symmetry; molecular dimensions are in Fig. 1. The molecule suffers from intramolecular overcrowding effects; there is a very short H...H distance 1.96 Å. Relief from steric strain is achieved by angle bending and by bond length extension. Only two intermolecular contact distances (Se...C = 3.48 and 3.60 Å) are less than van der Waals distances.

TETRAPHENYLTHIENO[3,4-c]THIOPHENE

$C_{30}H_{20}S_2$

M.D. GLICK and R.E. COOK, 1972. Acta Cryst., B28, 1336-1339.

Monoclinic, $P2_1/a$, a = 17.431, b = 5.953, c = 13.842 Å, β = 128.40°, Z = 2. Cu radiation, R = 0.061 for 1446 reflexions.

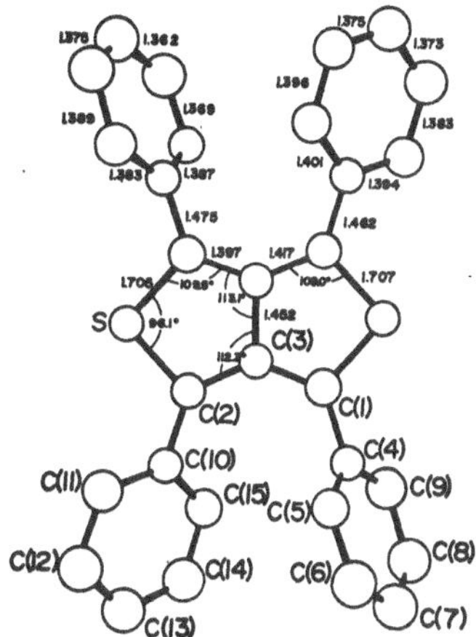

Fig. 1. Tetraphenylthieno[3,4-c]thiophene with bond distances (σ = 0.005 - 0.008 Å) and bond angles.

The molecule is located about a centre of symmetry and the thieno[3,4-c]-thiophene moiety is planar with approximate mmm (D_{2h})symmetry. The molecular geometry is in Fig. 1; the phenyl rings are rotated out of plane by 39.6 and 58.4°. There are no unusual contacts in the molecular packing.

2-OXAZOLIDINONE

$C_3H_5NO_2$

J.W. TURLEY, 1972. Acta Cryst., B28, 140-143.

Monoclinic, $P2_1/c$, a = 7.313, b = 5.672, c = 9.970 Å, β = 110.78°, D_m = 1.423, Z = 4. Cu radiation, R = 0.132 for 551 reflexions.

The molecule is essentially planar in the solid state with a root-mean-square deviation of 0.05 Å from the molecular plane (Fig. 1). Major resonance forms (I) and (II) are suggested for the structure from the observed bond lengths:

```
      H  H                    H  H
      |  |                    |  |
    H-C- C-H                H-C- C-H
     /    \                  /    \
(+)HN      O:              HN      O:(+)
     \    /                  \    /
       C                       C
       |                       |
      :O:                     :O:
      (-)                     (-)
      (I)                     (II)
```

C(1)-C(2) = 1.497(10), C(2)-O(1) = 1.453(8), O(1)-C(3) = 1.356(10), C(3)-N = 1.301(8), N-C(1) = 1.466(7), and C(3)-O(2) = 1.210(8) Å. In the crystal, hydrogen bonded chains are formed in the y direction by N-H...O(2), 2.862 Å, hydrogen bonds; there are no other close intermolecular contacts.

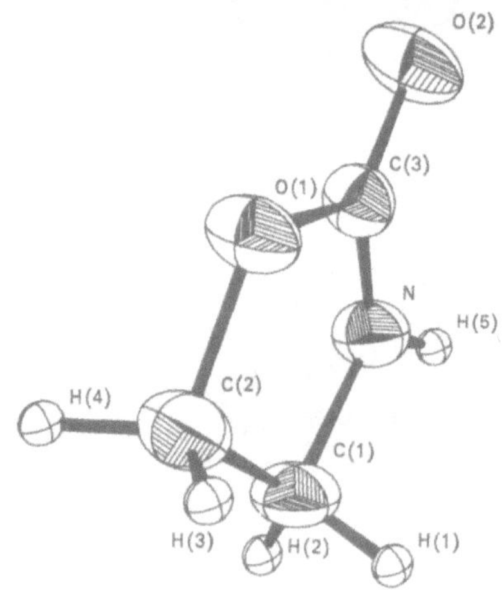

Fig. 1. Molecular structure of 2-oxazolidinone.

4-METHYL-3-FUROXANCARBOXYLIC ACID HYDRAZIDE

$C_4H_6N_4O_3$

I. M. CALLERI, G. CHIARI and D. VITERBO, 1972. Cryst. Struct. Comm., 1, 407-410.

Monoclinic, $P2_1/c$, a = 12.151, b = 7.531, c = 7.236 Å, β = 94.11°, Z = 4. Mo radiation, R = 0.049 for 932 reflexions, for isomer with m.p. 129-131°C.

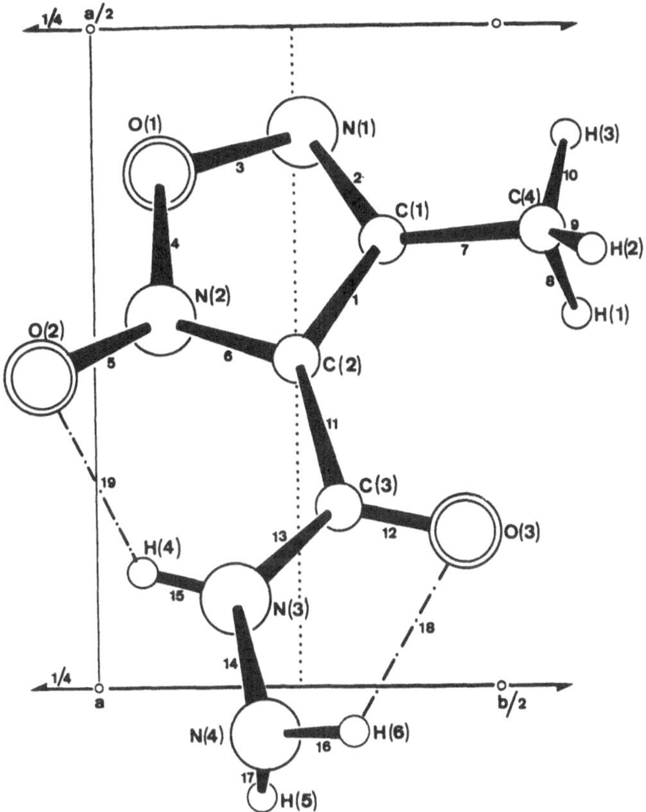

Fig. 1. The 4-methyl-3-furoxancarboxylic acid hydrazide molecule (m.p. 129-131°C) projected onto the (001) plane.

The molecule and crystallographic numbering scheme are shown in Fig. 1. Among the non-hydrogen atoms only N(3) and O(3) deviate slightly and N(4) significantly from planarity. Bond O(1)-N(2) (1.427(3) Å) is much longer than bond O(1)-N(1) (1.388(4) Å), and bond N(2)-C(2) (1.326(4) Å) significantly longer than N(1)-C(1) (1.307(4) Å). The hydrazine N(3)-N(4) bond is 1.416(4) Å, whereas N(3)-C(3) is 1.329(4) Å. The presence of intramolecular NH...O bonds is discounted on geometrical grounds.

II. G. GERMAIN and D. VITERBO, 1972. Cryst. Struct. Comm., 1, 411-414.

Triclinic, $P\bar{1}$, a = 9.092, b = 6.733, c = 6.987 Å, α = 69.91°, β = 117.72°, γ = 117.40°, Z = 2. Mo radiation, R = 0.09 for 1205 reflexions, for isomer with m.p. 124-126°C.

The structure of this lower melting isomer (Fig. 2) confirms assignments made from n.m.r. studies. Principal bond lengths are O(1)-N(1) = 1.358(3), O(1)-N(2) = 1.450(5), C(1)-N(2) = 1.317(4), C(2)-N(1) = 1.306(5), N(3)-N(4) = 1.395(4), and C(3)-N(3) = 1.328(6) Å.

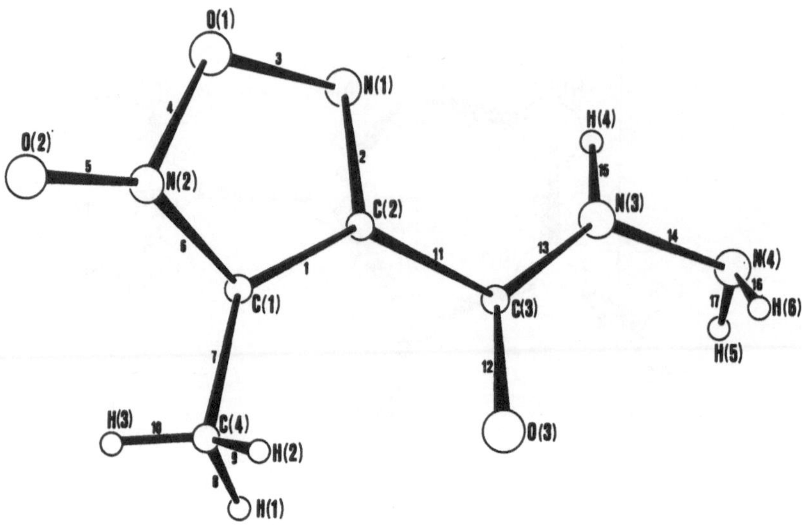

Fig. 2. Projection of the lower melting isomer of $C_4H_6N_4O_3$.

4,6-DINITROBENZFUROXAN

$C_6H_2N_4O_6$

C.K. PROUT, O.J.R. HODDER and D. VITERBO, 1972. Acta Cryst., B28, 1523-1526.

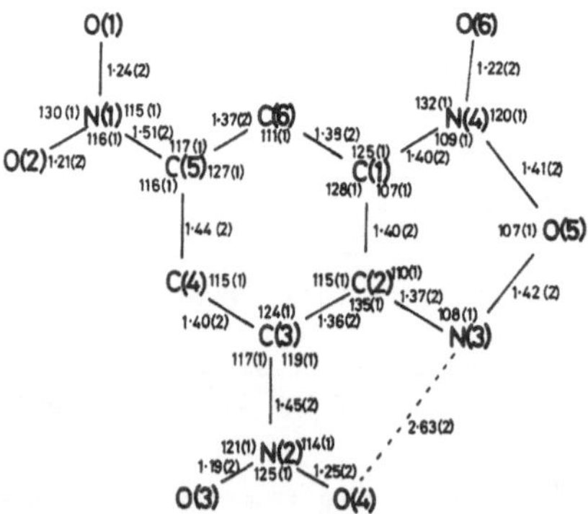

Fig. 1. Interatomic distances and bond angles in 4,6-dinitrobenzfuroxan.

Monoclinic $P2_1$, a = 7.40, b = 9.78, c = 6.19 Å, γ = 108.0°, D_m = 1.752, Z = 2. Cu radiation, R = 0.063 for 656 reflexions.

The crystal structure contains isolated, essentially planar molecules; molecular dimensions are in Fig. 1. The planes of the N(1) and N(2) nitro groups are inclined at angles of 8.9 and 2.7° respectively to the benzene ring plane.

5,6-DICHLOROBENZFURAZAN-1-OXIDE

$C_6H_2C\ell_2N_2O_2$

D. BRITTON, J. KONNERT, J. HAMER and L.M. TREFONAS, 1972. Acta Cryst., B28, 1123-1126.

Monoclinic, C2/c, a = 11.609, b = 8.956, c = 7.355 Å, β = 96.2°, D_m = 1.80, Z = 4. Mo radiation, R = 0.135 for 482 reflexions.

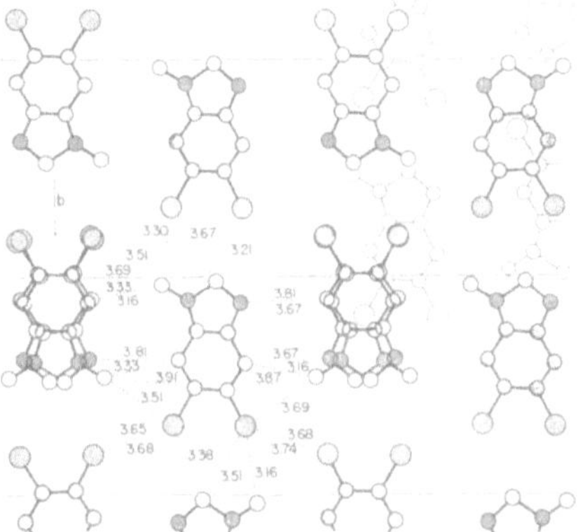

Fig. 1. The crystal structure of 5,6-dichlorobenzfurazan-1-oxide viewed perpendicular to the molecular planes.

The structure is shown in Fig. 1. The molecules lie approximately in planes parallel to (10$\bar{1}$). The average distance between layers is 3.19 Å. The molecules are disordered with the pseudo-twofold axis of the molecule approximately aligned with the twofold axis in the crystal. The O...Cℓ contacts within the rows of molecules are normal (3.2 Å). There is one short intermolecular distance in the structure, an O...C distance of 3.16 or 3.33 Å depending on the disorder.

5-BROMOBENZFURAZAN-1-OXIDE

$C_6H_3BrN_2O_2$

D. BRITTON, G.L. HARDGROVE, R. HEGSTROM and G.V. NELSON, 1972. Acta Cryst., B28, 1121-1123.

Monoclinic, $P2_1/n$, a = 12.67, b = 7.35, c = 7.79 Å, β = 90.0°, Z = 4. Cu radiation, R = 0.20 for 262 reflexions.

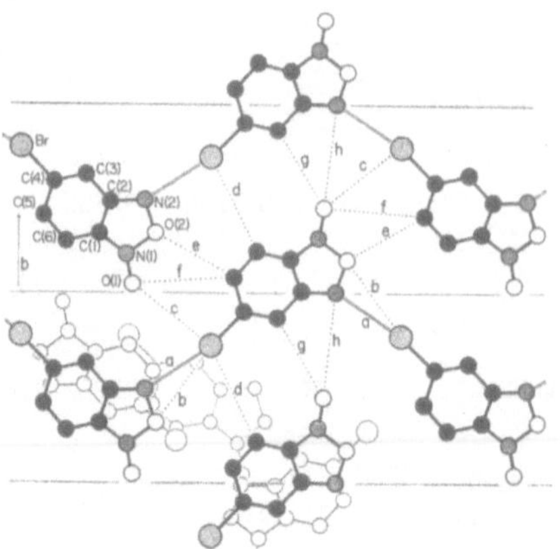

Fig. 1. The crystal structure of 5-bromobenzfurazan-1-oxide perpendicular to (101). The dashed line is the short N(2)...Br contact.

The structure was refined as a rigid group using the structure of the previously refined 5-methylbenzfurazan-1-oxide (1) with the bromine replacing the methyl group 1.870 Å from the ring, and the molecule given exact planarity. The crystal structure is shown in Fig. 1. The molecules form layers parallel to (101), the average distance between layers being 3.37 Å. All intermolecular contacts correspond to van der Waals interactions except a N(2)...Br distance of 3.14 Å. This is assigned to a specific acid-base interaction to explain the difference in packing compared with the methyl compound (1). A similar effect is suggested for the chloro compound.

1. This volume, p. 371.

5-IODOBENZFURAZAN-1-OXIDE

$C_6H_3IN_2O_2$

R.C. GEHRZ and D. BRITTON, 1972. Acta Cryst., B28, 1126-1130.

Monoclinic, $P2_1/c$, a = 10.413, b = 19.787, c = 7.596 Å, β = 103.4°, Z = 8. Mo radiation, R = 0.134 for 483 reflexions.

The crystal structure is shown in Fig. 1. The molecules occur in two independent chains parallel to b. All of the intermolecular distances are near the expected van der Waals contact distances except for two values: O(11)...I(1) is 3.11 Å and O(12)...I(2) is 3.08 Å, both of which are significantly shorter than the expected value of 3.65 Å. These are interpreted as Lewis acid-base interactions between the atoms.

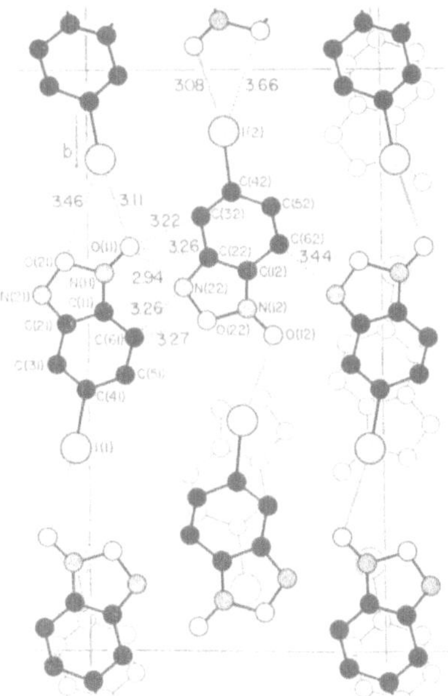

Fig. 1. The crystal structure of 5-iodobenzfurazan-1-oxide viewed along c.

5-METHYLBENZFURAZAN-1-OXIDE

$C_7H_6N_2O_2$

D. BRITTON and W.E. NOLAND, 1972. Acta Cryst., B28, 1116-1121.

Monoclinic, $P2_1/c$, a = 4.031, b = 15.042, c = 11.425 Å, β = 92.82°, D_m = 1.430, Z = 4. Mo radiation, R = 0.059 for 702 reflexions.

The bond lengths and angles corrected for libration are shown in Fig. 1. There appear to be three resonance forms that contribute appreciably to the structure. The bond lengths suggest that I makes the largest contribution.

I II III

The C(3)-C(4) and C(5)-C(6) distances are shorter than the remaining benzene ring distances, and the two C-N distances are about halfway between single and double bond values. The inequality of the 2 ring N-O distances suggests an appreciable contribution from II, and the N(1)-O(1) distance of 1.240 Å suggests III also contributes significantly. Atoms O(1), N(2), and C(7) are significantly out of the molecular plane. The intermolecular contacts are normal van der Waals interactions.

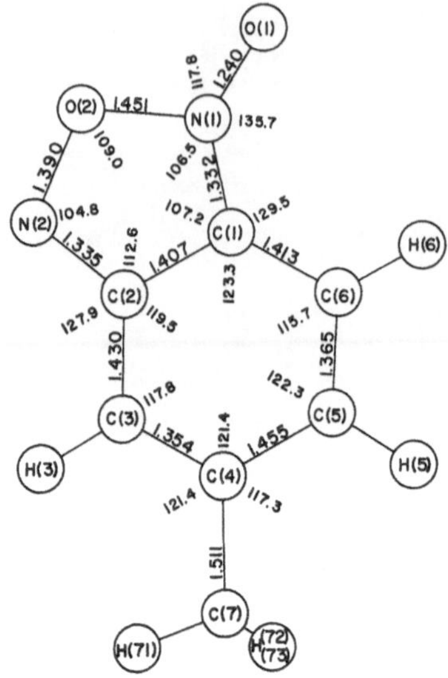

Fig. 1. Bond distances and angles in 5-methylbenzfurazan-1-oxide (σ = 0.004 - 0.005 Å and 0.3°).

3-METHYL-BENZOXAZOLINE-2-THIONE

C_8H_7NOS

P. GROTH, K. DAVIDKOV and D. SIMOV, 1972. Acta Chem. Scand., 26, 1931-1937.

Monoclinic, $P2_1/c$, a = 9.24, b = 6.90, c = 12.86 Å, β = 99.0°, D_m = 1.34, Z = 4. R = 0.077 for 674 reflexions.

The crystal structure contains well separated molecules. The molecule is planar to within 0.02 Å. The C = S distance is 1.629(7) Å, and the C(thionyl)-O distance is 1.393(9) Å. Other distances have normal values.

cis-2-ISOPROPYL-3-(4-NITROPHENYL)OXAZIRIDINE

$C_{10}H_{12}N_2O_3$

J.F. CANNON, J. DALY, J.V. SILVERTON, D.R. BOYD and D.M. JERINA, 1972. J. Chem. Soc., Perkin II, 1137-1140.

Monoclinic. $P2_1/n$, a = 11.206, b = 6.516, c = 14.757 Å, β = 98.50°, D_m = 1.29, Z = 4. Mo radiation, R = 0.005 for 993 reflexions.

Fig. 1. Molecular conformation in cis-2-isopropyl-3-(4-nitrophenyl) oxaziridine.

In this molecule (Fig. 1) the oxaziridine ring contains a long N-O bond (1.500(6) Å), a fairly short C-N bond (1.434(6) Å), and a C-O bond, 1.405(5) Å, of normal length. Atoms C(4), C(7), N(1) and C(8) are nearly coplanar and this plane makes an angle of 102.1° with that of the oxaziridine ring. Other bond lengths and angles have their expected values.

DECAHYDROPYRAZINO[2,3-b]PYRAZINO-1,6'.4,2'.5,3'.8,5-DIOXANE

$C_{10}H_{14}N_4O_2$

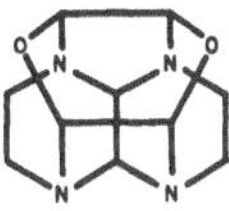

R.D. GILARDI, 1972. Acta Cryst., B28, 742-746.

Monoclinic, C2/c, a = 11.271, b = 5.983, c = 14.755 Å, β = 106.1°, Z = 4. Cu radiation, R = 0.053 for 840 reflexions.

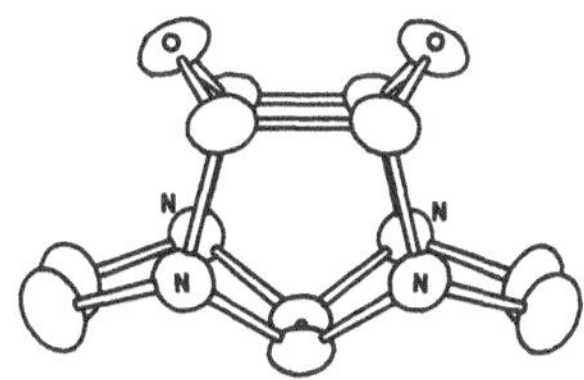

Fig. 1. A view of the $C_{10}H_{14}N_4O_2$ cage structure.

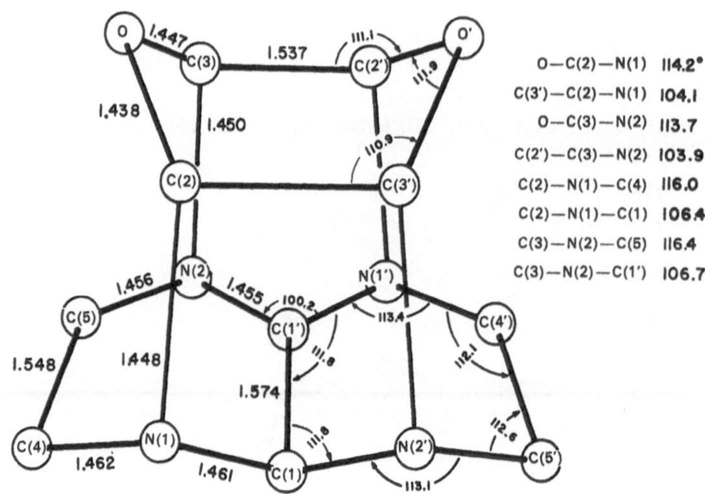

Fig. 2. Bond lengths (σ = 0.0025 - 0.0035 Å) and angles (σ = 0.15 - 0.20°) for $C_{10}H_{14}N_4O_2$.

The cage structure of this molecule can be considered as two layers of boat-shaped six-membered rings (Fig. 1). The carbon-carbon bond which forms the junction of the piperazine rings is the only one which appears to be significantly lengthened by cage formation (1.574(3) Å, Fig. 2). The molecules are held in the crystals by van der Waals forces.

3-PHENYL-5-CHLORO-7-BROMOISOXAZOLO[4,5-d]PYRIMIDINE

$C_{11}H_5BrC\ell N_3O$

B. BOVIO and S. LOCCHI, 1972. J. Cryst. Mol. Struct., 2, 251-257.

Orthorhombic, Pnam, a = 15.72, b = 10.65, c = 6.72 Å, D_m = 1.78, Z = 4. Cu radiation, R = 0.102 for 689 reflexions (films, densitometer intensities).

The molecules lie on crystallographic mirror planes and are planar. Bond distances and angles are shown in Fig. 1.

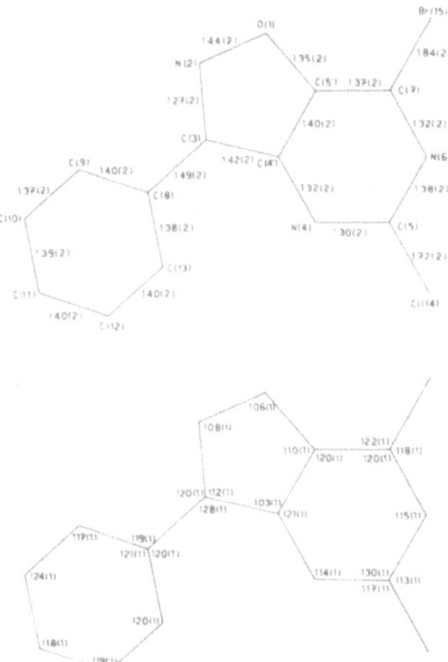

Fig. 1. Bond lengths (Å) and angles (degrees) with their standard deviations in the $C_{11}H_5BrC\ell N_3O$ molecule.

3-PHENYL-7-BROMOISOXAZOLO[4,5-d]PYRIDAZIN-4(5H)-ONE

$C_{11}H_6BrN_3O_2$

B. BOVIO and S. LOCCHI, 1972. J. Cryst. Mol. Struct., 2, 89-97.

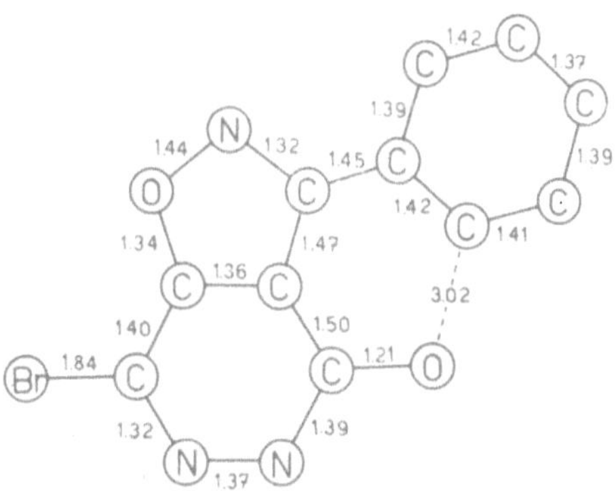

Fig. 1. Bond distances in $C_{11}H_6BrN_3O_2$; σ(C-Br) = 0.01, σ(other bonds) = 0.02 - 0.03 Å.

Monoclinic, $P2_1/c$, a = 7.67, b = 6.84, c = 20.52 Å, β = 91.5°, Z = 4. Cu radiation, R = 0.101 for 1187 reflexions (films, visual intensities).

The benzene ring makes an angle of 4·5° with the plane of the isoxazole ring, and the other part of the molecule is slightly non-planar. The bond distances (Fig. 1) and angles compare well with those of other similar molecules. Two N-H...O hydrogen bonds (N...O = 2.80 Å) join two molecules related by a centre of symmetry. Close molecular packing is found in layers parallel to (100).

3-METHYL-4-BENZYLIDENEISOXAZOLINE-5-ONE

$C_{11}H_9NO_2$

J. MEUNIER-PIRET, P. PIRET, G. GERMAIN, J.-P. PUTZEYS and M. van MEERSSCHE, 1972. Acta Cryst., B28, 1308-1311.

Monoclinic, $P2_1/n$, a = 12.140, b = 6.748, c = 12.325 Å, β = 114.27°, Z = 4. Cu radiation, R = 0.061 for 698 reflexions.

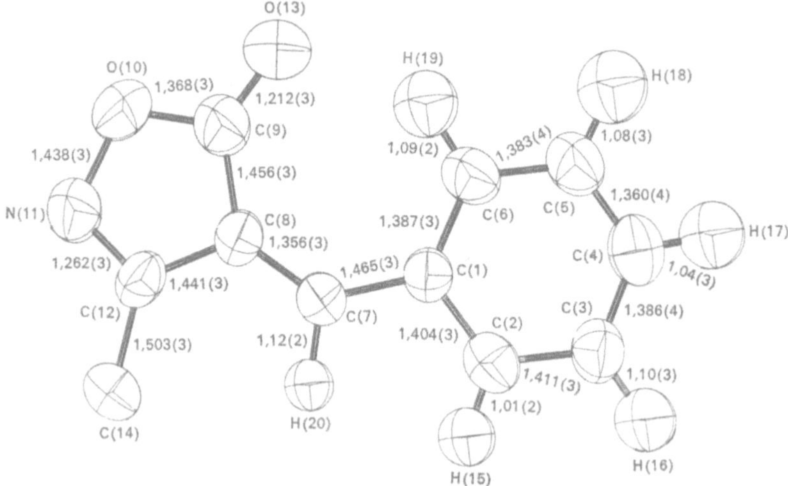

Fig. 1. Bond lengths in 3-methyl-4-benzylideneisoxazoline-5-one.

The bond lengths in the molecule are shown in Fig. 1. C(8)-C(9) and C(8)-C(12) are shorter, because of conjugation with the adjacent C(8)-C(7) double bond, than the corresponding values found in 3-phenyl-isoxazoline-5-one (1). The molecule is planar resulting in the opening of the angles C(6)-C(1)-C(7) to 126°, C(1)-C(7)-C(8) to 132°, C(7)-C(8)-C(9) to 134°, and C(8)-C(9)-O(13) to 133°.

1. M. CANNAS, S. BIAGINI and G. MARONGIU, 1969. Acta Cryst., B25, 1050.

2-METHYL-3-PHENYL-4-(N-METHYL-N-HYDROXYAMIDIN)-ISOXAZOLIN-5-ONE HYDROBROMIDE

$C_{12}H_{14}BrN_3O_3$ $[C_{12}H_{14}N_3O_3]^+.Br^-$

L. FANFANI, A.NUNZI, P.F. ZANAZZI and A.R. ZANZARI, 1972. Acta Cryst., B28, 2598-2604.

Orthorhombic, Pbca, a = 19.539, b = 10.676, c = 13.418 Å, D_m = 1.53, Z = 8. Cu radiation, R = 0.075 for 1452 reflexions.

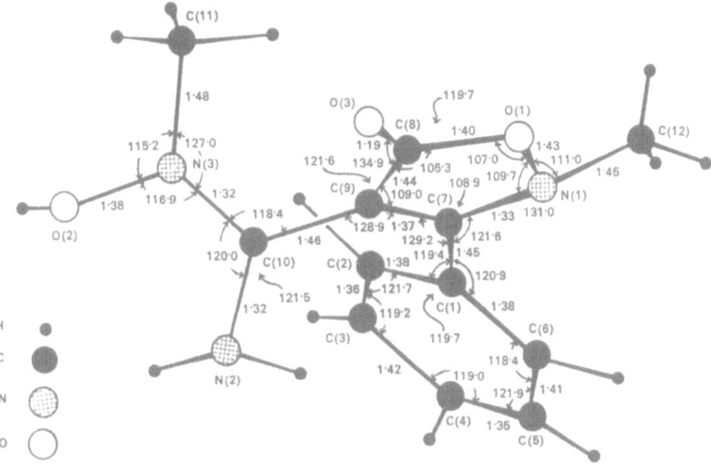

Fig. 1. Bond lengths (σ = 0.01 - 0.02 Å) and angles in the $[C_{12}H_{14}N_3O_3]^+$ ion.

The heterocyclic ring is planar and the bond lengths (Fig. 1) are consistent with two mesomeric forms containing $\gt N-\overset{|}{C}=\overset{+}{N}H_2$ and $\gt\overset{+}{N}=\overset{|}{C}-NH_2$ side chains. The iminic nitrogen, N(2), is protonated. Each Br^- ion links two molecules by hydrogen bonding (NH...Br = 3.31 and OH...Br = 3.09 Å).

1,7,10,16-TETRAOXA-4,13-DIAZACYCLOOCTADECANE

$C_{12}H_{26}N_2O_4$

M. HERCEG and R. WEISS, 1972. Bull. Soc. Chim. Fr., 549-551.

Monoclinic, $P2_1/c$, a = 11.155, b = 8.556, c = 8.999 Å, β = 119°43', D_m = 1.11, Z = 2. Cu radiation, R = 0.044 for 366 reflexions.

In the molecule (Fig. 1) the C-O and C-N dihedral angles are all close to 180° and the C-C dihedral angles are near 60°. The crystallographic symmetry of the molecule is C_i but it is experimentally very close to C_{2h} (2/m) with the twofold axis through the mid-points of the C(8)-C(9) and C(17)-C(18) bonds. Bond lengths are C-O = 1.421 - 1.471(5), C-N = 1.440 - 1.451(9), and C-C = 1.477 - 1.515(9) Å. The molecules are held in the crystal by van der Waals forces.

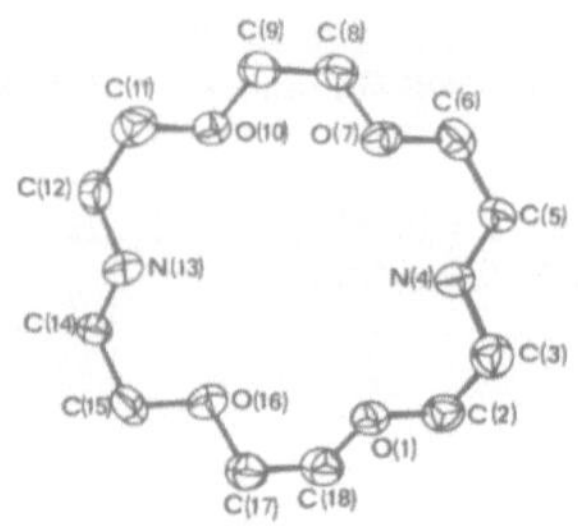

Fig. 1. The conformation of 1,7,10,16-tetraoxa-4,13-diazacyclooctadecane.

1,7,10,16-TETRAOXA-4,13-DIAZACYCLOOCTADECANE: POTASSIUM THIOCYANATE COMPLEX

$C_{13}H_{26}KN_3O_4S$ $[C_{12}H_{26}N_2O_4K]^+SCN^-$

D. MORAS, P. METZ, M. HERCEG and R. WEISS, 1972. Bull. Soc. Chim. Fr., 551-555.

Triclinic, $P\overline{1}$, a = 8.027, b = 8.271, c = 7.504 Å, α = 91.45°, β = 104.66°, γ = 101.82°, D_m = 1.30, Z = 1. Cu radiation, R = 0.065 for 949 reflexions.

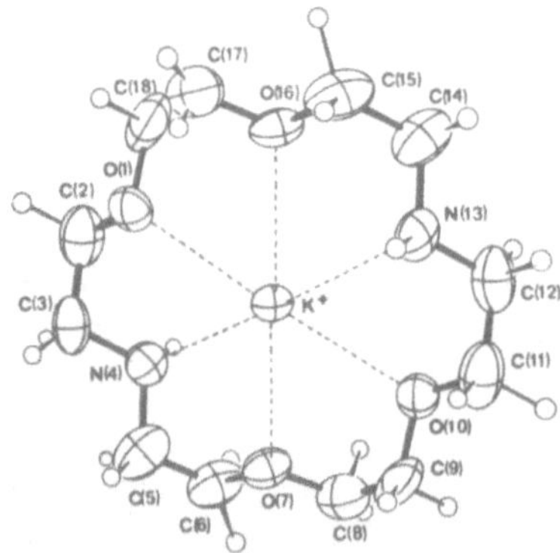

Fig. 1. Conformation of the $[C_{12}H_{26}N_2O_4K]^+$ ion.

The crystal structure contains $[C_{12}H_{26}N_2O_4K]^+$ ions and SCN^- ions centred on different sets of inversion centres; thus the SCN^- ion is disordered. The macrocyclic cation (Fig. 1) has experimental symmetry C_{2h} (2/m). The K^+ ion is at the centre of the macrocycle with K...O(1) = 2.836, K...O(2) = 2.825, and K...N(4) = 2.856(3) Å; there is only a small disturbance of the conformation of the free molecule (1) on complex formation.

1. Preceding report.

4-CHLOROANTHRAQUINONOXADIAZOLE

$C_{14}H_5C\ell N_2O_3$

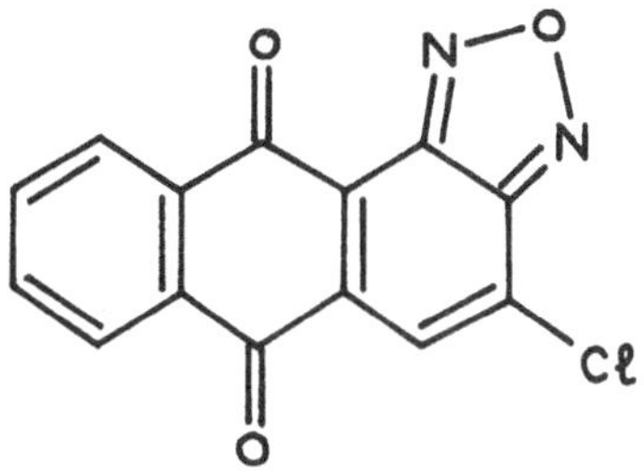

O.A. MIKHNO, Z.I. EŽKOVA, L.A. CHETKINA, S.L. GINZBURG, M.G. NEIGAUZ and L.A. NOVAKOVSKAJA, 1972. Kristallografija 17, 1140-1145 [Soviet Physics - Crystallography, 17, 1009-1012].

Monoclinic, $P2_1/c$, a = 12.10, b = 17.05, c = 5.90 Å, β = 104°20', D_m = 1.58, Z = 4. Cu radiation, R = 0.145 for 1160 reflexions (Weissenberg films).

In this non-planar molecule the oxygen atom remote from the heterocyclic ring is reported to be 0.34 Å from the plane of the central ring. The rings are all turned one relative to the other by about 3°. The angle between the end rings is 6.5°. There appear to be no unusual intermolecular contacts.

ANTHRA[1,2-c][1,2,5]OXADIAZOLE-6,11-DIONE

$C_{14}H_6O_3N_2$

L.A. CHETKINA, S.L. GINZBURG, M.G. NEIGAUZ and G.A. GOL'DER, 1972. Ž. Strukt. Khim., 13, 91-98 [J. Struct. Chem., 13, 79-84].

Monoclinic, $P2_1/a$, a = 20.53, b = 7.19, c = 7.56 Å, β = 104.6°, D_m = 1.56, Z = 4. Cu radiation, R = 0.198 for 1368 reflexions (films, visual intensities).

The addition of the five-membered heterocycle to the anthraquinone ring leads to a change in the bond length and bond angles in the ring nearest to the heterocycle (Fig. 1) and to destruction of the planarity of the molecule. The carbonyl groups project 6° out of the mean plane of the molecule. In the ring adjacent to the heterocycle one of the carbon bonds C(13)-C(14) has a length of 1.32 Å, and the C(13) atom deviates from the plane of the ring by 0.10 Å.

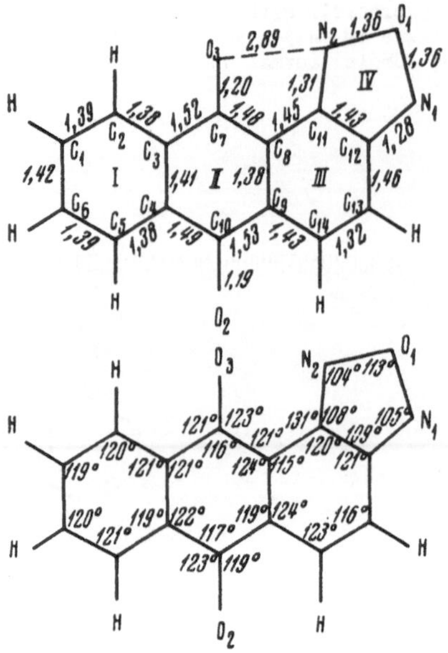

Fig. 1. Bond lengths and angles in the anthraquinoneoxadiazole molecule; σ(bonds) = 0.02 Å, σ(angles) = 1°.

2-PHENYL-7-BROMO-BENZ[d][1,3]OXAZEPINE

$C_{15}H_{10}BrNO$

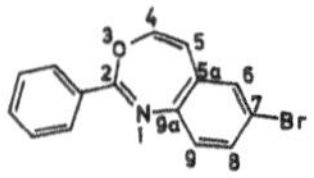

B. JENSEN, 1972. Acta Cryst., B28, 771-774.

Orthorhombic, $P2_12_12_1$, a = 11.09, b = 24.83, c = 4.565 Å, D_m = 1.57, Z = 4. Mo radiation, R = 0.086 for 706 reflexions.

A two-dimensional analysis of this structure has appeared previously (1). The results of this three-dimensional analysis show that the oxazepine ring adopts a boat conformation with N(1), C(2), C(4), and C(5) coplanar. Although standard deviations in bond length are high (0.03 Å) it is clear that the bond length data support the formulation shown above.

1. O. BUCHARDT, B. JENSEN and I.K. LARSEN, 1967. Acta Chem. Scand., 21, 1841.

1-p-NITROPHENYL-3-METHYLPERHYDRO-2,9-PYRIDOXAZINE

$C_{15}H_{20}N_2O_3$

C.S. HUBER, 1972. Acta Cryst., B28, 37-44.

Monoclinic, C2, a = 22.506, b = 8.089, c = 8.443 Å, β = 108.05°, D_m = 1.23, Z = 4. Cu radiation, R = 0.038 for 1654 reflexions.

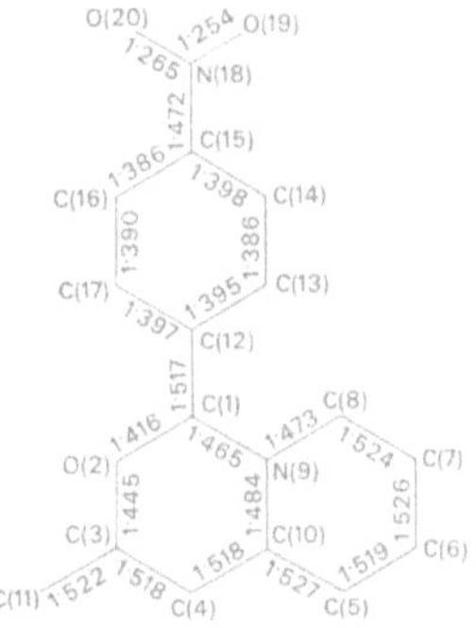

Fig. 1. Bond lengths corrected for libration in $C_{15}H_{20}N_2O_3$.

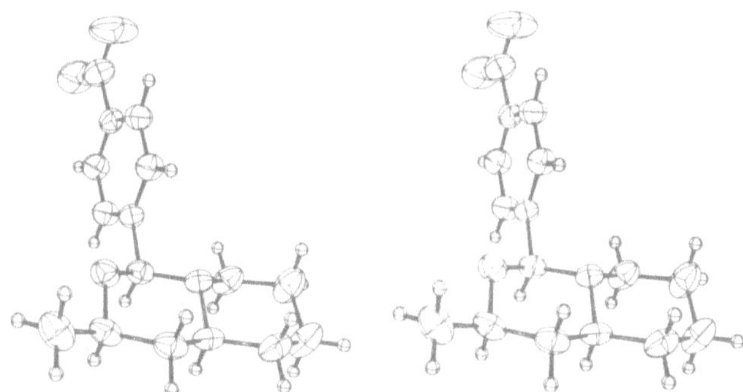

Fig. 2. View of the $C_{15}H_{20}N_2O_3$ molecule showing the absolute configuration.

The trans-fused heterocyclic rings are in flattened chair conformation with the methyl and p-nitrophenyl groups equatorial. In general, the bond lengths (Fig. 1) seem to be normal. The angles within the heterocyclic rings have average value 110.5(2)°. The absolute stereochemistry as determined from anomalous scattering of the oxygen atoms is in Fig. 2. The molecular packing (Fig. 3) shows that the p-nitrophenyl groups are closely interleaved between 2-fold axes.

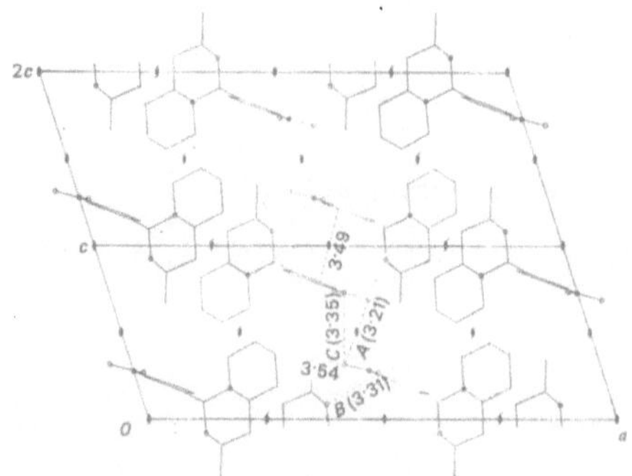

Fig. 3. Projection of the crystal structure of $C_{15}H_{20}N_2O_3$ on (010), showing some of the intermolecular contacts.

10-CHLORO-2,3,5,6,7,11b-HEXAHYDRO-3-HYDROXIMINO-7-METHYL-11b-PHENYL-OXAZOLO[3,2-d]BENZO-1,4-DIAZEPINE

$C_{18}H_{18}C\ell N_3O_2$

R. JAUNIN, W.E. OBERHÄNSLI and J. HELLERBACH, 1972. Helv. Chim. Acta, 55, 2975-2990.

Monoclinic, $P2_1/c$, a = 15.972, b = 17.606, c = 7.102 Å, β = 94.25°, Z = 4. Mo radiation, R = 0.108.

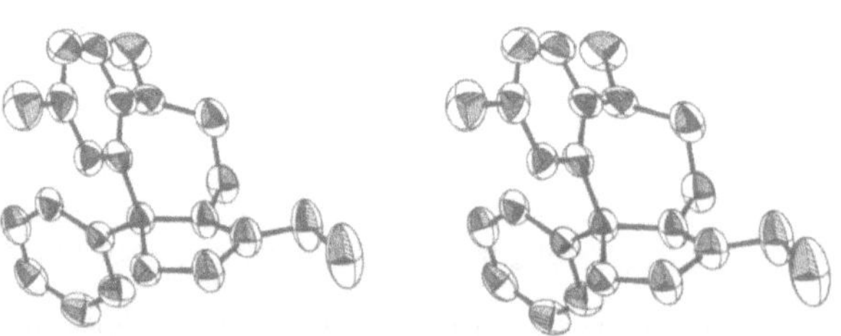

Fig. 1. Diagram of the $C_{18}H_{18}C\ell N_3O_2$ molecule.

This analysis has confirmed the structure proposed on chemical grounds and establishes that the OH group attached to the double-bonded nitrogen has the anti configuration (Fig. 1). Bond lengths and angles are normal. Solvent of crystallisation is trapped in the crystal lattice but the nature of the solvent is not determined.

TRIPHENYL-4,4,6-TRIMETHYL-1-OXA-3-AZACYCLOHEX-2-ENYLCARBENYL-PHOSPHONIUM CHLORIDE MONOHYDRATE

$C_{26}H_{30}ClNO_2P$

L.M. TREFONAS and J.N. BROWN, 1972. J. Heterocyclic Chem., 9, 985-989.

Monoclinic, $P2_1/c$, a = 9.832, b = 15.398, c = 19.762 Å, β = 123.06°, D_m = 1.18, Z = 4. Cu radiation, R = 0.053 for 1633 reflexions.

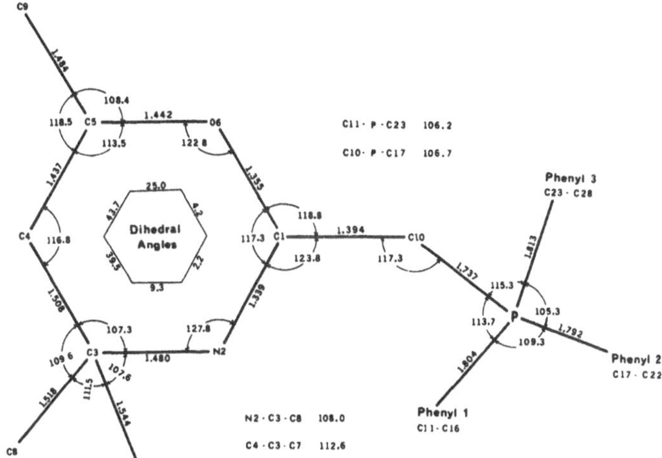

Fig. 1. Details of molecular geometry of the triphenyl-4,4,6-trimethyl-1-oxa-3-azacyclohex-2-enylcarbenyl-phosphonium cation.

The molecular geometry details (Fig. 1) suggest a partial hybridisation of the P atom to accommodate a C(10)-P bond order greater than one. Delocalisation also involves bonds C(1)-N(2) and C(1)-C(10), i.e., the exocyclic double bond shown in the title formula is delocalised through a major part of the oxazinyl system.

2,4,5,7-TETRAPHENYL-6-(4-BROMOPHENYL)-1,3-OXAZEPINE

$C_{35}H_{24}BrNO$

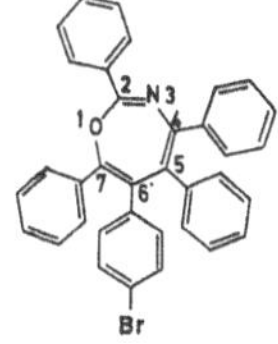

B. JENSEN, 1972. Acta Cryst., B28, 774-781.

Triclinic, $P\bar{1}$, a = 11.403, b = 6.394, c = 19.376 Å, α = 86.52, β = 100.36, γ = 98.97°, D_m = 1.34, Z = 2. Mo radiation, R = 0.068 for 3077 reflexions.

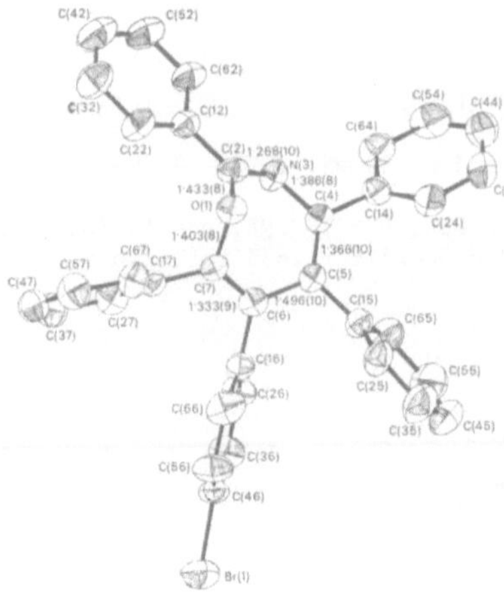

Fig. 1. The 2,4,5,7-tetraphenyl-6-(4-bromophenyl)-1,3-oxazepine molecule, with bond lengths in the oxazepine ring.

The molecule (Fig. 1) has the seven-membered oxazepine ring in a boat conformation. The bond lengths are consistent with the double bond character of C(2)-N(3), C(4)-C(5), and C(6)-C(7). The molecules are held in the crystal lattice by van der Waals forces.

5-AMINO-2-THIOL-1,3,4-THIADIAZOLE

$C_2H_3N_3S_2$

T.C. DOWNIE, W. HARRISON, E.S. RAPER and M.A. HEPWORTH, 1972. Acta Cryst., B28, 1584-1590.

Monoclinic, $P2_1/c$, a = 6.167, b = 10.048, c = 9.007 Å, β = 116.8°, D_m = 1.75, Z = 4. Cu radiation, R = 0.106 for 796 reflexions.

Fig. 1. Tautomeric forms of 5-amino-2-thiol-1,3,4-thiadiazole.

The molecule is essentially planar and exists in the thione form (I) rather than the thiol form (II) (Fig. 1), as evidenced by the bond length data (Fig. 2).

There is a reasonable probability that hydrogen bonds N(2)H...S(2) (3.32 Å) exist and join pairs of molecules across centres of symmetry.

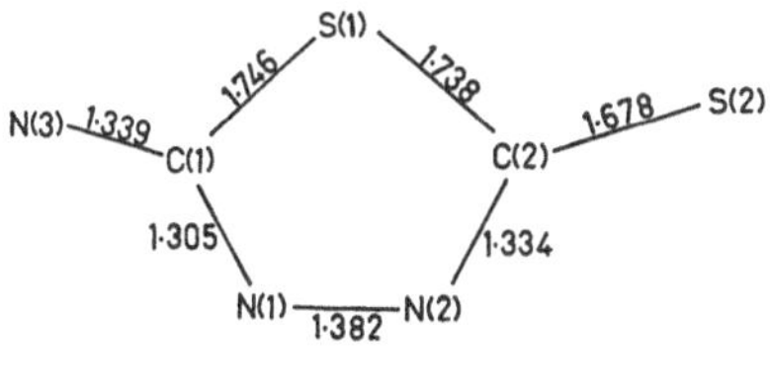

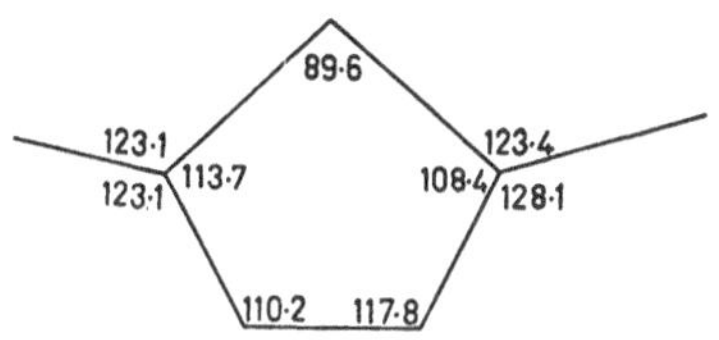

Fig. 2. 5-Amino-2-thiol-1,3,4-thiadiazole: bond lengths (standard deviations are 0.005 - 0.007 for S-C, 0.006 - 0.009 Å for remainder) and bond angles. The intramolecular S...S contact (3.008(3) Å) is very short.

2-IMINO-4-THIAZOLIDINONE

$C_3H_4N_2OS$

V. AMIRTHALINGAM and K.V. MURALIDHARAN, 1972. Acta Cryst., B28, 2421-2423.

Monoclinic, $P2_1/n$, a = 4.09, b = 9.05, c = 13.15 Å, β = 93.0°, D_m = 1.56, Z = 4. Cu radiation, R = 0.115 for 467 reflexions.

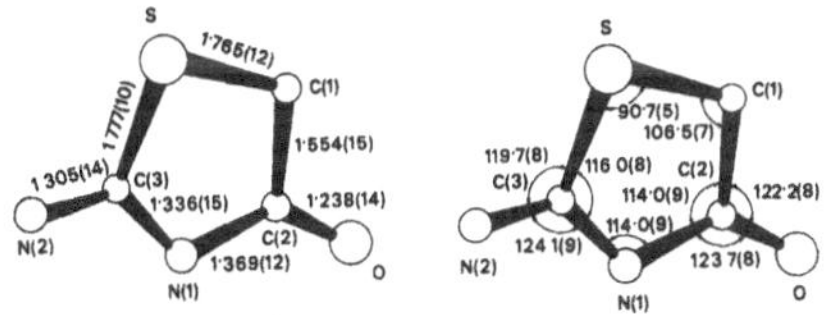

Fig. 1. Bond lengths and angles in 2-imino-4-thiazolidinone.

Bond length and angle data for this planar molecule are in Fig. 1. The exocyclic C(3)-N(2) bond (1.305(14) Å) is suggested to be a double bond, consistent with the molecule being in the 2-imino form. Molecules are hydrogen bonded about centres of inversion, N(1)...N(2) (3.00(1) Å); these dimers are linked into chains by NH...O (2.96(1) Å) hydrogen bonds.

2-AMINO-4-THIAZOLIDINONE-5-ACETIC ACID

$C_5H_6N_2O_3S$

V. AMIRTHALINGAM and K.V. MURALIDHARAN, 1972. Acta Cryst., B28, 2417-2421.

Monoclinic, $P2_1/c$, a = 10.30, b = 7.32, c = 12.64 Å, β = 128°, D_m = 1.50, Z = 4. Cu radiation, R = 0.11 for 957 reflexions.

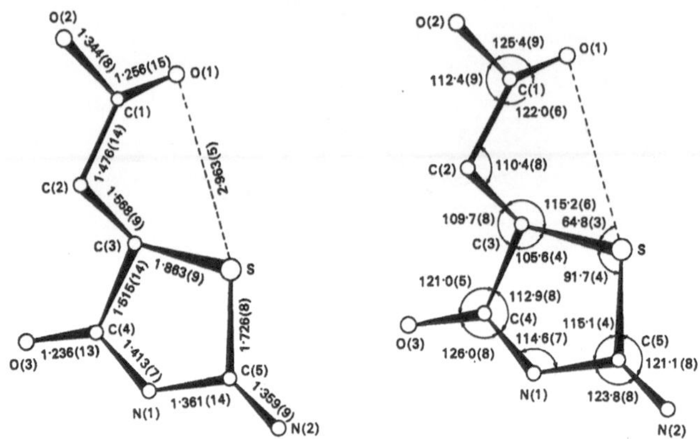

Fig. 1. Bond lengths and angles in 2-amino-4-thiazolidinone-5-acetic acid.

On the basis of the bond length data (Fig. 1) bonds S-C(5), N(1)-C(5), and N(2)-C(5) are suggested to possess partial double bond character. The five-membered ring and exocyclic atoms N(2) and O(3) are coplanar. The intramolecular contact O(1)...S, 2.963(5) Å, is proposed as an intramolecular hydrogen bond, but hydrogen atoms were not located. The molecules are linked in the crystal by OH...N (2.60 Å) and NH...O (2.84, 2.88(1) Å) hydrogen bonds.

3-METHYL-4-OXO-1,3-THIAZINE-2-THIONE

$C_5H_7NOS_2$

V. AMIRTHALINGAM and V.S. JAKKAL, 1972. Acta Cryst., B28, 2612-2614.

Monoclinic, $P2_1/c$, a = 8.02, b = 7.11, c = 15.47 Å, β = 124°, D_m = 1.45, Z = 4. Cu radiation, R = 0.108 for 623 reflexions.

Bond lengths and angles in the molecule are shown in Fig. 1, and indicate some double bond character for the bonds at C(2). The six-membered ring is in boat conformation with C(1) 0.723 Å and N 0.138 Å from the plane of the other four ring atoms. The molecules are held in the crystal by normal van der Waals forces.

Fig. 1. Bond lengths and angles in 3-methyl-4-oxo-1,3-thiazine-2-thione.

CHONDRINE

$C_5H_9NO_3S$

K.J. PALMER, K.S. LEE, R.Y. WONG and J.F. CARSON, 1972. Acta Cryst., B28, 2789-2793.

Orthorhombic, $P2_12_12_1$, a = 9.694, b = 7.640, c = 9.249 Å, D_m = 1.58, Z = 4. Cu radiation, R = 0.034 for 658 reflexions.

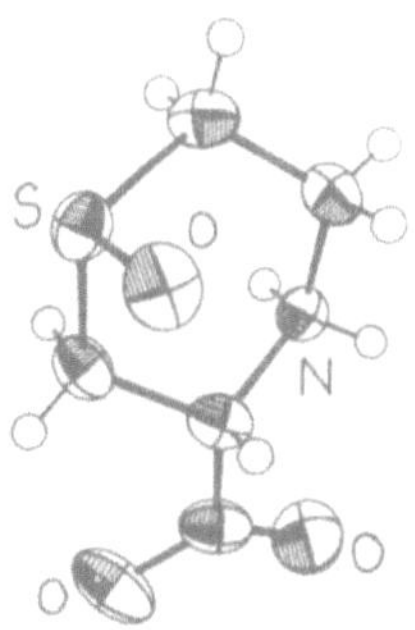

Fig. 1. The chondrine molecular structure.

In this molecule (Fig. 1) the six-membered ring is in a chair conformation with the S-O bond axial and the carboxyl group equatorial. The molecules occur as zwitterions in the crystal and are held together by a network of N-H...O hydrogen bonds, 2.660 and 2.770(5) (Fig. 2). Bond lengths and angles have values close to those found in related molecules (e.g., S-C = 1.795 and 1.806(5), N-C = 1.483(5) Å).

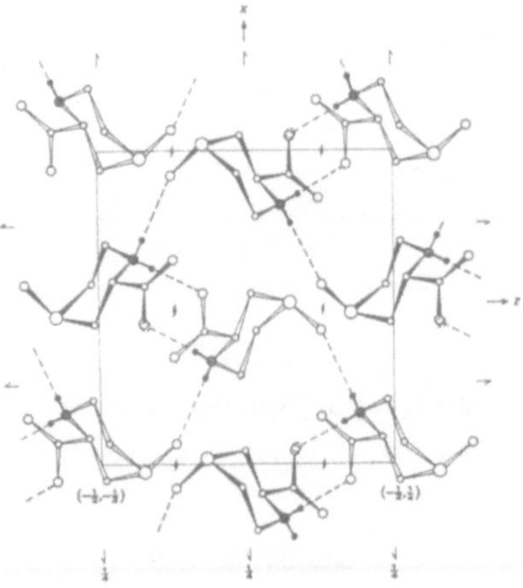

Fig. 2. Projection of the structure of chondrine onto the (101) plane.

2,3,4-TRIMETHYLTHIAZOLIUM BROMIDE

$C_6H_{10}NSBr$

G. PÈPE and M. PIERROT, 1972. Acta Cryst., B28, 2118-2123.

Triclinic, $P\bar{1}$, $a = 8.76$, $b = 7.57$, $c = 6.94$ Å, $\alpha = 106.4°$, $\beta = 77.4°$, $\gamma = 111.8°$, $D_m = 1.70$, $Z = 2$. Mo radiation, $R = 0.087$ for 990 reflexions.

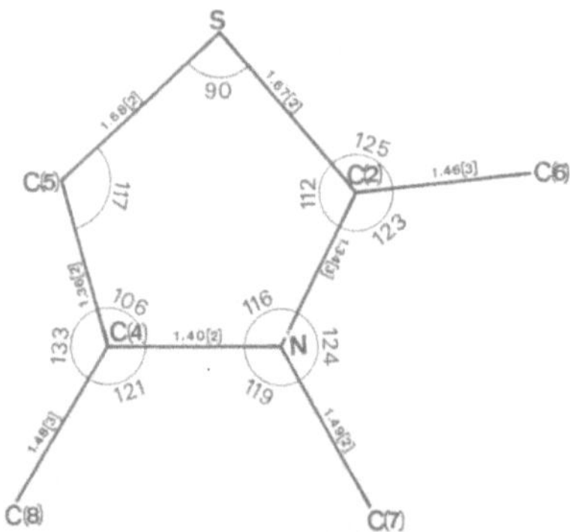

Fig. 1. The geometry of the thiazolium ring of 2,3,4-trimethylthiazolium bromide.

The molecule is disordered in the crystal lattice but this was allowed for and details of the molecular geometry obtained are shown in Fig. 1. There are short intermolecular contacts between bromine, sulphur, and the CH groups, with Br...S = 3.42 Å and Br...CH = 3.18 Å. The corresponding isopropyl derivative is also disordered.

5-CHLORO-2,1-BENZISOTHIAZOLE

C_7H_5ClNS

M. DAVIS, W.A. DENNE and M.F. MACKAY, 1972. J. Chem. Soc., Perkin II, 565-568.

Orthorhombic, $P2_12_12_1$, a = 3.926, b = 9.928, c = 18.204 Å, D_m = 1.586, Z = 4. Cu radiation, R = 0.035 for 475 reflexions.

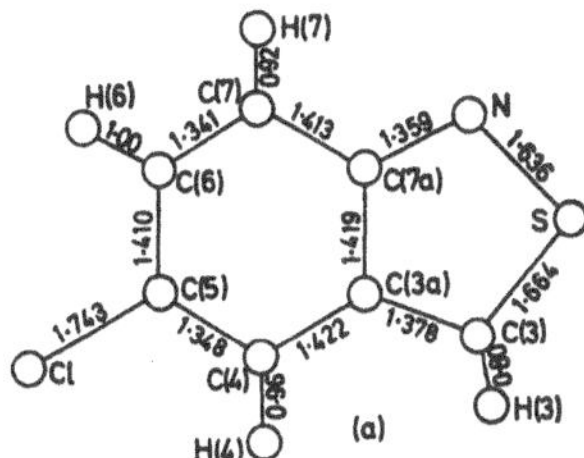

Fig. 1. Bond lengths in 5-chloro-2,1-benzisothiazole; estimated standard deviations for lengths involving Cl and S 0.005 Å, for those involving H 0.04 Å, and for all others 0.007 Å.

The molecule is planar and the bond lengths (Fig. 1), while indicating some small degree of bond fixation in the benzene ring, are interpreted in terms of considerable π-electron delocalisation extending throughout the two fused rings as in naphthalene. In the crystal structure the molecules pack with their molecular planes inclined at about 26° to (100). There are no unusual intermolecular contacts.

HYDROCHLOROTHIAZIDE

$C_7H_8ClN_3O_4S_2$

L. DUPONT and O. DIDEBERG, 1972. Acta Cryst., B28, 2340-2347.

Monoclinic, $P2_1$, a = 7.419, b = 8.521, c = 10.003 Å, β = 111.72°, D_m = 1.68, Z = 2. Mo radiation, R = 0.066 for 956 reflexions.

The bond lengths are in Fig. 1. The short bond N(2)-C(2) is explained in terms of partial donation of the nitrogen lone pair to the benzene ring π-system. The fused ring system is planar with the exception of N(1) which lies 0.69 Å from the ring plane. The molecules are linked in the crystal by NH...O hydrogen bonds (2.88 - 2.94 Å). Other intermolecular distances correspond to van der Waals contacts.

Fig. 1. Bond lengths (σ = 0.010 - 0.019 Å) in hydrochlorothiazide.

PERHYDRO-trans-THIENO[3,4-d]IMIDAZOL-2-ONE SS-DIOXIDE

$C_7H_{12}N_2O_3S$

F. ELLIS, M.B. HURSTHOUSE, S. NEIDLE and P.G. SAMMES, 1972. J. Chem. Soc., Perkin I, 1560-1563.

Monoclinic, C2/c, a = 8.616, b = 15.280, c = 7.409 Å, β = 102.62°, D_m = 1.42, Z = 4. Cu radiation, R = 0.062 for 519 reflexions.

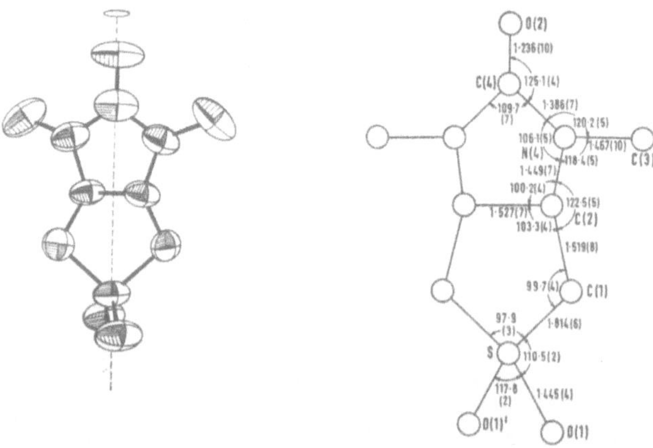

Fig. 1. The (001) projection of the $C_7H_{12}N_2O_3S$ molecule, and bond lengths and angles with standard deviations in parentheses. O(1)' refers to the two-fold axis atom related to O(1). The C(1)-(S)-O(1)' angle is 109·2(3)°.

The twofold molecular symmetry demanded by the space group is indicative of a trans-fused [3.3.0] system (Fig. 1). Both five-membered rings are highly twisted and do not have envelope conformations.

N-CHLOROACETYLCHLOROTHIAZIDE

$C_9H_7Cl_2N_3O_5S_2$

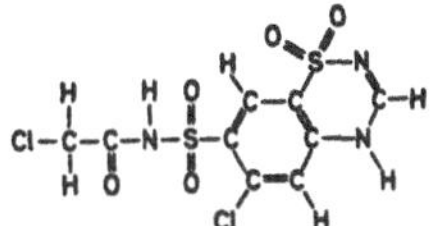

B. MAGNUSSON and O. LINDQVIST, 1972. Acta Chem. Scand., 26, 1411-1422.

Orthorhombic, $P2_12_12_1$, a = 17.577, b = 10.153, c = 7.867 Å, Z = 4. Cu radiation, R = 0.093 for 1123 reflexions.

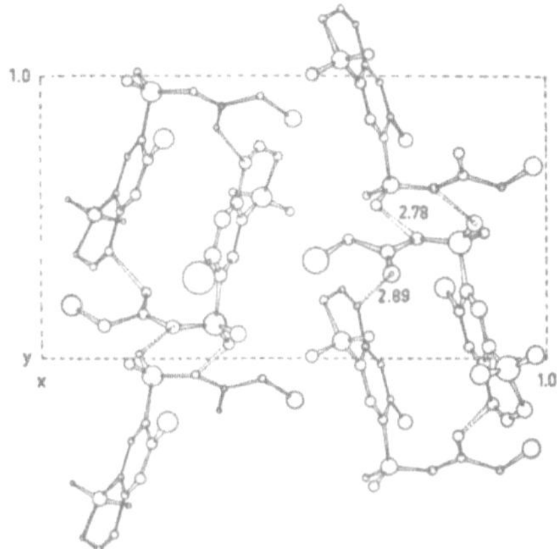

Fig. 1. Molecular packing of N-chloroacetylchlorothiazide, viewed along c. Hydrogen bonds are indicated by broken lines.

The structure of the molecule is as shown above. In the crystal each molecule is connected by hydrogen bonds to four others, to form two dimensional sheets perpendicular to the a-axis (Fig. 1); the sheets are held together by van der Waals forces only. Bond distances in the sulphonamide groups agree with those found previously (1). In the ring system there is a C-N double bond (1.241(16) Å) and in the chloroacetyl group the N-C distance (1.385(16) Å) is longer than in other chloroacetyl amides. The two rings are inclined at an angle of 6°.

1. Structure Reports, 35B, 85.

2-AMINO-5-PHENYL-4-THIAZOLINONE

$C_9H_8N_2OS$

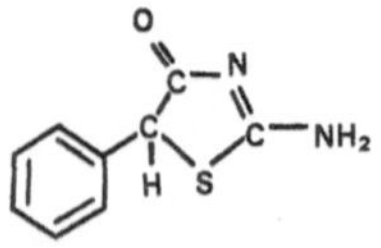

J.-P. MORNON and R. BALLY, 1972. Acta Cryst., B28, 2074-2079.

Monoclinic, $P2_1/c$, a = 12.60, b = 5.665, c = 29.50 Å, β = 121°, Z = 8. Cu radiation, R = 0.065 for 2517 reflexions.

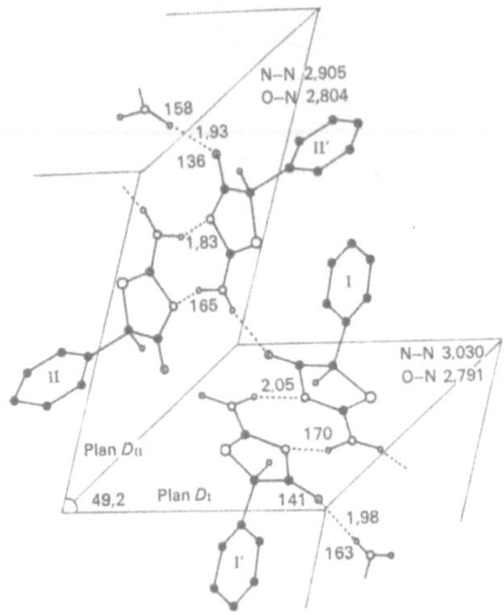

Fig. 1. Hydrogen bonding in 2-amino-5-phenyl-4-thiazolinone.

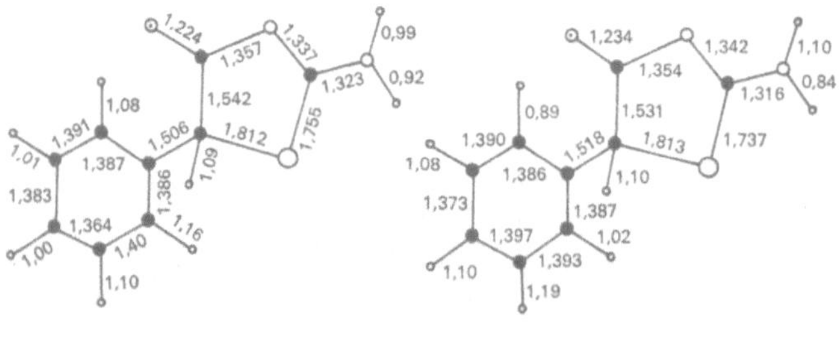

Fig. 2. Bond lengths in the two independent 2-amino-5-phenyl-4-thiazolinone molecules.

This molecule has been described previously in a different crystal modification (1). The results from this form are in accord with those found earlier. The hydrogen atoms were located (Fig. 1) and the molecule exists as the amino rather than the imino tautomer. Two crystallographically independent pairs of molecules form dimers by way of N-H...N hydrogen bonds (2.91 and 3.03 Å) about centres of symmetry. The dimers are then linked to form endless chains by N-H...O bonds (2.79 and 2.80 Å). Bond distances in the two molecules are shown in Fig. 2. The dihedral angles between the heterocyclic and phenyl rings are 83.3 and 87.6°.

1. L.A. PLASTAS and J.M. STEWART, 1969. Chem. Comm., 811; J.-P. MORNON and B. RAVEAU, 1971. Acta Cryst., B27, 95.

SULPHATHIAZOLE

(POLYMORPHS I and III)

$C_9H_9N_3O_2S_2$ $NH_2.C_6H_4.SO_2.N.C_3H_3NS$

G.J. KRUGER and G. GAFNER, 1972. Acta Cryst., B28, 272-283.

Polymorph I
Monoclinic, $P2_1/c$, a = 10.554, b = 13.220, c = 17.050 Å, β = 108.06°, D_m = 1.50, Z = 8. Mo radiation, R = 0.080 for 1843 reflexions.

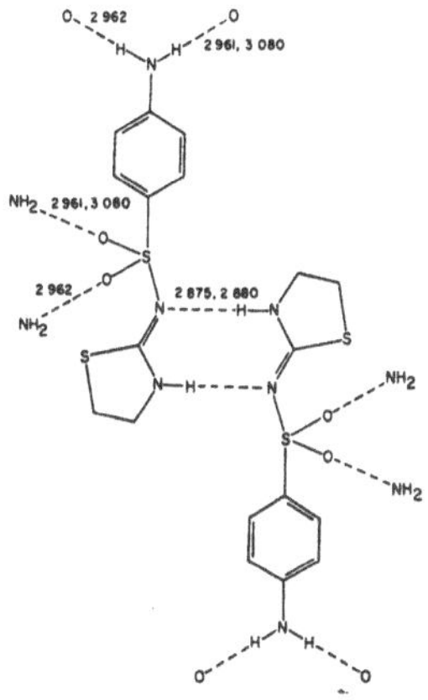

Fig. 1. Hydrogen bonding pattern and hydrogen bond distances in polymorph I of sulphathiazole.

Sulphathiazole is found in three polymorphs. In modification I there are two independent molecules in the asymmetric unit. Centrosymmetric dimers are formed by pairs of molecules and are further hydrogen bonded into chains (Fig. 1). Only symmetry-related molecules are bonded, i.e., no hydrogen bonding exists between the independent molecules. This means that two intermeshed but independent systems of hydrogen bonded molecules exist in the crystal. The systems are the same except that in one set the NH_2-group bonds to one oxygen of a related molecule, while in the other the NH_2-group bonds to two oxygens in related molecules. In both molecules, the five-membered ring is non-planar with the sulphur atom 0.102 and 0.037 Å respectively from the appropriate best-planes through the carbon and nitrogen atoms; these planes make dihedral angles of 101 and 99° with their respective benzene ring planes. Bond lengths are normal

(C=N = 1.306(10), C-S = 1.729(11), N-S = 1.614(8) Å).

Polymorph III
Monoclinic $P2_1/c$, a = 17.570, b = 8.574, c = 15.583 Å, β = 112.93°, D_m = 1.57, Z = 8. Mo radiation, R = 0.063 for 2729 reflexions.

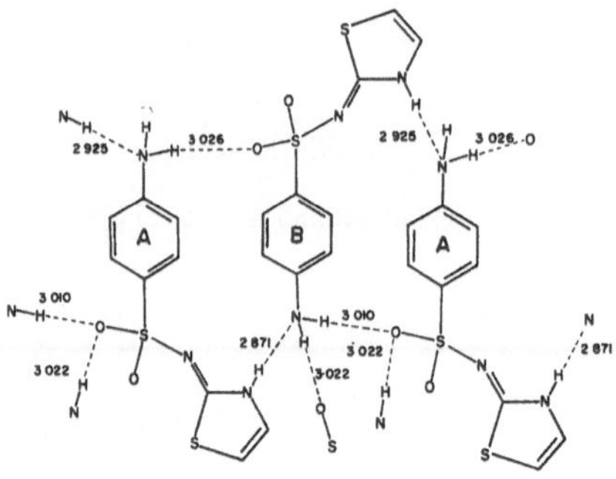

Fig. 2. Hydrogen bond pattern and hydrogen bond distances in polymorph III of sulphathiazole.

Just as in form I there are two independent molecules in the asymmetric unit but the molecular packing is quite different. The hydrogen bonding in form III is illustrated in Fig. 2. The planes of the benzene ring atoms make angles of 92 and 100° with the planes through the nitrogen and carbon atoms of the appropriate five-membered rings. Bonded distances agree with those found in polymorph I (C=N = 1.318(7), C-S = 1.735(6), N-S = 1.606(5) Å).

1,6-DIMETHYL-3,4-TRIMETHYLENE-6a-THIA-AZOPHTHENE

$C_{10}H_{14}N_2S$

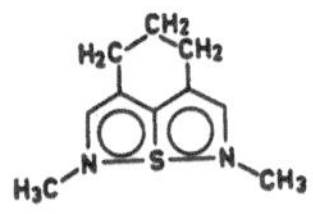

A. HORDVIK and K. JULSHAMN, 1972. Acta Chem. Scand., 26, 343-354.

Orthorhombic, Pbca, a = 16.115, b = 21.676, c = 5.825 Å, D_m = 1.29, Z = 8. Cu radiation, R = 0.073 for 1671 reflexions.

Bond lengths and angles in the molecule are shown in Fig. 1. The 6a-thia-azophthene system is planar within experimental error. Methyl carbons C(7) and C(8) lie -0.069 and -0.077 Å off the plane, while methylene carbons, C(9), C(10), and C(11) lie 0.075, -0.556 and -0.024 Å from the plane. There are no intermolecular distances shorter than appropriate van der Waals distances.

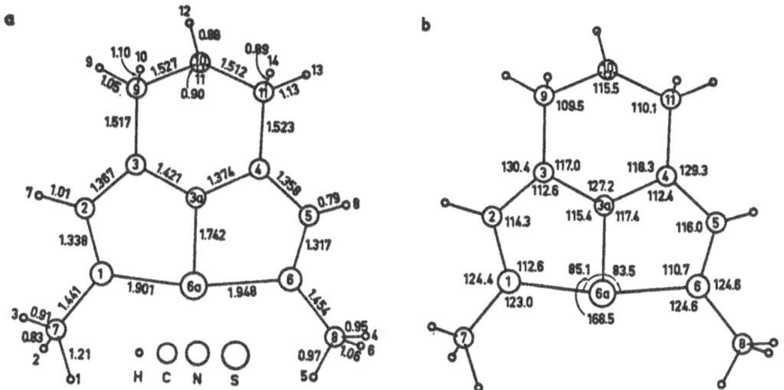

Fig. 1. Bond lengths (σ(N-S) = 0.005, σ(N-C) = 0.008, σ(C-C) = 0.007 - 0.013 Å) and bond angles in the thia-azophthene $C_{10}H_{14}N_2S$.

6-PHENYL-IMIDAZO[2,1-b]THIAZOLE

$C_{11}H_8N_2S$

L. CAVALCA, P. DOMIANO, and A. MUSATTI, 1972. Cryst. Struct. Comm., 1, 345-348.

Orthorhombic, $Pna2_1$, a = 15.934, b = 10.360, c = 5.871 Å, Z = 4. Cu radiation, R = 0.033 for 685 reflexions.

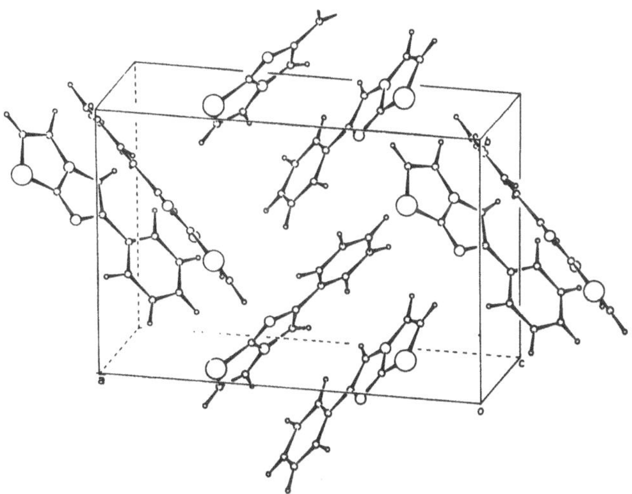

Fig. 1. Clinographic projection of the 6-phenyl-imidazo[2,1-b]thiazole structure.

The molecule is planar as a consequence of conjugation effects. The crystal structure is shown in Fig. 1. Bond lengths and angles are normal. The molecules are held together only by van der Waals forces.

2,3,5,6-TETRAHYDRO-6-PHENYL-IMIDAZO[2,1-b]THIAZOLE

(TETRAMISOLE)

$C_{11}H_{12}N_2S$

A.L. SPEK, 1972. Cryst. Struct. Comm., 1, 309-312.

Monoclinic, $P2_1/n$, a = 11.256, b = 9.772, c = 9.550 Å, β = 105.54°, Z = 4. Mo radiation, R = 0.087 for 1644 reflexions.

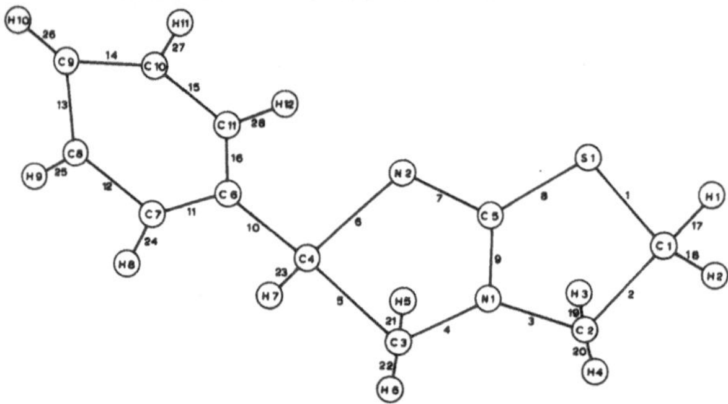

Fig. 1. Tetramisole.

The molecule and numbering scheme are in Fig. 1. The fragment C(1), S(1), C(5), N(1), N(2), C(4) is planar (to within 0.03 Å). The phenyl group and carbon atoms C(2) and C(3) are positioned on the same side of this plane. Bonded distances include: S(1)-C(1) = 1.834(6), S(1)-C(5) = 1.752(4), C(5)-N(2) = 1.267 (6), C(5)-N(1) = 1.381(6) Å.

THIAMINE PYROPHOSPHATE HYDROCHLORIDE

$C_{12}H_{19}C\ell N_4O_7P_2S$

J. PLETCHER and M. SAX, 1972. J. Amer. Chem. Soc., 94, 3998-4005.

Monoclinic, $P2_1/c$, a = 6.815, b = 11.171, c = 24.541 Å, β = 94.00°, D_m = 1.656, Z = 4. Cu radiation, R = 0.037 for 2824 reflexions.

Thiamine pyrophosphate hydrochloride contains a zwitterion with one negative and two positive charges and a chloride ion. One positive charge is associated with the aromatic thiazolium ring while the other results from protonation of

N(11) in the pyrimidine ring. The single negative charge in the zwitterion is produced by ionization of the acidic hydrogen from the inner phosphate group of the pyrophosphate moiety. In the crystal the net positive charge on the zwitterion is balanced by a chloride ion. One significant feature of the molecular packing is the short contact distance between the chloride ion and the C(2) of the thiazolium ring (C-H...Cℓ = 3.302 Å). The normal absence of hydrogen bonding to the bridging oxygen atoms is also a feature of this structure. Details of interatomic distances and angles within a molecule are given in Fig. 1.

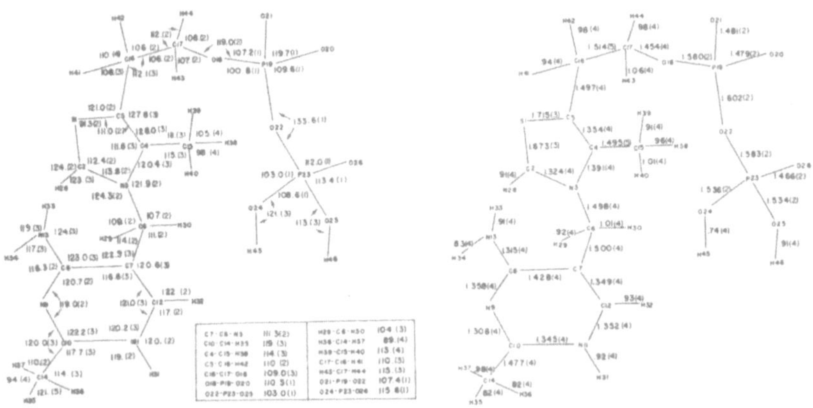

C7-C6-H3	111.3(2)	H29-C6-H30	104(3)
C10-C14-H35	119(3)	H36-C14-H37	89(4)
C4-C15-H38	114(3)	H39-C15-H40	113(4)
C5-C16-H42	110(2)	C17-C16-H41	110(3)
C16-C17-O18	109.0(3)	H43-C17-H44	115(3)
O18-P19-O20	110.5(1)	O21-P19-O22	107.4(1)
O22-P23-O25	103.0(1)	O24-P23-O26	115.6(1)

Fig. 1. Bond distances and angles in thiamine pyrophosphate hydrochloride.

DIMETHYL 1,3-DIMETHYL-8-THIA-2-AZABICYCLO[3,2,1]OCT-3-ENE-4,7-DICARBOXYLATE

$C_{12}H_{17}NO_4S$

J.L. FLIPPEN, 1972. Acta Cryst., B28, 1519-1522.

Monoclinic, $P2_1/c$, a = 11.076, b = 8.341, c = 14.799 Å, β = 103.07°, D_m not given, Z = 4. Cu radiation, R = 0.053 for 2148 reflexions.

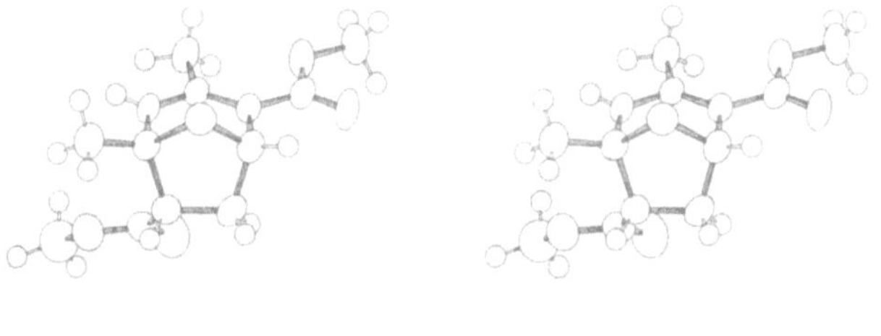

Fig. 1. Drawing of the $C_{12}H_{17}NO_4S$ molecule, including refined positions of the hydrogen atoms.

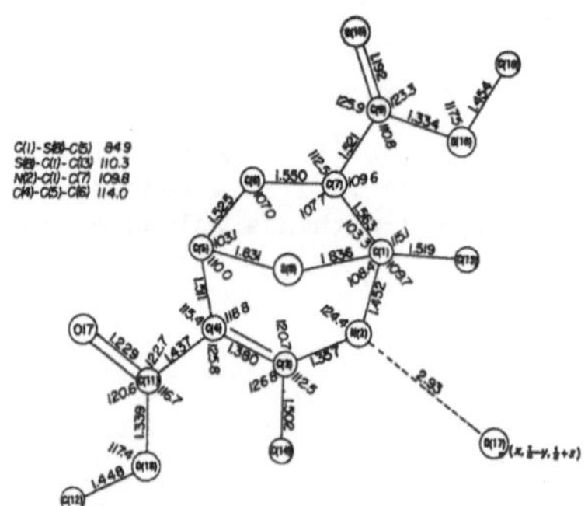

Fig. 2. Bond distances (σ = 0.002 - 0.004 Å) and angles (σ = 0.1 - 0.2°) in the $C_{12}H_{17}NO_4S$ molecule.

The molecular configuration is shown in Fig. 1, the ester group at C(7) being trans to the sulphur bridge as predicted by chemical studies. Bond lengths and angles are illustrated in Fig. 2. The O(17)-C(11)-C(4)-C(3)-N(2) grouping is somewhat conjugated as evidenced by the bond length data. The segments C(5)-C(6)-C(7)-C(1) and C(1)-N(2)-C(3)-C(4)-C(5) are individually planar; the dihedral angle between them is 66.5°. There is one intermolecular N-H...O hydrogen bond which links the molecules into infinite chains along <u>c</u>.

RACEMIC 7-CHLORO-1,3,5-TRIMETHYL-5H-PYRIMIDO[5,4-b]-[1,4]BENZOTHIAZINE-2,4(1H,3H)-DIONE

$C_{13}H_{12}ClN_3O_2S$

J.P. SCHAEFER and L.L. REED, 1972. J. Amer. Chem. Soc., <u>94</u>, 908-913.

Monoclinic, $P2_1/n$, a = 12.514, b = 8.029, c = 13.730 Å, β = 99.61°, D_m = 1.51, Z = 4. Cu radiation, R = 0.059 for 1699 reflexions.

In this stable sulphonium ylide the sulphur atom is in a pyramidal configuration and from the molecular geometry details it is estimated that the canonical form shown above contributes 85% to the overall bonding. The shortest intermolecular contact (3.15 Å) occurs between one sulphur methyl group and the negatively charged oxygen of a neighbouring molecule.

1,1'-DIETHYL-2,2'-THIAZOLINECARBOCYANINE IODIDE

$C_{13}H_{21}IN_2S_2$

T.E. BOROWIAK, N.G. BOKIJ, and Ju.T. STRUCHKOV, 1972. Ž. Strukt. Khim.,

13, 480-485 [J. Struct. Chem., 13, 447-451].

Monoclinic, $P2_1/a$, a = 12.40, b = 16.72, c = 7.99 Å, β = 107.6°, D_m = 1.61, Z = 4. Cu radiation, R = 0.089 for ∿650 reflexions (films, visual intensities).

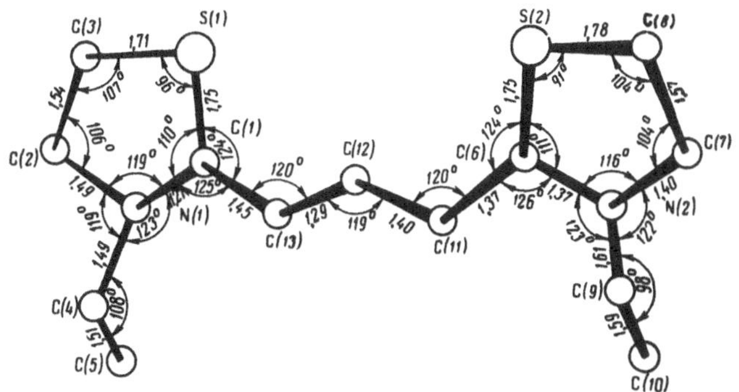

Fig. 1. Bond lengths and angles in the $C_{13}H_{21}N_2S_2^+$ ion; σ(bond lengths) = 0.02 - 0.05 Å.

The carbocyanine cation is non-planar, the dihedral angle between the mean planes of the five-membered rings being 13°. The thiazoline rings are also non-planar. One ring has an envelope conformation (C(1), N(1), C(2), C(3), S(1)) while the other has a half-chair conformation. Bond distances and angles are shown in Fig. 1.

cis-2,3-(3'-CYCLOHEXANON-1',2'-YLENE)-5-METHYL-8-ETHOXYDIHYDROTHIAZOLO[3,2-a]PYRIDINIUM BROMIDE

$C_{14}H_{18}BrNO_2S$

P.GROTH, 1972. Acta Chem. Scand., 26, 3131-3140.

Triclinic, $P\bar{1}$, a = 6.895, b = 10.196, c = 11.259 Å, α = 103.14, β = 92.50, γ = 107.52°, D_m = 1.55, Z = 2. Mo radiation, R = 0.045 for 1785 reflexions.

The molecular structure and dimensions are shown in Fig. 1. The five-membered ring has the envelope conformation with one of the atoms, C(9), 0.60 Å out of the plane through the four other atoms. The cyclohexanone ring has the chair form. There are no unusual intermolecular contacts.

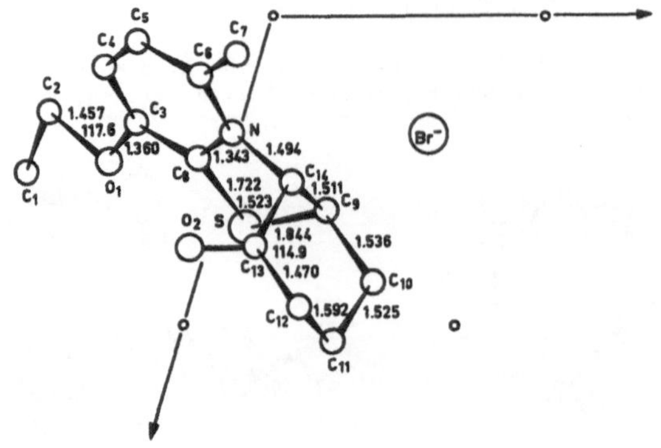

Fig. 1. Schematical drawing of the $C_{14}H_{18}NO_2S^+$ ion viewed along [100]. For the bond lengths σ values are: C-S = 0.005, C-O, C-N = 0.006, C-C = 0.007 - 0.008 Å.

PHENOTHIAZINE-10-PROPIONIC ACID

$C_{15}H_{13}NO_2S$

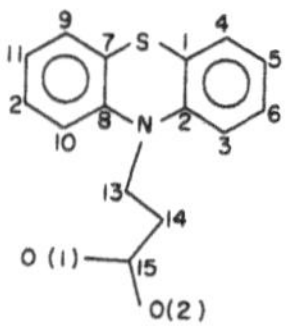

M.C. MALMSTROM and A.W. CORDES, 1972. J. Heterocyclic Chem., 9, 325-330.

Monoclinic, $P2_1/c$, a = 7.888, b = 8.703, c = 19.700 Å, β = 101.42°, D_m = 1.35, Z = 4. Mo radiation, R = 0.072 for 500 reflexions.

The molecule is folded along the S-N axis and the dihedral angle is 136.5°. The C-S-C angle is 98.5(7)° and the average C-S bond is 1.77(2) Å. The shortening of the C-S bond, the small value of the C-S-C angle, and the folding of the molecule are typical of the phenothiazine class of compounds and are assumed to be due to sulphur d orbital participation in ring bonding. In the crystal carboxyl groups are linked via O-H...O hydrogen bonds (2.65 Å) about centres of symmetry. All other intermolecular distances correspond to or are greater than van der Waals distances.

exo-7,11-METHANO-5,6a,7,7a,10,11,11a-OCTAHYDROBENZ-[c]INDENO[5,6-e]THIAZINE 6-OXIDE

$C_{16}H_{17}NOS$

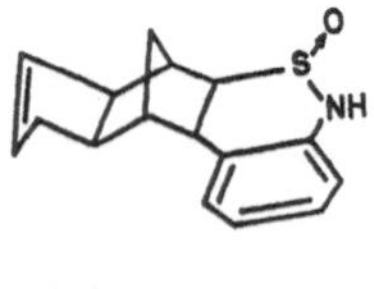

D.J. POINTER, J.B. WILFORD and K.M. CHUI, 1972. J. Chem. Soc., Perkin II, 1134-1136.

Monoclinic, $P2_1/c$, a = 7.018, b = 12.86, c = 14.35 Å, β = 92.04°, D_m = 1.39, Z = 4. Cu radiation, R = 0.103 for 2201 reflexions.

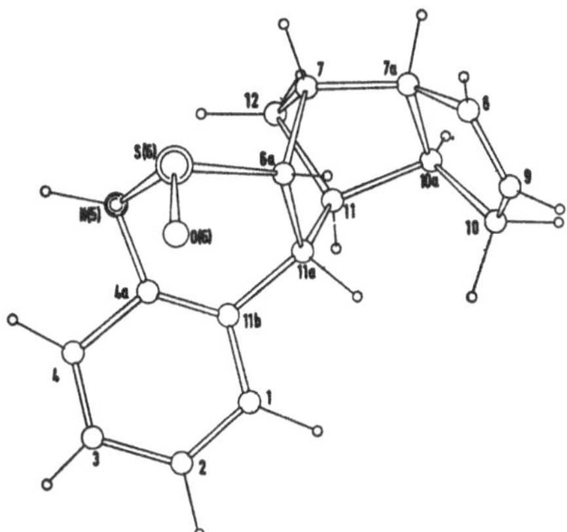

Fig. 1. The structure of the octahydrobenz[c]indeno[5,6-e]thiazine derivative, $C_{16}H_{17}NOS$.

The molecule (Fig. 1) has exo-stereochemistry. Bond lengths are for the most part normal, except for the S-O bond length (1.501(4) Å) which is longer by 0.05 Å than expected. The C(7)-C(12)-C(11) bond angle (94.5(5)°) is as expected much less than tetrahedral. In the crystal structure two enantiomeric molecules form a centrosymmetric dimer by N-H...O (2.975(6) Å) hydrogen bonds.

PHENOXYMETHYLANHYDROPENICILLIN ETHANOL SOLVATE

$C_{18}H_{22}N_2O_5S$ $C_{16}H_{16}N_2O_4S.C_2H_5OH$

G.L. SIMON, R.B. MORIN and L.F. DAHL, 1972. J. Amer. Chem. Soc., 94, 8557-8563.

Orthorhombic, $P2_12_12_1$, a = 8.4565, b = 6.9243, c = 32.5349 Å, D_m = 1.33, Z = 4. Cu radiation, R = 0.054 for 1532 reflexions.

The structure is analogous to that of the active penicillins with respect to the resulting stereochemistry caused by the fusion of the four-membered β-lactam ring to the five-membered ring; namely, a thiazolidinone ring for the anhydropenicillins versus a thiazolidine ring for the penicillins. As in the penicillins, the large pyramidal character of the nitrogen atom in the β-lactam ring of phenoxymethylanhydropenicillin (the nitrogen atom is 0.41 Å out of the plane of its three bonded constituents) is likewise assumed to inhibit considerably any delocalization of its lone pair of electrons. Hence, the chemical stability of the anhydropenicillins to hydrolysis of the β-lactam ring (in contrast to that of the penicillins) cannot apparently be simply ascribed to extensive delocalization in the ground electronic state of the electron pair on the nitrogen atom into the adjacent α,β-unsaturated system. This conclusion is in accord with the observed N-C and C-O bond lengths of 1.415(7) and 1.183(7) Å, respectively, in the β-lactam ring and the N-C bond length of 1.431(7) Å in the thiazolidinone ring. The ethanol molecule of crystallization is involved in hydrogen bonding with the carbonyl oxygen atom of the peptide linkage in the side chain.

5-BENZOYLIMINO-2-(4-BROMOPHENYL)-3-PHENYL-2,5-DIHYDRO-1,2,4-THIAZOLE

$C_{21}H_{14}BrOSN_3$

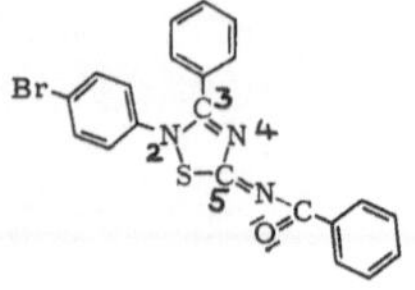

A. KUTOGLU and H. JEPSEN, 1972. Chem. Ber., 105, 125-131.

Triclinic, P$\bar{1}$, a = 10.065, b = 9.858, c = 10.115 Å, α = 104.82, β = 77.15, γ = 90.91°, Z = 2. Cu radiation, R = 0.077 for 2157 reflexions.

The bromophenyl group is bonded to N(2) adjacent to the sulphur atom. The observed bond lengths are not inconsistent with a delocalised system rather than that of the formal double and single bonds shown above, e.g., S-N(2) = 1.759(15), N(2)-C(3) = 1.31(2), C(3)-N(4) = 1.35(2), N(4)-C(5) = 1.33(2), C(5)-S = 1.712(17), C(5)-N(benzoylimino) = 1.35(2) Å.

2,4-BIS-(1,2-BENZISOTHIAZOL-3-YL)3-AMINOCROTONONITRILE DIMETHYLFORMAMIDE SOLVATE

$C_{21}H_{19}N_5OS_2$ $C_{18}H_{12}N_4S_2.HCON(CH_3)_2$

E. GAETANI, T. VITALI, A.MANGIA, M.NARDELLI and G. PELIZZI, 1972. J. Chem. Soc., Perkin II, 2125-2129.

Monoclinic, $P2_1/c$, a = 8.85, b = 12.08, c = 19.52 Å, β = 95.9°, Z = 4. Cu radiation, R = 0.082 for 2622 reflexions.

The material obtained on hydrolysis of ethyl 2-(1,2-benzisothiazolin-3-ylidene)-cyanoacetate was established as (I) by this crystal structure analysis.

I

The two benzthiazole systems are individually planar and form a dihedral angle of 86°. The bond lengths and angles are consistent with formulation (I) but there may also be a small contribution from another tautomeric form in which the NH_2-C=C-C=N- moiety is replaced by NH=C-C=C-NH-. The packing of the molecule is determined by normal van der Waals forces. The dimethylformamide molecule is hydrogen bonded to the NH_2 group (N-H...O 2.88 Å).

3,5-EPIDITHIO-2,5-DIPHENYL-2,4-PENTADIENYLIDENEANILINE

$C_{23}H_{17}NS_2$

F. LEUNG and S.C. NYBURG, 1972. Canad. J. Chem., 50, 324-332.

Monoclinic, $P2_1/c$, a = 7.350, b = 12.309, c = 23.268 Å, β = 118.35°, D_m = 1.32, Z = 4. Cu radiation, R = 0.07 for 2377 reflexions.

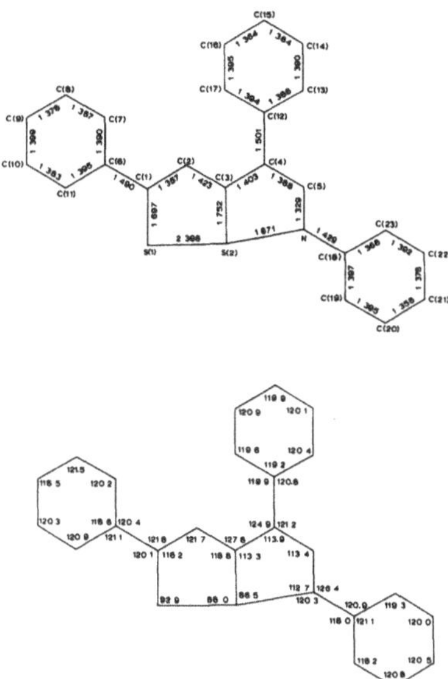

Fig. 1. Bond lengths and angles in $C_{23}H_{17}NS_2$ (σ(S-S) 0.012, σ(S-C) and σ(N-C) 0.007, σ(S-N) 0.010, σ(C-C) 0.008 - 0.010 Å)

The bond lengths in this molecule (Fig. 1) are in good agreement with those obtained for the related thiathiophthen system in 3,5-epidithio-2,5-diphenyl-2,4-pentadienylidene-3-aminoquinolone (1). Of the three substituent phenyl rings, that at C(4) is most twisted out of planarity with the central ring (46.3°). The structure is compared with that of the related thiathiophthen nitrogen isostere 2,4-diphenyl-5-(3'-phenyl-2'-thiabutene-4'-yl)-iso-thiazolium iodide (2).

1. F. LEUNG and S.C. NYBURG, 1971. Canad. J. Chem., 49, 167; Structure Reports, 37B, 191.
2. Following report.

2,4-DIPHENYL-5-(3'-PHENYL-2'-THIABUTENE-4'-YL)-ISOTHIAZOLIUM IODIDE 6

$C_{24}H_{20}INS_2$

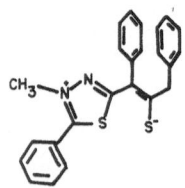

F. LEUNG and S.C. NYBURG, 1972. Canad. J. Chem., 50, 324-332.

Monoclinic, $P2_1/c$, a = 9.34, b = 9.58, c = 25.65 Å, β = 108.08°, D_m = 1.57, Z = 4. Mo radiation, R = 0.071 for 3073 reflexions.

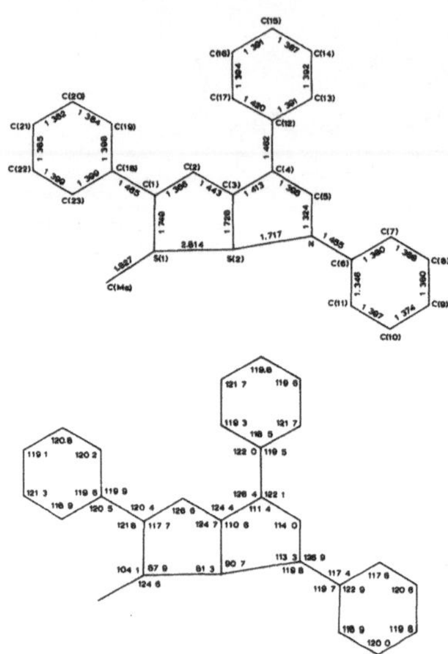

Fig. 1. Bond lengths and angles in $C_{24}H_{20}INS_2$.

The bond lengths and angles for this molecule are in Fig. 1, and indicate that the resonance form shown above predominates. The phenyl ring attached to nitrogen is virtually coplanar (angle 4.2°) with the central molecular plane, unlike the situation in the preceding report where the corresponding angle is 32°.

MESIONIC THIADIAZOLE DERIVATIVE

$C_{24}H_{20}N_2S_2$

J.L. FLIPPEN, 1972. Acta Cryst., B28, 2749-2754.

Monoclinic, $P2_1/n$, a = 14.673, b = 11.037, c = 12.995 Å, β = 102.1°, Z = 4. R = 0.061 for 3325 reflexions.

Fig. 1. Diagram of the thiadiazole derivative $C_{24}H_{20}N_2S_2$.

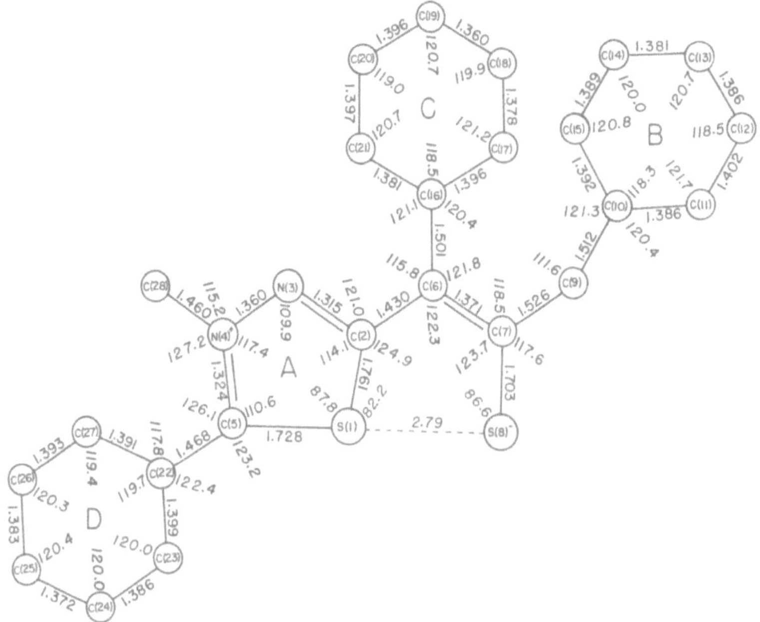

Fig. 2. Bond lengths (σ = 0.004 Å for S or N bonds, 0.008 Å for C-C) and angles in the thiadiazole derivative $C_{24}H_{20}N_2S_2$. The C(16)-C(10) intramolecular distance is only 3.23 Å.

The conformation of this new mesionic thiadiazole derivative is shown in Fig. 1 and bond lengths and angles are in Fig. 2. The two sulphur atoms, which are in a syn conformation with respect to each other are only 2.79 Å apart; this distance is too long to be considered analogous to the S-S bond lengths (2.23 - 2.57 Å) in thiathiophthene systems. The entire central portion of the molecule (atoms 1 - 8 and C(28)) is essentially planar (±0.2 Å). Intermolecular contacts are close to van der Waals distances.

3-(2-DIETHYLAMMONIUMETHOXY)-1,2-BENZISOTHIAZOLE TETRACHLOROCOBALTATE

$C_{26}H_{38}C\ell_4CoN_4O_2S_2$ $(C_{13}H_{19}N_2OS^+)_2[CoC\ell_4]^{2-}$

A.C. BONAMARTINI, M. NARDELLI and C. PALMIERI, 1972. Acta Cryst., B28, 1207-1210.

Triclinic, $P\bar{1}$, a = 8.92, b = 13.57, c = 14.47 Å, α = 88.1, β = 84.6, γ = 69.2°, D_m = 1.44, Z = 2. Co radiation, R = 0.087 for three-dimensional Weissenberg data.

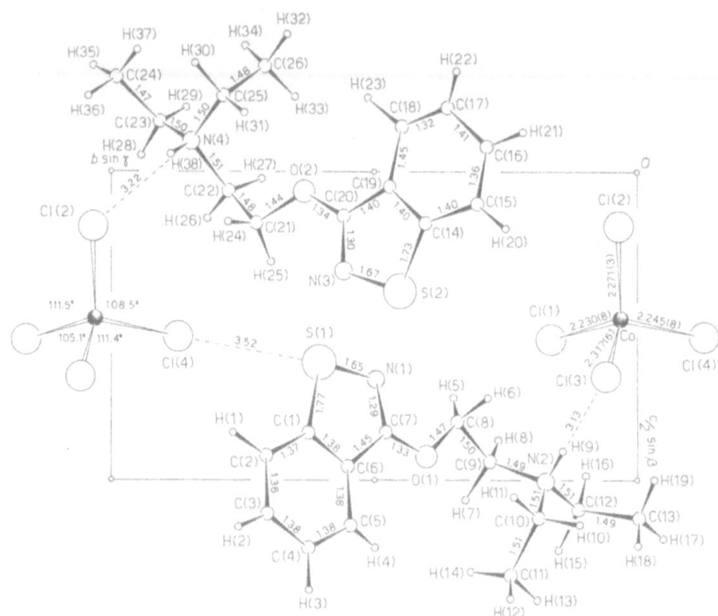

Fig. 1. Projection of the 3-(2-diethylammoniumethoxy)-1,2-benzisothiazole tetrachlorocobaltate structure along [100].

The anion has a slightly distorted tetrahedral structure (Co-Cℓ = 2.230, 2.271, 2.317, 2.245 Å; Cℓ - Co-Cℓ = 105.1 - 111.7°, mean 109.5°). Two Co-Cℓ distances are significantly longer than the other two, and correspond to the Cℓ(2) and Cℓ(3) atoms respectively which are involved in hydrogen bonds with the ammonium nitrogens N(4) and N(3) of two independent organic cations. The two independent 3-(2-diethylammoniumethoxy)-1,2-benzisothiazole systems (Fig. 1). are practically identical and bond distances and angles in them are normal. Packing is mainly determined by the hydrogen bonding: N(2)H...Cℓ(3) = 3.13 Å and N(4)H...Cℓ(2) = 3.22 Å.

1,5-DIHYDROXY-9-OXA-3,7-DITHIABICYCLO[3.3.1]NONANE-3-OXIDE

$C_6H_{10}O_4S_2$

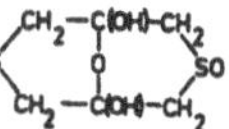

S. ABRAHAMSSON and G. REHNBERG, 1972. Acta Chem. Scand., 26, 3309-3318.

Orthorhombic, $Pna2_1$, a = 9.756, b = 11.996, c = 7.212 Å, Z = 4. Cu radiation, R = 0.038 for 749 reflexions.

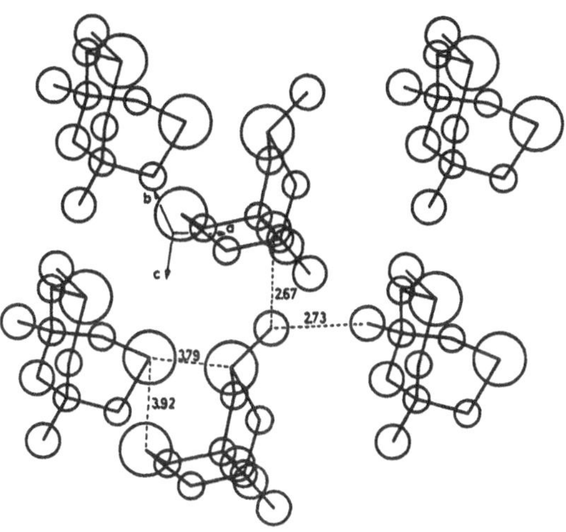

Fig. 1. The packing of $C_6H_{10}O_4S_2$. The O-H...O distances are 2.67 and 2.73 Å; the other distances shown are S...S intermolecular contacts.

The molecule has a twin-chair conformation. Hydrogen bonds between hydroxyl groups of two different molecules and the sulphoxide oxygen link the molecules together (Fig. 1). Principal bond lengths are C-S(sulphoxide) = 1.817(5), C-S = 1.807(5), S-O = 1.508(4), C-O(hydroxyl) = 1.378(6), C-O = 1.416(6), and C-C = 1.528(7) Å. The intramolecular S...S contact is short (3.17 Å) and leads to deformation of valence angles at two carbon atoms (by +5°).

triptych-BOROXAZOLIDINE

$C_6H_{12}BNO_3$

R. MATTES, D. FENSKE and K.-F. TEBBE, 1972. Chem. Ber., 105, 2089-2094.

Orthorhombic, $Pca2_1$, a = 11.386, b = 6.586, c = 9.649 Å, D_m = 1.423, Z = 4. Mo radiation, R = 0.049 for 777 reflexions.

The molecule (Fig. 1) possesses approximate (non-crystallographic) threefold symmetry along the intramolecular B-N bond. Principal bond lengths are B-N = 1.677(6), B-O = 1.428(6) - 1.448(6), mean C-C = 1.519(6), mean C-O = 1.423(5), mean C-N = 1.487(5) Å; for the angles, mean O-B-O = 115.5(4), mean N-B-O = 103.0(4)°.

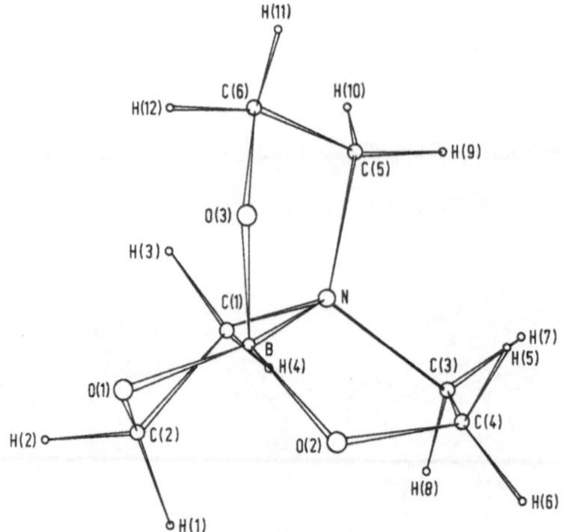

Fig. 1. Molecular structure of triptych-boroxazolidine.

4-PHENYL-1,3,2-OXATHIAZOL-5-ONE

$C_8H_5NO_2S$

G.D. ANDREETTI, G. BOCELLI, L. CAVALCA and P. SGARABOTTO, 1972. Gazz. Chim. Ital., 102, 23-35.

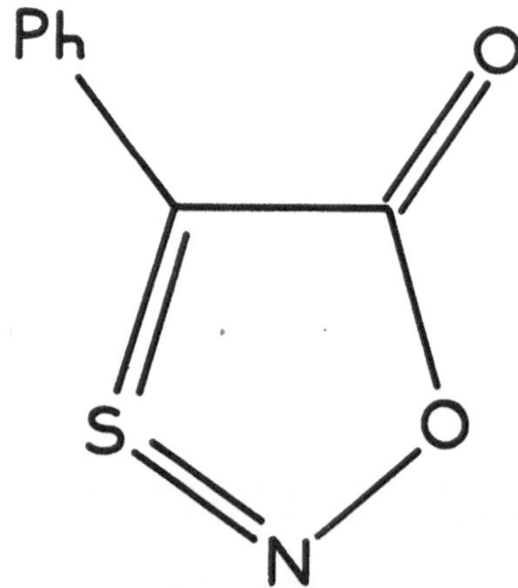

Fig. 1. 4-Phenyl-1,3,2-oxathiazol-5-one.

Monoclinic, $P2_1/c$, a = 9.55, b = 7.07, c = 12.50 Å, β = 114.2°, D_m = 1.55, Z = 4. Mo radiation, R = 0.041 for 1325 reflexions.

Bond lengths in the sydnone ring are S-N = 1.604, N-O = 1.375, O-C = 1.405, C-C = 1.443, C-S = 1.654, exocyclic C=O = 1.195 Å. These distances favour the structure in Fig. 1. The sydnone ring is slightly non-planar, with the carbonyl carbon atom 0.026 Å out of the plane of the other four atoms; the angle between the sydnone and phenyl planes is 17°.

[1,2,5]OXASELENAZOLO[2,3-b][1,2,5]OXASELENAZOLE-7-Se^{IV}

$C_8H_{10}N_2O_2Se$

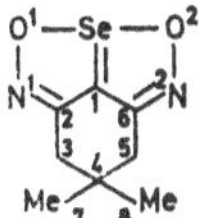

E.C. LLAGUNO and I.C. PAUL, 1972. J. Chem. Soc., Perkin II, 2001-2006.

Monoclinic, $P2_1/n$, a = 5.837, b = 20.337, c = 7.971 Å, β = 97°55', D_m = 1.76, Z = 4. Cu radiation, R = 0.056 for 1154 reflexions.

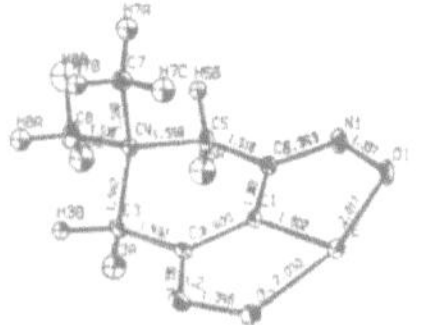

Fig. 1. Drawing of the selenium derivative $C_8H_{10}N_2O_2Se$ showing bond lengths (Å). The mean C-Se-O angle is 80.4(4)°.

The non-hydrogen atoms in the molecule are almost coplanar except for the atoms of the $C(CH_3)_2$ group in the six-membered ring (Fig. 1). The selenium atom interacts with the oxygen atoms of the two nitroso-groups in the molecule to give Se-O 2.017(9) and 2.030(9) Å; O-Se-O is 199.2(4)°. The lengths of the Se-O, C-Se, C-C, C-N, and N-O bonds (Fig. 1) all support a bicyclic structure for the O-Se-O-N-C-C-C-N moeity involving three-coordinate selenium. In the crystal, the molecules are arranged as centrosymmetric dimers with two Se...O(2) intermolecular distances of 3.20 Å.

1-ACETOXY-1,2-BENZIODOXOLIN-3-ONE

$C_9H_7IO_4$

J.Z. GOUGOUTAS and J.C. CLARDY, 1972. J. Solid State Chem., 4, 226-229.

Monoclinic, $P2_1/a$, a = 16.191, b = 7.762, c = 7.705 Å, β = 82.0°, Z = 4. Cu radiation, R = 0.09 for 777 reflexions.

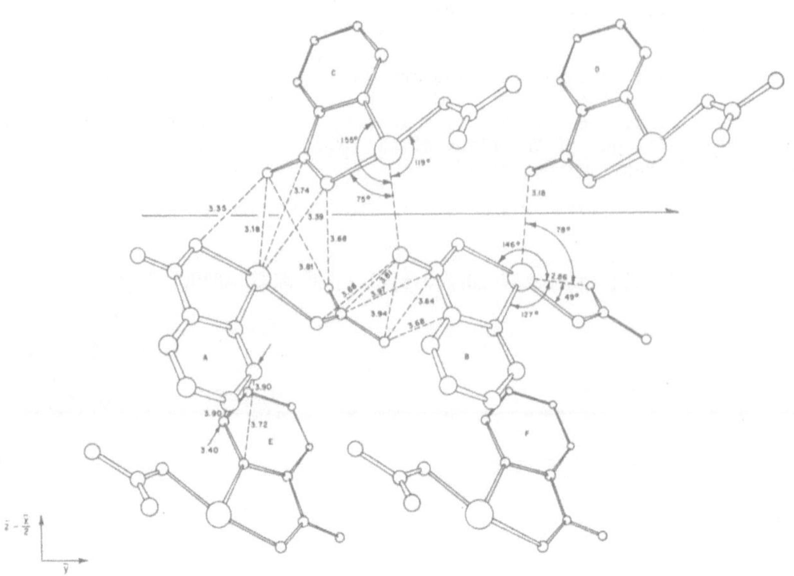

Fig. 1. The crystal structure of 1-acetoxy-1,2-benziodoxolin-3-one in its (201) projection, showing intermolecular contacts less than 4 Å.

The molecular conformation and packing are shown in Fig. 1. The packing is dominated by an intermolecular I...O interaction 3.18 Å which is appreciably less than the sum of van der Waals radii (~3.46 Å). The intramolecular distances involving iodine are I-O(ring) = 2.13(2), I-C = 2.00(3), I-O(acetate) = 2.11(2) Å. The C-I-O(ring) and C-I-O(acetate) angles are 81(1) and 83(1)° respectively. The benziodoxolinone moiety is planar. The planar acetoxy group is rotated about the exocyclic I-O bond through 24°, resulting in an intramolecular I...O (carbonyl) distance of 2.86(2) Å.

1-(2'-IODOBENZOYLOXY)-1,2-BENZIODOXOLIN-3-ONE

(α and β forms)

$C_{14}H_8I_2O_4$

J.Z. GOUGOUTAS and J.C. CLARDY, 1972. J. Solid State Chem., 4, 230-242.

α Form
Monoclinic, Cc, a = 4.21, b = 30.86, c = 22.52 Å, β = 93.3°, Z = 8. Cu radiation, R = 0.13 for 1547 reflexions.

This α crystal structure is formed directly in the three-dimensionally ordered topotactic transformation of the isomeric bis-(iodobenzoyl) peroxide crystal structure, and further is hydrolytically transformed to the single crystal phase of 2-iodobenzoic acid. The asymmetric unit (Fig. 1) consists of two independent molecules which are mutually associated through two strong I-O intermolecular coordination (2.9 Å) bonds across a pseudo inversion centre. In both molecules, the monovalent 2'-iodo substituent and the carbonyl oxygen atom of the adjacent carboxyl group are oriented in an unusual transoid relationship as a consequence of the mutual coordination bonds.

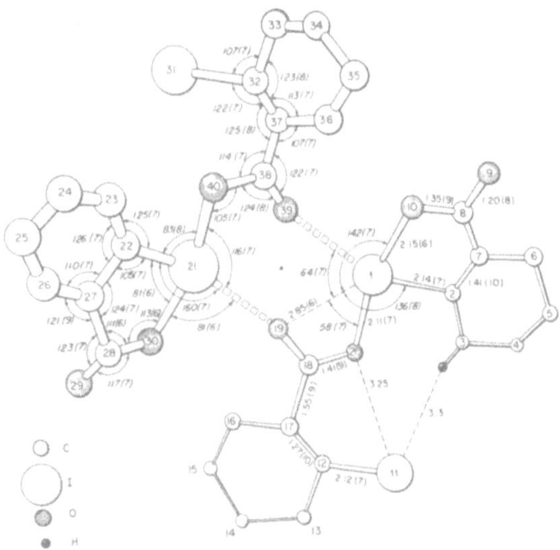

Fig. 1. α-Form of $C_{14}H_8I_2O_4$: the two independent molecules in the asymmetric unit.

β-Form
Monoclinic, $P2_1/c$, a = 8.03, b = 12.58, c = 13.74 Å, β = 91.6°, Z = 4. Cu radiation, R = 0.09 for 1053 reflexions.

The molecular conformation is shown in Fig. 2 along with details of molecular geometry. The monovalent 2'-iodo substituent and the carbonyl oxygen of the adjacent carbonyl group are oriented in a cisoid manner. The crystal structure does not contain the common I-O coordination bonds. However, closely analogous coordination bonds are indicated between the monovalent and trivalent iodine atoms in the β form. A sluggish polymorphic transformation (single crystal→ polycrystal) of the α structure to the β structure occurs at ∿110°C.

A schematic representation of a bond delocalisation process which could eventually result in the transformation of the α- to the β-form is shown in Fig. 3.

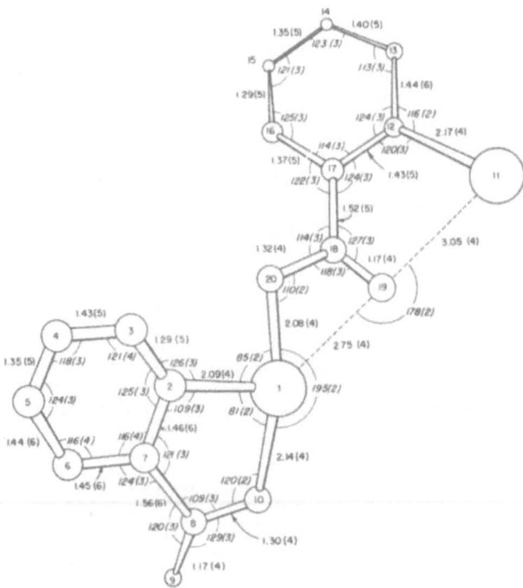

Fig. 2. β-Form of $C_{14}H_8I_2O_4$.

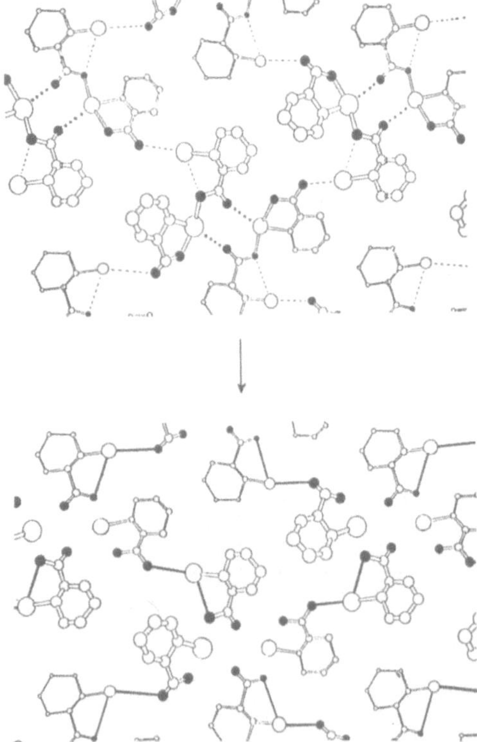

Fig. 3. A possible bond delocalization process in the α- to β-$C_{14}H_8I_2O_4$ transformation. New covalent I-O bonds are shown by heavy lines in the lower half of the diagram.

2-p-NITROPHENYL-1,3-OXATHIANE

$C_{10}H_{11}NO_3S$

N. deWOLF, G.C. VERSCHOOR and C. ROMERS, 1972. Acta Cryst., B28, 2424-2429.

Monoclinic, $P2_1/n$, a = 17.248, b = 11.563, c = 10.782 Å, β = 100.24°, D_m = 1.38, Z = 8. Cu radiation, R = 0.069 for 1774 reflexions.

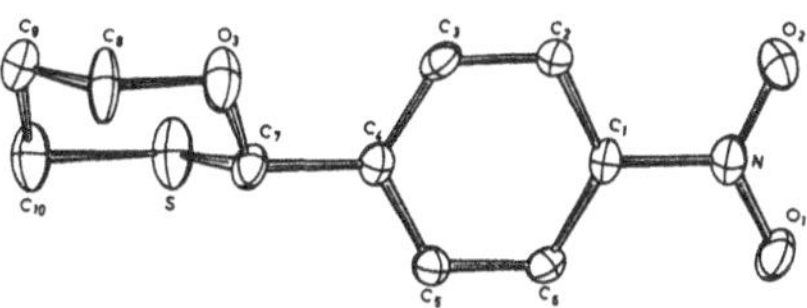

Fig. 1. Atomic numbering and conformation of 2-p-nitrophenyl-1,3-oxathiane.

The oxathiane ring and nitro group in the molecule (Fig. 1) are apparently disordered and the observed structure is an average of at least two conformations. The chair form oxathiane rings have mean endocyclic torsion angles of 61°, greater than the value 55.9° found for cyclohexane. The two independent molecules, apart from the orientation of their nitro groups, are approximate mirror images. One nitro group is twisted by +7°, and the other by -8°, out of the appropriate phenyl ring plane. Mean values of bond lengths include: S-C = 1.80(2), C-O = 1.40(3) Å, the low accuracy arising from the disorder. Nearly all the intermolecular contacts between 3.2 and 3.5 Å involve the nitro group and atoms C(8), C(9), and O(3) of the oxathiane ring.

2H-THIOPYRAN p-BROMOBENZYLESTER

$C_{20}H_{22}BrNO_4S_2$

A.E. SMITH, R. KALISH and E.J. SMUTNY, 1972. Acta Cryst., B28, 3494-3502.

Monoclinic, $P2_1$, a = 11.36, b = 8.052, c = 11.77 Å, β = 99.86°, D_m = 1.52, Z = 2. Mo radiation, R = 0.057 for 1429 reflexions.

The molecular structure and bond distances are shown in Fig. 1. The bond distances are normal excepting the C(1)-C(6) bond (1.33(2) Å) the shortness of which is suggested to be possibly "related to the proximity of the Br atom...". The molecules are separated by van der Waals distances. The thiopyran ring has a flattened boat conformation, whereas the morpholinium ring has a chair form.

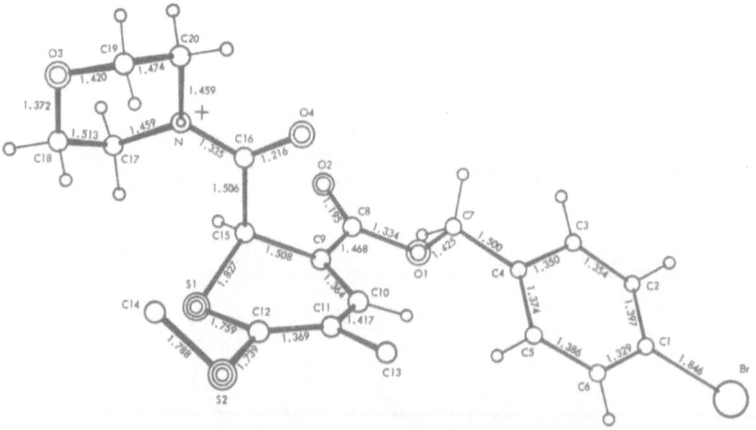

Fig. 1. The structure of 2H-thiopyran p-bromobenzylester.

3,3,5,6-TETRAPHENYL-2,3-DIHYDRO-1,4-OXATHIIN-2-ONE-4,4-DIOXIDE

$C_{28}H_{20}O_4S$

N. YASUOKA, N. KASAI, M. TANAKA, T. NAGAI and N. TOKURA, 1972. Acta Cryst., B28, 3393-3399.

Monoclinic, $P2_1/n$, a = 18.692, b = 9.840, c = 12.962 Å, β = 104.44°, D_m = 1.29, Z = 4. Cu radiation, R = 0.057 for 2045 reflexions.

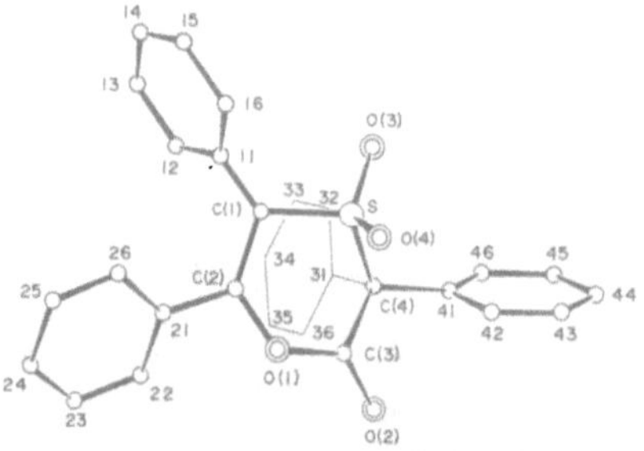

Fig. 1. The molecular structure of 3,3,5,6-tetraphenyl-2,3-dihydro-1,4-oxathiin-2-one-4,4-dioxide projected onto the best plane through S, C(1), C(2), and O(1).

The molecular structure (Fig. 1) has a non-planar six-membered ring containing the sulphur atom; the four atoms S,C(1), C(2), and O(1) almost lie on a plane with C(3) and C(4) located 0.46 and 1.19 Å above the plane, respectively. The C(sp^2)-S and C(sp^3)-S bonds have distances of 1.761(6) and 1.838(6) Å respectively and there is no evidence to indicate any conjugation between the sulphone group and the adjacent unsaturated bond. The C(1)-C(2) double bond has a value of 1.341(8) Å while the two S-O distances of 1.426(5) and 1.436(5) Å are similar values to those found in other sulphones. The bond lengths and angles in the phenyl groups are quite normal (C-C = 1.361 - 1.395(12) Å). There are two probable intermolecular C-H...O hydrogen bonds, C(15)-H(15)...O(2) = 3.339 and C(22)-H(22)...O(2) = 3.252 Å.

LITHIUM PURPURATE DIHYDRATE

$C_8H_8LiN_5O_8$

Li^+. $.2H_2O$

H.B. BÜRGI, S. DJURIĆ, M. DOBLER and J.D. DUNITZ, 1972. Helv. Chim. Acta, 55, 1771-1782.

Triclinic, $P\bar{1}$, a = 6.715, b = 6.822, c = 14.038 Å, α = 96.85, β = 100.94, γ = 67.41°, D_m = 1.75, Z = 2. Mo radiation, R = 0.056 for 3175 reflexions.

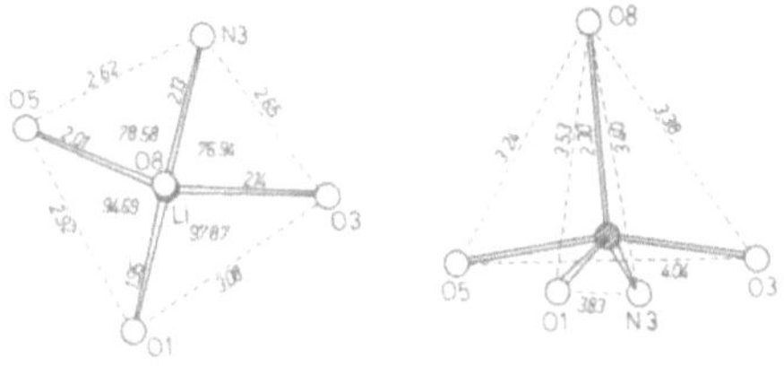

Fig. 1. Coordination of lithium ion in lithium purpurate dihydrate.

Fig. 2. View of crystal structure of lithium purpurate dihydrate.

In this crystal structure the lithium ion is bonded to two carbonyl oxygen atoms and to the central nitrogen atom of one purpurate anion (Li...O, 2.01, 2.14 Å; Li...N, 2.13 Å). It is further bonded to a carbonyl oxygen atom of a second purpurate anion (Li...O, 1.95 Å) and to a water molecule (Li...O, 2.30 Å) to complete a roughly square-pyramidal 5-coordination with the water molecule as apex (Fig. 1). The purpurate anion (Fig. 2) deviates markedly from planarity. It has virtual (non-crystallographic) C_2 symmetry, with the two six-membered rings twisted by 22° (in the same sense) out of the plane of the central three atoms. Some of the bond angles show large deviations from the standard value of 120° at trigonal atoms; in particular, the central C-N-C angle is increased to 135° and its two adjacent N-C-C angles to 128°.

SODIUM 1-METHYL-5,5-DIETHYLBARBITURATE

(SODIUM METHARBITAL)

$C_9H_{13}NaN_2O_3$

B. BERKING, 1972. Acta Cryst., B28, 1539-1545.

Orthorhombic, $Pna2_1$, a = 12.670, b = 10.796, c = 7.802 Å, D_m = 1.369, Z = 4. Cu radiation, R = 0.023 for 889 reflexions.

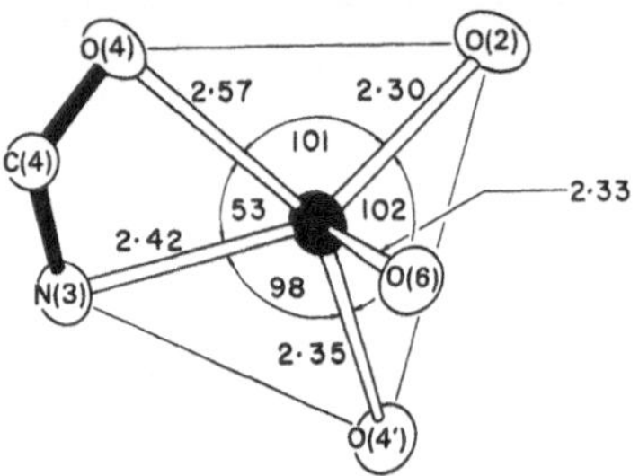

Fig. 1. Coordination at the Na^+ ion in sodium metharbital. Atoms N(3), C(4) and O(4) belong to the same metharbital ion.

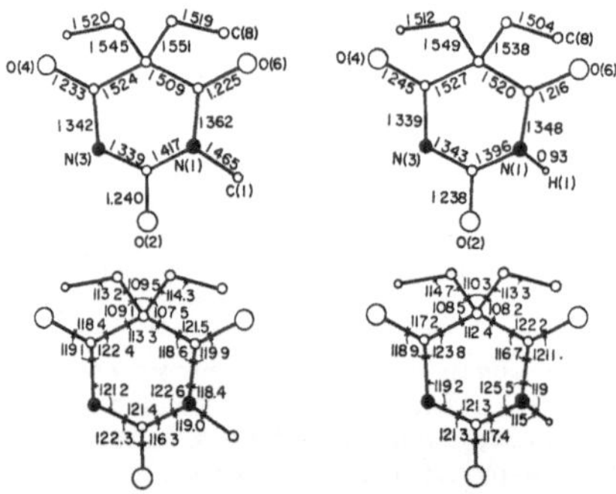

Fig. 2. Bond lengths (σ= 0.002 - 0.004 Å) and angles in the metharbital ion.

In the absence of hydrogen bonding the dominant role in the crystal packing is played by ionic coordination interactions. The sodium ion is fivefold coordinated (Fig. 1). The extremely short Na-N(3) distance (2.42 Å) suggests that N(3) carries a great part of the formal negative charge of the metharbital ion. Bond lengths and angles in the anion are shown in Fig. 2, and compare where relevant with those of the barbital ion in sodium barbital (1). The pyrimidine ring is slightly puckered with maximum deviations from planarity occurring at

C(6) (-0.040 Å) and N(1) (+0.044 Å).

1. B. BERKING and B.M. CRAVEN, 1971. Acta Cryst., B27, 1107; Structure Reports, 37B, 217.

CALCIUM 5,5-DIETHYLBARBITURATE TRIHYDRATE

(CALCIUM BARBITAL TRIHYDRATE)

$C_{16}H_{28}CaN_4O_9$ $Ca(C_8H_{11}N_2O_3)_2.3H_2O$

B. BERKING, 1972. Acta Cryst., B28, 98-113.

Triclinic, $P\bar{1}$, a = 16.999, b = 9.122, c = 16.780 Å, α = 77.29, β = 89.36, γ = 74.86°, Z = 4. Cu radiation. R = 0.046 for 8237 reflexions.

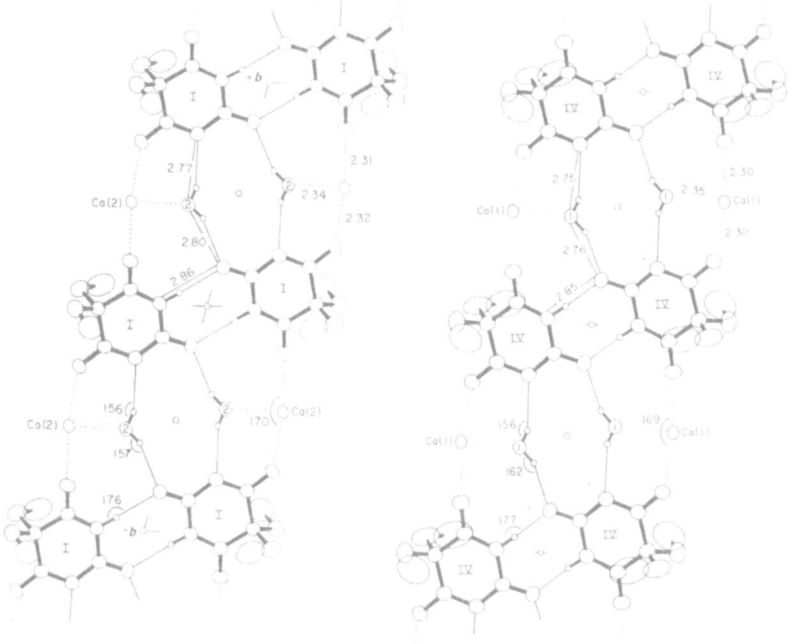

Fig. 1. Projection of part of the calcium barbital structure along c. Only two of the dimer pairs are shown.

Deprotonation of the four crystallographically independent pyrimidine rings causes changes in bond lengths and internal bond angles from those found in barbital (1). The greater part of the formal negative charge seems to be equally distributed between the two oxygen atoms adjacent to the deprotonated nitrogen atom. There are conformational differences in the barbiturate rings: one is flat and three are puckered with the quaternary carbon, C(5), out of plane.

Deprotonated nitrogen atoms are involved either in hydrogen bonds or in coordination with the calcium ions. Centrosymmetrically related barbital ions form doubly hydrogen-bonded dimers (NH...O 2.78 - 2.86 Å) which are isolated with regard to having no hydrogen bonds with other barbital ions (Fig. 1). Dimers are linked by hydrogen bonds via water molecules and calcium ions to form ribbons along b and sheets perpendicular to b*. The two calcium ions (Fig. 2) are 7-fold coordinated by six oxygen and one nitrogen atom at the corners of irregular pentagonal bipyramids with Ca-O in the range 2.30 - 2.59, Ca-N 2.52 - 2.55 Å; the polyhedra are not interconnected, but they interact with the anions in the ribbons and sheets establishing a three-dimensional packing.

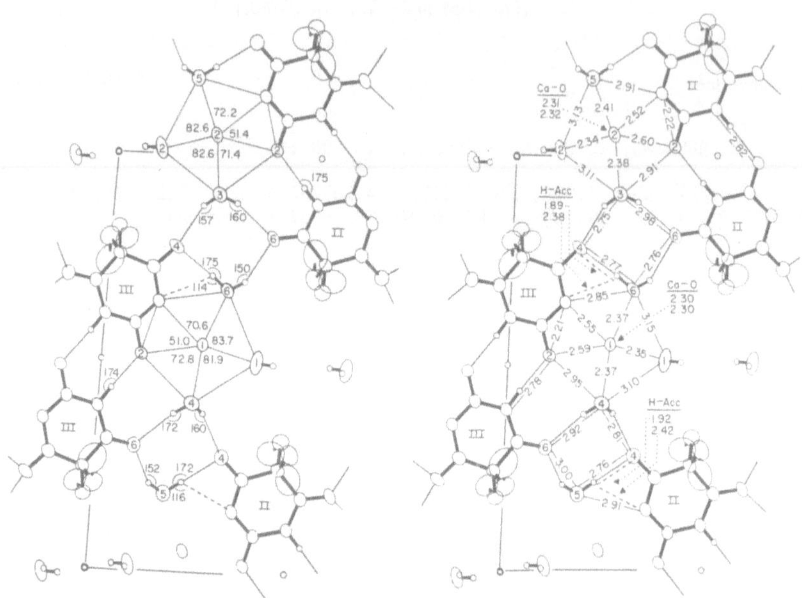

Fig. 2. The Ca coordination and hydrogen bonding of barbital ions (II) and (III).

1. B.M. CRAVEN, E.A. VIZZINI and M.M. RODRIGUES, 1969. Acta Cryst., B25, 1978; Structure Reports, 34B.

5-FLUOROPYRIMIDINE-2-ONE

$C_4H_3FN_2O$

S. FURBERG and S. PETERSEN, 1972. Acta Chem. Scand., 26, 760-768.

Monoclinic, $P2_1/c$, a = 11.118, b = 6.012, c = 7.147 Å, β = 104.31°, D_m = 1.64, Z = 4. Mo radiation, R = 0.069 for 527 reflexions.

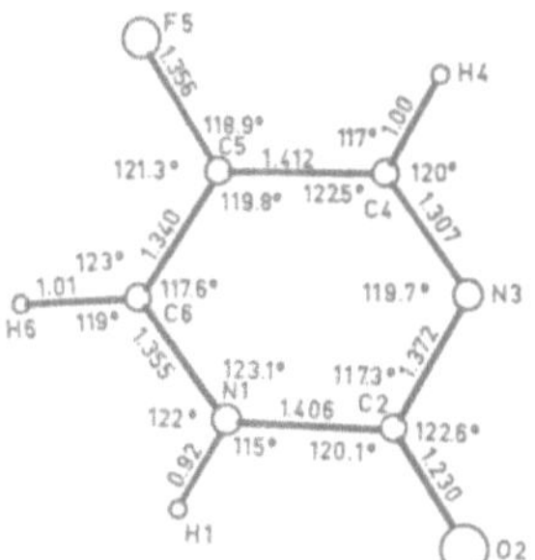

Fig. 1. Bond lengths (corrected for libration, σ = 0.004 - 0.005 Å) and angles in 5-fluoropyrimidine-2-one.

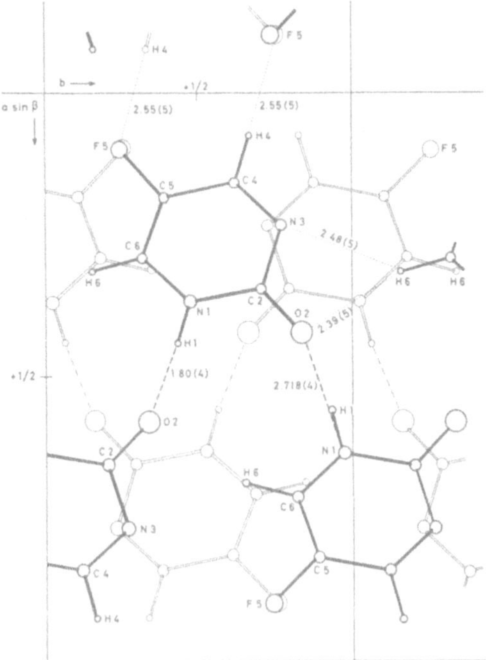

Fig. 2. The c-projection of the structure of 5-fluoropyrimidine-2-one. Broken lines are hydrogen bonds; dotted lines are short intermolecular contacts.

The main effects of fluorine substitution of pyrimidine-2-one (1) are to increase the bond angle at C(5) (Fig. 1) by 2.9° and to cause small but probably significant changes in bond lengths (the double bonds become shorter, the single bonds longer, but the sum of bond lengths remains the same). The molecules are linked by nearly planar zigzag chains of N-H...O hydrogen bonds (2.718 Å) (Fig. 2).

1. Structure Reports, 35B, 248.

ALLOPURINOL

$C_5H_4N_4O$

P. PRUSINER and M. SUNDARALINGAM, 1972. Acta Cryst., B28, 2148-2152.

Monoclinic, $P2_1/c$, a = 3.683, b = 14.685, c = 10.318 Å, β = 97.47°, D_m = 1.635, Z = 4. Cu radiation, R = 0.045 for 647 reflexions.

The pyrimidine portion of the allopurinol ring has normal distances (Fig. 1) while the five-membered diazole ring exhibits considerable distortions of bond angles compared to imidazole rings. The purine ring is planar. Each base ring is surrounded in the same plane by six bases (but forms hydrogen bonds (N-H...N = 2.880, 2.875 Å) with only five of them (Fig. 2)). There are also two base rings above and below the base plane held by van der Waals forces, with interplanar spacing of 3.301 Å between the bases. The carbonyl O(6) is involved in an intermolecular C-H...O hydrogen bond (3.208 Å) with C(2). The packing consists of stacks of parallel hydrogen-bonded sheets of bases with partial overlap of rings.

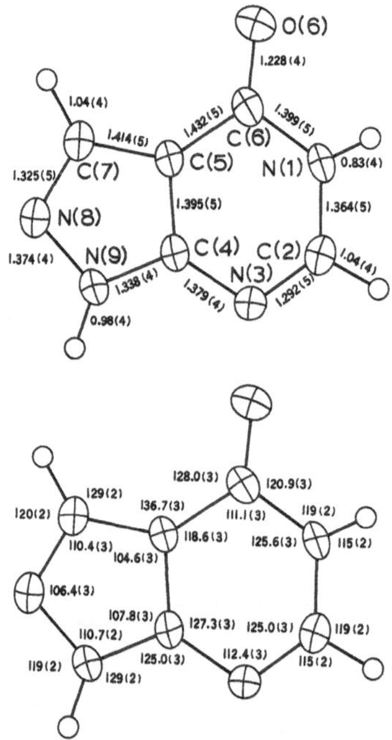

Fig. 1. Bond distances and angles in allopurinol.

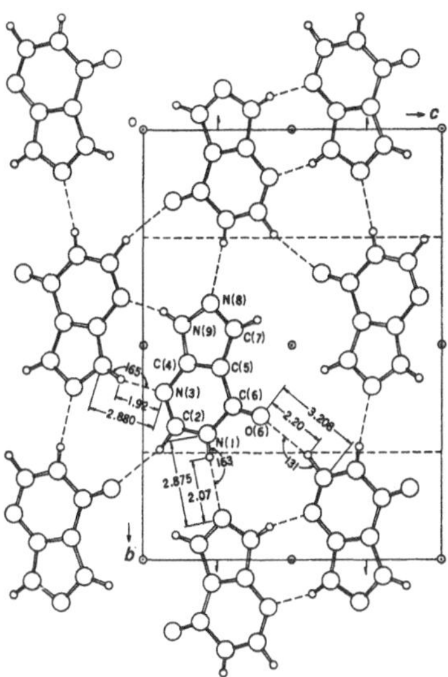

Fig. 2. A view along the a* axis, showing the hydrogen bonding in allopurinol.

ADENINE N[1]-OXIDE-SULPHURIC ACID

$C_5H_7N_5O_5S$

P. PRUSINER and M. SUNDARALINGAM , 1972. Acta Cryst., B28, 2142-2148.

Orthorhombic, $P2_12_12_1$, a = 10.224, b = 14.944, c = 5.639 Å, D_m = 1.910, Z = 4. Cu radiation, R = 0.042 for 838 reflexions.

The molecule (Fig. 1) is the protonated form of adenine N[1]-oxide with protonated sites O(1), N(7), and N(9) on the base. The purine ring system is planar except for displacements of N(1) (-0.02 Å) and the substituent O(1) (0.015 Å). The S-O bond distances in the sulphate group range from 1.455 - 1.494(4) Å. The crystal packing (Fig. 2) consists of seven negatively charged sulphate groups surrounding each positively charged base. Five of the sulphates are involved in hydrogen bonding (N-H...O = 2.644 - 2.908, O-H...O = 2.546 Å) whereas the other two sulphates above and below the plane of the purine ring are involved in ionic interactions (N...O = 3.143 - 3.808 Å). There is a C-H...O interaction involving C(8) of the base and the sulphate oxygen (C-H...O = 3.189 Å).

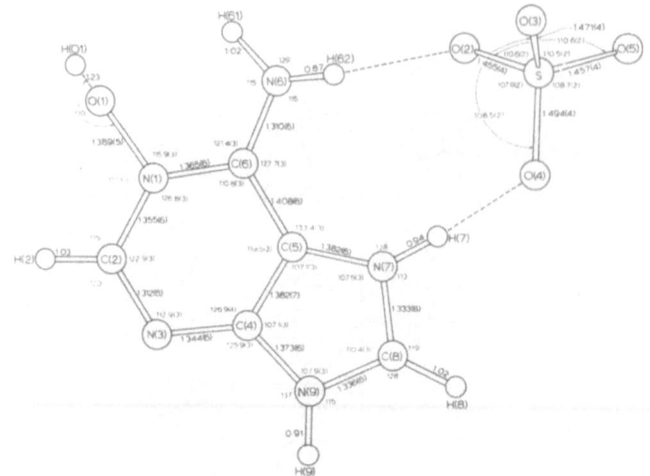

Fig. 1. Bond distances and angles for adenine N^1-oxide-sulphuric acid.

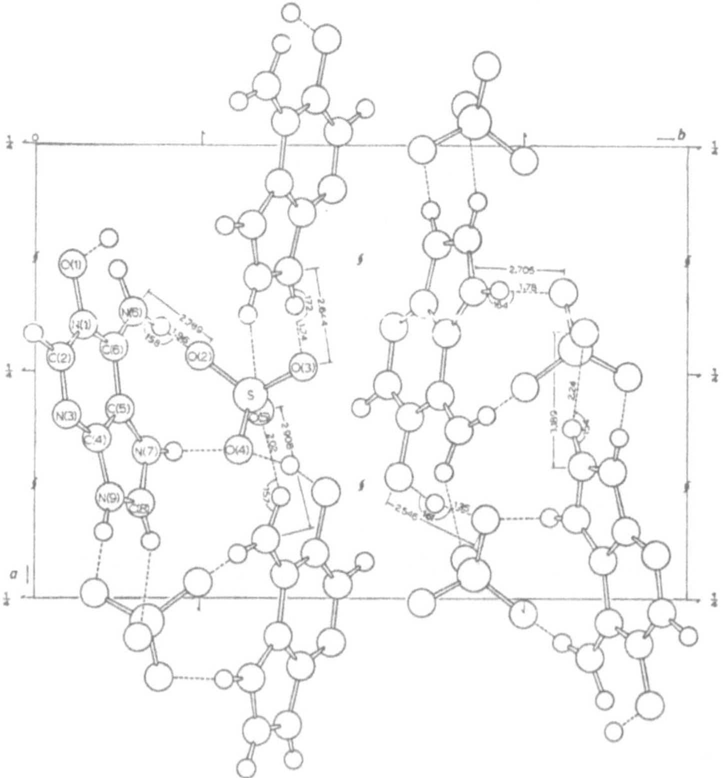

Fig. 2. The molecular packing of adenine N^1-oxide-sulphuric acid viewed down c.

1-METHYLCYTOSINE HYDROCHLORIDE

$C_5H_8C\ell N_3O$ $C_5H_8N_3O^+.C\ell^-$

B.L. TRUS and R.E. MARSH, 1972. Acta Cryst., B28, 1834-1840.

Monoclinic, $P2_1/n$, a = 6.695, b = 32.30, c = 6.912 Å, β = 104.22°, D_m = 1.483, Z = 8. Cu radiation, R = 0.055 for 2609 reflexions.

Bond distances and angles in the two independent molecules in the asymmetric unit (Fig. 1) are in statistical agreement. The molecular packing (Fig. 2) features layers of molecules joined by N-H...O and N-H...$C\ell^-$ hydrogen bonds, relatively short C-H...O interactions, and closely meshing methyl groups.

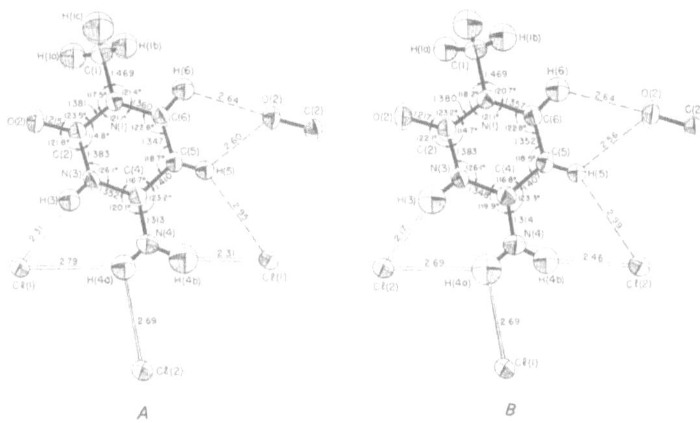

Fig. 1. Distances and angles in 1-methylcytosine hydrochloride. Estimated standard deviations are 0.004 Å and 0.15° for bonds and angles not involving hydrogen (for which the values are about ten times larger).

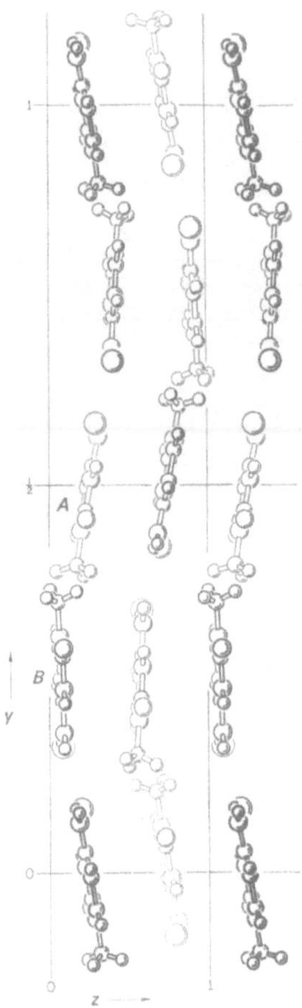

Fig. 2. View up the a-axis in 1-methylcytosine hydrochloride.

5,6-DIHYDRO-1-METHYL-4-THIOURACIL

$C_5H_8N_2OS$

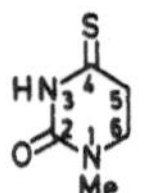

M.J.E. HEWLINS, 1972. J. Chem. Soc., Perkin II, 275-279.

Monoclinic, $P2_1/c$, a = 8.086, b = 19.57, c = 8.864 Å, β = 100.53°, D_m = 1.38, Z = 8. Cu radiation, R = 0.107 for 1476 reflexions.

The two independent molecules in the asymmetric unit are linked by two hydrogen bonds between N(3) and oxygen, with NH...O 2.829 and 2.840(10) Å.

There is a twist of 33° between the planes of the two molecules. Both molecules have the ketothione form (C=S = 1.658(8), C=O = 1.229(14) Å), and a half-chair conformation with atoms C(5) and C(6) respectively above (0.36 Å) and below (0.25 Å) the planes formed by the other non-hydrogen atoms. The N(3)-C(2) distance (mean 1.438(11) Å) is slightly greater than values found for this bond in related non-sulphur containing derivatives (1.383 - 1.395 Å).

1-ETHYL-5-BROMOURACIL

$C_6H_7BrN_2O_2$

I. H. MIZUNO, T. FUJIWARA and K. TOMITA, 1972. Bull. Chem. Soc. Japan, 45, 905-908.

II. T. TSUKIHARA, T. ASHIDA and M. KAKUDO, 1972. Ibid., 45, 909-912.

	Form I	Form II
Crystal system	Monoclinic	Tetragonal
Space group	$P2_1/c$	$P4_2/n$
a (Å)	7.89	17.13
b	12.01	
c	9.71	5.36
β	121.0°	
D_m	1.84	1.84
Z	4	8

Mo, Cu radiations, R = 0.052 and 0.064 for forms I and II respectively.

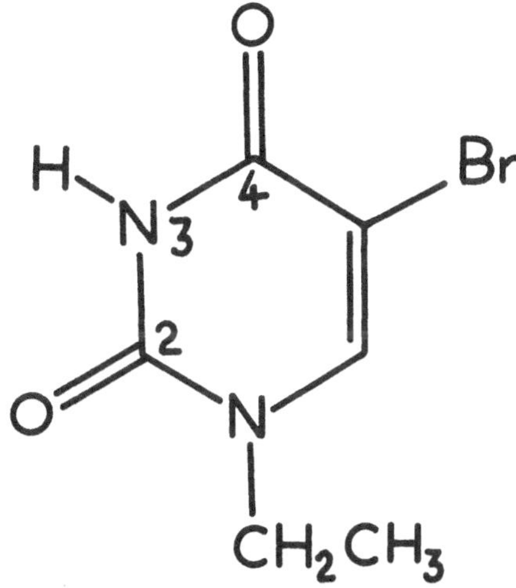

Fig. 1. 1-Ethyl-5-bromouracil.

Molecular dimensions are normal for the molecules (Fig. 1) in both forms. In form I the molecules are joined into centrosymmetrical dimers by N(3)-H...O(4) hydrogen bonds, 2.85 Å, and there is in addition a possible weak C-H...O bond. In form II centrosymmetrical dimers are formed by N(3)-H...O(2) hydrogen bonding, and again a C-H...O bond is possible; in this form the ethyl group is disordered.

6-METHYLURACIL-5-ACETIC ACID

$C_7H_8N_2O_4$

R. DESTRO and R.E. MARSH, 1972. Acta Cryst. B28, 2971-2977.

Monoclinic, $P2_1/n$, a = 4.8929, b = 12.6368, c = 12.7455 Å, β = 99.174°, D_m = 1.57, Z = 4. Cu radiation, R = 0.036 for 1600 reflexions.

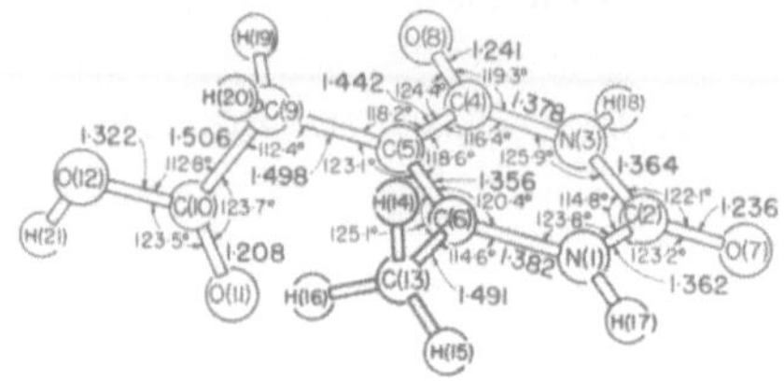

Fig. 1. Bond lengths and angles in 6-methyluracil-5-acetic acid. Standard deviations are about 0.002 Å and 0.10°.

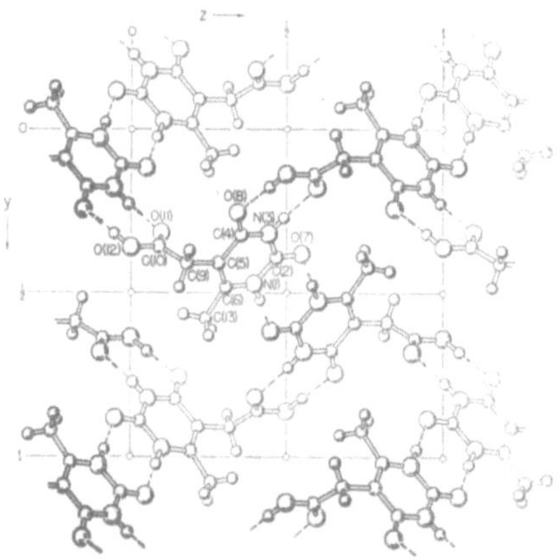

Fig. 2. The crystal structure of 6-methyluracil-5-acetic acid viewed down a. Dashed lines represent hydrogen bonds.

Details of bond lengths and angles are in Fig. 1. The pyrimidine ring is slightly non-planar being folded about the C(2)...C(5) axis to relieve strain between exocyclic substituents. The crystal structure (Fig. 2) features an off-set stacking of parallel pyrimidine rings at a separation of about 3.29 Å. There are two types of hydrogen bonding; across centres of symmetry (N(1)H...O(7) = 2.798 Å) to form base pairs, and between the carboxyl group and the ring (O(12)H...O(8) = 2.667 Å and O(11)...HN(3) = 2.804 Å).

N^6-(Δ^2-ISOPENTENYL)ADENINE

$C_{10}H_{13}N_5$

C.E. BUGG and U. THEWALT, 1972. Biochem. Biophys. Res. Commun., 46, 779-784.

Triclinic, P$\bar{1}$, a = 4.983, b = 13.427, c = 7.869 Å, α = 97.57, β = 100.52, γ = 92.70°, Z = 2. Cu radiation, R = 0.048 for 1684 reflexions.

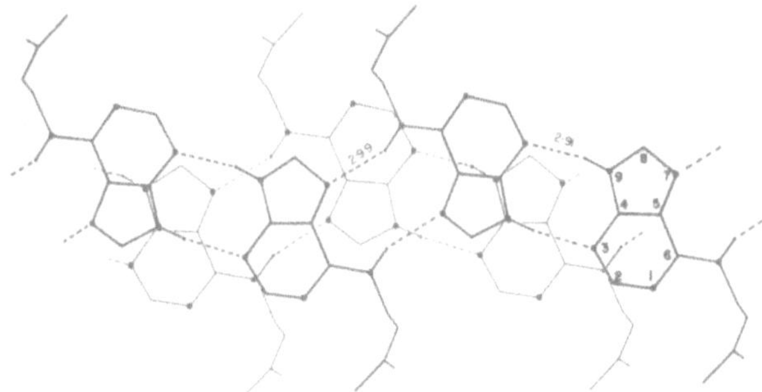

Fig. 1. The hydrogen bonding scheme and base stacking pattern in isopentenyladenine. Only those hydrogen atoms involved in hydrogen bonding are shown. The planes of purine bases are about 3.3 Å apart.

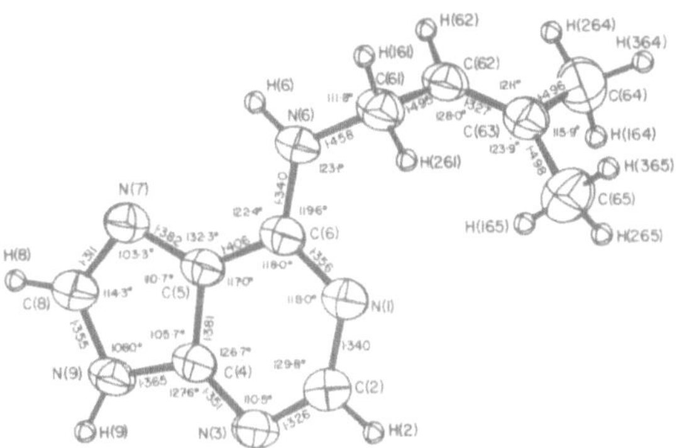

Fig. 2. Conformation of isopentenyladenine, with bond lengths (σ = 0.004 Å) and angles (σ = 0.2°).

In the crystal structure (Fig. 1) the bases are linked across crystallographic inversion centres by N(9)-H...N(3) and N(6)-H...N(7) hydrogen bonds resulting in continuous ribbons of purine bases. The isopentenyl moiety is pointing away from the adenine imidazole ring, and assumes a conformation (Fig. 2) that prevents the N(1) position of adenine from accepting hydrogen bonds;

consequently, the two sites, N(6)-H and N(1), that are normally utilized by adenine bases for complementary pairing within double helical regions of nucleic acids are blocked by the isopentenyl sidechain. The observed structure is consistent with the hypothesis that this purine derivative plays a role in maintaining the anticodon loop in a single stranded conformation that enhances codon-anticodon interactions.

6-HISTAMINOPURINE DIHYDRATE

$C_{10}H_{15}N_7O_2$ $C_{10}H_{11}N_7 . 2H_2O$

U. THEWALT and C.E. BUGG, 1972. Acta Cryst., B28, 1767-1773.

Monoclinic, $P2_1/c$, a = 5.507, b = 31.839, c = 7.336 Å, β = 105.58°, D_m = 1.42, Z = 4. Cu radiation, R = 0.136 for 2075 reflexions.

The crystal structure is shown in Fig. 1. All the hydrogen atoms that are covalently bonded to N or O participate in hydrogen bonding resulting in ribbons parallel to a and stacking of adenine moieties along c. The stacked base planes are approximately parallel, with an interplanar spacing of 3.4 Å. The molecular conformation and details of geometry are in Fig. 2; the adenine moiety is in the N(7)-H tautomer form.

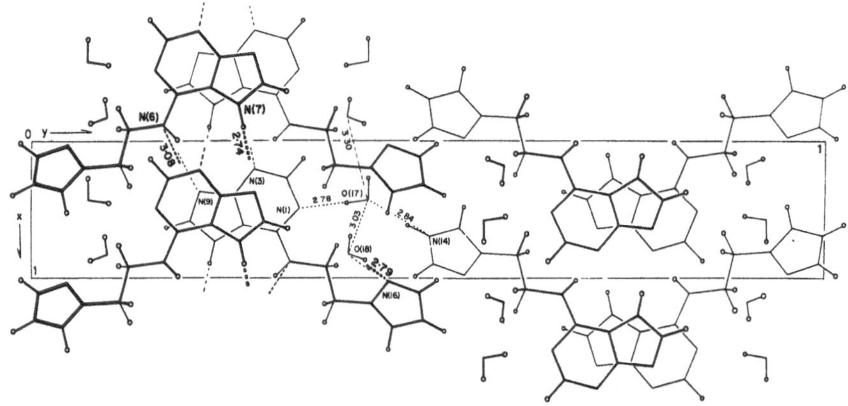

Fig. 1. The structure of 6-histaminopurine dihydrate viewed down the c-axis. Hydrogen bonds are shown by dashed lines.

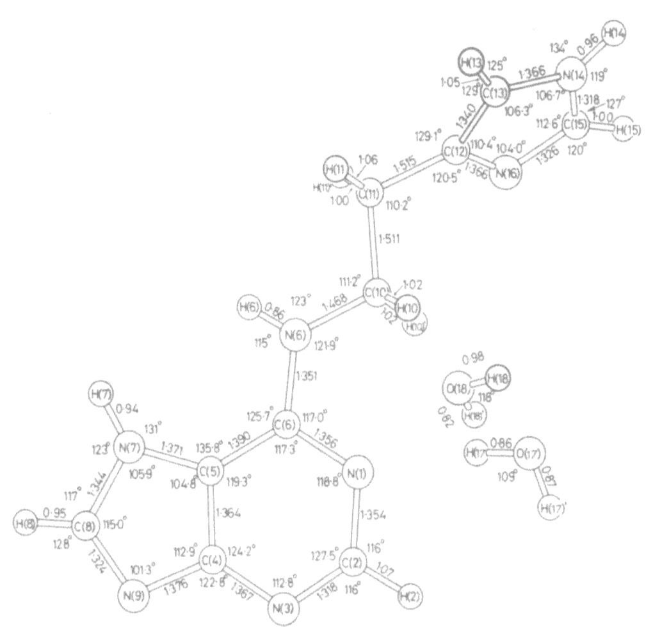

Fig. 2. Bond distances (Å) and angles (°) in 6-histaminopurine dihydrate. Average e.s.d.'s are 0.008 Å for bond lengths between heavy atoms and 0.07 Å for those involving hydrogen atoms. Corresponding e.s.d.'s in bond angles are 0.5° and 6° respectively.

THIAMINE CHLORIDE MONOHYDRATE

$C_{12}H_{19}C\ell N_4O_2S$ $C_{12}H_{17}C\ell N_4OS.H_2O$

J. PLETCHER, M. SAX, S. SENGUPTA, J. CHU and C.S. YOO, 1972. Acta Cryst., B28, 2928-2935.

Monoclinic, $P2_1/c$, a = 8.914, b = 16.775, c = 10.405 Å, β = 94.28°, D_m = 1.361, Z = 4. Cu radiation, R = 0.033 for 2368 reflexions.

Details of molecular geometry are in Fig. 1. The pyrimidine and thiazolium rings are individually planar. The conformation found for the free base (Fig. 2) is very similar to that which is observed for protonated forms. The absence of the proton on the pyrimidine ring reduces the latter's "quinoidal" character but the bond to the amino group, although longer in the unprotonated compound, still exhibits considerable double bond character. In the crystal structure the hydrogen bond from C(2) (C(2)H...Cℓ(20) = 3.305 Å Fig. 2) is a feature observed in a number of structures containing a thiazolium ring which is unsubstituted at C(2). Another feature is the absence of hydrogen bonding to the N(11) position of the pyrimidine ring.

Fig. 1. Bond lengths (σ = 0.002 - 0.004 Å for bonds between non-hydrogen atoms) and angles (σ = 0.1 - 0.2° for angles not involving hydrogen atoms) in thiamine chloride monohydrate.

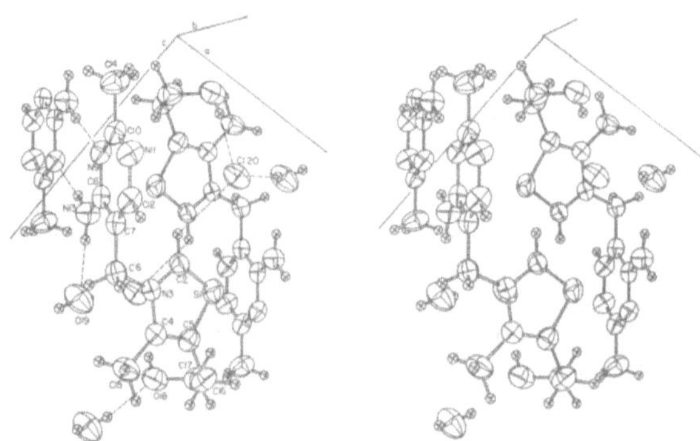

Fig. 2. View showing the conformation of thiamine chloride and hydrogen bonding between molecules.

D-iso-ASCORBIC ACID

$C_6H_8O_6$

O=C
HO-C
HO-C
H-C*
H-C*-OH
CH_2OH

N. AZARNIA, H.M. BERMAN and R.D. ROSENSTEIN, 1972. Acta Cryst., B28, 2157-2161. [See also Acta Cryst., B29, 1170 for minor corrections.]

Monoclinic, $P2_1$, a = 5.165, b = 14.504, c = 4.724 Å, β = 99.50°, D_m = 1.654, Z = 2. Cu radiation, R = 0.037 for 620 reflexions.

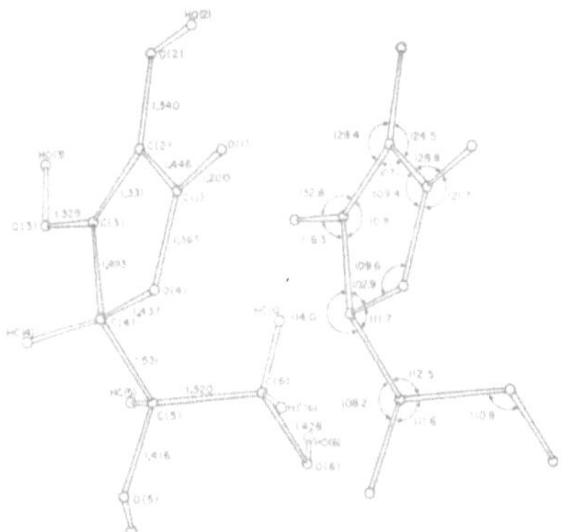

Fig. 1. Bond distances (σ = 0.005 Å) and angles (σ = 0.3 °) in D-iso-ascorbic acid.

The molecule (Fig. 1),which is a stereoisomer of L-ascorbic acid (vitamin C) with inversion at C(5), consists of a lactone ring and an acyclic side chain. The ring system which comprises C(1), C(2), C(3), C(4), O(4), O(1), O(2), O(3), HO(2) and HO(3) is roughly planar while the lactone group C(1), C(2), C(4), O(4), O(1), the endiol group HO(2), C(2), C(3), HO(3) and the ring C(1), C(2), C(3), C(4), O(4) are all individually planar. The four atoms C(4), C(5), C(6), O(6) form an extended roughly planar zigzag forming a dihedral angle of 102.4° with the ring plane. Atom O(6) is in a staggered conformation that takes it farthest from the ring. The molecules which are arranged head-to-tail along b, are linked together by a zigzag chain of hydrogen bonds between hydroxyls (O-H...O = 2.584 - 2.827 Å) with a single hydrogen bond (O-H...O = 2.643 Å) to the carbonyl oxygen atom along c. The ring oxygen atom is not hydrogen-bonded.

1,6-ANHYDRO-β-D-MANNOFURANOSE

$C_6H_{10}O_5$

J. LECHAT and G.A. JEFFREY, 1972. Acta Cryst., B28, 3410-3415.

Orthorhombic, $P2_12_12_1$, a = 5.877, b = 11.396, c = 9.915 Å, D_m = 1.613, Z = 4. Cu radiation, R = 0.032 for 589 reflexions.

The molecular structure (Fig. 1) has a fused-ring system which consists of a chair-shaped six-membered ring and an envelope furanose ring having the furanose ring oxygen -0.676 Å out of the plane C(1), C(2), C(3), C(4), exo to the anhydro ring. Four of the C-C bond lengths lie within a narrow range from 1.536 to 1.544(4) Å, and C(4) - C(5), 1.499(4) Å, is 0.040 Å shorter than their mean. The C-O bond lengths show a variation from 1.409 to 1.466(4) Å, which is similar to that observed in the vicinity of the ring oxygen and anomeric carbon in most pyranose sugars. The hydrogen bonding (Fig. 2) is intermolecular and consists of infinite spirals and isolated links (O-H...O = 2.818 - 3.140 Å). The hydroxyls O(2)H and O(5)H form bifurcated hydrogen bonds to O(4), O(5) and O(2), O(3) respectively. O(3)H forms a normal hydrogen bond to the anhydride ring oxygen O(1).

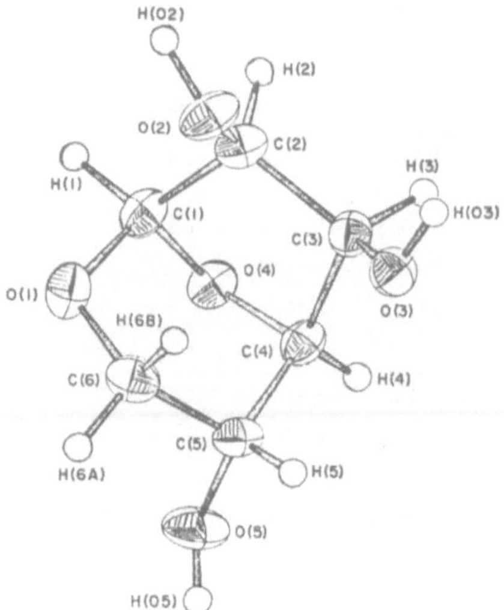

Fig. 1. 1,6-Anhydro-β-D-mannofuranose.

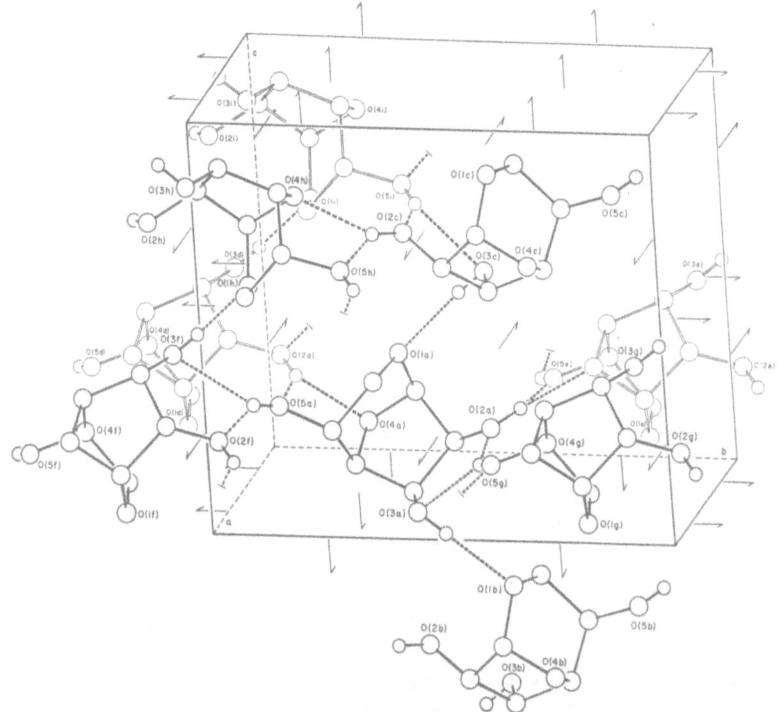

Fig. 2. Clinographic representation of the unit cell of 1,6-anhydro-β-D-mannofuranose.

POTASSIUM D-GLUCONATE MONOHYDRATE

$C_6H_{13}KO_8$ $C_6H_{11}O_7K.H_2O$

G.A. JEFFREY and E.J. FASISKA, 1972. Carbohyd. Res., 21, 187-199.

Orthorhombic, $P2_12_12_1$, a = 8.220, b = 17.840, c = 6.717 Å, D_m = 1.743, Z = 4. Cu radiation, R = 0.061 for 1632 reflexions.

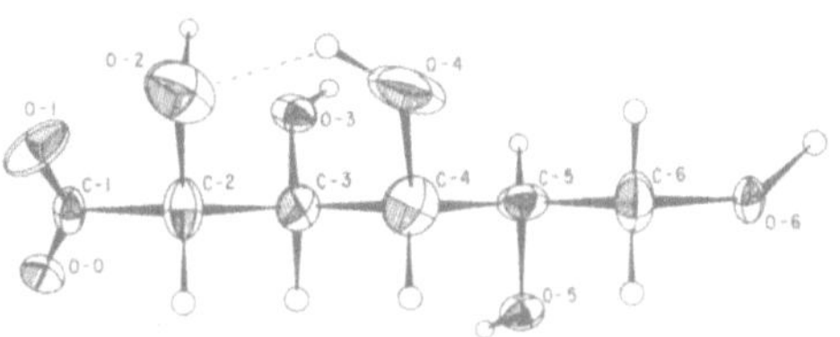

Fig. 1. The D-gluconate ion in potassium D-gluconate monohydrate.

Fig. 2. Schematic diagram of the puckered, hydrogen-bonded sheets of D-gluconate ions and water molecules in projection. The centre of the D-gluconate ions is at approximately x = 0; the x coordinates of the K^+ ions are shown.

The D-gluconate ion (Fig. 1) has a planar extended chain conformation with an intramolecular hydrogen bond O(4)-H...O(2), 2.62 Å, which stabilizes this conformation. The D-gluconate ions and water molecules are linked in puckered sheets (Fig. 2) by a series of hydrogen bonds (OH...O = 2.71 - 3.04 Å) that involve the water molecules, the carboxylate groups, and pairs of hydroxyl groups. One hydroxyl group in the ion does not form a hydrogen bond. The potassium ions lie between the puckered sheets, with an eight-fold coordination of six D-gluconate groups and two water oxygen atoms; K...O distances are in the range 2.613 - 3.229 Å. For details of the anhydrous structure see (1).

1. Structure Reports, 17, 636.

β-D-GALACTOSAMINE HYDROCHLORIDE

$C_6H_{14}C\ell NO_5$

M. TAKAI, S. WATANABE, T. ASHIDA and M. KAKUDO, 1972. Acta Cryst., B28, 2370-2376.

Orthorhombic, $P2_12_12_1$, a = 9.794, b = 19.686, c = 9.385 Å, D_m = 1.591, Z = 8. Cu radiation, R = 0.045 for 1248 reflexions.

The two galactosamine molecules in the asymmetric unit have almost the same conformation. Each ring has the normal Sachse trans configuration. The rings have the C1 chair form with conformation 1e2e3e4a5e. Details of molecular geometry are in Fig. 1. The bond lengths for equatorial bonds C(1)-O(1) and C(7)-O(7) are shortened by about 0.03 Å relative to the other C-O bonds; this is a common finding in sugar molecules. The molecules are held in the crystal by a network of hydrogen bonds. There are thirteen hydrogen bonds in the asymmetric unit (Fig. 2). Only H(O9) (bonded to O(9)) of the hydrogen atoms which belong to functional groups, does not participate in hydrogen bonding.

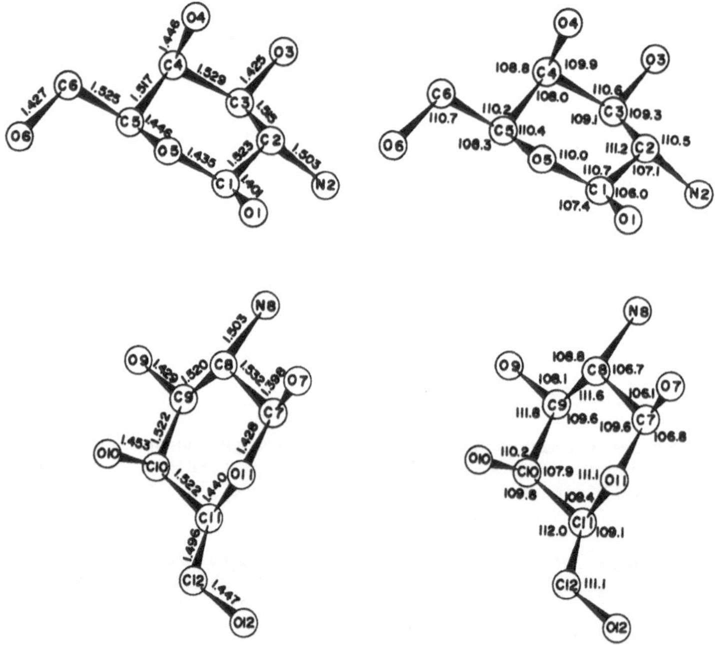

Fig. 1. Bond lengths (σ = 0.009 Å) and angles (σ = 0.6°) in β-D-galactosamine.

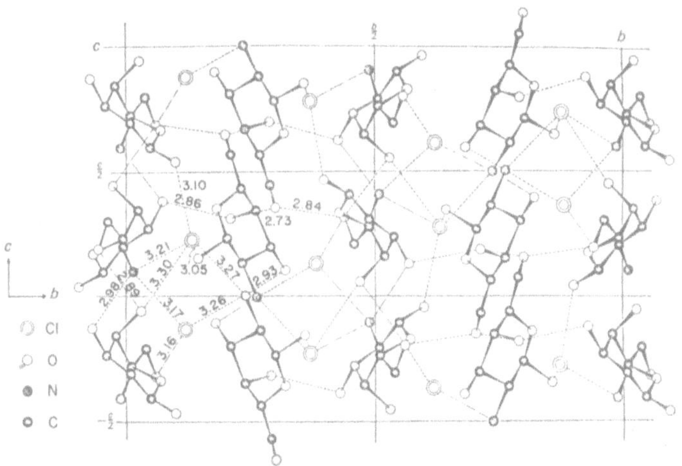

Fig. 2. The β-D-galactosamine structure viewed along the a axis. Hydrogen bonds are indicated by broken lines.

ALLITOL

$C_6H_{14}O_6$

N. AZARNIA, G.A. JEFFREY and M.S. SHEN, 1972. Acta Cryst., B28, 1007-1013.

Monoclinic, $P2_1/c$, a = 4.708, b = 13.408, c = 6.616 Å, β = 100.1°, D_m = 1.482, Z = 2. Cu radiation, R = 0.04 for 587 reflexions.

Fig. 1. Allitol.

The molecule possesses a centre of symmetry midway between C(3) and C(4) (Fig. 1). The molecule has the only possible bent-chain conformation with no parallel C(n)-O/C(n+2)-O interaction. O(1), C(1), C(2), and C(3) lie within 0.06 Å of one plane, and the centrosymmetrically related atoms C(4), C(5), C(6), and O(6) lie near a parallel plane separated from the first by 1.49 Å. The four central carbon atoms [C(2), C(3), C(4), and C(5)] and the two oxygens O(2) and O(5) are coplanar to within ±0.1 Å. The bond lengths and angles are normal. The molecules are linked together by a system of hydrogen bonds in which each hydroxyl group is involved in 2 bonds, in one as donor and in one as acceptor. The mean O(H)...O distance is 2.714 Å.

D-IDITOL

$C_6H_{14}O_6$

N. AZARNIA, G.A. JEFFREY and M.S. SHEN, 1972. Acta Cryst., B28, 1007-1013.

Monoclinic, $P2_1$, a = 8.124, b = 8.386, c = 5.870 Å, β = 93.8°, D_m = 1.516, Z = 2. Cu radiation, R = 0.08 for 697 reflexions.

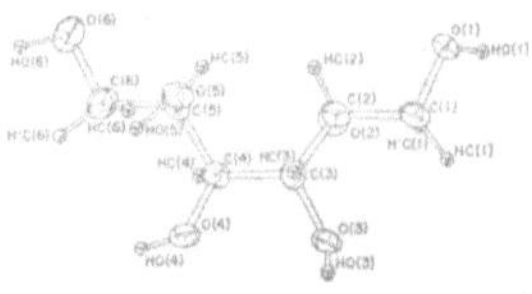

Fig. 1. D-Iditol

The conformation observed in the crystal structure is a bent-chain one without parallel C(n)-O/C(n+2)-O interaction and a pseudo C_2 axis normal to the C(3)-C(4) bond (Fig. 1). With the exception of the hydroxyl hydrogens involved in the intermolecular hydrogen bonding, the deviations from this noncrystallographic molecular axis are less than 0.024 Å. O(1), C(1), C(2), C(3), and C(4) are coplanar within 0.053 Å, while C(3), C(4), C(5), C(6), and O(6) are coplanar within 0.05 Å. The dihedral angle between these two planes is 49°. The bond lengths and angles in the molecule are normal. The molecules are joined together by a system of intermolecular hydrogen bonds in which one hydroxyl, O(4)H, is a donor only, and one O(3)H, is a donor for one and an acceptor for three hydrogen bonds. The mean O(H)...O distance is 2.786 Å.

METHYL 3,6-ANHYDRO-α-D-GALACTOSIDE

$C_7H_{12}O_5$

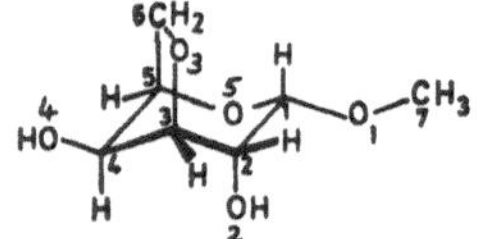

J.W. CAMPBELL and M.M. HARDING, 1972. J. Chem. Soc., Perkin II, 1721-1723.

Orthorhombic, $P2_12_12_1$, a = 9.46, b = 12.05, c = 6.93 Å, D_m = 1.48, Z = 4. Mo radiation, R = 0.097 for 1926 reflexions.

The geometry of the five-membered ring of the molecule (C(3)-(6), O(3)) is similar to that found in furanose rings and that of the rest of the molecule is typical of a pyranose sugar. Apart from C(1)-O(1) the bond lengths do not differ significantly from the expected values. C(1)-O(1), 1.374 Å, is 0.067 Å (>18σ) shorter than the average C-O bond length. There is no significant asymmetry in the lengths of the C-O(5) bonds. The bond angles, also, have typical values though the angle C(3)-C(4)-C(5) (97.7°) is rather smaller than average. The distance across the pyranose ring from C(3) to C(5) is 2.30 Å which is 0.2 Å less than that found for an unbridged pyranose ring. Atoms C(1) and C(4) are 0.543 and 0.921 Å respectively out of the best plane through atoms C(2), C(3), C(5) and O(5). Atom C(4) is 0.706 Å from the best plane through atoms C(3), O(3), C(5) and C(6). In the crystal structure there are two sets of hydrogen bonds, O(2)... O(4') 2.845 and O(2)...O(4") 2.744 Å, making double chains of molecules parallel to the c axis.

ETHYL 2-S-ETHYL-1,2-DITHIO-α-D-MANNOFURANOSIDE

$C_{10}H_{20}O_4S_2$

A. DUCRUIX and C. PASCARD-BILLY, 1972. Acta Cryst., B28, 1195-1201.

Orthorhombic, $P2_12_12_1$, a = 5.070, b = 8.113, c = 32.18 Å, D_m = 1.36, Z = 4. Cu radiation, R = 0.11 for 552 reflexions.

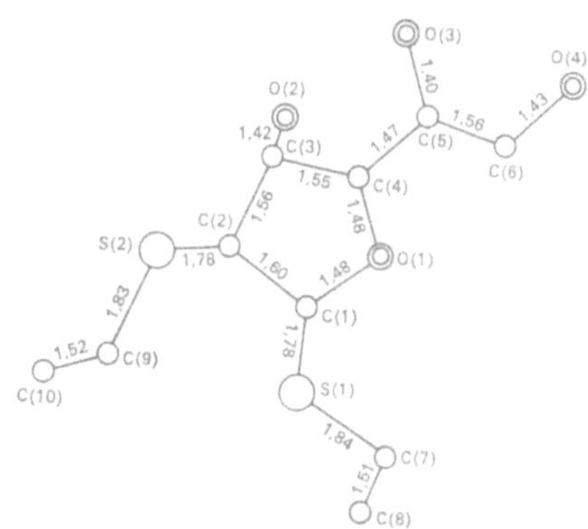

Fig. 1. Bond lengths in ethyl 2-S-ethyl-1,2-dithio-α-D-mannofuranoside (σ = 0.03 Å).

The conformation of the furanose ring is intermediate between an envelope and a twist, and is best described by the torsion angles around the ring: C(1)-C(2) -8.6°, C(2)-C(3) 30.3°, C(3)-C(4) -40.7°, C(4)-O(1) 36.9°, and O(1)-C(1) -17.9°. The two S-alkyl groups are in trans positions with the manno configuration for the S-ethyl substituent at C(2). The bond lengths (Fig. 1) are close to accepted values. The molecules are linked together in the crystal by hydrogen bonding. O(3) is hydrogen bonded to O(2) of one molecule and O(4) of a second molecule (O(3)...O(2), 2.86; O(3)...O(4), 2.78 Å).

3,6-(ACETYLEPIMINO)-3,6-DIDEOXY-1,2-O-ISOPROPYLIDENE-β-L-IDOFURANOSE

$C_{11}H_{17}NO_5$

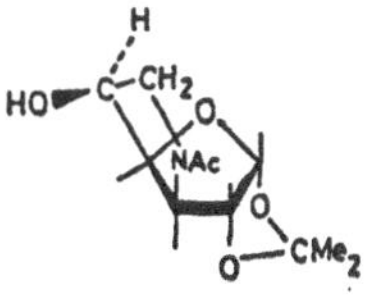

J.S. BRIMACOMBE, J. IBALL and J.N. LOW, 1972. J. Chem. Soc., Perkin II, 937-942.

Monoclinic, $P2_1$, a = 12.13, b = 9.51, c = 5.32 Å, β = 93.74°, Z = 2. Mo radiation, R = 0.078 for 946 reflexions.

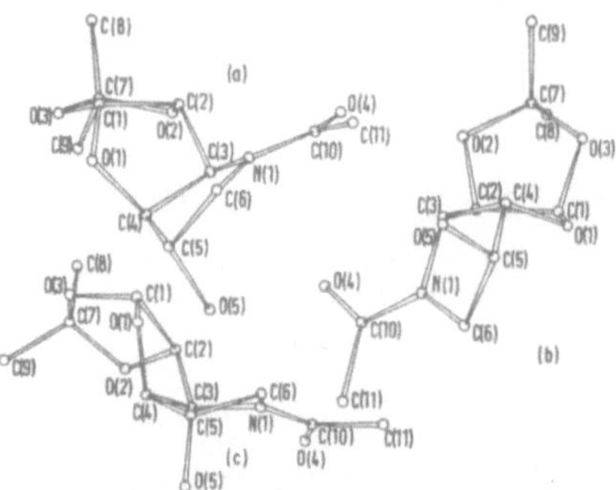

Fig. 1. View of the epimine, $C_{11}H_{17}NO_5$, looking along the mean plane of (a) ring A, (b) ring B and (c) ring C.

The structure of this molecule is shown above and in Fig. 1. There are no unusual bond lengths or angles. The two cis-fused, five-membered rings of the trioxabicyclo[3,3,0]octane system (rings A and B) each adopt envelope-like conformations. The furanose ring (Fig. 1b) has C(4) displaced by 0.459 Å in an exo-direction (with respect to the bicyclic system) from the mean plane through the other four atoms and the acetal carbon atom, C(7), of the dioxolan ring is also displaced in an exo-direction by 0.38 Å (Fig. 1a). The five-membered ring containing the nitrogen atom (ring C) likewise assumes an envelope conformation having C(5) displaced by 0.538 Å in an endo-direction with respect to the oxa-azabicyclo[3,3,0]octane ring-system. These values are comparable with out-of-plane displacements commonly found for furanose rings adopting envelope conformations, although in many other instances the out-of-plane atom of the furanose ring has been either C(2) or C(3). In the crystal structure molecules are linked by O(4)H...O(5) hydrogen bonds, 2.723 Å.

The signs of the hydrogen atom y/b coordinates are wrongly given in the paper and should be changed (1).

1. J. IBALL, 1973. Personal communication.

DEHYDRO-L-ASCORBIC ACID DIMER

$C_{12}H_{12}O_{12}$

J. HVOSLEF, 1972. Acta Cryst., B28, 916-923.

Monoclinic, C2, a = 15.728, b = 5.530, c = 9.453 Å, β = 130.56°, D_m = 1.8, Z = 2. Cu radiation, R = 0.061 for 730 reflexions.

Dehydro-L-ascorbic acid, the first oxidation product of vitamin C is found as a dimer in the crystalline state. The molecule has twofold symmetry and comprises a system of five fused rings (Fig. 1). Non-planar γ-lactone and furanose rings are attached to a central dioxan ring which is in a twisted boat conformation. The external axially bonded C-OH groups have relatively short C-O distances and contain, in addition to the carbonyl groups, the only oxygen atoms taking part in hydrogen bonding (Fig. 2). Each molecule is linked to eight others by one O-H...O bond to each of them (O...O = 2.808, 2.880(7) Å).

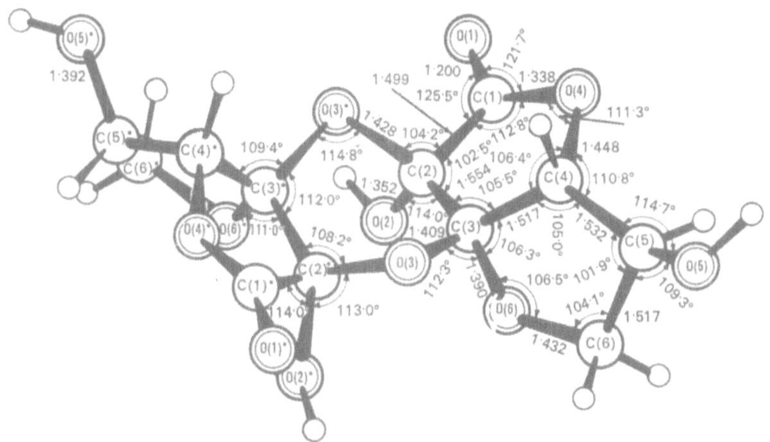

Fig. 1. Bond distances (corrected for thermal motion) and angles in dehydro-L-ascorbic acid dimer.

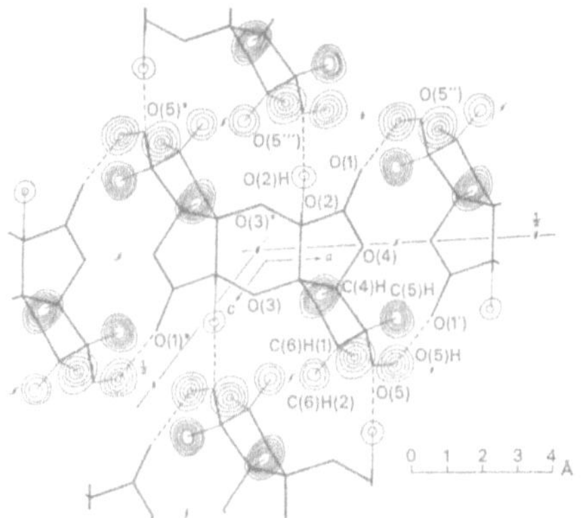

Fig. 2. Hydrogen bonding in dehydro-L-ascorbic acid dimer.

3,6-ANHYDRO-α-D-GLUCOSYL-1,4:3,6-DIANHYDRO-β-D-FRUCTOSIDE

$C_{12}H_{16}O_8$

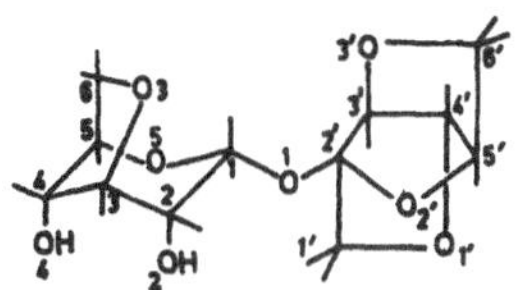

N.W. ISAACS and C.H.L. KENNARD, 1972. J. Chem. Soc., Perkin II, 582-585.

Orthorhombic, $P2_12_12_1$, a = 6.437, b = 11.298, c = 16.754 Å, D_m = 1.57, Z = 4. Cu radiation, R = 0.071 for 991 reflexions.

In this compound, which is a dehydrated form of sucrose, the most interesting feature is the tricyclic furanose moiety. The strain involved in the formation of this system results in a "tightening" of the angles about C(3') and C(4'), lengthening of the C-O distances, and increased puckering of the ring compared with the corresponding data for sucrose itself (1). The closure of two additional rings on the furanose part causes the sign of the torsion angle for each of the bonds in the 'primary' ring C(2'), C(3'), C(4'), C(5'),O(2') to be opposite to that in the sucrose molecule and forces the ring into the 'half-chair' conformation. The pyranose ring is in the chair form, with the hydroxy-groups axial, and the hydrogen atoms equatorial. This conformation is opposite to that observed in sucrose. As in sucrose, the sign of the torsion angle alternates round the ring, but the magnitudes of the angles are much greater (49 - 70°) than in sucrose (54.8 - 56.0°). The packing of the molecules in the crystal is governed by a hydrogen bonding scheme involving both hydroxyl groups with O-H...O 2.788 and 2.864(7) A.

1. G.M. BROWN and H.A. LEVY, 1963. Science, 141, 921.

3-CHLORO-3-DEOXY-1,2:5,6-DI-O-ISOPROPYLIDENE-β-D-IDOSE

$C_{12}H_{19}C\ell O_5$

F.W.B. EINSTEIN and K.N. SLESSOR, 1972. Canad. J. Chem., 50, 93-98.

Orthorhombic, $P2_12_12_1$, a = 9.744, b = 26.76, c = 5.403 Å, D_m = 1.27, Z = 4. Mo radiation, R = 0.057 for 817 reflexions.

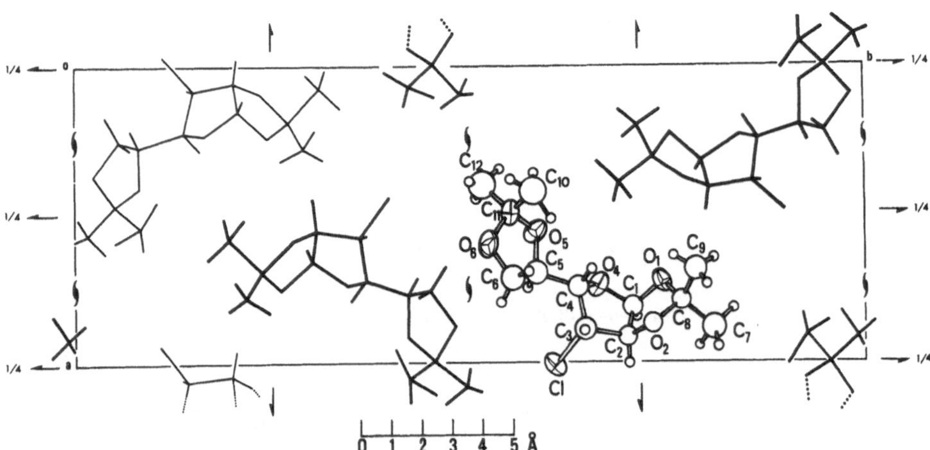

Fig. 1. The crystal structure of 3-chloro-3-deoxy-1,2:5,6-di-O-isopropylidene-β-D-idose viewed down the c axis.

This structure analysis served to confirm that chlorination of the sugar occurred stereospecifically at position 3 (Fig. 1). Bond lengths and angles are in good agreement with accepted values. There are no significant intermolecular interactions. The conformation of the furan ring is an envelope arrangement with C(4) displaced from the mean plane; the dioxolane rings both adopt twist conformations but to different extents.

α,α-TREHALOSE DIHYDRATE

$C_{12}H_{26}O_{13}$

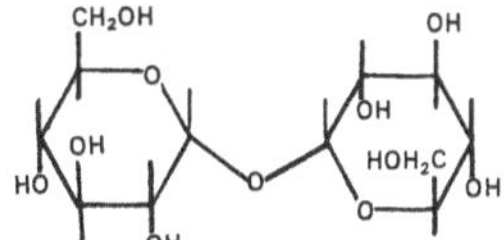

I. G.M. BROWN, D.C. ROHRER, B. BERKING, C.A. BEEVERS, R.O. GOULD and R. SIMPSON, 1972. Acta Cryst., B28, 3145-3158.

II. T. TAGA, M. SENMA and K. OSAKI, 1972. Acta Cryst., B28, 3258-3263.

Orthorhombic, $P2_12_12_1$, a = 12.230, b = 17.890, c = 7.596 Å, D_m = 1.512, Z = 4. Mo radiation, R = 0.057 for 3293 reflexions; Cu radiation, R = 0.041 for 1611 reflexions; Cu radiation, R = 0.055 for 2149 reflexions; and Cu radiation, R = 0.057 for 1892 reflexions.

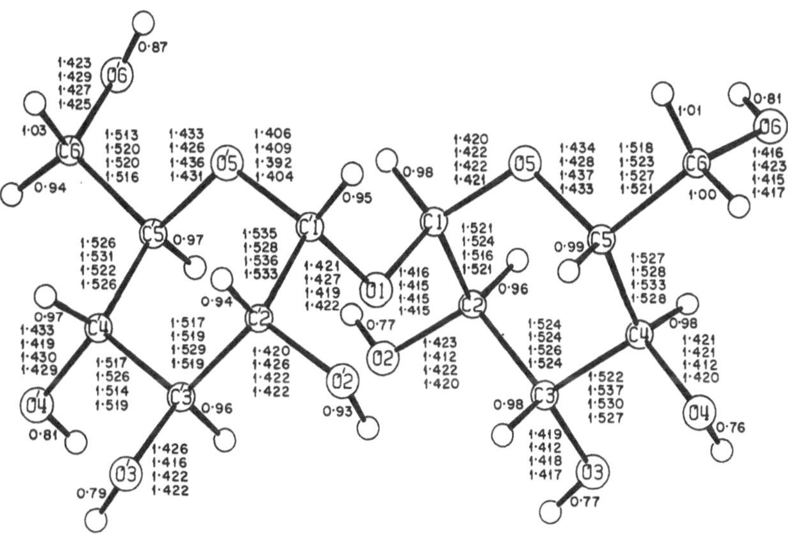

Fig. 1. Bond lengths in the α,α-trehalose molecule. The bottom value corresponds to the average value of the three determinations reported in I. Results from II are in accord with these.

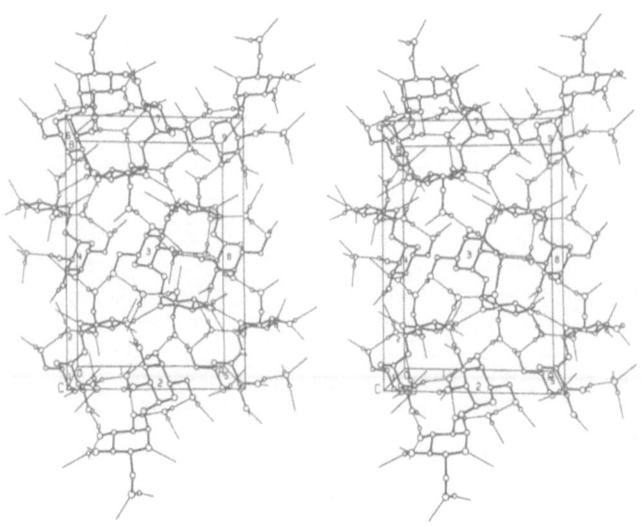

Fig. 2. View of the crystal structure of α,α-trehalose dihydrate. Thin lines denote hydrogen bonds.

Four independent structural determinations are reported on α,α-trehalose which is the first carbohydrate oligomer to be studied with a bridging oxygen atom axial to two rings. The two glucopyranosyl rings have the chair form and are linked through O(1) oxygen in approximate twofold symmetry; there are some significant structural differences between the two chemically equivalent halves (Fig. 1). The angles at the ring oxygen atoms are 114.0 and 114.2°; the angle at the glycosidic oxygen is 115.8°. The trehalose and water molecules are held together by twelve O-H...O hydrogen bonds as shown in the packing diagram (Fig. 2), with O...O = 2.697 - 2.906 Å.

ETHYL 2,3:4,5-DI-O-ISOPROPYLIDENE-1-THIO-β-D-GLUCOSEPTANOSIDE

$C_{14}H_{24}O_5S$

J.P. BEALE, N.C. STEPHENSON and J.D. STEVENS, 1972. Acta Cryst., B28, 3115-3121.

Orthorhombic, $P2_12_12_1$, a = 16.70, b = 11.70, c = 8.78 Å, D_m = 1.20, Z = 4. Cu radiation, R = 0.103 for 1322 reflexions.

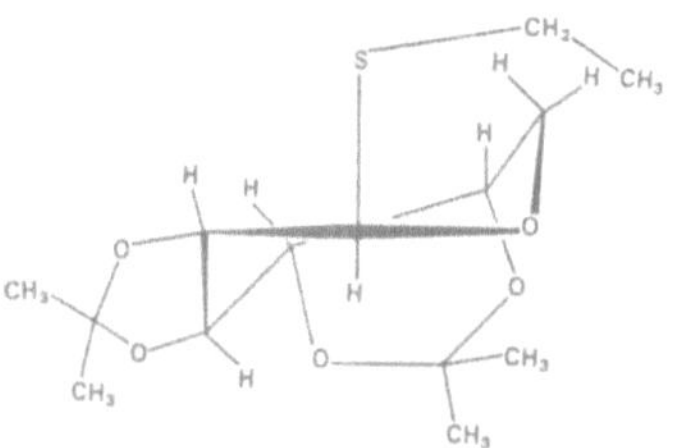

Fig. 1. Diagrammatic representation of ethyl 2,3:4,5-di-O-isopropylidene-1-thio-β-D-glucoseptanoside.

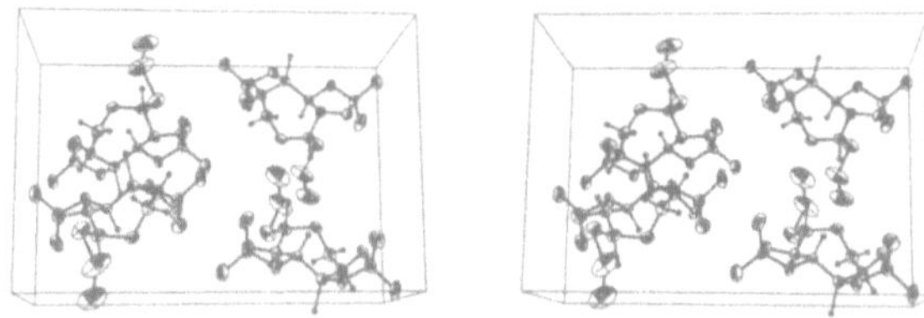

Fig. 2. Molecular packing diagram viewed along c for ethyl 2,3:4,5-di-O-isopropylidene-1-thio-β-D-glucoseptanoside.

The central seven-membered septanose ring is in a twist-chair conformation (Fig. 1), and the two trans-fused dioxolan rings are in twist conformations. Fig. 2 shows the molecular packing in the unit cell. S-C = 1.79(2), 1.76(3) Å, mean C-O = 1.44 Å. The C(2)-C(3) (1.54(2)) and C(4)-C(5) (1.57(2) Å) bonds which form part of the dioxolan ring are significantly longer than the other ring C-C bonds, C(1)-C(2) = 1.49(2), C(3)-C(4) = 1.50(2), and c(5)-C(6) = 1.49(2) Å. Bond angles within the seven-membered ring are all greater than the tetrahedral value.

1-KESTOSE

$C_{18}H_{32}O_{16}$

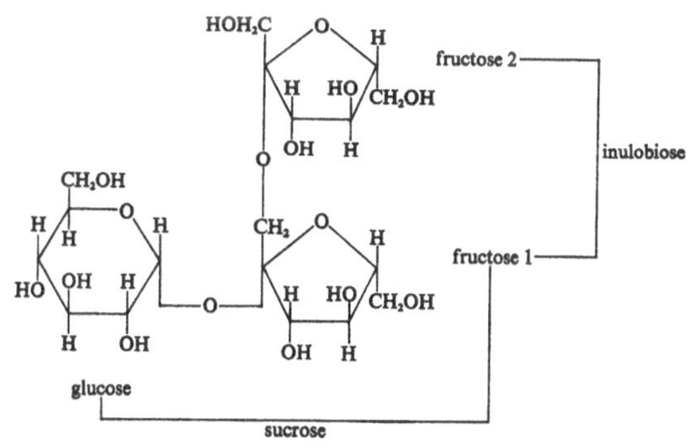

G.A. JEFFREY and Y.J. PARK, 1972. Acta Cryst., B28, 257-267.

Orthorhombic, $P2_12_12_1$, a = 7.935, b = 9.994, c = 26.699 Å, D_m = 1.574, Z = 4. Cu radiation, R = 0.039 for 2089 reflexions.

In crystals of 1-kestose, O-α-D-glucopyranosyl-(1→2)-O-β-D-fructofuranosyl-(1→2)-β-D-fructofuranoside, the α-D-glucopyranoside unit has the normal C1 chair conformation, with a primary alcohol group that is disordered over two orientations (Fig. 1). Fructofuranoside units have puckered rings with different conformations (Fig. 2). Bond lengths and angles are in Fig. 3. All hydrogen bonding is intermolecular with the majority of hydroxyls functioning as both donor and acceptor atoms (Fig. 4).

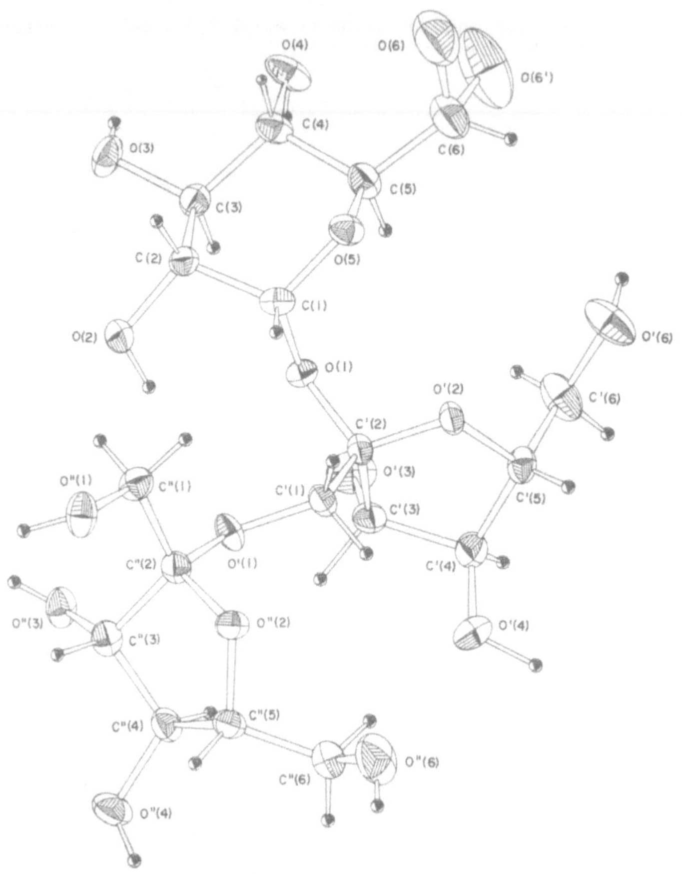

Fig. 1. Molecular conformation and numbering scheme for 1-kestose.

Fig. 2. Conformations of the two fructofuranoside rings in 1-kestose.

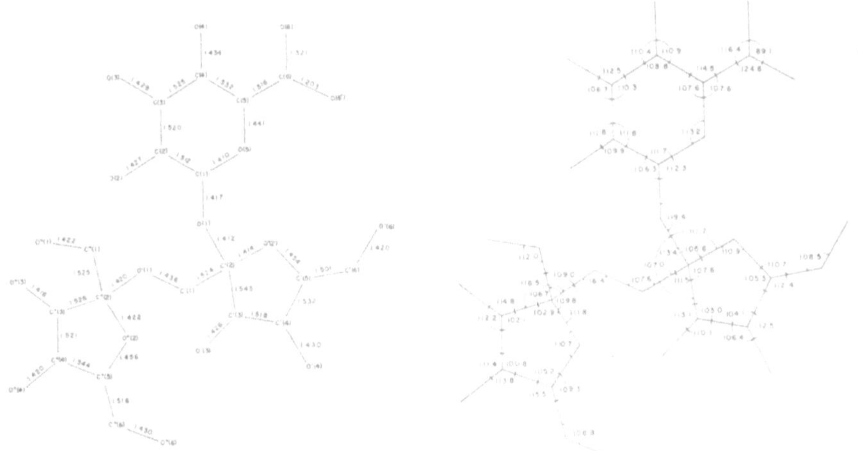

Fig. 3. Bond lengths (σ = 0.005 Å) and angles (σ ≑ 0.3°) in 1-kestose.

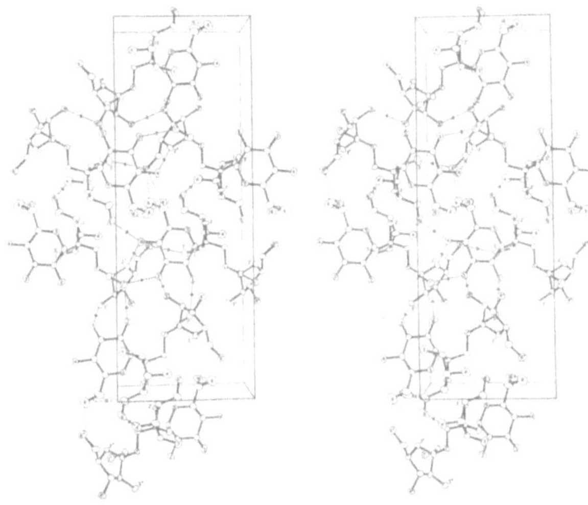

Fig. 4. Diagram showing hydrogen bonding as thin lines with arrows indicating donor direction in 1-kestose.

PLANTEOSE DIHYDRATE

$C_{18}H_{36}O_{18}$ $C_{18}H_{32}O_{16}.2H_2O$

D.C. ROHRER, 1972. Acta Cryst., B28, 425-433.

Orthorhombic, $P2_12_12_1$, a = 32.43, b = 8.152, c = 8.711 Å, D_m = 1.547, Z = 4. Cu radiation, R = 0.036 for 2197 reflexions.

The structure of planteose is shown in (I).

OH OH OH
O CH₂ O OH
OH OH HO
HO O O CH₂OH O
CH₂OH CH₂OH
(galactose) (fructose) (glucose)
planteobiose sucrose

(I)

The molecule has a circular conformation (Fig. 1) with O(6) of the glucose and O(6") of the galactose at opposite ends, both hydrogen bonded to the same hydroxyl, O(2)H, of a neighbouring molecule. Bond lengths and angles are in Fig. 2. Most of the hydroxyl groups are involved in intermolecular hydrogen-bonding (Fig. 3). Planteose and raffinose (1) are structural isomers and a detailed comparison is given of the conformations of the saccharide units in the two molecules and in sucrose (2) and kestose (3).

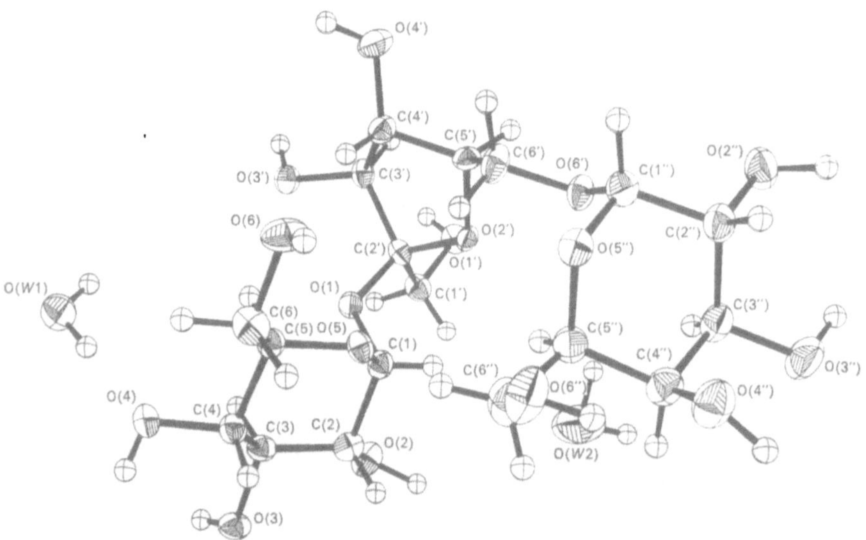

Fig. 1. Molecular conformation and numbering scheme in planteose dihydrate.

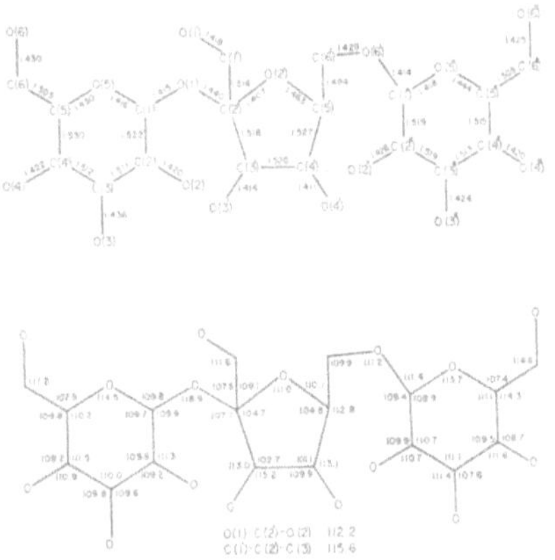

Fig. 2. Bond lengths and angles in planteose dihydrate (σ = 0.003 (C-O) and 0.004 Å (C-C)).

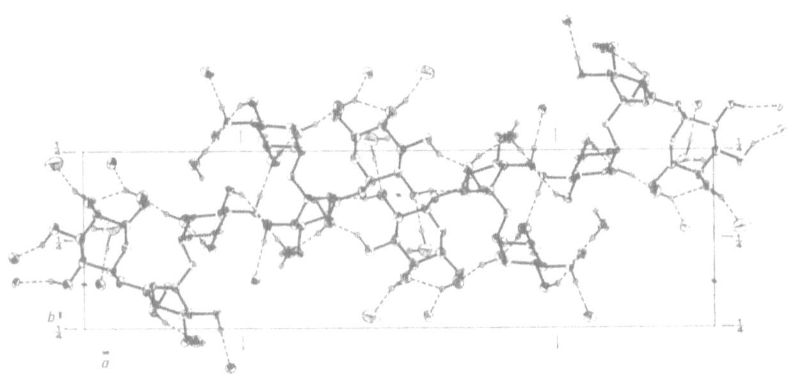

Fig. 3. Molecular packing and hydrogen bonding in planteose dihydrate.

1. Structure Reports, 35B, 270.
2. G.M. BROWN and H.A. LEVY, 1963. Science, 141, 921.
3. Preceding report.

METHYL 2,3-ANHYDRO-4,6-BENZYLIDENE-α-D-MANNOPYRANOSIDE

$C_{14}H_{16}O_5$

A.-M. PILOTTI and B. STENSLAND, 1972. Acta Cryst., B28, 2821-2825.

Monoclinic, $P2_1$, a = 8.565, b = 4.531, c = 16.711 Å, β = 90.42°, Z = 2. Cu radiation, R = 0.039 for 1283 reflexions.

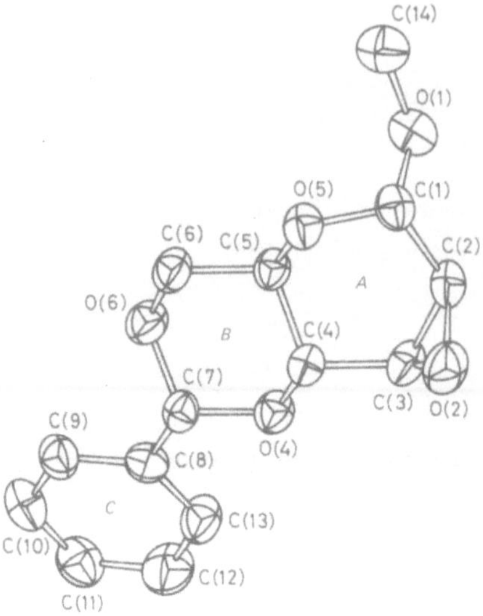

Fig. 1. A perspective view of the mannopyranoside derivative $C_{14}H_{16}O_5$.

In the molecule (Fig. 1) the pyranoside ring A is flattened, relative to the ordinary chair form, in the presence of the epoxide ring. Ring B has a slightly distorted chair conformation with C(7) -0.705 and C(5) +0.666 Å from the best plane through C(6), O(6), C(4), O(4). The C(1)-O(1) bond at the pyranoside ring is 1.395(4) Å and the average of the other C-O bonds is 1.427 Å. The C(2)-C(3) bond in the epoxide ring is 1.456(3) Å. In the crystal structure there are two intermolecular contacts (O(2)...C(3') = 3.289 and C(2)...O(2") = 3.321 Å) a little shorter than the normal van der Waals separation of about 3.4 Å.

5-BROSYL-3-DEOXY-3-C-(R)-(ETHOXYCARBONYLFORMAMIDO)-METHYL-2-O-ISOPROPYLIDENE-α-D-RIBOFURANOSE

$C_{19}H_{24}BrNO_9S$

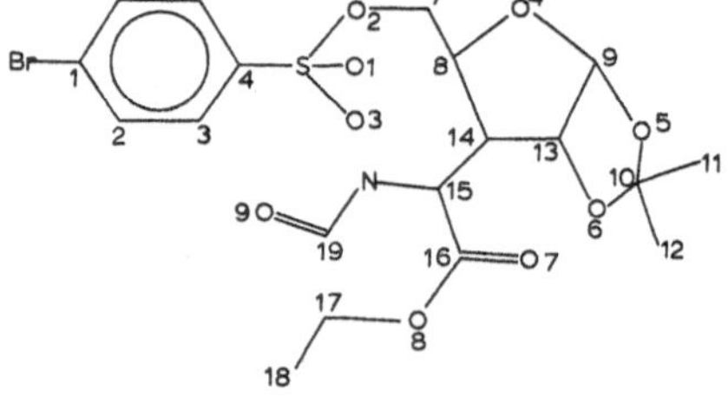

J. COETZER, A. JORDAAN, G.J. LOURENS and M.J. NOLTE, 1972. Acta Cryst., B28, 3537-3542.

Monoclinic, $P2_1$, a = 10.381, b = 9.142, c = 11.850 Å, β = 99.92°, D_m = 1.55, Z = 2. Mo radiation, R = 0.055 for 1474 reflexions.

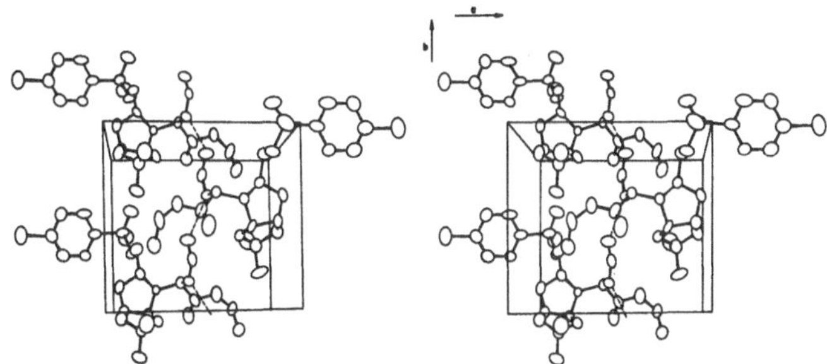

Fig. 1. Diagram of the molecular packing of the ribofuranose derivative $C_{19}H_{24}BrNO_9S$.

The molecules (Fig. 1) are linked by NH...O hydrogen bonds (3.05(2) Å). Bond lengths and angles are close to accepted values; the molecular conformation, in particular the puckering of the two fused five-membered rings, can be seen in Fig. 1. The analysis establishes absolute stereochemistry R for C(3).

1,2:5,6-DI-O-ISOPROPYLIDENE-3,4-DI-O-TOSYL-L-chiro-INOSITOL

$C_{26}H_{32}O_{10}S_2$

J.F. McCONNELL, S.J. ANGYAL and J.D. STEVENS, 1972. J. Chem. Soc.,Perkin II, 2039-2044.

Tetragonal, $P4_12_12$, a = 9.304, c = 32.008 Å, D_m = 1.40, Z = 4. Cu radiation, R = 0.047 for 1317 reflexions.

The molecule has twofold symmetry (space group requirement). The cyclohexane ring is in a skew conformation as indicated above, with the four substituents axial. The dioxolan rings take up twist conformations: the acetal carbon atom is above, and O(1) below, the plane formed by the other three atoms of the ring. The torsion angle at the bridgehead, -12°, is surprisingly small. The cyclohexane ring is under some strain; all the angles within it are greater (115 - 118°) than expected.

METHYL DIAMMONIUM PHOSPHATE DIHYDRATE

$CH_{15}N_2O_6P$ $[NH_4^+]_2[CH_3PO_4^{2-}].2H_2O$

F. GARBASSI, L. GIARDA and G. FAGHERAZZI, 1972. Acta Cryst., B28, 1665-1670.

Orthorhombic, $P2_12_12_1$, a = 12.39, b = 11.37, c = 6.10 Å, D_m = 1.47, Z = 4. Cu radiation, R = 0.090 for 787 reflexions.

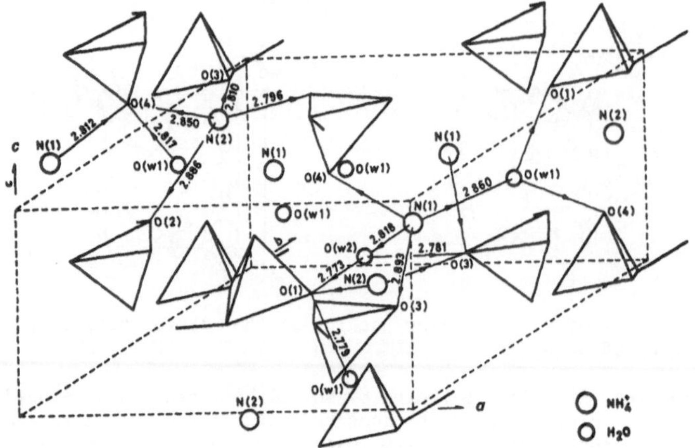

Fig. 1. A general view of the $[NH_4^+]_2[CH_3PO_4^{2-}].2H_2O$ structure. Tetrahedra represent the PO_4 groups with P at the centre; circles represent N and O atoms belonging to NH_4^+ ions or H_2O molecules. Arrows indicate hydrogen bonds.

In the phosphate tetrahedron there are three short P-O bonds with an average 1.507(7) Å; the longest bond is P-O(2) = 1.597(6) Å. The O-P-O angles fall into two classes, ∿105° and ∿113°, the long P-O(2) bond being involved in all the smaller O-P-O angles. The PO-C distance is 1.464(12) Å. A complex system of hydrogen bonds (Fig. 1) holds the ions and water molecules together.

DISODIUM DL-GLYCEROL 3-PHOSPHATE HEXAHYDRATE

$C_3H_{18}Na_2O_{12}P$ $(Na^+)_2(CH_2OH.CHOH.CHOPO_3)^{2-}.6H_2O$

R.H. FENN and G.E. MARSHALL, 1972. Biochem. J., 130, 1-10.

Monoclinic, C2/c, a = 24.10, b = 8.35, c = 13.80 Å, β = 105.38°, D_m = 1.606, Z = 8. Cu radiation, R = 0.125 for 1700 reflexions.

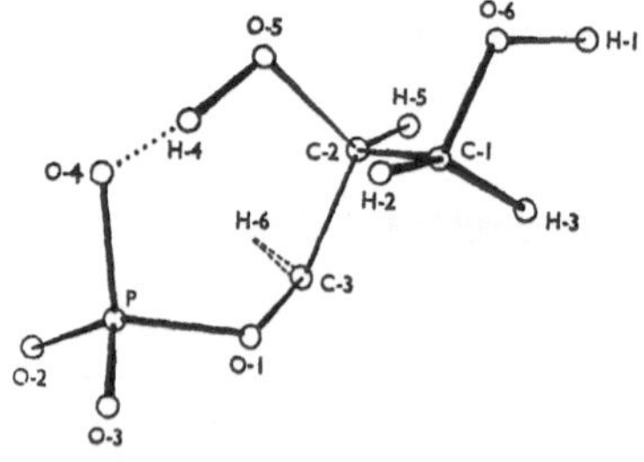

Fig. 1. Projection of D-glycerol 3-phosphate molecule.

The conformation of the molecule is shown in Fig. 1. Bond lengths are for the most part in good agreement with those found in similar molecules but the P-O(1) bond, 1.637(7) Å, is notably long for an ester bond. The other P-O distances are in the range 1.493 -1.516(7) Å. An intramolecular O(5)H...O(4) hydrogen bond results in the formation of puckered seven-membered ring holding C(2), C(3) and the phosphate group in a stable configuration. The Na^+ ions are six-coordinate. The crystal structure is held together by an intricate hydrogen bond network involving water and glycerol phosphate molecules, and by the Na^+ coordination.

SILVER DIETHYL PHOSPHATE

$C_4H_{10}AgO_4P$ $AgPO_2(OC_2H_5)_2$

J.P. HAZEL and R.L. COLLIN, 1972. Acta Cryst., B28, 2951-2957.

Orthorhombic, Pccn, a = 20.19, b = 14.35, c = 5.86 Å, D_m = 1.997, Z = 8. Mo radiation, R = 0.043 for 853 reflexions.

The silver ion is coordinated to four oxygen atoms in a distorted tetrahedral arrangement with mean Ag-O 2.36 Å. Thermal motion in the diethylphosphate anion is high and is qualitatively explained as a rigid body motion of the phosphate group coupled with a librational motion of the ester groups about the P-O bonds. Uncorrected P=O distances are 1.459 and 1.497(6) Å; the P-O distances are 1.548, 1.549(8) Å. The O-P-O angles are in the range 102(O-P-O) to 117° (O=P=O).

CATECHOL CYCLIC PHOSPHATE

$C_6H_5O_4P$

F.P. BOER, 1972. Acta Cryst., B28, 1201-1207.

Orthorhombic, Pbca, a = 8.549, b = 15.041, c = 11.053 Å, Z = 8. Cu radiation, R = 0.063 for 851 reflexions.

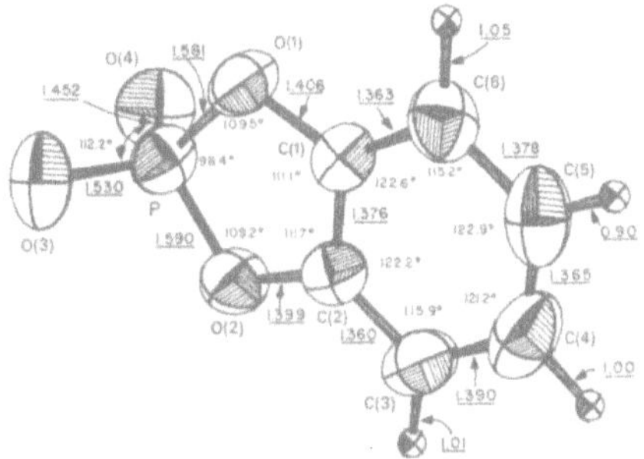

Fig. 1. The catechol cyclic phosphate molecule; σ = 0.003, 0.005, 0.006 Å for P-O, C-O, C-C.

The ring system in the molecule (Fig. 1) is essentially planar, the maximum deviation being -0.017 Å for the phosphorus atom. The O-P-O angle in the highly strained five-membered ring is 98.4(2)°. The endocyclic P-O-C and O-C-C angles average 109.3 and 111.4° respectively. The six-membered ring is evidently distorted from an ideal hexagon by the presence of the strained five-membered system which increases the C(2)-C(1)-C(6) and C(1)-C(2)-C(3) angles (to 122.6 (5) and 122.2(5)° respectively). The internal angles at C(4) and C(5) are also larger than 120°, resulting in the angles at C(3) and C(6) decreasing to 115.9(5) and 115.2(5)° respectively. The bond lengths are normal. The crystal is held together by infinite hydrogen-bonded chains linking the phosphoryl and hydroxyl oxygen atoms of adjacent molecules (O-H...O = 2.464 Å).

PROPYLGUANIDINIUM DIETHYLPHOSPHATE

$C_8H_{22}N_3O_4P$ $CH_3(CH_2)_2NHC(NH_2)_2^+.(C_2H_5O)_2P(O)_2^-$

S. FURBERG and J. SOLBAKK, 1972. Acta Chem. Scand., 26, 3699-3707.

Monoclinic, $P2_1/c$, a = 8.292, b = 19.527, c = 12.231 Å, β = 131.97°, D_m = 1.15, Z = 4. Mo radiation, R = 0.089 for 1798 reflexions.

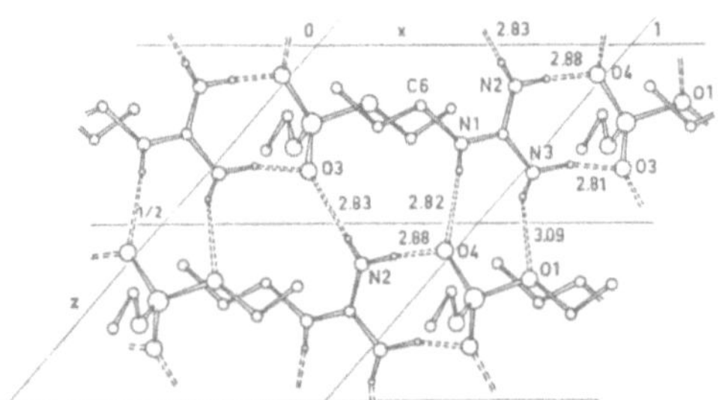

Fig. 1. The structure of propylguanidinium diethylphosphate projected along b. The N-H...O hydrogen bonds are indicated.

The crystal structure is shown in Fig. 1. The propylguanidinium cation is nearly planar (to within 0.06 Å if H atoms are ignored) and the conformation of the diethylphosphate moiety is gauche at both P-O bonds. The two amino groups of the guanidinium residues are linked by hydrogen bonds to oxo oxygen atoms in the same phosphate group (Fig. 1). Bond lengths and angles at the P atom are normal. In the cation the $C(sp^2)$-NH_2 corrected distances are 1.373 and 1.376 Å, and the $C(sp^2)$-NH corrected distance is 1.302 Å.

(-)-EPHEDRINE DIHYDROGEN PHOSPHATE

$C_{10}H_{18}NO_5P$ $[C_6H_5CH(OH)CH(CH_3)NH_2CH_3]^+H_2PO_4^-$

R.A. HEARN and C.E. BUGG, 1972. Acta Cryst., B28, 3662-3667.

Monoclinic, C2, a = 14.738, b = 5.710, c = 15.302 Å, β = 97.17°, D_m = 1.37, Z = 4. Cu radiation, R = 0.026 for 1180 reflexions.

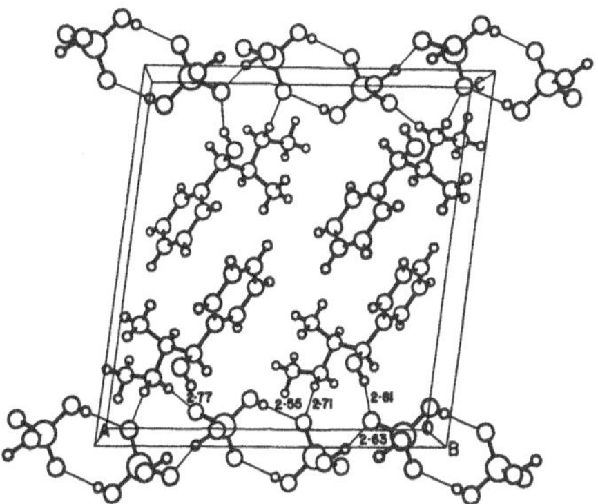

Fig. 1. Crystal structure of (-)-ephedrine dihydrogen phosphate. Thin lines show hydrogen bonds with distances in Å.

In the crystal structure (Fig. 1) the hydrophobic phenyl and methyl groups are clustered in the central region of the unit cell and the polar amino, hydroxyl, and phosphate moieties are arranged in layers parallel to the ab plane. All hydrogen atoms covalently bonded to N and O atoms participate in hydrogen bonding (Fig. 1). Phosphate ions are hydrogen bonded around twofold axes to form dimers that involve two short (2.55 Å) hydrogen bonds. Bond length and angle data for the ephedrine cation agree well with those found for the hydrochloride derivative (1); the major difference between the cation conformations is a 47° difference in the torsion angle at the C-O bond. In the $H_2PO_4^-$ ion the bond lengths are P-OH = 1.555, 1.578(5), P-O = 1.498, 1.505(5) Å. The ephedrine cations are linked to the non-protonated phosphate oxygen atoms by hydrogen bonds from the β-ethanolamine moiety.

1. R. BERGIN, 1971. Acta Cryst., B27, 381; Structure Reports, 37B, 318.

TETRAMETHYLFORMAMIDINIUM PHOSPHONIC ANHYDRIDE

$C_{10}H_{24}N_4O_5P_2$

$(Me_2N)_2C^{+}$–P(=O)(O^-)–O–P(=O)(O^-)–$C^{+}(NMe_2)_2$

F. SANZ and J.J. DALY, 1972. J. Chem. Soc., Dalton, 2267-2269.

Monoclinic, C2/c, a = 11.170, b = 10.613, c = 14.077 Å, β = 103.4°, Z = 4. Mo radiation, R = 0.065 for 2092 reflexions.

In the crystal structure the molecule (Fig. 1) lies on a 2-fold symmetry axis which passes through the central oxygen atom. Important bond lengths are: mean P-O(terminal) = 1.469(2), P-O(bridge) = 1.619(2), P-C = 1.880(2), mean N-C(sp^2) = 1.331(3), mean N-C(sp^3) = 1.463(4) Å; P-O-P is 125.9(1)°. There are no unusual intermolecular contacts.

Fig. 1. The molecule of tetramethylformamidinium phosphonic anhydride viewed down b.

PUTRESCINIUM DI-(DIETHYL PHOSPHATE)

$C_{12}H_{34}N_2O_8P_2$ $^+NH_3(CH_2)_4NH_3^+.2[(C_2H_5O)_2P(O)_2^-]$

S. FURBERG and J. SOLBAKK, 1972. Acta Chem. Scand., 26, 2855-2862.

Monoclinic, C2/c, a = 27.821, b = 9.067, c = 8.451 Å, β = 91.66°, D_m = 1.21, Z = 4. Mo radiation, R = 0.064 for 1741 reflexions.

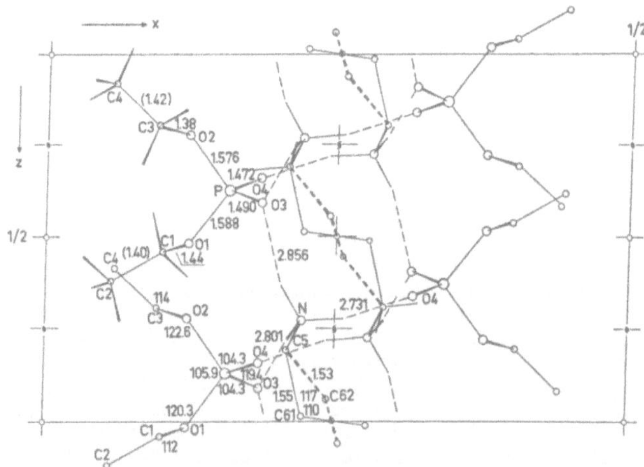

Fig. 1. The structure of putrescinium di-(diethyl phosphate) viewed along b. One of the positions of the putrescinium ion is shown by broken lines.

The crystal structure is shown in Fig. 1. The asymmetric unit contains one diethylphosphate ion and half a putrescinium ion. The conformation of the diethylphosphate is gauche at both P-O bonds. In the cation, which is disordered, the conformation is anti at the central C-C bond and gauche at the others. Each NH_3^+ group is linked by three N-H...O hydrogen bonds (2.731 - 2.856(4) Å) to adjacent diethylphosphate groups.

2'-CHLORO-2'-DEOXYURIDINE

$C_9H_{11}ClN_2O_5$

D. SUCK, W. SAENGER and J. HOBBS, 1972. Biochem. Biophys. Acta, 259, 157-163.

Orthorhombic, $P2_12_12_1$, a = 12.569, b = 4.830, c = 17.409 Å, D_m = 1.55, Z = 4. Mo radiation, R = 0.036 for 1379 reflexions.

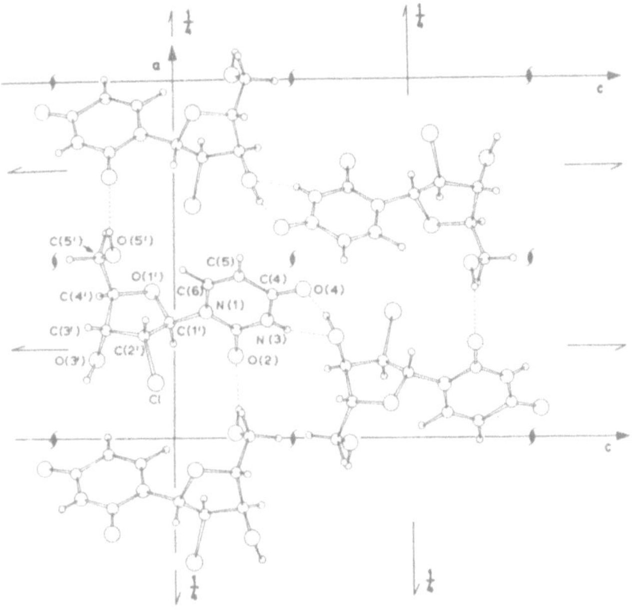

Fig. 1. Projection of the crystal structure of 2'-chloro-2'-deoxyuridine along b. Hydrogen bonds are indicated by dotted lines.

The crystal structure is shown in projection in Fig. 1. All potential proton acceptor and donor atoms are involved in hydrogen bonds connecting atoms O(5') and O(2) (2.782 Å), atoms N(3) and O(3') (2.951 Å) and atoms O(3') and O(4) (2.756 Å). The conformation of the molecule and its dimensions are similar to those found in other pyrimidine nucleoside structures: the orientation of the base with respect to the sugar is anti, and the conformation about the C(4')-C(5') bond is gauche, gauche. The 2'-chloro-2'-deoxyribose is puckered with C(2')-

endo and despite the chloro substitution shows no significant differences in bond lengths, bond angles or torsional angles in comparison with the average values for unmodified ribose residues in the same C(2')-endo conformation.

2'-DEOXYURIDINE

$C_9H_{12}N_2O_5$

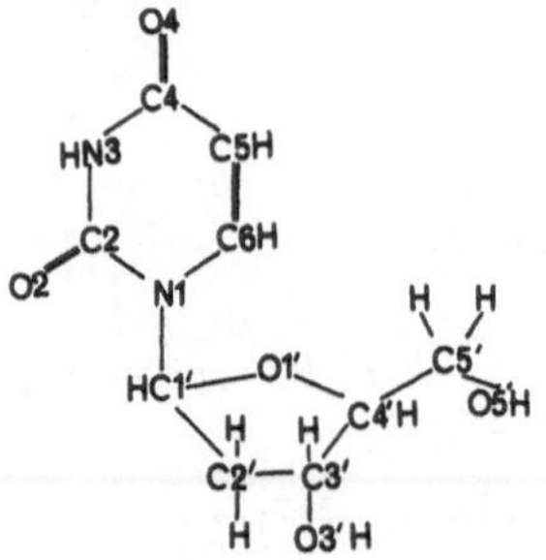

A. RAHMAN and H.R. WILSON, 1972. Acta Cryst., B28, 2260-2270.

Monoclinic, $P2_1$, a = 7.91, b = 6.710, C = 18.77 Å, β = 96.6°, Z = 4. Mo radiation, R = 0.084 for 1611 reflexions.

There are two molecules in the asymmetric unit; the dihedral angle in molecule (I) between the base and the sugar is 53° while in molecule (II) it is 47° with glycosidic torsion angle ϕ_{CN} -26° and -25°, respectively. Both pyrimidine bases are planar while the sugar rings are puckered so that C(2') is *endo* and C(3') is *exo* relative to the plane through atoms C(1'), O(1'), and C(4'). The orientation of the C(5')-O(5') bond is *trans* to the C(4')-O(1') bond and *gauche* to C(4')-C(3') in both molecules. Bond distances and angles are given in Fig. 1. The bases form an infinite sheet parallel to (001), with hydrogen bonds in a zigzag arrangement within the sheet (N-H...O = 2.77, 2.8 and O-H...O = 2.68 - 2.75 Å).

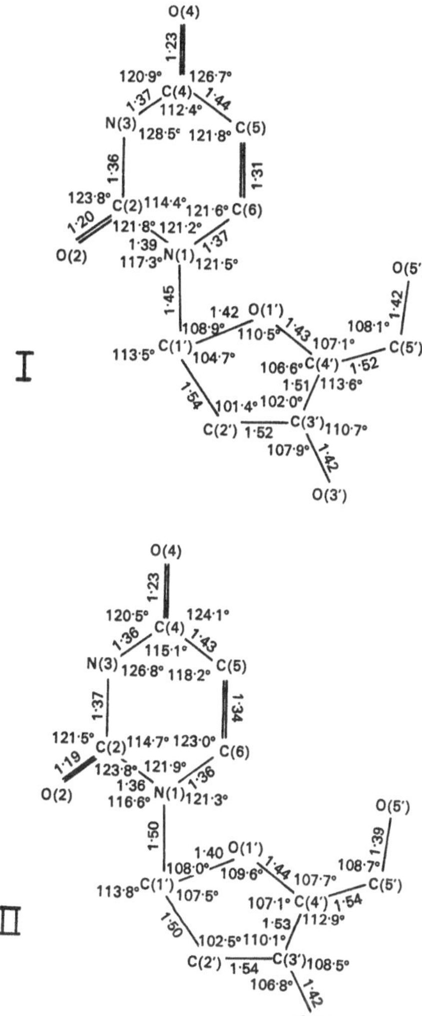

Fig. 1. Bond lengths and angles in molecules (I) and (II) of 2'-deoxyuridine.

DIHYDROURIDINE HEMIHYDRATE

$C_9H_{14}N_2O_6.1/2H_2O$

D. SUCK, W. SAENGER and K. ZECHMEISTER, 1972. Acta Cryst., B28, 596-605.

Orthorhombic, $P2_12_12_1$, a = 11.779, b = 8.150, c = 23.068 Å, D_m = 1.51, Z = 8. Cu radiation, R = 0.051 for 2143 reflexions.

The material crystallises with two independent molecules of nucleoside plus one water molecule in the asymmetric unit. The chemical structure and numbering scheme are in Fig. 1. The oxygen atom O(5') of one molecule is threefold disordered, but generally bond lengths and angles are similar in both molecules; mean values are shown in Fig. 2. The two nucleosides within the asymmetric unit are conformational isomers with atom C(6) on opposite sides of the plane through N(1), C(2), N(3), C(4) (Fig. 3). Both molecules exhibit the usual anti conformation; the conformation of the ribose residues is C(2') endo. The packing of the molecules in the crystals is determined by an extensive network of OH...O and NH...O hydrogen bonds and shows no base stacking.

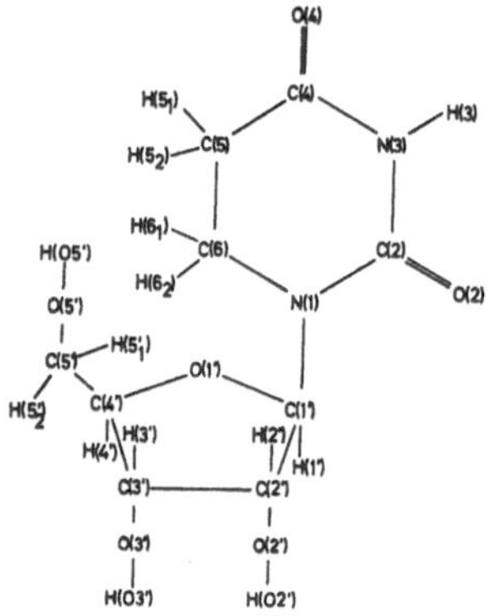

Fig. 1. Chemical structure and numbering scheme for dihydrouridine.

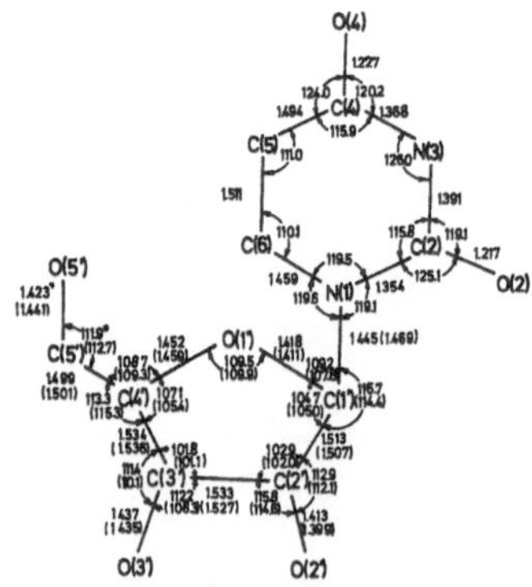

Fig. 2. Mean bond distances (σ = 0.005 Å) and angles for dihydrouridine (figures in parentheses are average values from several C(2')-endo ribose units (1).)

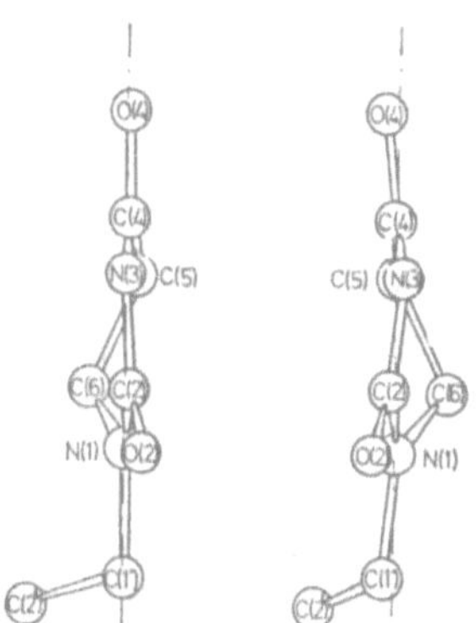

Fig. 3. Schematic view along the N(1), C(2), N(3), C(4) plane through the two dihydrouracil residues in dihydrouridine hemihydrate.

1. W. SAENGER and F. ECKSTEIN, 1970. J. Amer. Chem. Soc., 92, 4712; Structure Reports, 35B, 292.

1-β-D-ARABINOFURANOSYL-4-THIOURACIL MONOHYDRATE

$C_9H_{14}N_2O_6S$ $C_9H_{12}N_2O_5S.H_2O$

W. SAENGER, 1972. J. Amer. Chem. Soc., 94, 621-626.

Monoclinic, $P2_1$, a = 4.787, b = 15.060, c = 8.661 Å, β = 73.0°, D_m = 1.54, Z = 2. Mo radiation, R = 0.046 for 1444 reflexions.

The conformation of the arabino nucleoside is similar to the conformation observed for ribonucleosides, the preferred *anti* conformation. The pucker of the arabinose moiety is C(3') *endo* and the *conformation* about the C(4')-C(5') bond is such that O(5') is *above* the sugar. Molecular dimensions are compared in Fig. 1 with data for the average C(3') *endo* ribose unit (1). Some short intramolecular contacts are depicted in Fig. 2. Hydrogen bonds O-H...S = 3.311 and 3.343, O-H...O = 2.669, and N-H...O = 2.792 Å link the molecules in the crystal structure.

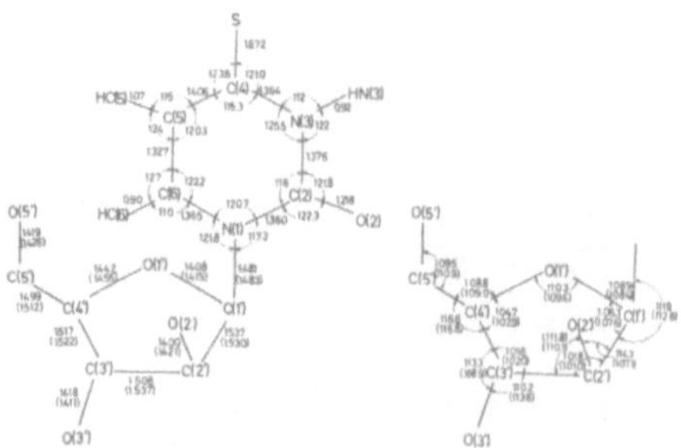

Fig. 1. Intramolecular distances and angles in $C_9H_{12}N_2O_5S.H_2O$. The numbers in parentheses refer to the data obtained for the average C(3') endo ribose unit. The standard deviations σ for data not involving hydrogen atoms are 0.003 Å for C(4)-S and 0.005 Å for the other bond lengths while bond angles are uncertain to 0.3°. Bond angles involving hydrogen atoms are 107 ± 6°, bond distances are 0.98 ± 0.07 Å.

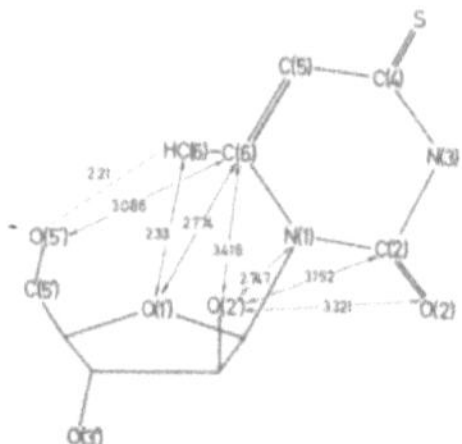

Fig. 2. Intramolecular short nonbonded contacts in $C_9H_{12}N_2O_5S.H_2O$; distances O(2')...N(1), O(1)...C(6), and O(5')...C(6) are shorter than the corresponding van der Waals contacts (O...N, 2.9 Å; O...aromatic system, 3.2 Å).

1. W. SAENGER and F. ECKSTEIN, 1970. J. Amer. Chem. Soc., 92, 4712; Structure Reports, 35B, 292.

DISODIUM URIDINE 3'-PHOSPHATE TETRAHYDRATE

$C_9H_{19}N_2Na_2O_{13}P$

M.A. VISWAMITRA, B.S. REDDY, M.N.G. JAMES and G.J.B. WILLIAMS, 1972. Acta Cryst., B28, 1108-1116.

Orthorhombic, $P2_12_12$, a = 6.746, b = 39.653, c = 6.483 Å, D_m = 1.689, Z = 4. Mo radiation, R = 0.043 for 2426 reflexions.

Fig. 1. Bond lengths in the uridine-3'-phosphate anion (σ = 0.005 Å).

A view of the uridine-3'-phosphate anion is shown in Fig. 1. The torsional angle about the glycosidic bond N(1)-C(1') is 45.1°, corresponding to the *anti* conformation. The ribose conformation can be described as C(2')-*endo*, C(3')-*endo*, with C(2') and C(3') displaced by 0.683 and 0.147 Å [σ(r) = 0.006 Å] from the plane defined by C(4')-O(1')-C(1'). The conformation about the C(4')-C(5') bond is the commonly observed *gauche-gauche* with ϕ_{OO} = -77.4° and ϕ_{OC} = 42.9°. The pyrimidine ring is not strictly planar. The exocyclic atoms O(2) and O(4) are displaced in opposite directions from the mean base plane by -0.057 and 0.078 Å respectively. Each of the phosphate oxygen atoms, except O(3') in the ester linkage, participates in forming three hydrogen bonds. The Na(1) ion has an octahedral coordination of six oxygen atoms, all from water molecules. The Na(2) ion has a capped octahedral coordination of seven neighbours, two of which are phosphate oxygen atoms. The sodium polyhedra share faces as well as corners. All the available hydrogen atoms participate in hydrogen bonding. Both the carbonyl oxygen atoms of the base participate in hydrogen bonding, giving rise to an infinite chain of nucleotide molecules parallel to the *a* axis. These chains are stacked against the sodium coordination polyhedra as alternate channels in the structure.

ADENOSINE

$C_{10}H_{13}N_5O_4$

T.F. LAI and R.E. MARSH, 1972. Acta Cryst., B28, 1982-1989.

Monoclinic, $P2_1$, a = 4.825, b = 10.282, c = 11.823 Å, β = 99.30°, D_m = 1.54, Z = 2. Cu radiation, R = 0.024 for 1333 reflexions.

The absolute configuration of the molecule and the bond distances and angles are shown in Fig. 1. The furanose ring is puckered with C(3') endo. The small value of the torsion angle about the glycosidic bond, 9.9°, is characteristic of purine nucleosides. Fig. 2 shows the structure. There is a parallel stacking of adenine rings with an interplanar spacing of 3.7 Å. Each nucleoside is hydrogen bonded to six neighbouring molecules with all nitrogen and oxygen atoms except N(9) and O(1') participating (N-H...O = 2.774 - 2.947, N-H...N = 3.133, and O-H...O = 2.749 Å). A weak C-H...O hydrogen bond between C(2) and O(2'), 3.09 Å, has been postulated.

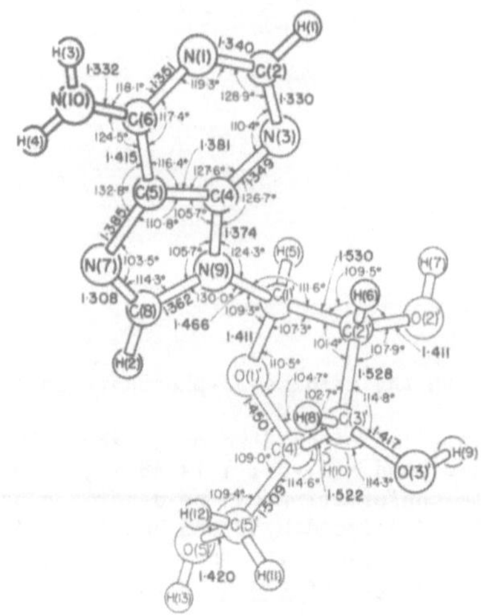

Fig. 1. Bond distances and angles for adenosine (σ = 0.003 Å and 0.2°).

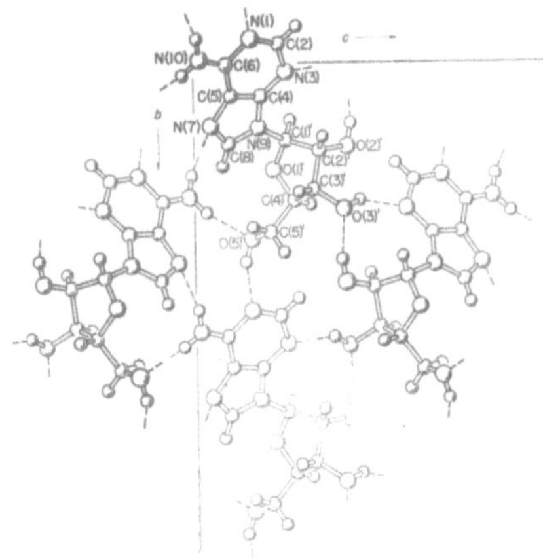

Fig. 2. The adenosine structure viewed down a.

6-METHYLURIDINE

$C_{10}H_{14}N_2O_6$

D. SUCK and W. SAENGER, 1972. J. Amer. Chem. Soc., 94, 6520-6526.

Monoclinic, $P2_1$, a = 19.895, b = 8.201, c = 6.786 Å, β = 92.70°, D_m = 1.540, Z = 4. Mo radiation, R = 0.049 for 2670 reflexions.

The molecular geometry is summarized in Fig. 1, and a stereoscopic view of the two molecules in the asymmetric unit is shown in Fig. 2. 6-Methyluridine exhibits the syn conformation which seems to be stabilized mainly by the influence of the bulky methyl group in the 6 position of the base. The C(2')-endo pucker of the ribose unit offers optimal steric conditions for this syn conformation nucleoside, which is obvious from the bond distances and angles showing no significant distortion compared to the corresponding values in nucleosides in the normal anti conformation. Due to the different orientation of the C(5')-O(5') bond in the two crystallographic independent molecules (gauche, gauche and trans, gauche, respectively) only one of the 6-methyluridine molecules shows an intramolecular hydrogen bond between O(5') of the sugar and O(2) of the base (Fig. 3). The 6-methyluracil bases are stacked along the c axis parallel to each other at interplanar spacings of 3.24 Å.

Fig. 1. Bond angles and distances in molecules A and B of 6-methyluridine (data for molecule B in parentheses). The average estimated standard deviations are 0.005 Å and 0.3°, respectively.

Fig. 2. View of the asymmetric unit of 6-methyluridine.

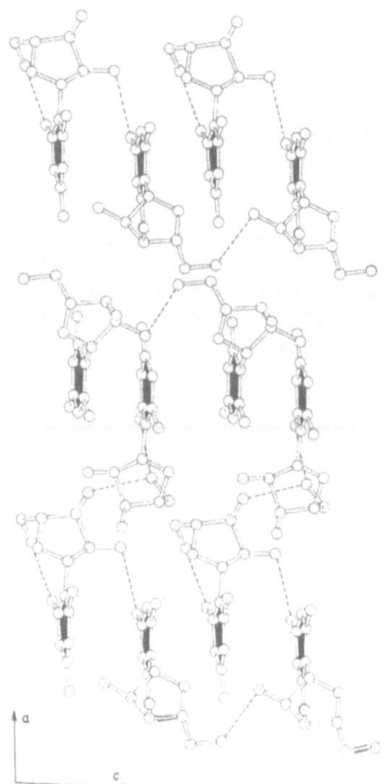

Fig. 3. Projection of the crystal structure of 6-methyluridine along the b axis (rotated by 10° about the a axis). Hydrogen bonds are indicated by broken lines.

6-THIOGUANOSINE MONOHYDRATE

$C_{10}H_{15}N_5O_5S$ $C_{10}H_{13}N_5O_4S.H_2O$

U. THEWALT and C.E. BUGG, 1972. J. Amer. Chem. Soc., 94, 8892-8898.

Orthorhombic, $C222_1$, a = 12.272, b = 6.904, c = 32.466 Å, D_m = 1.53, Z = 8. Cu radiation, R = 0.073 for 1285 reflexions.

The conformation around the glycosidic linkage (Fig. 1) is anti; the torsion angle χ_{CN} (O(1')-C(1')-N(9)-C(8)) is 65.7°. The ribose ring assumes the C(2') endo conformation; atoms C(1'), O(1'), C(4'), and C(3') are coplanar with no deviations from the plane in excess of 0.02 Å, and atom C(2') is displaced 0.56 Å from the plane of the four other ring atoms. The conformation about the C(4')-C(5') bond is gauche-trans; the torsion angle (O(5')-C(5')-C(4')-O(1')) is 64.3°, and the torsion angle (O(5')-C(5')-C(4')-C(3')) is -175.4°. The nine atoms of the purine ring are coplanar within experimental error; S, N(2), and C(1') are displaced from the least-squares purine plane by 0.01. 0.01, and 0.05 Å, respectively. Bond lengths and angles are normal. Apparently, replacing the carbonyl oxygen atom of guanosine with a sulphur substituent has little effect on the bond lengths and angles within the purine ring. The crystal packing is depicted in Fig. 2. Thioguanine moieties are hydrogen bonded around screw axes to form planar ribbons of bases running in the *a* direction and lying nearly parallel to the *ac* plane. The ribbons of bases are stacked in the *b* direction. The ribose moieties and the water molecules are arranged in channels between stacks of bases. Thus, the crystal structure consists of successive layers of bases, sugars, and water molecules, with the layers running parallel to the *ab* plane. N(1)-H...N(7) and N(2)-H...S hydrogen bonds have lengths of 2.96 and 3.27 Å, respectively. The parallel ribbons of bases are stacked with an interplanar separation of 3.4 Å. The base-stacking pattern involves a slight base overlap, with the sulphur atoms in close contact with purine rings of adjacent bases.

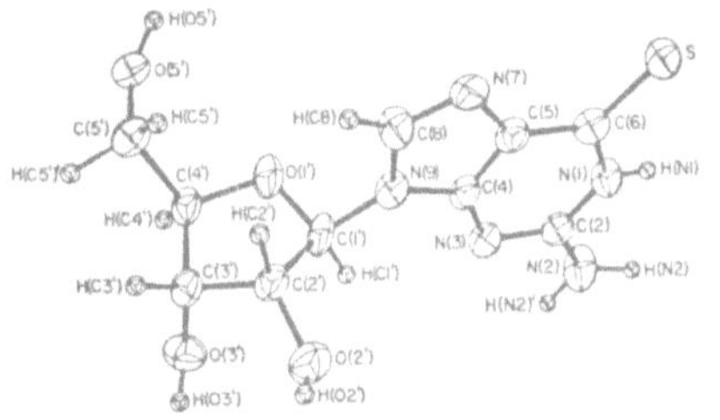

Fig. 1. The conformation of 6-thioguanosine.

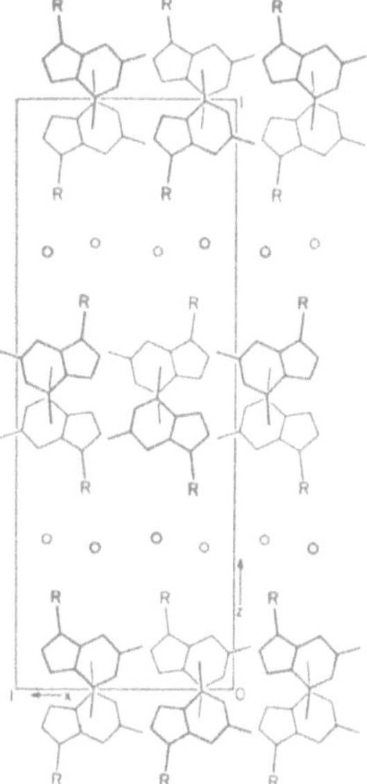

Fig. 2. The crystal packing in 6-thioguanosine monohydrate as viewed down *b*. The molecules represented by the heavy lines are at y = 0.75 and those represented by the light lines are at y = 0.25. Ribose moieties are represented by the letter R and water molecules are represented by circles.

5'-METHYLENEADENOSINE 3',5'-CYCLIC MONOPHOSPHONATE MONOHYDRATE

$C_{11}H_{16}N_5O_6P$ $C_{11}H_{14}N_5O_5P.H_2O$

M. SUNDARALINGAM and J. ABOLA, 1972. J. Amer. Chem. Soc., 94, 5070-5076.

Orthorhombic, $P2_12_12_1$, a = 10.755, b = 8.869, c = 15.198 Å, D_m = 1.578, Z = 4. R = 0.04 for 1199 reflexions.

In the crystal structure of this isosteric phosphonate analogue of cyclic AMP the base displays the *syn* conformation with respect to rotation about the glycosyl bond, χ being -125.8°. The constraints of the cyclization freeze the ribose into the conformation, $^3T_4$(C(3')-*endo*-C(4')-*exo*), which appears to be characteristic of cyclic nucleotides. For a similar reason the bond distances and bond angles in the ribose ring (Fig. 1) show marked differences from those of the nucleotides themselves. The phosphonate ring is in the chair conformation and flattened at the phosphorus end. The molecule is a zwitterion with the base site N(1) protonated by a phosphonate hydrogen. The nucleotides are packed in a head-to-tail fashion and are involved in an intricate scheme of hydrogen bonding both to themselves and the water of crystallization (Fig. 2). A comparison of the conformation of the analogue is made with the available data on cyclic AMP and cyclic UMP.

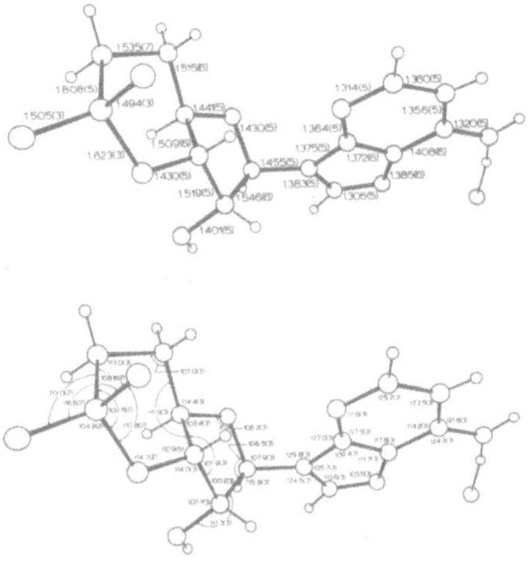

Fig. 1. Bond distances and angles in the 5'-methylene analogue of cyclic AMP.

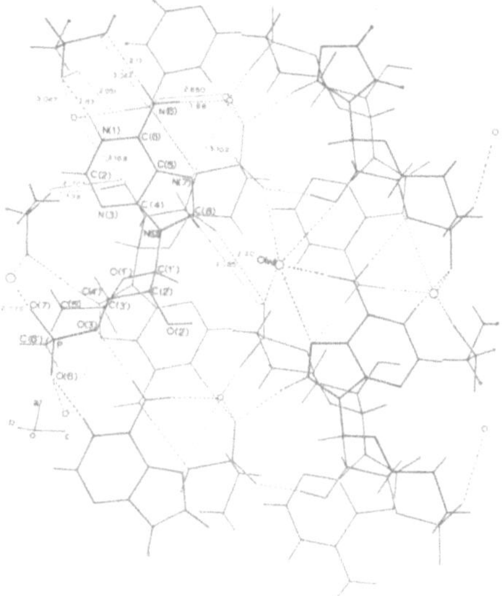

Fig. 2. Molecular packing and hydrogen bonding scheme in the 5'-methylene analogue of cyclic AMP.

N^2-DIMETHYLGUANOSINE

$C_{12}H_{17}N_5O_5$

T. BRENNAN, C. WEEKS, E. SHEFTER, S.T. RAO and M. SUNDARALINGAM, 1972. J. Amer. Chem. Soc., 94, 8548-8553.

Orthorhombic, $P2_12_12_1$, a = 6.927, b = 11.792, c = 16.362 Å, D_m = 1.558, Z = 4. Cu radiation, R = 0.046 for 1125 reflexions.

N^2-Dimethylguanosine (DMG) is in the <u>syn</u> conformation with a χ_{CN} value of -103.9°. Bond lengths and angles are in Fig. 1. The sugar moiety exhibits $^2T_3$ puckering and the conformation about the C(4')-C(5') bond is gauche-trans. A view of the crystal structure is shown in Fig. 2, and while there is no base-pairing, there is extensive intermolecular hydrogen bonding between base and sugar groups. Bases related by 2_1 screw axes stack over each other in such a way that the N-2 dimethylamine groups interact extensively with the adjacent pyrimidine rings (with the shortest interaction, 3.35 Å, between methyl C(10) and pyrimidine C(6)).

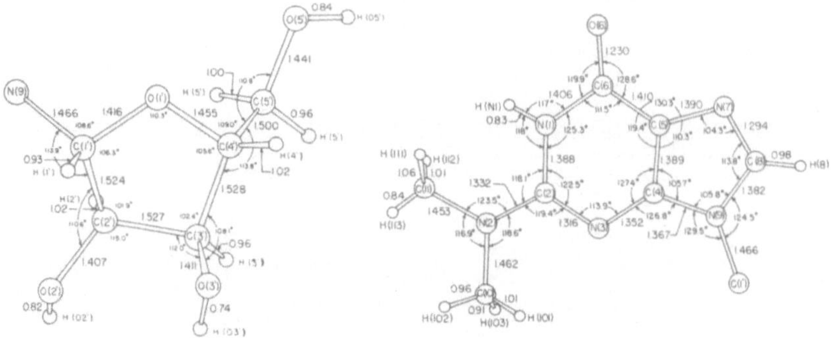

Fig. 1. Bond distances and bond angles in the base and ribose components of dimethylguanosine (for non-hydrogen atoms σ = 0.005 Å for bond lengths and 0.3° for angles).

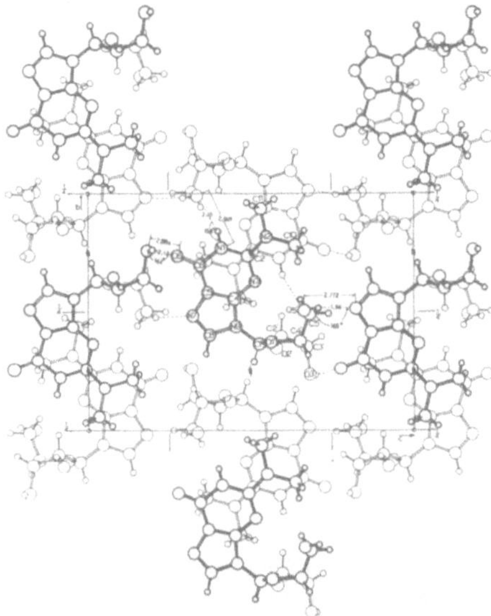

Fig. 2. The a axis projection of the crystal structure of dimethylguanosine, showing the intermolecular hydrogen bonding and the base stacking pattern.

3'-DEOXY-3'-(DIHYDROXYPHOSPHINYLMETHYL)ADENOSINE ETHANOL SOLVATE

$C_{13}H_{22}N_5O_7P$ $C_{11}H_{16}N_5O_6P.C_2H_5OH$

S.M. HECHT and M. SUNDARALINGAM, 1972. J. Amer. Chem. Soc., 94, 4314-4319.

Orthorhombic, $P2_12_12_1$, a = 5.592, b = 20.287, c = 15.198 Å, D_m = 1.503, Z = 4. Cu radiation, R = 0.044 for 1417 reflexions.

This molecule (Fig. 1) is in the anti conformation with respect to the glycosyl torsion angle, having χ = 28.1°. The conformation about the C(4')-C(5') bond is gauche-gauche, where O(1')-C(4')-C(5')-O(5') = -69.0° and C(3')-C(4')-C(5')-O(5') = 49.2°. The conformation of the sugar ring is $^3T_2$, the torsion angles about the ring bonds being O(1')-C(1') = 8.9°, C(1')-C(2') = -29.5°, C(2')-C(3') = 37.4°, C(3')-C(4') = -33.1°, C(4')-O(1') = 15.4°. These conformational parameters are similar to those found in adenosine 3'-phosphate dihydrate. The molecule is a zwitterion, N(1) of the base being protonated by an adjacent phosphonate hydrogen. The same phosphonate group is also hydrogen bonded to N(6) of the base. Similarly, the second hydrogen on N(6) and the site N(7) are involved in a hydrogen-bonded pair to a symmetry related phosphonate group. The base-phosphonate (or phosphate) hydrogen bonding is a characteristic feature of the crystal chemistry of adenine and cytidine nucleotides. The alcohol of solvation is hydrogen bonded to an adjacent ribose O(5') atom. The remaining potential hydrogen bonding sites are also involved in hydrogen bonding (Fig. 2).

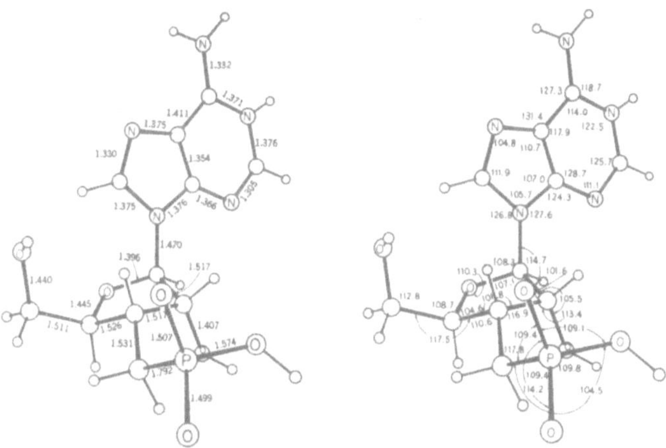

Fig. 1. Bond distances (σ = 0.008°) and angles (σ = 0.4°) in $C_{11}H_{16}N_5O_6P$.

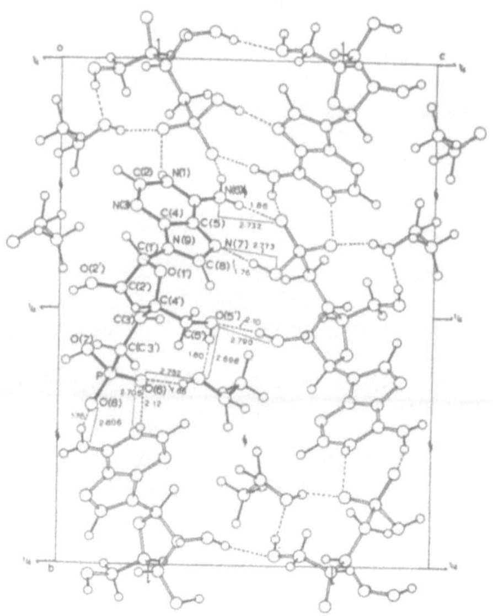

Fig. 2. Hydrogen bonding scheme seen in projection down a and the hydrogen bond distances in $C_{11}H_{16}N_5O_6P.C_2H_5OH$.

cis-1-(6-ACETOXYMETHYLTETRAHYDRO-2-PYRANYL)-5,6-DICHLOROBENZOTRIAZOLE

$C_{14}H_{15}Cl_2N_3O_3$

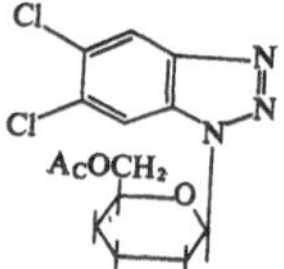

J. FAYOS and S. GARCÍA-BLANCO, 1972. Acta Cryst., B28, 2863-2868.

Triclinic, $P\bar{1}$, a = 12.178, b = 11.680, c = 5.686 Å, α = 93.2, β = 98.4, γ = 95.1°, D_m = 1.40, Z = 2. Cu radiation, R = 0.064 for 2997 reflexions.

Bond distances and angles in this analogue of a purine nucleoside are shown in Fig. 1. The benzotriazole and acetoxymethyl groups are individually planar, and the tetrahydropyranyl group has a regular chair conformation (Fig. 2). In the crystal structure the molecules pack around centres of symmetry such that a hydrogen atom on C(4) is 2.53 Å from N(3) of a centrosymmetrically related molecule; it is suggested that this is a weak C-H...N bond (C...N = 3.446 Å).

Fig. 1. Bond distances (σ = 0.004 Å) and angles in the benzotriazole derivative $C_{14}H_{15}C\ell_2N_3O_3$.

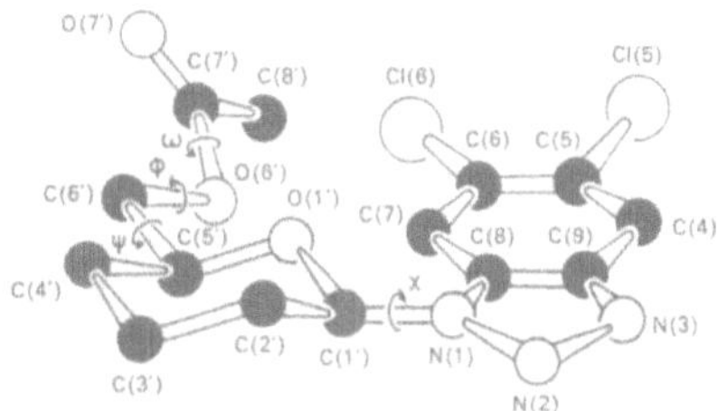

Fig. 2. A view of the $C_{14}H_{15}C\ell_2N_3O_3$ molecular conformation. The torsion angles χ, ψ, φ, and ω are -38.8 -170.3, 129.8, and -178.4° respectively.

URIDYLYL-(3'-5')-ADENOSINE HEMIHYDRATE

$C_{19}H_{24}N_7O_{12}P.1/2H_2O$

J. RUBIN, T. BRENNAN and M. SUNDARALINGAM, 1972. Biochemistry, 11, 3112-3128.

Monoclinic, $P2_1$, a = 16.961, b = 12.350, c = 11.245 Å, β = 95.88°, D_m = 1.625, Z = 4. Cu radiation, R = 0.040 for 3179 reflexions.

There are two independent molecules of UpA and a molecule of water per asymmetric unit. The UpA molecules exist as zwitterions; the adenine base is protonated at N(1) while the phosphate is negatively charged. The molecular dimensions of the nucleoside and nucleotide units are generally in good agreement with the values found in the corresponding monomer structures. Similarly, the nucleotide moieties are in the preferred conformation as found for the mononucleotide structures: anti glycosyl torsion angle and gauche, gauche conformation about the C(4')-C(5') bond. All four riboses exhibit the same puckering, C(3')-endo,-C(2')-exo ($^3T_2$), which is one of the two preferred conformations found for the sugars. However, the two UpA molecules (Fig. 1) are markedly different in their overall conformations, due mainly to the differences in the rotations about the P-O(5') and P-O(3') ester bonds. UpA1 exhibits the trans,(-)gauche conformation for the phosphodiester group while UpA2 exhibits the (+)gauche, (+)-gauche conformation. Both conformations are different from the (-)gauche, (-)-gauche conformation found for the phosphodiester group in the accurately analyzed double-helical polynucleotide structures. The conformation of UpA2 provides a model for the formation of hairpin loops in polynucleotides as, for example, the anticodon loop of tRNAs. The structure of UpA taken in conjunction with the available 5'-nucleotide structures indicates that tertiary folding of a polynucleotide chain is achieved primarily through conformational changes in the phosphodiester function. The Watson and Crick complementary base pairing is not observed in this structure. Rather, the crystal structure may be considered as made up of alternate sheets of self-paired uracil and adenine bases (Fig. 2), lying on (402) with an average separation of 3.4 Å between planes. Both base-base and base-ribose O(1') interactions are found in UpA. The water molecule, which lies close to the plane of the uracil bases, is trapped in a pocket where it is hydrogen bonded to five oxygen atoms of adjacent UpA molecules. The structure demonstrates both intramolecular and intermolecular C-H...O hydrogen bonding. An independent study of the same crystal structure has also been published (1).

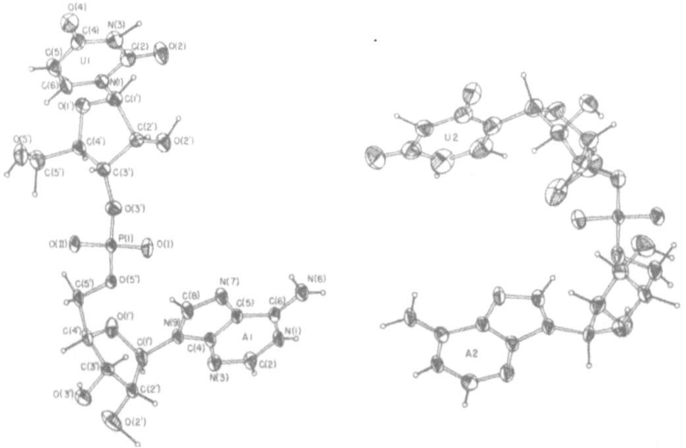

Fig. 1. The conformation of molecules UpA1 and UpA2.

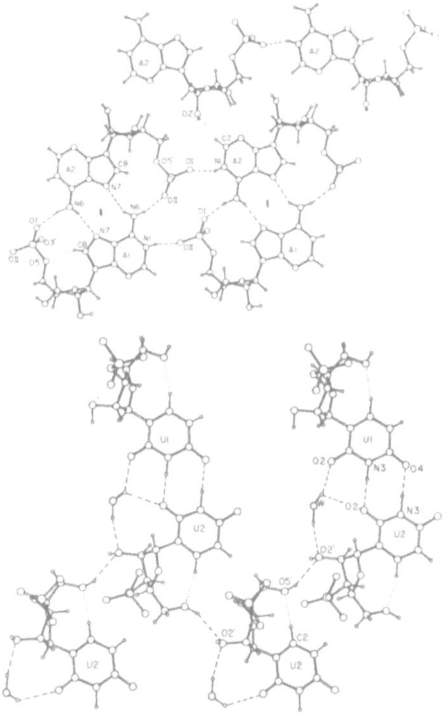

Fig. 2. The intralayer hydrogen bonding in (a, upper) adenine sheets and (b, lower) uracil sheets. The intra- and intermolecular C-H...O hydrogen bonds are shown by dotted lines. Other hydrogen bonds are shown by broken lines.

1. N.C. SEEMAN, J.L. SUSSMAN, H.M. BERMAN and S.H. KIM, 1971. Nature, New Biol., 223, 90.

α-GLYCINE (AMINOACETIC ACID)

(NEUTRON STUDY)

$C_2H_5NO_2$ $^{+}H_3N-CH_2-COO^{-}$

P.-G. JÖNSSON and Å. KVICK, 1972. Acta Cryst., B28, 1827-1833.

Monoclinic, $P2_1/n$, a = 5.1054, b = 11.9688, c = 5.4645 Å, β = 111.697°, D_m = 1.606, Z = 4. Neutron diffraction, R = 0.030 for 757 reflexions.

The hydrogen bonds postulated from X-ray work (1) are confirmed by this neutron study (Fig. 1) except that the difference in strength between the two H...O contacts of the bifurcated N-H...O hydrogen bond is larger than previously assumed. One bond (N-H(3)...O(2)) with an H...O distance of 2.121 Å is clearly a hydrogen bond and the other (N-H(3)...O(1)), with H...O 2.365 Å, is probably better classified as a short van der Waals contact. Intramolecular bond distances are C-O = 1.250 and 1.251(1), C-C = 1.526(1), C-N = 1.476(1), C-H = 1.089 and 1.090(2), N-H = 1.025, 1.037, and 1.054 Å.

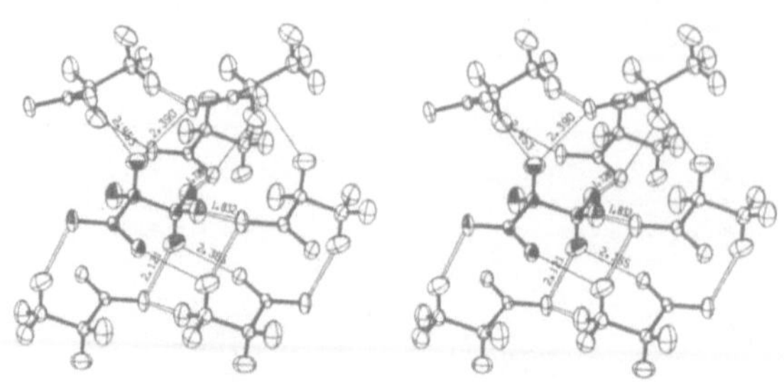

Fig. 1. Hydrogen bonding around a glycine molecule. All contacts with H...O distances less than 2.5 Å are included.

1. Structure Reports, 22, 644.

L-ALANINE

(NEUTRON STUDY)

$C_3H_7NO_2$ $CH_3CH(\overset{+}{N}H_3)COO^-$

M.S. LEHMANN, T.F. KOETZLE and W.C. HAMILTON, 1972. J. Amer. Chem. Soc., 94, 2657-2660.

Orthorhombic, $P2_12_12_1$, a = 6.025, b = 12.324, c = 5.783 Å, Z = 4. Neutron radiation, R = 0.02 for 786 reflexions.

The atomic positions found by X-ray methods (1, 2) are confirmed and the hydrogen atoms located precisely. The molecule is in the zwitterion form and the three N-H...O hydrogen bonds are nearly equal in lengths and angles (2.813 - 2.853(1) Å and 160.9 - 168.1(2)°). The methyl C-H distance is 1.096(3) Å, whereas the N-H bond length is 1.039(6). Average angles in the methyl group are C-C-H = 110.2(1)° and H-C-H = 108.7(2)°. The barriers to rotation for the methyl group and the ammonium group are estimated to be 5.6 and 20 kcal/mol, respectively. The molecules are linked together by hydrogen bonds to form a three-dimensional network (Fig. 1). The methyl groups are situated in channels formed by this network.

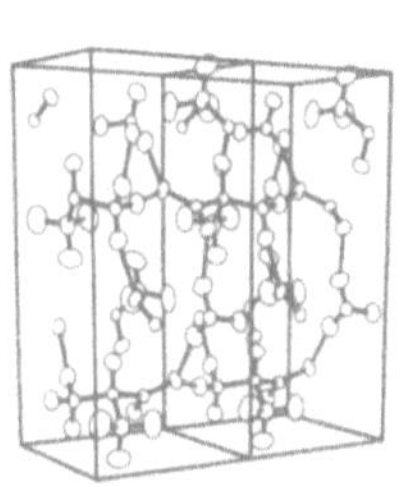
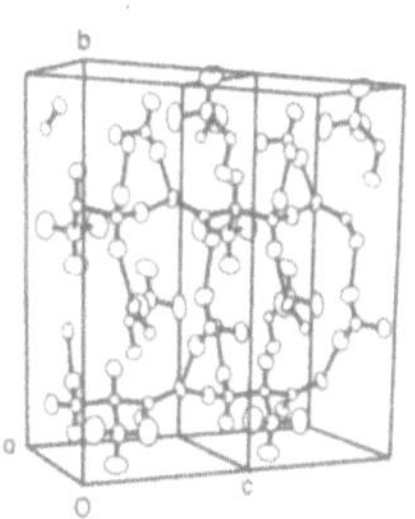

Fig. 1. View of two unit cells of L-alanine; molecular bonds solid, hydrogen bonds open.

1. H.J. SIMPSON and R.E. MARSH, 1966. Acta Cryst., 20, 550.
2. J.D. DUNITZ and R.R. RYAN, 1966. Acta Cryst., 20, 617.

L-(-)-SERINE

$C_3H_7NO_3$ $HO.CH_2CH(\overset{+}{N}H_3).COO^-$

E. BENEDETTI, C. PEDONE and A. SIRIGU, 1972. Cryst. Struct. Comm., 1, 35-37.

Orthorhombic, $P2_12_12_1$, a = 8.571, b = 9.325, c = 5.615 A, Z = 4. Cu radiation, R = 0.044 for 502 reflexions.

The amino acid is present in the solid state in the zwitterionic form. A rigid three-dimensional network of intermolecular hydrogen bonds is found in the crystal. The nitrogen atom, the carboxylic oxygens O(1) and O(2), and the hydroxylic oxygen O(3) experience three, two, one, and two hydrogen bonds, respectively. The conformation of the molecule is similar to that found for the DL mixture (1), the internal rotation angles 3,5,6 and 4,5,6 being G_+ (+62°) and G_- (-61°), respectively (trans-conformation = 180°). The absolute configuration is assigned with respect to that of the L-(-)-isoleucine hydrobromide (2).

1. Structure Reports, 17, 655.
2. Ibid., 18, 666.

DL-α-AMINO-n-BUTYRIC-ACID

(C-FORM AT -90°C)

$C_4H_9NO_2$ $CH_3CH_2CH(NH_2)CO_2H$

T. AKIMOTO and Y. IITAKA, 1972. Acta Cryst., B28, 3106-3107.

Monoclinic, I2/a, a = 9.8, b = 4.7, c = 23.1 Å, β = 94.0°, Z = 8. R = 0.247 for 114 h0ℓ and 52 0kℓ reflexions (film data).

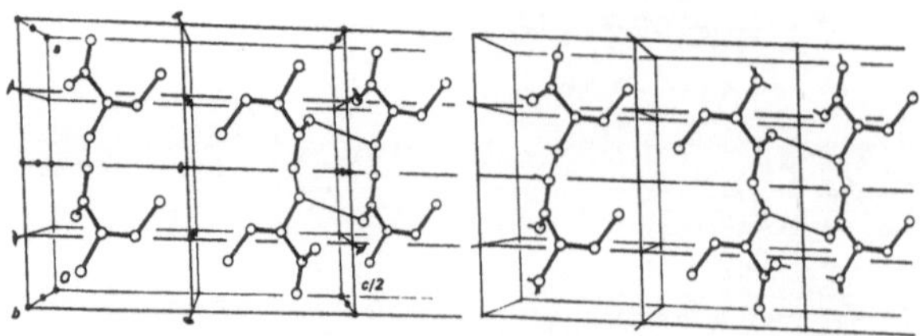

Fig. 1. Drawing of the structure of the C-form of DL-α-amino-n-butyric acid.

The room temperature (A-form) modification has been reported previously (1). In this C-form the unit cell has doubled c-length as well as changing space group from $P2_1/a$ in the A-form. The change in structure has involved the shift of every second double layer of molecules by (a + b)/2. This gives enough space for the C-form molecules to take a trans-conformation about N-Cα-Cβ-Cγ (Fig. 1). In the A-form the Cγ atom is disordered over three positions.

1. Structure Reports, 33B, 180.

L-ASPARAGINE MONOHYDRATE

(NEUTRON STUDY)

$C_4H_{10}N_2O_4$ $NH_2COCH_2CH\overset{+}{N}H_3COO^-.H_2O$

I. J.J. VERBIST, M.S. LEHMANN, T.F. KOETZLE and W.C. HAMILTON, 1972. Acta Cryst., B28, 3006-3013.

II. M. RAMANADHAM, S.K. SIKKA and R. CHIDAMBARAM, 1972. Acta Cryst., B28, 3000-3005.

Orthorhombic, $P2_12_12_1$, a = 5.593, b = 9.827, c = 11.808 Å, D_m = 1.543, Z = 4. Neutron radiation, R = 0.026 for 1787 reflexions (from I), R = 0.096 for 749 reflexions (from II).

The two independent neutron studies are in good agreement, confirm the correctness of the structure as determined by X-ray methods (1), and locate all hydrogen atoms. The molecule is in the zwitterion structure shown above. There is good agreement between non-hydrogen atom bond lengths in the X-ray and neutron structures. In the crystal structure there are seven hydrogen bonds, involving the seven hydrogen atoms bonded to N or O per asymmetric unit of the crystal. N-H...O distances are in the range 2.806(1) - 3.026(1) Å, and the two O-H...O distances are 2.812(2) and 2.843(2) Å.

1. G. KARTHA and A. de VRIES, 1961. Nature, 192, 862.

GLYCYLGLYCINE MONOHYDROCHLORIDE MONOHYDRATE (NEUTRON STUDY)

$C_4H_{11}C\ell N_2O_4$ $[NH_3CH_2CONHCH_2COOH]^+.C\ell^-.H_2O$

T.F. KOETZLE, W.C. HAMILTON and R. PARTHASARATHY, 1972. Acta Cryst., B28, 2083-2090.

Monoclinic, $P2_1/c$, a = 8.813, b = 9.755, c = 9.788 Å, β = 104.10°, D_m = 1.513, Z = 4. Neutron radiation, R = 0.068 for 2243 reflexions.

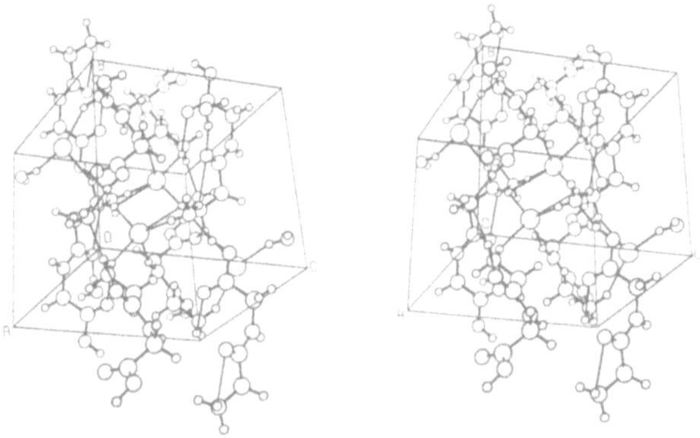

Fig. 1. View of the packing in a unit cell of glycylglycine monohydrochloride monohydrate.

All features involving heavy atoms are nearly identical with those found from X-ray studies (1). This neutron study indicates the presence of a network (Fig. 1) of seven distinct hydrogen bonds. The molecular conformation and numbering scheme are shown in Fig. 2. There is one hydrogen bond for each hydrogen atom covalently bonded to nitrogen or oxygen. The $C\ell^-$ ion (Fig. 3) is involved in four strong hydrogen bonds; the surroundings of the terminal nitrogen atoms are shown in Fig. 4. The water oxygen O(4) has a tetrahedral environment, accepting two and donating two hydrogen bond systems (O(4)-H...O(1') = 2.728, O(4)-H...Cℓ = 3.086, O(3)-H...O(4) = 2.644 and N(1)-H...O(4) = 3.025(4) Å). Other N-H...O and N-H...Cℓ contacts shown in Figs. 3 and 4 should probably not be considered as hydrogen bonds.

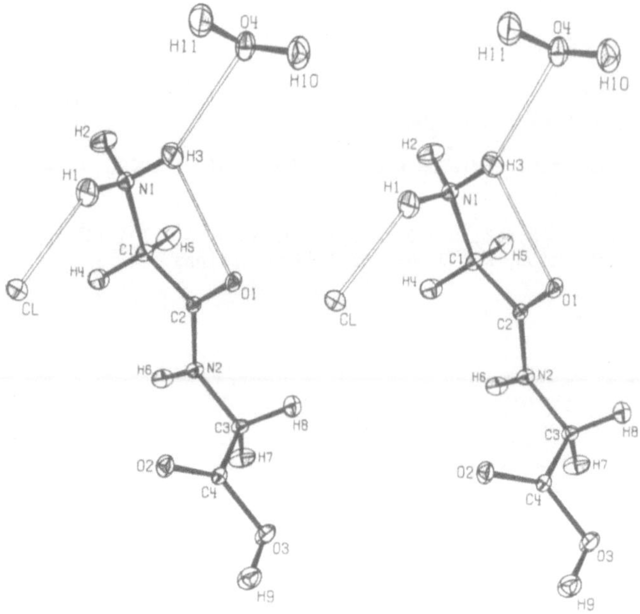

Fig. 2. View of the asymmetric unit of $[NH_3CH_2CONHCH_2COOH]^+.C\ell^-.H_2O$.

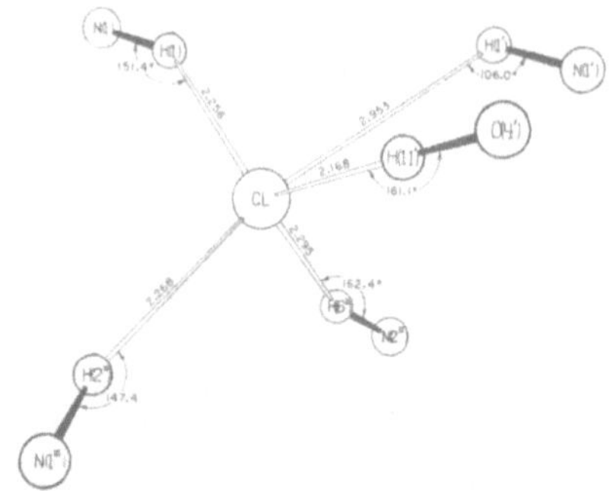

Fig. 3. Surroundings of the $C\ell^-$ ion in glycylglycine monohydrochloride monohydrate.

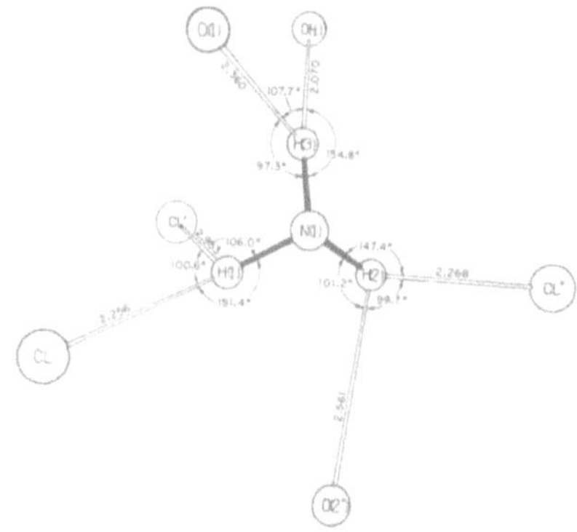

Fig. 4. Environment of the terminal nitrogen atom in glycylclycine monohydrochloride monohydrate.

1. R. PARTHASARATHY, 1969. Acta Cryst., B25, 509.

DIGLYCINE HYDROIODIDE

$C_4H_{11}IN_2O_4$

P. PIRET, J.M. PIRET, J. VERBIST and M. VAN MEERSSCHE, 1972. Bull. Chem. Soc. Belges, 81, 539-546.

Monoclinic, C2, a = 18.592, b = 5.160, c = 20.372 Å, β = 110.64°, D_m = 2.02, Z = 8. Mo radiation, R = 0.045 for 1350 reflexions.

The asymmetric unit contains two independent iodide ions, two glycinium ions $^+NH_3CH_2CO_2H$ and two zwitterions $^+NH_3CH_2CO_2^-$. The bond length data indicate that one of the glycinium ions may have a disordered carboxylic proton. In the crystal the different ions are connected by hydrogen bonds $N-H^+...I^-$, $N-H^+...O^-$ and $O-H...O^-$.

GLYCYLGLYCINE PHOSPHATE MONOHYDRATE

$C_4H_{13}N_2O_7P$ $(C_4H_9N_2O_2)^+.(H_2PO_4)^-.H_2O$

G.R. FREEMAN, R.A. HEARN and C.E. BUGG, 1972. Acta Cryst., B28, 2906-2912.

Orthorhombic, $Pna2_1$, a = 17.879, b = 12.130, c = 4.621 Å, D_m = 1.63, Z = 4. Cu radiation, R = 0.024 for 954 reflexions.

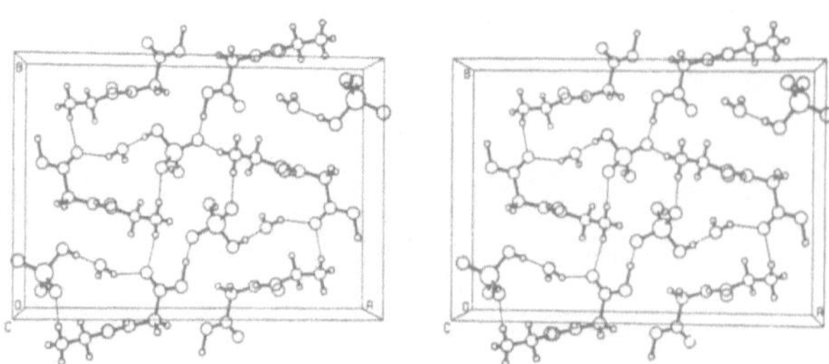

Fig. 1. View of the crystal structure of glycylglycine phosphate monohydrate viewed along c.

Bond lengths and angles in the peptide moiety are in good agreement with other glycylclycine structures and with the structures of other peptides. The conformation of the dipeptide, the crystal packing, and hydrogen bonding are depicted in Fig. 1. The dihedral angle between the planar peptide moiety ($^{+}NH_3CH_2CONHCH_2$) and the carboxyl plane is 85.2°. The crystal structure features a short hydrogen bond (2.503(3) Å) between the phosphate moiety and the carboxyl group of the dipeptide. Although the least-squares refinement suggested that the hydrogen atom involved in this bond might be positioned symmetrically between the two oxygen atoms, difference Fourier maps and the observed C-O and P-O bond lengths indicate that it is bonded covalently to the carboxyl oxygen atom. Thus, the compound is properly formulated as (glycylglycine.H)$^{+}$(H_2PO_4)$^{-}$. The crystal structure also involves a short hydrogen bond of 2.571(3) Å between phosphate moieties related by the c translation. The peptide conformation is not significantly different from that found in other crystal structures.

PYROGLUTAMIC ACID

(5-OXO-PROLINE)

$C_5H_7NO_3$

V. PATTABHI and K. VENKATESAN, 1972. Cryst. Struct. Comm., 1, 87-89.

Monoclinic, $P2_1/c$, a = 8.14, b = 8.86, c = 9.32 Å, β = 116.5°, Z = 4. Cu radiation, R = 0.091 for 1138 reflexions.

The crystal structure (Fig. 1) is stabilized by N-H...O and O-H...O hydrogen bonds. The atoms C(3), O(3), C(4), N and C(2) correspond to the cis peptide group, which is significantly non-planar; the value of the torsion angle about the bond C(3)-N is 4.2°. The five-membered ring has an envelope conformation with atom C(5) deviating 0.40 Å from the plane defined by C(2), N, C(3), and C(4). Bond lengths are normal, e.g., C(2)-N = 1.472(4), C(3)-N = 1.354(5) Å.

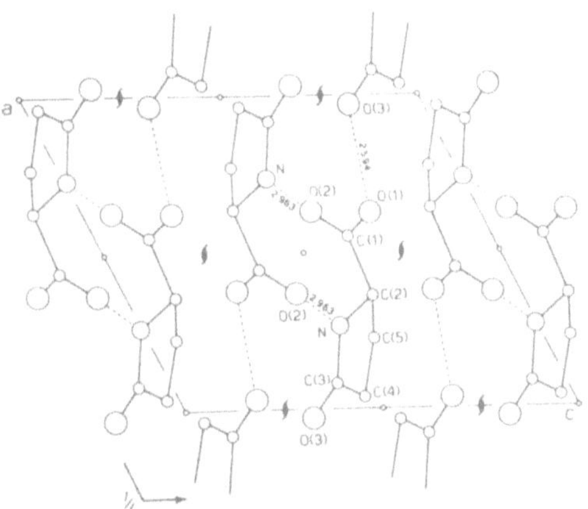

Fig. 1. The pyroglutamic acid crystal structure viewed down b.

L-GLUTAMIC ACID

$C_5H_9NO_4$ $^-OOCCH(NH_3^+)CH_2CH_2CO_2H$

M.S. LEHMANN, T.F. KOETZLE and W.C. HAMILTON, 1972. J. Cryst. Mol. Struct., 2, 225-233.

Orthorhombic, $P2_12_12_1$, a = 5.159, b = 17.30, c = 6.948 Å, D_m = 1.57, Z = 4. Neutron diffraction data, R = 0.026 for 803 reflexions.

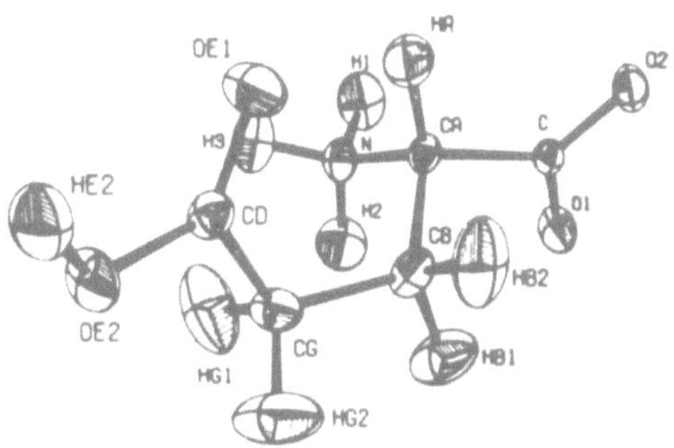

Fig. 1. A view of L-glutamic acid.

The structure has been determined earlier by X-ray diffraction (1). The molecule is in the zwitterion form (Fig. 1) and no intramolecular hydrogen bonds have been found. The hydrogen atom involved in a strong hydrogen bond between two carboxyl groups in adjacent molecules (O...O distance 2.519 Å) is covalently bonded to the carboxyl group belonging to the side chain of the amino acid. This side chain is buckled with Cδ gauche to Cα with respect to the Cβ-Cγ bond. The bond angles involving carbon atoms in the side chain are accordingly strained. Bond distances are: C-O = 1.219 - 1.312; C-N = 1.490; C-C = 1.501 - 1.533; O-H = 1.050; N-H = 1.027 - 1.038; C-H = 1.079 - 1.102 Å; σ(C-O, C-N, C-C) = 0.002; σ(O-H, N-H, C-H) = 0.003 - 0.004 Å.

1. Structure Reports, 19, 529.

L-GLUTAMIC ACID HYDROCHLORIDE

$C_5H_{10}C\ell NO_4$ $[CO_2HCH_2CH_2CH(NH_3)CO_2H]^+C\ell^-$

A. SEQUEIRA, H. RAJAGOPAL and R. CHIDAMBARAM, 1972. Acta Cryst., B28, 2514-2519.

Orthorhombic, $P2_12_12_1$, a = 5.151, b = 11.789, c = 13.347 Å, D_m = 1.524, Z = 4. Neutron radiation, R = 0.043 for 606 reflexions.

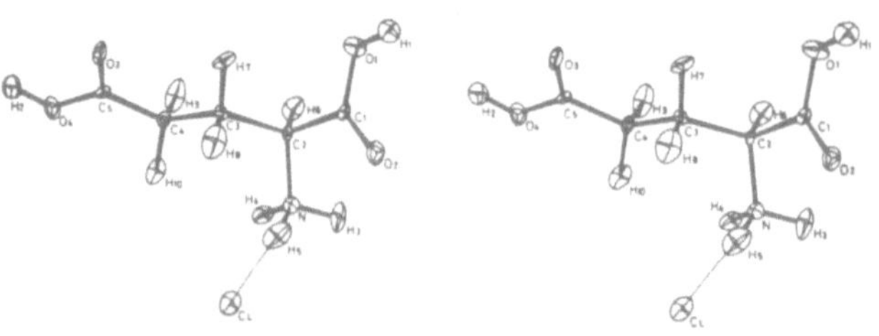

Fig. 1. Drawing of L-glutamic acid hydrochloride.

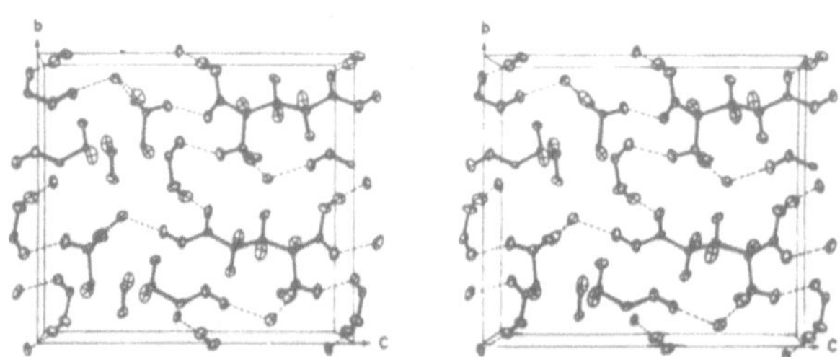

Fig. 2. Drawing of the crystal structure of L-glutamic acid hydrochloride viewed along a. Dotted lines represent hydrogen bonds.

This neutron study confirms the correctness of the earlier two-dimensional X-ray work (1). The molecular conformation and numbering scheme are shown in Fig. 1, and the crystal structure, which consists of molecules hydrogen bonded in zigzag chains along the c direction (O(1)H(1)...O(3) = 2.637(10) Å), is shown in Fig. 2. These chains are held together by N-H...O (2.904(8) Å), N-H...Cℓ (3.160 and 3.208(6) Å), and O-H...Cℓ (3.043(8) Å) hydrogen bonds. The bond lengths and angles are for the most part close to normal values with the exception of C(4)-C(5) (1.476(7) Å) which is somewhat shorter than a normal

C(sp^2)-C(sp^3) single bond.

1. Structure Reports, 17, 633.

L-GLUTAMINE HYDROCHLORIDE

$C_5H_{11}ClN_2O_3$

N. SHAMALA and K. VENKATESAN, 1972. Cryst. Struct. Comm., 1, 227-229.

Orthorhombic, $P2_12_12_1$, a = 13.32, b = 11.75, c = 5.16 Å, Z = 4. Cu radiation, R = 0.096 for 955 reflexions.

The crystal structure is stabilized by N-H...O, N-H...Cl⁻, and O-H...O hydrogen bonds. As in most amino acids, the torsion angle N-C-C=O is small (-18.6°). Hydrogen atom positions were not determined experimentally but the carboxyl oxygen O(3) is inferred to be protonated because of the short intermolecular contact (2.62(1) Å) between O(3) and the amide oxygen O(1) of a neighbouring molecule.

GLYCYL-L-ALANINE HYDROCHLORIDE

$C_5H_{11}ClN_2O_3$ $^+NH_3CH_2CONHCH(CH_3)COOH.Cl^-$

P.S. NAGANATHAN and K. VENKATESAN, 1972. Acta Cryst., B28, 552-556.

Monoclinic, $P2_1$, a = 8.21, b = 5.04, c = 10.82 Å, β = 100.1°, D_m = 1.383, Z = 2. Cu radiation, R = 0.098 for 895 reflexions.

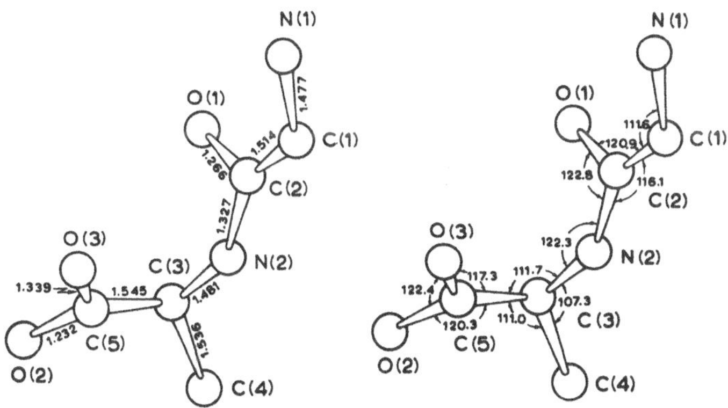

Fig. 1. Bond lengths (σ = 0.013 Å) and angles (σ = 1.0°) in glycyl-L-alanine hydrochloride.

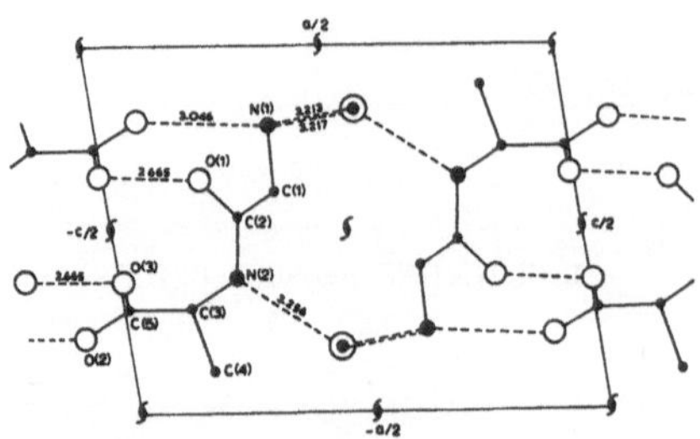

Fig. 2. The structure of glycyl-L-alanine hydrochloride viewed down b, showing intermolecular hydrogen bonds.

Bond distances and angles in the molecule (Fig. 1) are normal. The carboxyl group is un-ionized and the terminal NH_2 group is protonated. All N-H and O-H protons are involved in intermolecular hydrogen bonding; details are shown in Fig. 2. Torsional angles in the molecule are N(1)-C(1)-C(2)-O(1) = 11°, C(2)-N(2)-C(3)-C(5) = 108°.

γ-GUANIDINOBUTYRIC ACID HYDROCHLORIDE

γ-GUANIDINOBUTYRIC ACID HYDROBROMIDE

$C_5H_{12}ClN_3O_2$ $^+(H_2N)_2CNH(CH_2)_3CO_2H.\ Cl^-$
$C_5H_{12}BrN_3O_2$ $^+(H_2N)_2CNH(CH_2)_3CO_2H.\ Br^-$

T. MAEDA, T. FUJIWARA and K. TOMITA, 1972. Bull. Chem. Soc. Japan, 45, 3628-3633.

Triclinic, $P\bar{1}$, Z = 2.

	Hydrochloride	Hydrobromide
a(Å)	7.41	7.94
b	9.12	9.36
c	7.25	7.78
α	101.5°	103.2°
β	112.4	115.5
γ	65.2	61.5
D_m	1.429	1.664

Cu radiation, film data, R = 0.14 for 1435 reflexions for the hydrochloride.

The compounds are isostructural. The cation chain is extended, and bond lengths and angles are normal. The cation and Cl- anion lie in the $(20\bar{1})$ plane, and are joined by a system of six hydrogen bonds, N-H...O, N-H...Cl, and O-H...Cl.

L-HISTIDINE

$C_6H_9N_3O_2$

I. J.J. MADDEN, E.L. McGANDY and N.C. SEEMAN, 1972. Acta Cryst., B28, 2377-2382.

II. M.S. LEHMANN, T.F. KOETZLE and W.C. HAMILTON, 1972. Int. J. Peptide Protein Res., 4, 229-239.

III. J.J. MADDEN, E.L. McGANDY, N.C. SEEMAN, M.M. HARDING and A. HOY, 1972. Acta Cryst., B28, 2382-2389.

	I	II	III	
	Orthorhombic	Orthorhombic	Monoclinic	
Space group	$P2_12_12_1$	$P2_12_12_1$	$P2_1$	
a (Å)	5.177	5.175	5.172	
b	7.322	7.315	7.384	
c	18.87	18.75	9.474	
β	-	-	97.162°	
D_m	1.428	-	1.446	
Z	4	4	2	
Radiation	Mo	neutrons	Mo and Cu	
R	0.034	0.058	0.11	0.10
Reflexions	927	618	not given	492

The X-ray and neutron studies of the orthorhombic form are in good agreement. The molecule (Fig.1) is in the zwitterion form and the molecular conformation is stabilized by an intramolecular hydrogen bond between an ammonium group hydrogen and the imidazole nitrogen N(2) (Fig. 2) [N(1)...N(2) = 2.783, H(N(1))...N(2) = 1.90 Å]. The molecules are bonded together by hydrogen bonds to form chains in the c direction (Fig. 2) with NH...O = 2.773 - 2.851 Å. Bond lengths and angles in the molecule are in good agreement with those of other crystalline histidine forms of the free base. In this form the conformation is the open extended form. There is no protonation of the imidazole ring.

The monoclinic form yielded crystals which gave relatively poor data because of stacking faults. The bond lengths are in good agreement with those of the orthorhombic form. The molecules are extended in the c direction (Fig. 3) and linked in this direction by N(3)H...O(1) hydrogen bonds (2.73 Å). An intramolecular hydrogen bond N(1)H...N2 (2.75 Å) stabilizes the zwitterion form in a manner identical to that in the orthorhombic form. Two other intermolecular hydrogen bonds, N(1)H. ..O(2) = 2.76 and 2.85 Å, cross link the molecular chains (Fig. 3).

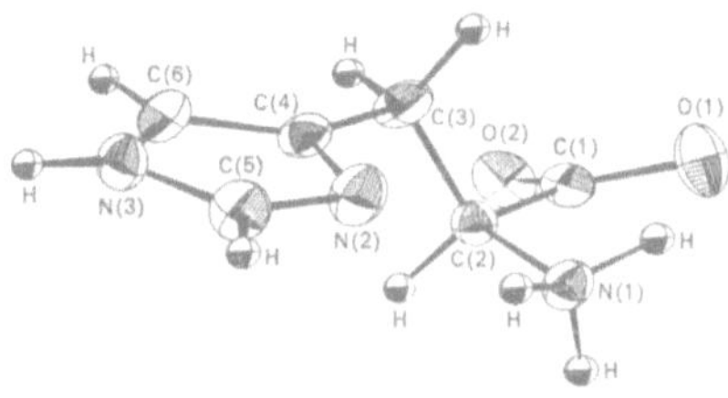

Fig. 1. Structure of orthorhombic L-histidine.

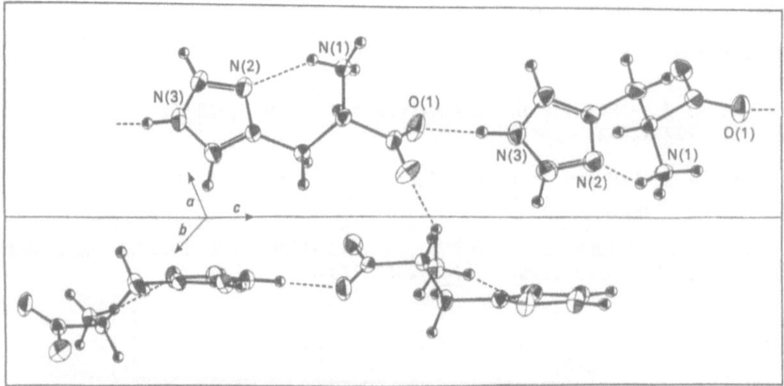

Fig. 2. Hydrogen bonding in orthorhombic L-histidine.

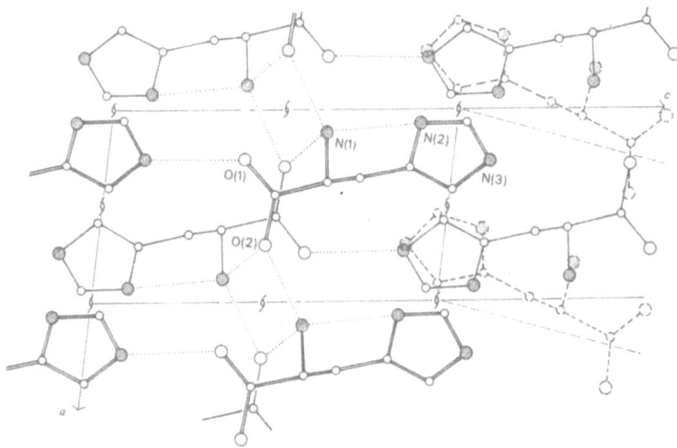

Fig. 3. Monoclinic L-histidine viewed along b. The full lines show the structure when only one orientation, P, is present. Broken lines show how molecules in an alternative orientation, Q, are stacked as a result of lamellar twinning. Dotted lines represent hydrogen bonds.

β-(PYRAZOLYL-3)-L-ALANINE

$C_6H_9N_3O_2$ $C_3H_3N_2.CH_2CH(NH_3^+)COO^-$

N.C. SEEMAN, E.L. McGANDY and R.D. ROSENSTEIN, 1972. J. Amer. Chem. Soc., 94, 1717-1720.

Monoclinic, $P2_1$, a = 4.62, b = 7.55, c = 10.06 Å, β = 98.61°, D_m = 1.419, Z = 2. Cu radiation, R = 0.078. The number of reflexions is not given in the paper.

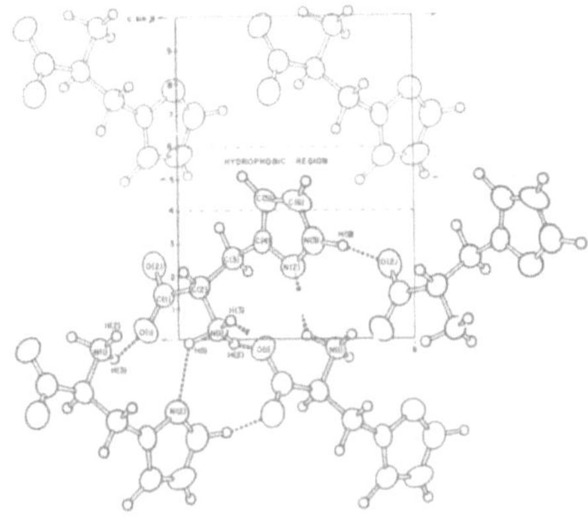

Fig. 1. Projection of the structure of β-(pyrazolyl-3)-L-alanine along a. The hydrogen bond lengths are N-H...N 2.925 Å, N-H...O 2.729 - 2.857 Å.

The molecule crystallised from a neutral solvent in the zwitterion form. The molecular packing (Fig. 1) has two separate regions: one hydrophilic which makes use of all four possible hydrogen bonds, and the other hydrophobic, involving a characteristic (perpendicular) stacking of bases. In β-(pyrazolyl-3)-L-alanine, N(3) and C(5) are reversed from their positions in histidine, but all other atoms are in identical positions, making this molecule a structure analogue of histidine. The molecular geometry is compared with values from four independent investigations of L- and DL-histidine, and demonstrates that this molecule is isosteric with histidine.

CYCLO-DI-β-ALANYL

$C_6H_{10}N_2O_2$

D.N.J. WHITE and J.D. DUNITZ, 1972. Israel J. Chem., 10, 249-256.

Monoclinic, C2/c, a = 11.85, b = 5.20, c = 10.66 Å, β = 94.83°, D_m = 1.43, Z = 4. Mo radiation, R = 0.042 for 731 reflexions. Molecular symmetry C_2.

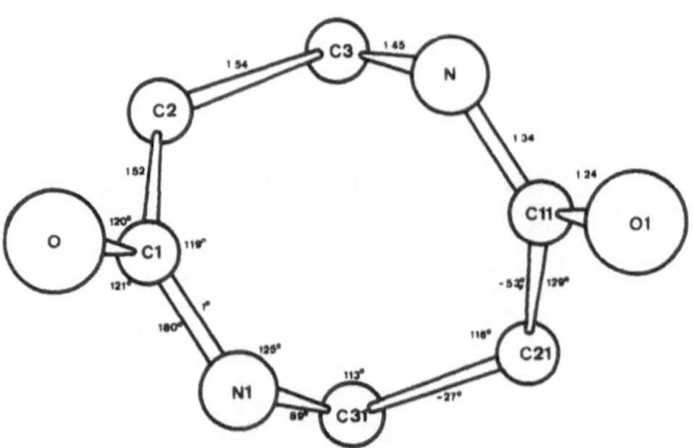

Fig. 1. Cyclo-di-β-alanyl. Structure projected on best plane of molecule. Bond lengths (σ = 0.002 Å), bond angles, and torsion angles are shown.

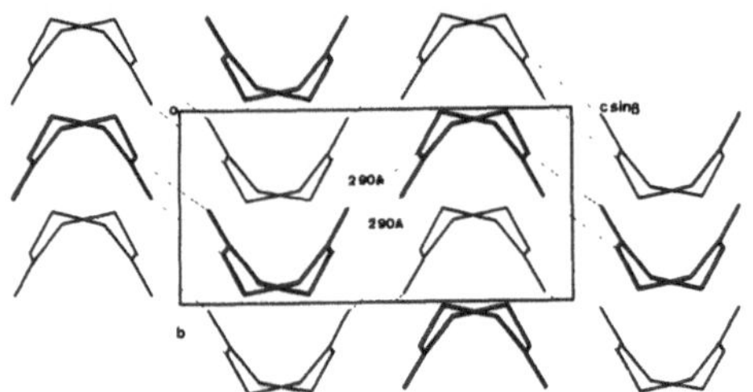

Fig. 2. Cyclo-di-β-alanyl molecular packing.

The molecular dimensions are summarized in Fig. 1. The two cis-peptide groups related by a twofold axis are almost exactly planar and the molecule occurs in a "flexible" chair conformation. In the crystal centrosymmetric pairs of hydrogen bonds link the molecules into infinite chains (Fig. 2) running parallel to c.

2S-CARBOXY-4R,5S-DIHYDROXYPIPERIDINE HYDROCHLORIDE

$C_6H_{12}C\ell NO_4$

G. EVRARD, F. DURANT and M. MARLIER, 1972. Cryst. Struct. Comm., 1, 215-217.

Orthorhombic, $P2_12_12_1$, a = 8.541, b = 14.816, c = 7.069 Å, D_m = 1.470, Z = 4. Cu radiation, R = 0.103 for 963 reflexions.

The piperidine ring has the chair conformation, and bond distances and angles are normal. The cohesion of the crystal is due to the intermolecular hydrogen bonds N-H...Cℓ (3.07, 3.16 Å), O-H...Cℓ (3.09, 3.16 Å) and O-H...O (2.64 Å). The absolute configuration was determined using the anomalous scattering of the chlorine atom.

HISTIDINE HYDROCHLORIDE MONOHYDRATE

$C_6H_{12}C\ell N_3O_3$ $C_6H_{10}O_2N_3^+.C\ell^-.H_2O$

K. ODA and H. KOYAMA, 1972. Acta Cryst., B28, 639-642.

Orthorhombic, $P2_12_12_1$, a = 15.301, b = 8.921, c = 6.846 Å, D_m = 1.483, Z = 4. Mo radiation, R = 0.037 for 1079 reflexions.

This crystal structure has been determined previously (1) and further refined (2). This report is of another refinement. The imidazole ring is perfectly planar. The atoms of the carboxyl group, the α-carbon atom, and the amino nitrogen are also planar; the dihedral angle between the planes is 72°. Mostly the corresponding bond lengths and angles are not significantly different from expected values. A summary of molecular dimensions in a number of histidine derivatives is also given.

1. Structure Reports, 20, 510.
2. Ibid., 29, 701.

L-CITRULLINE HYDROCHLORIDE

$C_6H_{14}C\ell N_3O_3$ $NH_2CONH(CH_2)_3CH(NH_3^+)COOH.C\ell^-$

T. ASHIDA, K. FUNAKOSHI, T. TSUKIHARA, T. UEKI and M. KAKUDO, 1972. Acta Cryst., B28, 1367-1374.

Monoclinic, C2, a = 18.089, B = 5.150, c = 11.918 Å, β = 106.9°, D_m = 1.32, Z = 4. Mo radiation, R = 0.042 for 1363 reflexions.

The structure has been determined previously (1) with photographic data. The present analysis utilised diffractometer data and agrees essentially with the previous work. The citrulline molecule consists of three planar groups, the carboxyl group, the aliphatic trans zig-zag chain, and the carbamylamino group. Differences between the two determinations occur in the cell data (the photographically determined cell data are about 1.2% smaller than those reported above) and as a consequence a number of the bonds reported previously as being shorter than usual, are found to have normal values in this analysis.

1. P.S. NAGANATHAN and K. VENKATESAN, 1971. Acta Cryst., B27, 1079; Structure Reports, 37B, 269.

L-HOMOCITRULLINE HYDROCHLORIDE

$C_7H_{16}C\ell N_3O_3$ $NH_2CONH(CH_2)_4CH(NH_3^+)COOH.C\ell^-$

T. ASHIDA, K. FUNAKOSHI, T. TSUKIHARA, T. UEKI and M. KAKUDO, 1972. Acta Cryst., B28, 1367-1374.

Orthorhombic, $P2_12_12_1$, a = 10.518, b = 21.627, c = 4.916 Å, D_m = 1.32, Z = 4. Mo radiation, R = 0.086 for 1691 reflexions.

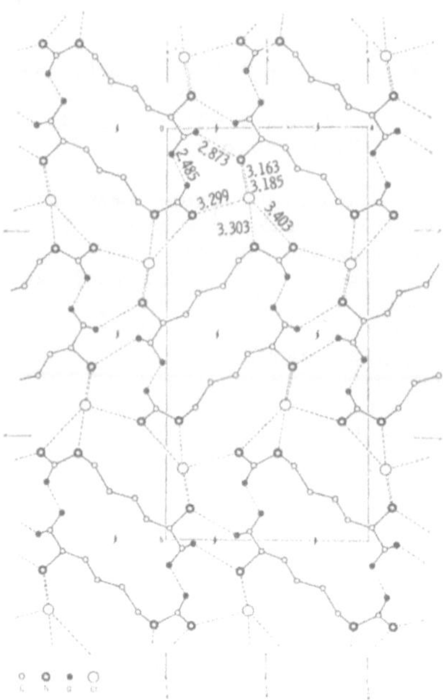

Fig. 1. Hydrogen bonding (broken lines) in homocitrulline hydrochloride.

Like citrulline (1) the homocitrulline molecule also consists of three planar groups, the carboxyl group, the trans zig-zag aliphatic chain, and the carbamylamino group. Bond distances are normal. The molecule and the chloride ion are held together in the crystal by seven hydrogen bonds (Fig. 1).

1. Preceding report.

L-ARGININE PHOSPHATE MONOHYDRATE

$C_6H_{19}N_4O_7P$ $C_6H_{15}O_2N_4{}^+H_2PO_4{}^-.H_2O$

W. SAENGER and K.G. WAGNER, 1972. Acta Cryst., B28, 2237-2244.

Monoclinic, $P2_1$, a = 10.898, b = 7.910, c = 7.339 Å, β = 97.97°, D_m = 1.53, Z = 2. Mo radiation, R = 0.039 for 1453 reflexions.

The crystal structure of L-arginine phosphate monohydrate which was previously reported (1) has been refined more completely. The complex (Fig. 1) is deprotonated at the carboxyl group and protonated at the guanidyl and at the Cα-amino groups while the phosphate ion carries one negative charge. The four atoms comprising the guanidinium group are coplanar, and C(5) is cis-planar with N(4). The bond lengths and angles are in Fig. 2. The carboxyl-Cα system is essentially planar. The crystal structure (Fig. 3) is interlaced by N-H...O = 2.796 - 3.372 and O-H...O = 2.551, 2.604 Å hydrogen bonds; the N(2)-H(11) group forms a bifurcated hydrogen bond with O(W) and O(6). The molecular packing indicates that screw-axis related L-arginine molecules form a hydrophobic zone along the b axis which is surrounded by hydrophilic phosphate groups and water molecules.

Fig. 1. Chemical formula of the L-arginine phosphate complex. The IUPAC nomenclature is Cα for C(2).

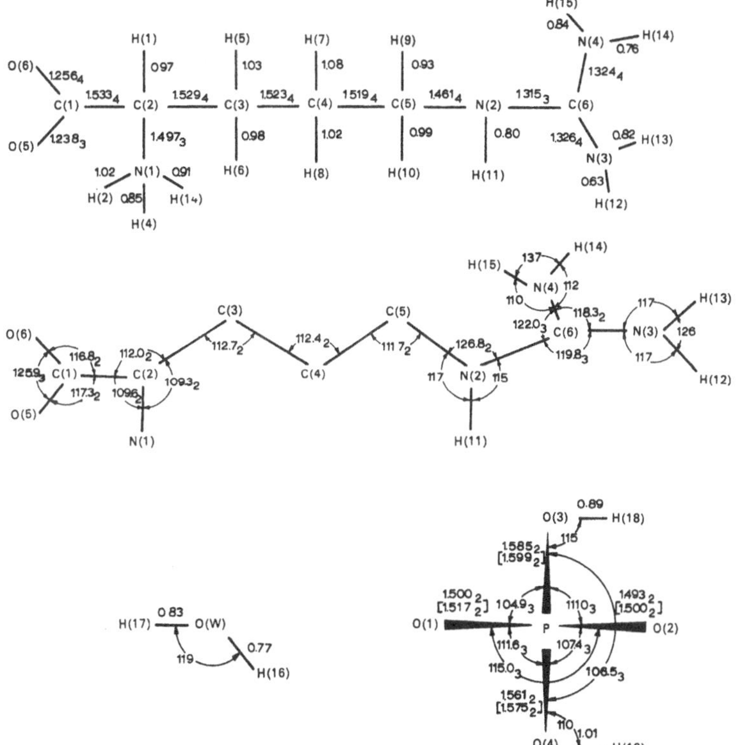

Fig. 2. Bond lengths and angles for L-arginine phosphate monohydrate.

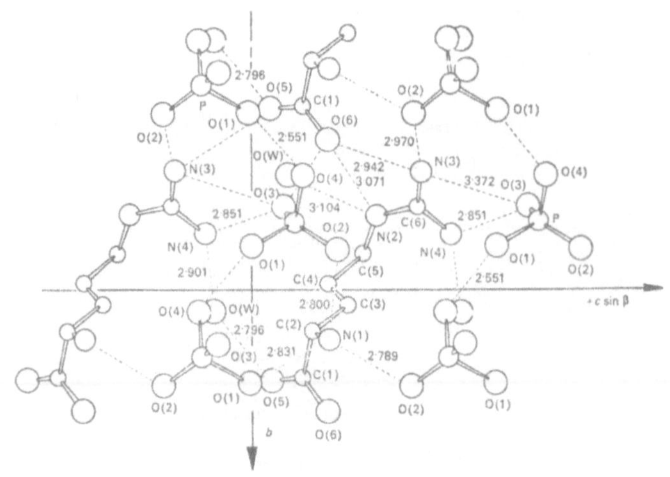

Fig. 3. Projection of the L-arginine phosphate monohydrate structure along a.

1. K. AOKI, K. NAGANO and Y. IITAKA, 1971. Acta Cryst., B27, 11; Structure Reports, 37B, 271.

L-LYSINE HYDROCHLORIDE DIHYDRATE

$C_6H_{19}C\ell N_2O_4$ $^+NH_3(CH_2)_4CH(NH_3^+)COO^-.C\ell^-.2H_2O$

I. T.F. KOETZLE, M.S. LEHMANN, J.J. VERBIST and W.C. HAMILTON, 1972. Acta Cryst., B28, 3207-3214.

II. R.R. BUGAYONG, A. SEQUEIRA and R. CHIDAMBARAM, 1972. Acta Cryst., B28, 3214-3219.

Monoclinic, $P2_1$, a = 7.492, b = 13.320, c = 5.879 Å, β = 97.79°, D_m = 1.249, Z = 2. Neutron radiation, R = 0.030 for 1240 reflexions (I) and R = 0.070 for 1275 reflexions (II).

The lysine molecule (Fig. 1) is a singly charged zwitterion with the extra charge carried by the terminal ammonium group as was found in two prior X-ray studies (1). The neutron diffraction studies confirm the X-ray determinations and locate the hydrogen atoms precisely. There are ten hydrogen bonds per asymmetric unit, one involving each hydrogen atom covalently bonded to nitrogen or oxygen and the packing scheme (Fig. 2) consists of chains of lysine cations inclined about 30° to the b axis and interconnected through the water molecules and chloride ions (N-H...O = 2.788 - 2.887(3), N-H...Cℓ = 3.173(2), 3.217(2), O-H...O = 2.718(6), and O-H...Cℓ = 3.254(4) Å).

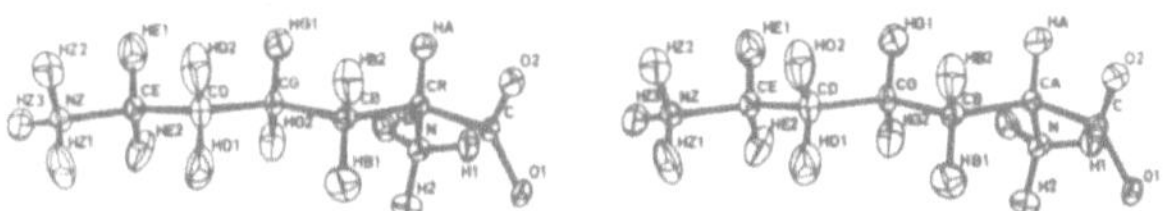

Fig. 1. View of the lysine molecule.

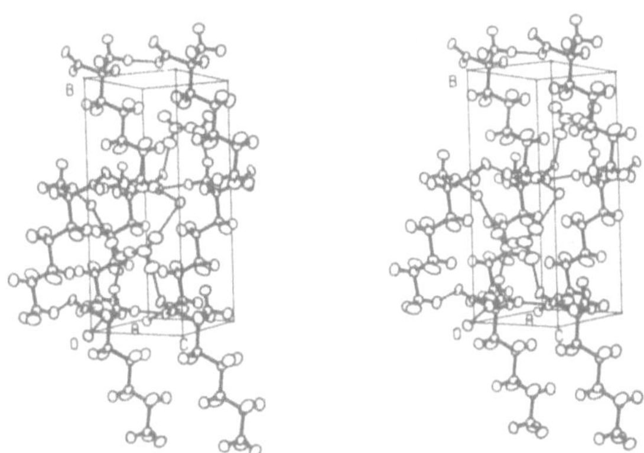

Fig. 2. View of the packing of one unit cell of lysine hydrochloride monohydrate. Molecular bonds are drawn thick and hydrogen bonds thin.

1. Structure Reports, 23, 596; 27, 767.

L-N-ACETYLHISTIDINE MONOHYDRATE

$C_8H_{13}N_3O_4$ $C_8H_{11}N_3O_3.H_2O$

T.J. KISTENMACHER, D.J. HUNT and R.E. MARSH, 1972. Acta Cryst., B28, 3352-3361.

Triclinic, P1, a = 8.865, b = 9.097, c = 7.346 Å, α = 102.24, β = 90.30, γ = 117.73°, D_m = 1.407, Z = 2. Cu radiation, R = 0.029 for 2152 reflexions.

The molecule exists in the crystal as a zwitterion with protonation of the imidazole ring. The two independent molecules have very different molecular conformations (Fig. 1) with molecule A adopting a 'closed' and molecule B an 'open' conformation. The differences are reflected in the torsion angle about C(2)-C(3); a small value corresponds to the closed form, in which the imidazole ring is folded back on top of the carboxylate group, while a larger value (near 180°) corresponds to an open or extended conformation. Bond lengths and angles (Fig. 1) are normal. All protons attached to water or nitrogen atoms are involved in hydrogen bonding (N-H...O = 2.664 - 3.207, O-H...O = 2.736 - 2.805 Å); a C(5)-H...O(4) intermolecular interaction, 3.199 Å, is shorter than a van der Waals contact.

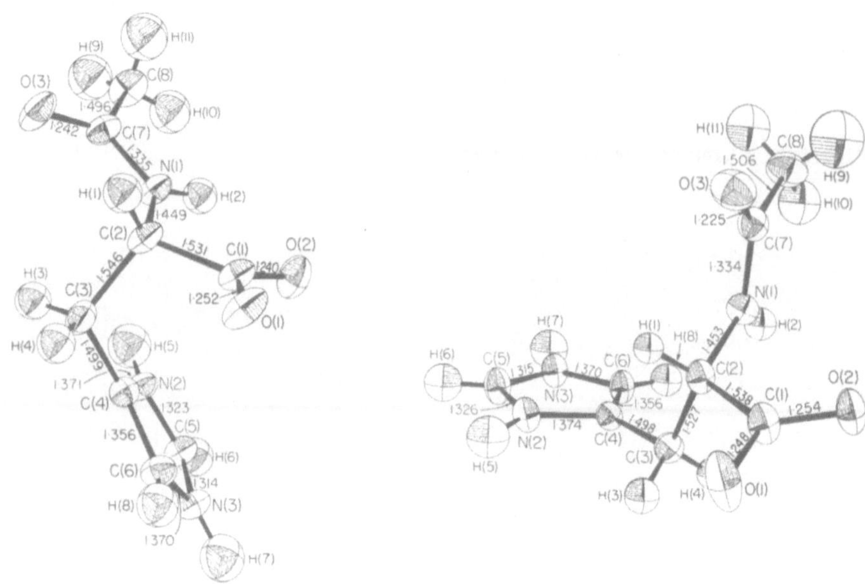

Fig. 1. Perspective view of the two independent molecules of L-N-acetylhistidine showing bond lengths (σ = 0.004 - 0.005 Å) and bond angles (σ = 0.1 - 0.2°).

HIPPURIC ACID

(N-BENZOYLGLYCINE)

$C_9H_9NO_3$ $C_6H_5CONHCH_2COOH$

W. HARRISON, S. RETTIG and J. TROTTER, 1972. J. Chem. Soc., Perkin II, 1036-1040.

Orthorhombic, $P2_12_12_1$, a = 10.586, b = 9.123, c = 8.880 Å, D_m = 1.371, Z = 4. Cu radiation, R = 0.034 for 750 reflexions.

This independent determination of the structure confirms that reported by Ringertz (1). The structure consists of molecules of hippuric acid linked by a three-dimensional network of intermolecular O-H...O (2.68 Å) and N-H...O (3.02 Å) hydrogen bonds. The molecule is characterized by three planes defined by the phenyl, amide, and carboxylic acid groups, which are twisted with respect to each other. There are errors in Figure 2 in the paper; bond angle C(2)-C(1)-O(1) should be 109.6(3)° and C(2)-C(1)-O(2) should be 125.6(2)° (2).

1. H. RINGERTZ, 1971. Acta Cryst., B27, 285; Structure Reports, 37B, 273.
2. J. TROTTER, 1972. Personal communication.

L-TYROSINE

(L-4-HYDROXYPHENYLALANINE)

$C_9H_{11}O_3N$ $HOC_6H_4CH_2CH(NH_3^+)COO^-$

A. MOSTAD, H.M. NISSEN and C. RØMMING, 1972. Acta Chem. Scand., 26, 3819-3833.

Orthorhombic, $P2_12_12_1$, a = 6.913, b = 21.116, c = 5.829 A, D_m = 1.41, Z = 4. Mo radiation, R = 0.049 for 2259 reflexions.

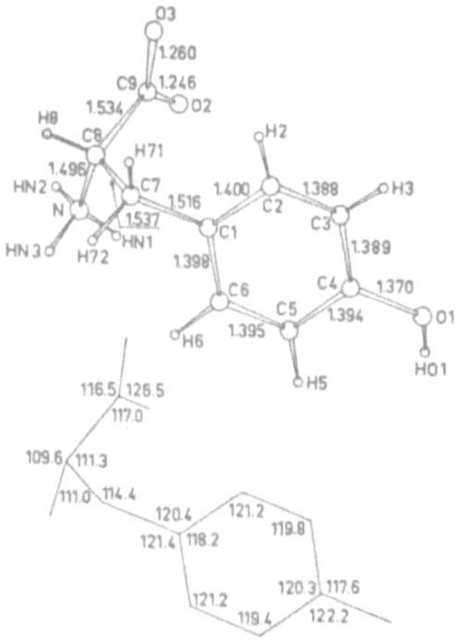

Fig. 1. Bond lengths (Å) corrected for thermal vibration effects and angles in L-tyrosine.

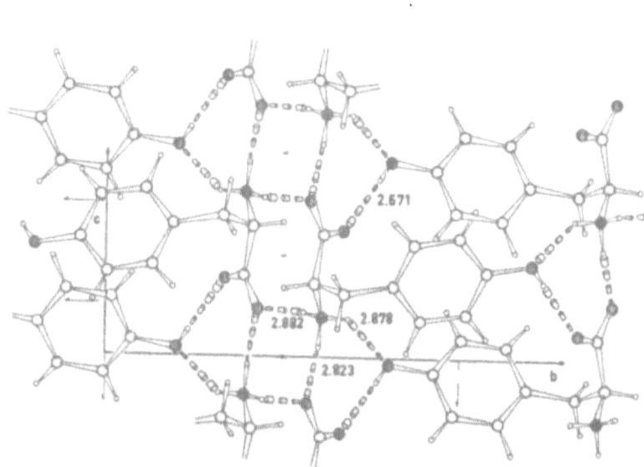

Fig. 2. The L-tyrosine crystal structure as seen down the a axis. The hydrogen bond lengths are indicated.

A drawing of the molecule showing bond lengths (corrected) and valency angles is shown in Fig. 1. In the alanine moiety the usual zwitterionic nature of a free amino acid is indicated both by the nearly equal C-O bond length in the carboxyl group and the tetrahedral arrangement of hydrogen atoms around the nitrogen atom. The conformation about the N-C(8) and C(7)-C(8) bonds is staggered with the C(8) hydrogen atom anti relative to the phenyl ring. The crystal structure is characterised by a three-dimensional hydrogen-bond network (Fig. 2). The nitrogen atom is hydrogen donor in three hydrogen bonds, two to carboxyl oxygen atoms of two different molecules (2.882 and 2.823 Å) and one to a phenolic oxygen atom of a third (2.878 Å). The phenolic hydrogen atom participates in a hydrogen bond to a carboxylic oxygen atom of a neighbouring molecule (2.671 Å). Each L-tyrosine molecule is hydrogen bonded to six neighbouring molecules.

cyclo-L-PROLYL-L-LEUCYL

$C_{11}H_{18}N_2O_2$

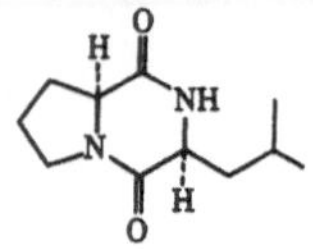

I.L. KARLE, 1972. J. Amer. Chem. Soc., 94, 81-84.

Orthorhombic, $P2_12_12_1$, a = 9.451, b = 19.587, c = 6.340 Å, D_m not given, Z = 4. Cu radiation, R = 0.064 for 1146 reflexions.

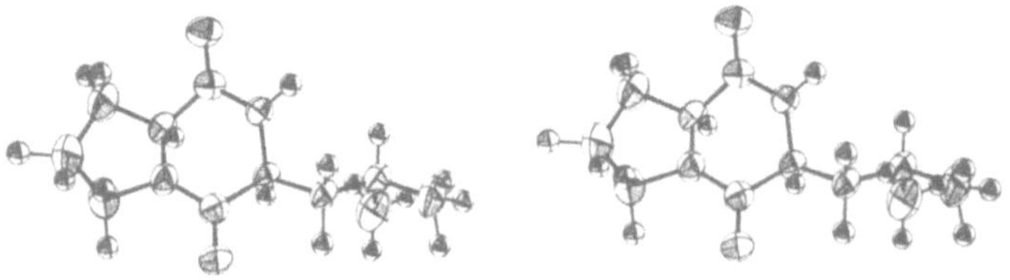

Fig. 1. Diagram of the conformation of cyclo-L-prolyl-L-leucyl.

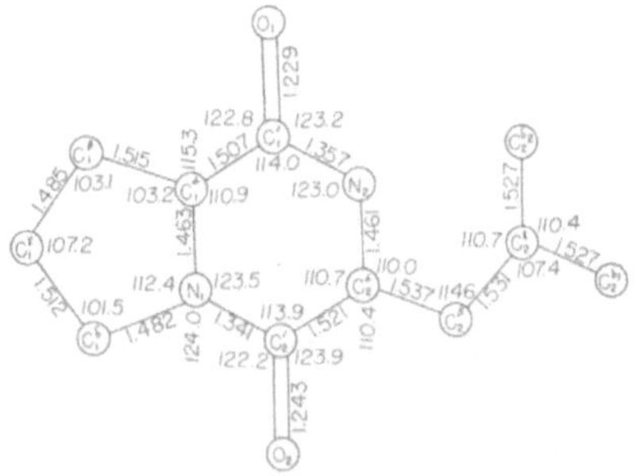

Fig. 2. Bond lengths and angles in cyclo-L-prolyl-L-leucyl; σ = 0.005 - 0.010 Å and 0.3 - 0.6°.

In this molecule the diketopiperazine ring has a folded conformation with a dihedral angle of 143° between the two planar peptide units; the fold occurs along the line joining the two tetrahedral Cα atoms (Fig. 1.). In the pyrolidine ring Cβ is 0.52 Å out of the plane of the other four atoms. The leucyl side chain is fully extended. Bond lengths and angles are in Fig. 2. The parameters and conformation for the diketopiperazine ring are the same within the standard deviations as found for 3,4-dehydroproline anhydride (1). Molecules form infinite chains parallel to a through N-H...O hydrogen bonds (N...O = 2.97 Å)

1. I.L. KARLE, 1972. In preparation.

DI-L-LEUCINE HYDROCHLORIDE

$C_{12}H_{27}C\ell N_2O_4$

L. GOLIC and W.C. HAMILTON, 1972. Acta Cryst., B28, 1265-1271.

Monoclinic, $P2_1$, a = 11.152, b = 5.116, c = 15.405 Å, β = 108.94°, D_m = 1.187, Z = 2. Cu radiation, R = 0.055 for 1360 reflexions.

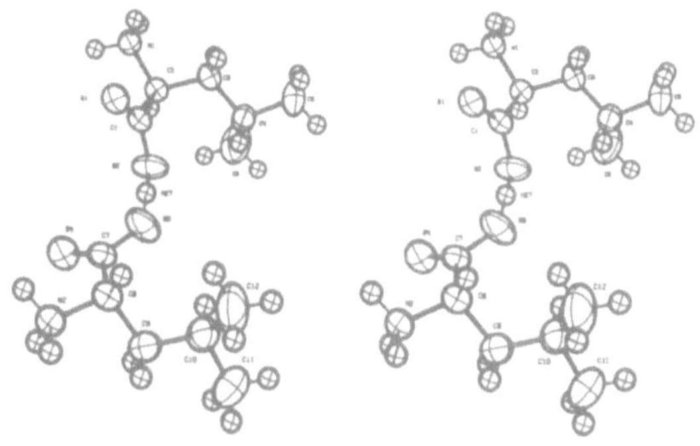

Fig. 1. The two leucine molecules of the dimeric cation.

The most striking feature of the structure is the existence of a hydrogen-bonded dimeric cation consisting of two leucine molecules sharing a proton in a very short (2.43 Å) hydrogen bond between the two carboxyl groups (Fig. 1). These dimers are more weakly hydrogen-bonded through the chloride ions to form double layers parallel to (001); NH...Cℓ = 3.204 - 3.370, mean 3.301(6) Å. There is also a single NH...O hydrogen bond of length 2.779(5) Å. The bond lengths in the two molecules of the dimer agree well. The L-leucine molecule is characterized by two planes: one through the carboxyl group, and the other through the side chain. Both carboxyl groups are nearly coplanar with the α-carbon atom. The mean values for the C-C bond lengths are 1.521(9) and 1.514(9) Å for the two molecules of the dimer. The mean values for the C-C-C bond angles are 111.0 (5)° and 113.5(5)° respectively.

N-ACETYL-L-TYROSINE ETHYL ESTER MONOHYDRATE

$C_{13}H_{19}NO_6$ $C_{13}H_{17}NO_4.H_2O$

A.F. PIERET and F. DURANT, 1972, Cryst. Struct. Comm., 1, 75-77.

Orthorhombic, $P2_12_12_1$, a = 13.132, b = 14.500, c = 7.369 Å, Z = 4. Mo radiation, R = 0.107 for 1036 reflexions.

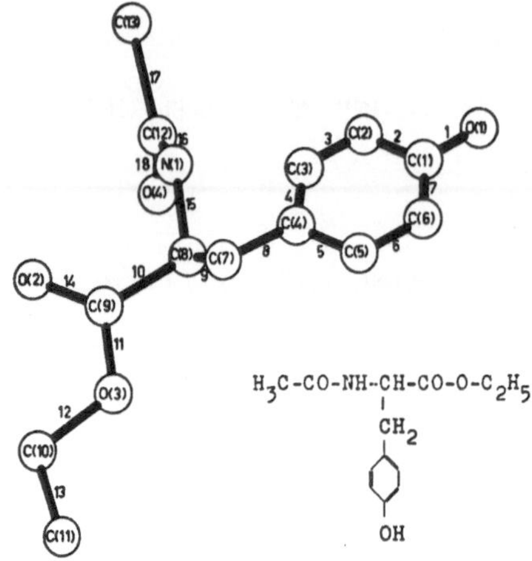

Fig. 1. Conformation of N-acetyl-L-tyrosine ethyl ester.

The molecular conformation (Fig. 1) reveals that this molecule does not adopt the "scorpion" arrangement found in tyrosine ethyl ester (1). The distances and angles correspond to the values found for most amino acid esters. The difference in activity of α-chymotripsin in the hydrolyses of tyrosine ethyl ester and of its N-acetyl derivative may be related to the different conformations of the two substrates.

1. Structure Reports, 35B, 325.

GLYCYL-L-PHENYLALANINE ARYLSULPHONATES

J.M. van der VEEN and B.W. LOW, 1972. Acta Cryst., B28, 3548-3559.

GLYCYL-L-PHENYLALANINE p-BROMOBENZENESULPHONATE

$C_{17}H_{19}BrN_2O_6S$

Monoclinic, C2, a = 35.91, b = 5.838, c = 9.679 Å, β = 93.59°, D_m = 1.498, Z = 4. Cu radiation, R = 0.12 for 1684 reflexions.

GLYCYL-L-PHENYLALANINE p-TOLUENESULPHONATE

$C_{18}H_{22}N_2O_6S$

Monoclinic, C2, a = 35.99, b = 6.005, c = 9.679 Å, β = 109.12°, D_m = 1.297, Z = 4. Cu radiation, R = 0.095 for 2073 reflexions.

The conformations of the peptides and of the sulphonates are nearly identical in these isomorphous structures. The molecular packing is characterized by alternating sheets of peptide (extended along a) and benzenesulphonate ions. The peptide consists of a planar region from the $\bar{N}(1)$ atom of glycine to the Cβ(2) atom of phenylalanine with the carboxyl and phenyl groups twisted 74 and 99° respectively from this plane (Fig. 1). The phenylalanine residue has the appropriate conformation for the right handed α-helix, while the glycine has an extended chain conformation. The molecules are linked by hydrogen bonds (Fig. 2). Within one peptide sheet a pair of peptide molecules is held together by two hydrogen bonds involving the terminal nitrogen (N(1), Fig. 1) of the peptide and the carboxyl oxygen O(1) of the glycyl residue of a neighbouring molecule (NH...O = 2.917(6) Å). There are also hydrogen bonds between N(1) and N(2) and the sulphonate groups (NH...O = 2.80 - 2.93 Å).

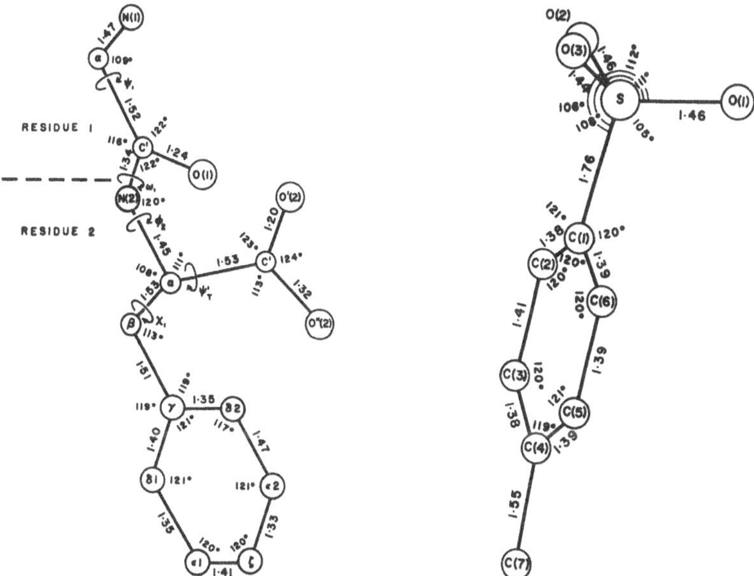

Fig. 1. Molecular structure, bond distances and angles of glycyl-L-phenylalanine p-toluenesulphonate, (a) peptide portion, (b) p-toluenesulphonate portion.

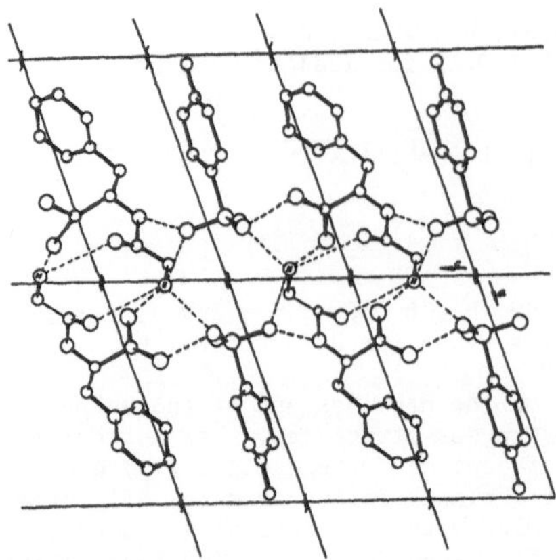

Fig. 2. Hydrogen bonding network in glycyl-L-phenylalanine p-toluenesulphonate viewed along a.

N-BENZOYL-L-TYROSINE ETHYL ESTER

$C_{18}H_{19}NO_4$ $C_6H_5CONHCH(CH_2C_6H_4OH)CO_2C_2H_5$

F. DURANT, A.F. PIERET, G. GERMAIN and M. KOCH, 1972. Cryst. Struct. Comm., 1, 435-438.

Orthorhombic, $P2_12_12_1$, a = 16.441, b = 17.053, c = 11.782 Å, Z = 8. Mo radiation, R = 0.080 for 2504 reflexions.

Bond distances and angles in the two molecules in the asymmetric unit are normal. The molecules are held together in the crystal by π-interactions between the phenyl rings of the benzoyl groups. The molecules in the different asymmetric units are linked by hydrogen bonds between hydroxyl and ester carbonyl groups, O-H...O = 2.654 and 2.724 Å.

3,5-DIIODO-L-THYRONINE N-METHYLACETAMIDE

(1:1 COMPLEX)

$C_{18}H_{21}I_2N_2O_5$

$.CH_3CONHCH_3$

V. CODY, W.L. DUAX and D.A. NORTON, 1972. Acta Cryst., B28, 2244-2252.

Monoclinic, $P2_1$, a = 7.998, b = 22.317, C = 5.995 Å, β = 95.54°, Z = 2. Cu radiation, R = 0.043 for 1808 reflexions.

The structure (Fig. 1) of the thyroxine analogue 3,5-diiodo-L-thyronine has been determined as a 1:1 complex with N-methylacetamide. The C-I distances, 2.112 and 2.089 Å, are longer than the average value (2.05 Å) in other aromatic iodinated structures. The amino acid backbone conformation, described by rotation about the Cα-Cβ bond, shows a preferred torsional angle of 300° about this bond, which is sterically favoured. The diphenylether system adopts a skewed conformation with the planes of the two phenyl rings mutually perpendicular to produce minimal interaction between the bulky 3,5-iodine atoms and the 2',6'-hydrogen atoms. The complex is held together by hydrogen bonds, with the amine nitrogen hydrogen bonded to three oxygen atoms in a tetrahedral manner (N-H...O = 2.73, 2.76, 2.96 and O-H...O = 2.62 Å). There is also an unusually short iodine-carbonyl (C=O...I) contact distance of 3.03 Å.

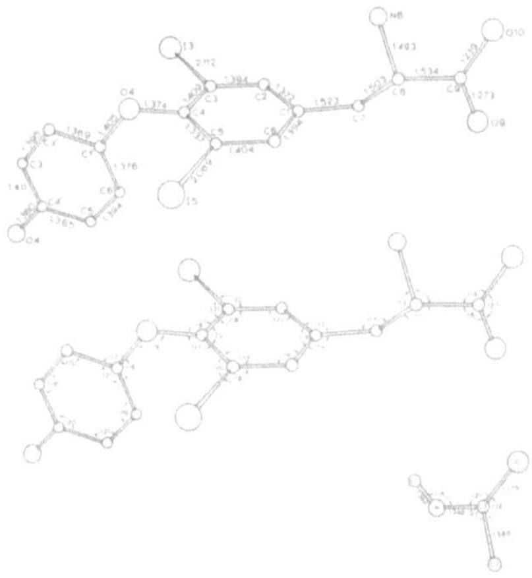

Fig. 1. Bond distances and angles for 3,5-diiodo-L-thyronine N-methylacetamide (σ = 0.016 Å and 1.2° for the diiodo-thyronine molecule, σ = 0.038 Å and 2.3° for the N-methylacetamide molecule).

N-(HALOACETYL)-L-PHENYLALANYL-L-PHENYLALANINE ETHYL ESTERS

C.H. WEI, D.G. DOHERTY and J.R. EINSTEIN, 1972. Acta Cryst., B28, 907-915.

N-(BROMOACETYL)-L-PHENYLALANYL-L-PHENYLALANINE ETHYL ESTER

$C_{22}H_{25}BrN_2O_4$

Monoclinic, $P2_1$, a = 12.951, b = 17.617, c = 4.951 Å, β = 103.23°, D_m = 1.39, Z = 2. Cu radiation, R = 0.090 for 522 reflexions.

N-(CHLOROACETYL)-L-PHENYLALANYL-L-PHENYLALANINE ETHYL ESTER

$C_{22}H_{25}C\ell N_2O_4$

Monoclinic, $P2_1$, a = 12.844, b = 17.677, c = 4.931 Å, β = 103.30°, D_m = 1.27, Z = 2. Cu radiation, R = 0.105 for 487 reflexions.

The compounds are isostructural. Details of their configuration and bond lengths are in Fig. 1; for both derivatives the standard deviations are high (0.03 - 0.05 Å) because of disorder. The molecular configuration is of the parallel-chain pleated-sheet type. Interchain hydrogen bonding is shown in Fig. 2. A complete description of the molecular conformations in terms of torsion angles is also given in the paper.

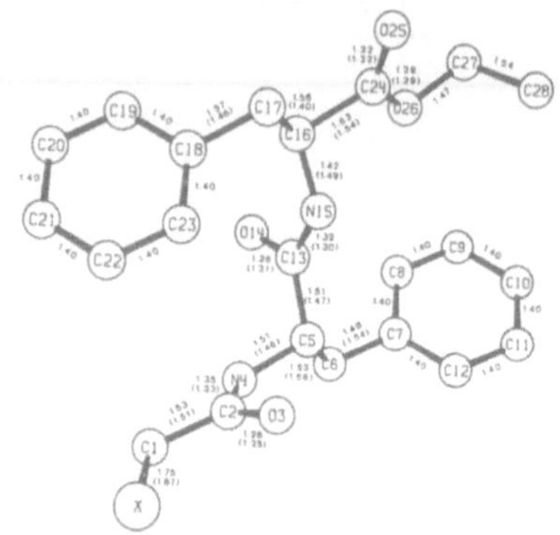

Fig. 1. Molecular configuration with bond distances for N-(haloacetyl)-L-phenylalanyl-L-phenylalanine ethyl ester. Values for the bromo-derivative are in parentheses.

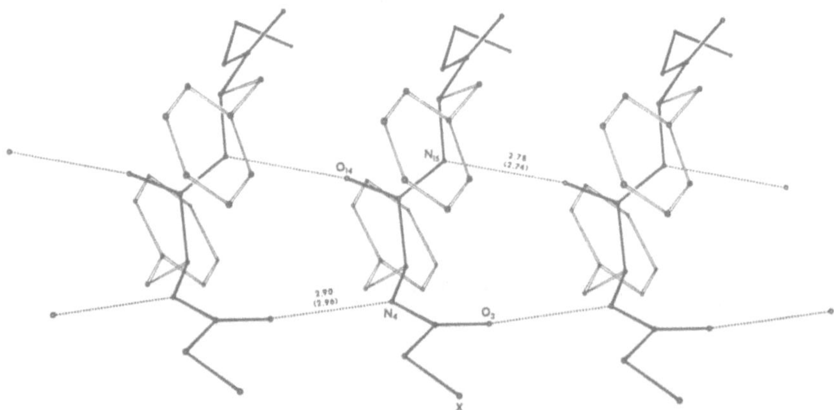

Fig. 2. Hydrogen bonding of parallel chains of N-(haloacetyl)-L-phenylalanyl-L-phenylalanine ethyl ester. The bromo derivative data are in parentheses.

PORPHINE

$C_{20}H_{14}N_4$

B.M.L. CHEN and A. TULINSKY, 1972. J. Amer. Chem. Soc., 94, 4144-4151.

Monoclinic, $P2_1/c$, a = 10.271, b = 12.089, c = 12.362 Å, β = 102.17°, D_m not given, Z = 4. Cu radiation, R = 0.059 for 1206 reflexions.

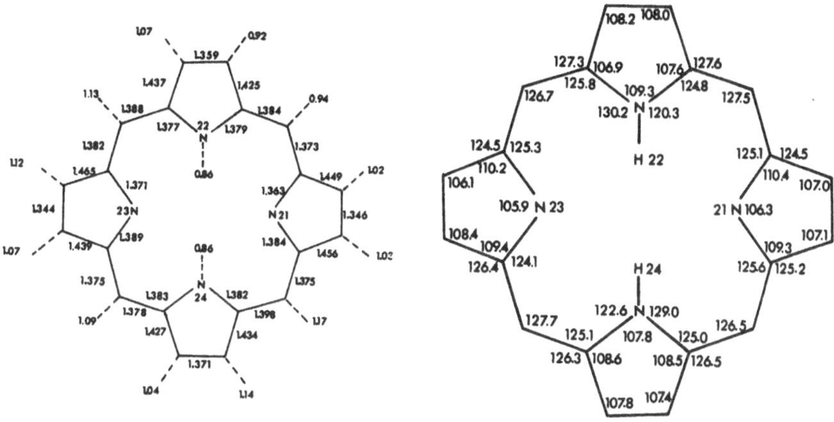

Fig. 1. Bond distances and angles in porphine (σ for C-C and C-N = 0.007 Å).

The structure of porphine has been redetermined with intensity measurements made in such a way as to optimize the reliability of the hydrogen atom structure, particularly that of the imino hydrogen atoms. The porphine molecule is composed of two like-opposite pairs of pyrroles rings, one pair of which carries imino hydrogen atoms. The different pairs of pyrroles show additional small but significant differences in bond distances and bond angles (Fig. 1).

The molecule is planar within about ±0.02 Å and possesses C_{2h} symmetry (approximate D_{2h} symmetry with the mirror planes bisecting the pyrrole rings). The different pyrroles are very similar to the two different pyrroles described in tetraphenylporphine (1). The structure of the free base macrocycle corresponds approximately with the hybrid of the two classical resonance forms of the porphine molecule. The new observations reported here were probably obscured in the original determination (2) by the presence of a metalloporphine impurity.

1. S.J. SILVERS and A. TULINSKY, 1967. J. Amer. Chem. Soc., 89, 3331.
2. L.E. WEBB and E.B. FLEISCHER, 1965. J. Chem. Phys., 43, 3100.

α,β,γ,δ-TETRA-n-PROPYLPORPHINE

$C_{32}H_{38}N_4$

P.W. CODDING and A. TULINSKY, 1972. J. Amer. Chem. Soc., 94, 4151-4157.

Monoclinic, $P2_1/c$, a = 5.078, b = 11.59, c = 22.39 Å, β = 99.50°, D_m = 1.22, Z = 2. Cu radiation, R = 0.054 for 1413 reflexions.

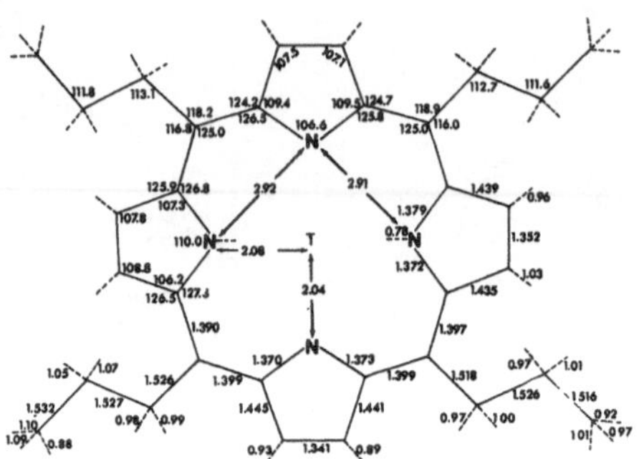

Fig. 1. Interatomic distances (for C-C and C-N, σ = 0.004 Å) and angles (σ = 0.2 - 0.4°) in tetrapropylporphine.

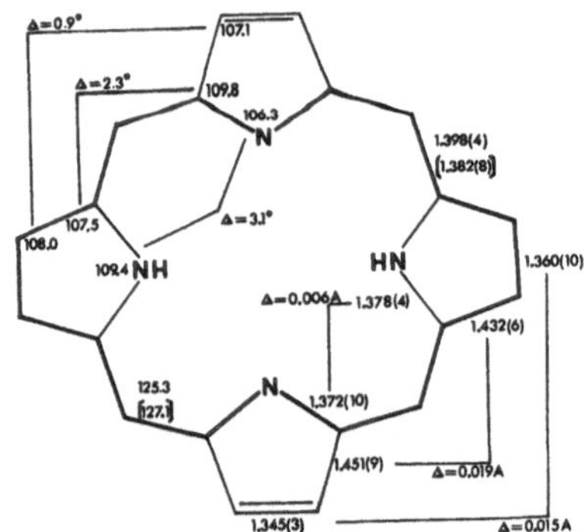

Fig. 2. Average structure for free base porphine. Standard deviations of distances in parentheses; standard deviations of angles 0.3 -0.7°. The average bridge distance and angle of porphine in brackets.

The molecule is centrosymmetric with two independent pyrrole rings. Details of molecular geometry are in Fig. 1. A common structure (Fig. 2) for the free base macrocycle was obtained by averaging results from tetraphenylporphine (1), porphine (2), and tetrapropylporphine structure analyses. The free base structure consists of opposite pyrrole rings possessing imino hydrogen atoms and the pyrrole rings differing in like pairs. Small differences which are localized at the bridge carbon atoms were found for the substituted compounds. In addition, the aliphatic substituent seems to decrease some of the differences between the two pyrrole rings. The individual pyrrole rings are planar to ±0.007 Å and the nucleus of the tetrapropylporphine ring is essentially planar (±0.04 Å). Some small deviations from planarity are probably due to packing effects with the closest intermolecular distance being 3.36 Å. The propyl groups are in an extended configuration efficiently filling the space between molecules in the crystal.

1. S.J. SILVERS and A. TULINSKY, 1967. J. Amer. Chem. Soc., 89, 3331.
2. Preceding report.

TETRABENZMONAZAPORPHIN

$C_{35}H_{21}N_5$

I.M. DAS and B. CHAUDHURI, 1972. Acta Cryst., B28, 579-585.

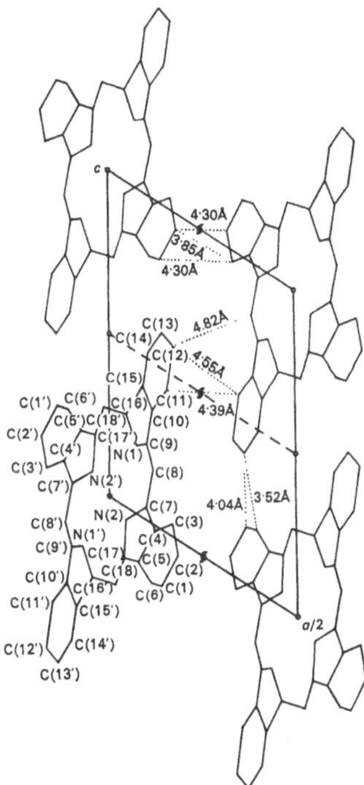

Fig. 1. The crystal structure of tetrabenzmonazaporphin viewed along b.

Monoclinic, $P2_1/a$, a = 17.63, b = 6.58, c = 12.50 Å, β = 123°24', D_m = 1.41, Z = 2, Cu radiation, R = 0.16 for some 200 h0ℓ reflexions, and 0.21 for some 80 0kℓ reflexions.

A view of the crystal structure is shown in Fig. 1. The molecules are centrosymmetric. The limited accuracy obtained with two-dimensional data precludes any detailed discussion of bond lengths or molecular conformation.

METHYL PHEOPHORBIDE a

$C_{36}H_{38}N_4O_5$

M.S. FISCHER, D.H. TEMPLETON, A. ZALKIN and M. CALVIN, 1972. J. Amer. Chem. Soc., 94, 3613-3619.

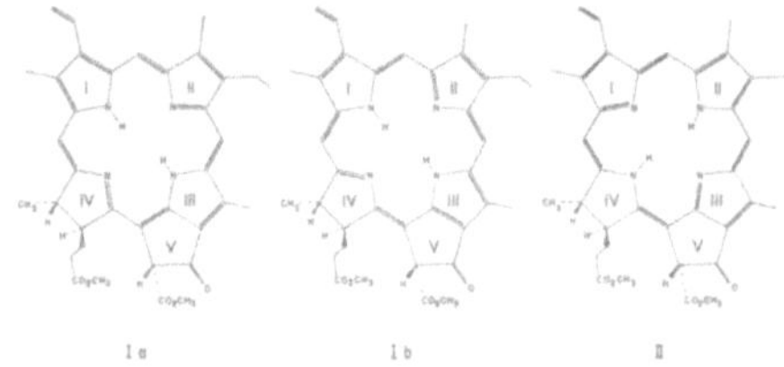

Fig. 1. Three resonance structures of two tautomers of methyl pheophorbide a.

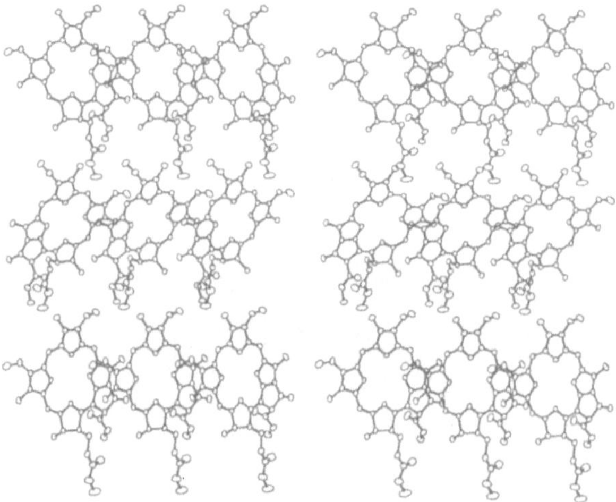

Fig. 2. View of the crystal structure of methyl pheophorbide a, viewed down the reciprocal c axis; a axis horizontal, b axis vertical.

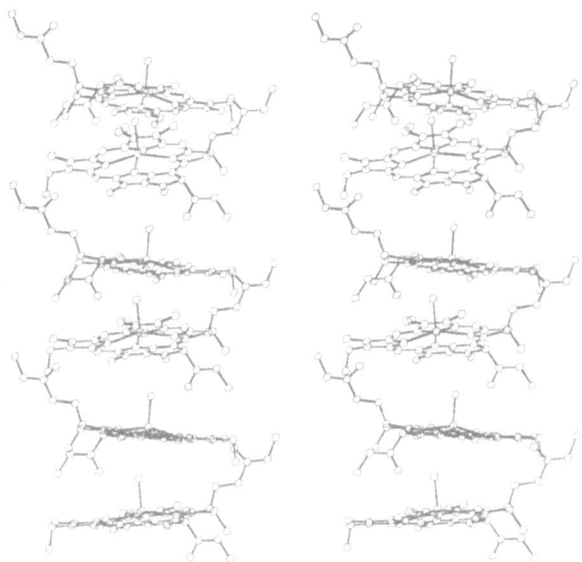

Fig. 3. View of proposed model of a chlorophyll-water polymer which might exist in photosynthetic lamellae. Only the first carbon atom of each phytyl group is shown. The region on each side of the depicted polymer contains phytyl chains, lipids, and, farther out, proteins.

Monoclinic, $P2_1$, a = 8.035, b = 28.531, c = 7.320 Å, β = 110.96°, D_m = 1.25, Z = 2. Cu radiation, R = 0.051 for 1616 reflexions.

Methyl pheophorbide a (Fig. 1) differs from chlorophyll a only by having two H atoms rather than a Mg atom in the centre and a methyl ester rather than a phytyl ($C_{20}H_{39}$) ester on the side chain off ring IV. The chlorin ring is nearly flat but the departures from planarity are statistically significant. The central hydrogen atoms are disordered but show a preference for rings I and III (Fig. 1). A view of the methyl pheophorbide a crystal structure is shown in Fig. 2. On the basis of this analysis a model for the arrangement of chlorophyll molecules in photosynthetic lamellae is proposed (Fig. 3) in which non-parallel chlorophyll molecules are related by a 2_1 screw axis and are linked by water molecules.

NICKEL(II) OCTAETHYLPORPHYRIN

$C_{36}H_{44}N_4Ni$

E.F. MEYER, 1972. Acta Cryst., B28, 2162-2167.

Tetragonal, $I4_1/a$, a = 14.93, c = 13.84 Å, D_m = 1.16, Z = 4. Mo radiation, R = 0.077 for 1205 reflexions. Molecular symmetry $\bar{4}$.

The five atoms in the pyrrole group N(1), C(1), C(2), C(3) and C(4) are planar while the macrocyclic ring of the porphyrin is nonplanar (Fig. 1). The nickel atom and the coordinated nitrogens form a normal square-planar coordination. The individual pyrrole groups lie at an angle of 14.2° to the nitrogen-nickel plane and therefore the groups are at an angle of 28.4° to each other. The nickel-pyrrole nitrogen atom bond distance is 1.929(3) Å, the shortest metal-nitrogen bond distance for a metalloporphyrin found so far. The independent bond lengths, angles, and internal geometry of the pyrrole group show a close relationship, so that the pseudo mirror plane perpendicular to the pyrrole ring and passing through the nitrogen atom is maintained. No unusual short intermolecular contact distances are observed.

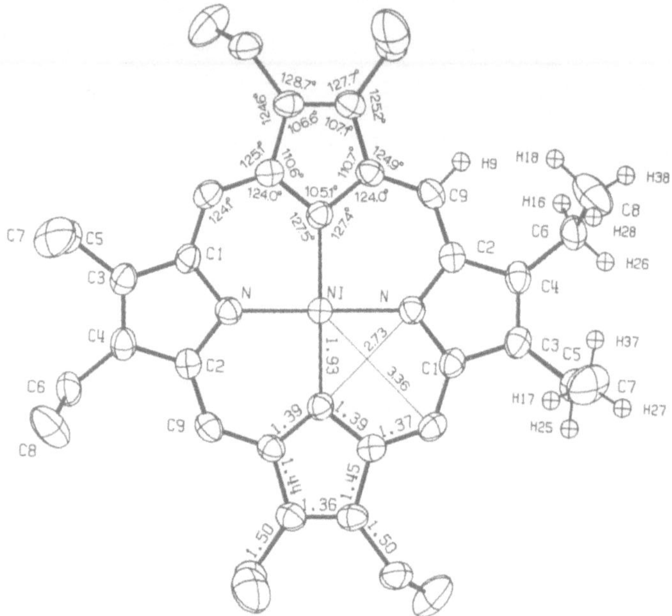

Fig. 1. The geometry of nickel(II) octaethylporphyrin as viewed down c.

(TETRAPHENYLPORPHINE)SILVER-TETRAPHENYLPORPHINE

$(C_{44}H_{28}AgN_4)_{0.54}(C_{44}H_{30}N_4)_{0.46}$

M.L. SCHNEIDER, 1972. J. Chem. Soc., Dalton, 1093-1096.

Tetragonal, I4/m, a = 13.384, c = 9.717 Å, D_m = 1.29, Z = 2. Mo radiation, R = 0.085 for 1221 reflexions.

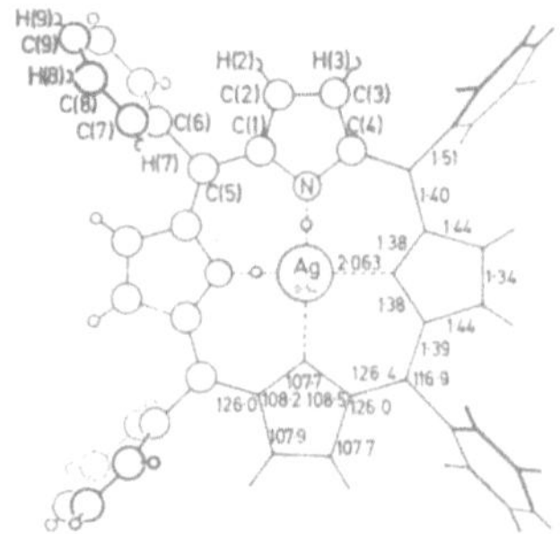

Fig. 1. View of the (tetraphenylporphine)silver-tetraphenylporphine system, showing selected bond lengths, angles, and atomic site-occupancy; two hydrogen atoms are not shown.

The free-base molecules are disordered in two orientations related to each other by a 90° rotation about an axis passing through the centre of the molecule and normal to the porphine ring. The averaged molecule is centred on a four-fold axis, its point symmetry is 4/m but it deviates only slightly from 4/mmm symmetry. The molecular geometry (Fig. 1) is consistent with that of related structures. A conformational feature is that phenyl rings are perpendicular to the macrocyclic ring.

α,β,γ,δ-TETRAPHENYLPORPHINATODICHLOROTIN(IV)

$C_{44}H_{28}C\ell_2N_4Sn$ ($C\ell_2SnTPP$)

D.M. COLLINS, W.R. SCHEIDT and J.L. HOARD, 1972. J. Amer. Chem. Soc., 94, 6689-6696.

Tetragonal, I4/m, a = 13.673, c = 9.961 Å, D_m = 1.43, Z = 2. Mo radiation, R = 0.04 for 3909 reflexions.

The large strongly complexing tin(IV) atom is centred in a highly expanded porphinato core of required C_{4h} symmetry (Fig. 1) with no significant departure from full D_{4h} symmetry. Main bond distances are Sn-Cℓ 2.420(1), Sn-N 2.098(2) Å. The porphinato core is compared with that of a nickel(II) porphyrin (1) (Fig. 2).

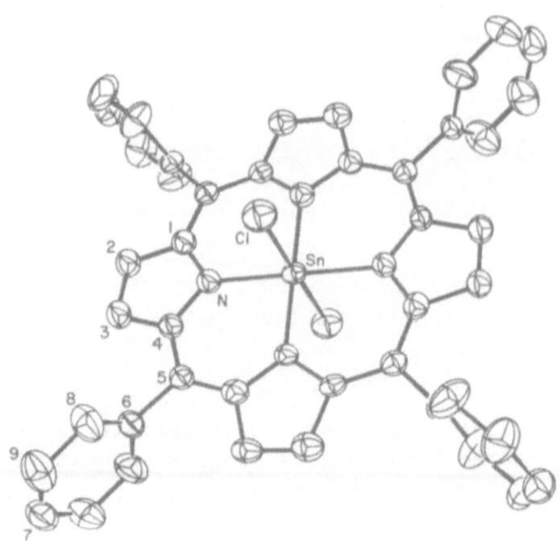

Fig. 1. The $C\ell_2SnTPP$ molecule.

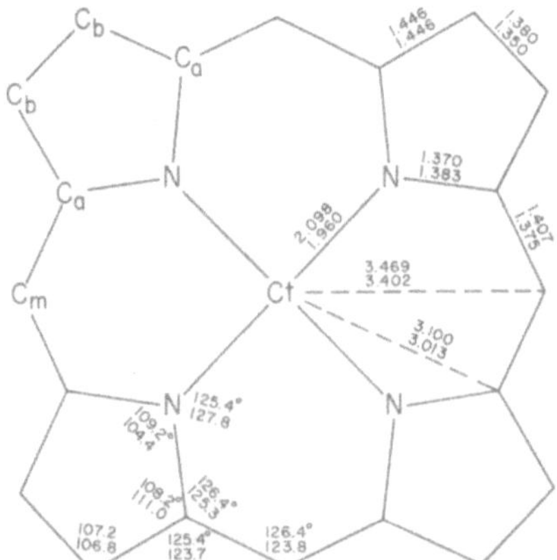

Fig. 2. Diagram of the carbon-nitrogen skeleton in porphinato core of a metalloporphyrin for the effective retention of D_{4h} symmetry. Values of the principal radii, bond lengths, and bond angles in $C\ell_2SnTPP$ and a nickel(II) porphyrin (1) are entered on the diagram; the upper datum in each pairing is the value in the tin(IV) porphyrin.

1. T.A. HAMOR, W.S. CAUGHEY and J.L. HOARD, 1965. J. Amer. Chem. Soc., 87, 2305.

BIS(IMIDAZOLE)-α,β,γ,δ-TETRAPHENYLPORPHINATOIRON(III) CHLORIDE METHANOL SOLVATE

$C_{51}H_{40}ClFeN_8O$ $(Im_2FeTPP)^+Cl^-.CH_3OH$

D.M. COLLINS, R. COUNTRYMAN and J.L. HOARD, 1972. J. Amer. Chem. Soc., 94, 2066-2072.

Monoclinic, $P2_1/n$, a = 14.265, b = 16.390, c = 18.121 Å, β = 90.83°, D_m = 1.36, Z = 4. Cu radiation, R = 0.076 for 5299 reflexions.

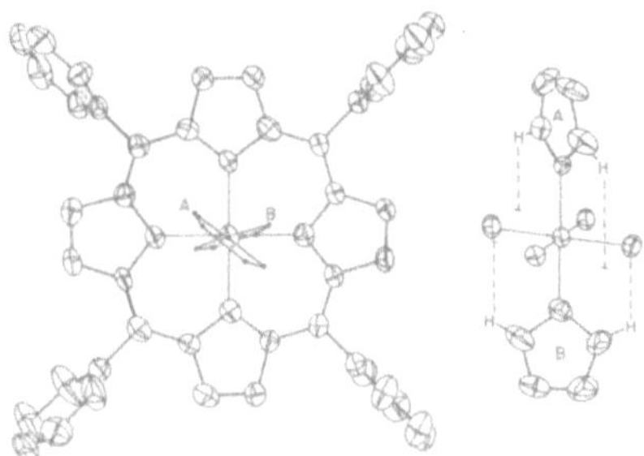

Fig. 1. Perspective drawing of the $[Im_2FeTPP]^+$ ion and at the right the coordination group.

The octahedral FeN_6 coordination group (Fig. 1) in the $(Im_2FeTPP)^+$ ion is dimensionally quasitetragonal. Equatorial Fe-N bond lengths are constrained by the porphinato core to the relatively high averaged value of 1.989 Å. Axial Fe-N bond lengths are 1.957(4) and 1.991(5) Å to, respectively, well oriented and badly oriented imidazole ligands; the latter is strongly hydrogen bonded through its N-H group to the chloride ion, and is involved in substantial steric repulsions with porphinato nitrogen atoms. Stereochemical parameters of the porphinato core are consistent with the structural data from other iron porphyrins.

BIS(PIPERIDINE)-α,β,γ,δ-TETRAPHENYLPORPHINATOIRON(II)

$C_{54}H_{50}FeN_6$

L.J. RADONOVICH, A. BLOOM and J.L. HOARD, 1972. J. Amer. Chem. Soc., 94, 2073-2078.

Triclinic, $P\bar{1}$, a = 11.113, b = 12.071, c = 9.797 Å, α = 105.67, β = 113.70, γ = 101.02°, D_m = 1.26, Z = 1. Mo radiation, R = 0.084 for 4267 reflexions.

The octahedral FeN_6 coordination group in the centrosymmetric molecule approximates dimensionally to full tetragonal symmetry. Equatorial Fe-N bond lengths average to 2.004 Å. Extension of the axial Fe-N bond length to 2.127(3) Å and the accompanying opening of the FeNC bond angles at piperidine nitrogen to 116.8(2)° are attributable to unavoidable steric interactions between nitrogen atoms of the porphinato core and the piperidine hydrogen atoms contiguous thereto. Bond distances and angles of any specified chemical class in the porphinato core display quite trivial departures from effective D_{4h} symmetry.

NEOVITAMIN B_{12}

$C_{63}H_{88}CoN_{14}O_{14}P.(probably)20H_2O$

H. STOECKLI-EVANS, E. EDMOND and D.C. HODGKIN, 1972. J. Chem. Soc., Perkin II, 605-614.

Orthorhombic, $P2_12_12_1$, a = 24.98, b = 22.06, c = 15.74 Å, D_m = 1.33, Z = 4. Cu radiation, R = 0.159 for 2986 reflexions.

The molecule (Fig. 1) is found to be cyano-13-epicobalamin, an isomer of cyanocobalamin, vitamin B_{12}. The differences in vitamin B_{12} and neovitamin B_{12} are illustrated in Fig. 2, which shows a cylindrical projection of the corrin rings as seen from the central metal atom outwards. Apart from the single propionamide chain at C(13) the position of the atoms in the neo-B_{12} molecule and many of the water molecules surrounding it are close to those found in the B_{12} crystal structures, rather nearer to those in air-dry (1) than wet B_{12} (2). The principal structural differences in neovitamin B_{12} compared to vitamin B_{12} are found to be a change in configuration at C(13) and a change in conformation in the tilt of the ββ' bond, C(12)-C(13). The propionamide side-chain attached to C(13) now rises vertically at right angles to the plane of the corrin nucleus in the same direction as the cyanide group attached to the cobalt atom. In B_{12}, this chain is also directed perpendicularly to the corrin nucleus but in the opposite direction.

The distances and angles in the molecule are shown in Fig. 3; those in the nucleotide, phosphate and propanolamine groups show little difference from those in B_{12} itself. Density measurements suggested the presence of twenty water molecules but only seventeen could be located (ten very clearly and a further seven affected by disorder).

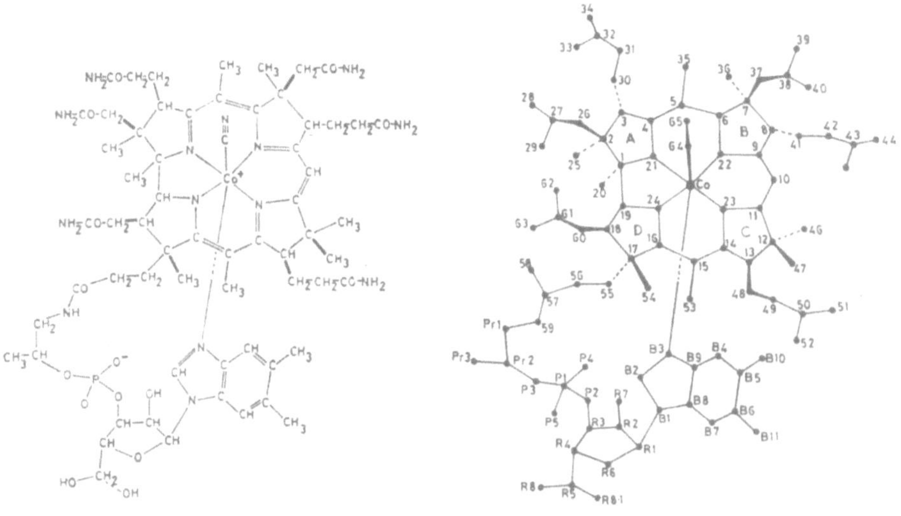

Fig. 1. Formula and numbering system used for neovitamin B_{12}.

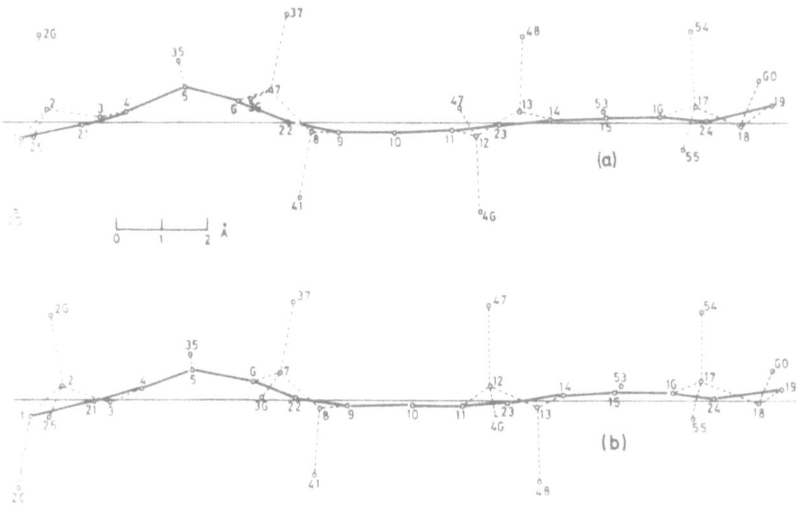

Fig. 2. A cylindrical projection of (a) neovitamin B_{12} and (b) vitamin B_{12}: the molecules are shown as they would be seen when viewed from the central metal atom outwards. The atoms are projected on a cylinder of radius 2·8 Å. Vertical displacement of each atom corresponds to the distance from the least-squares plane through the four nitrogen atoms (21)-(24).

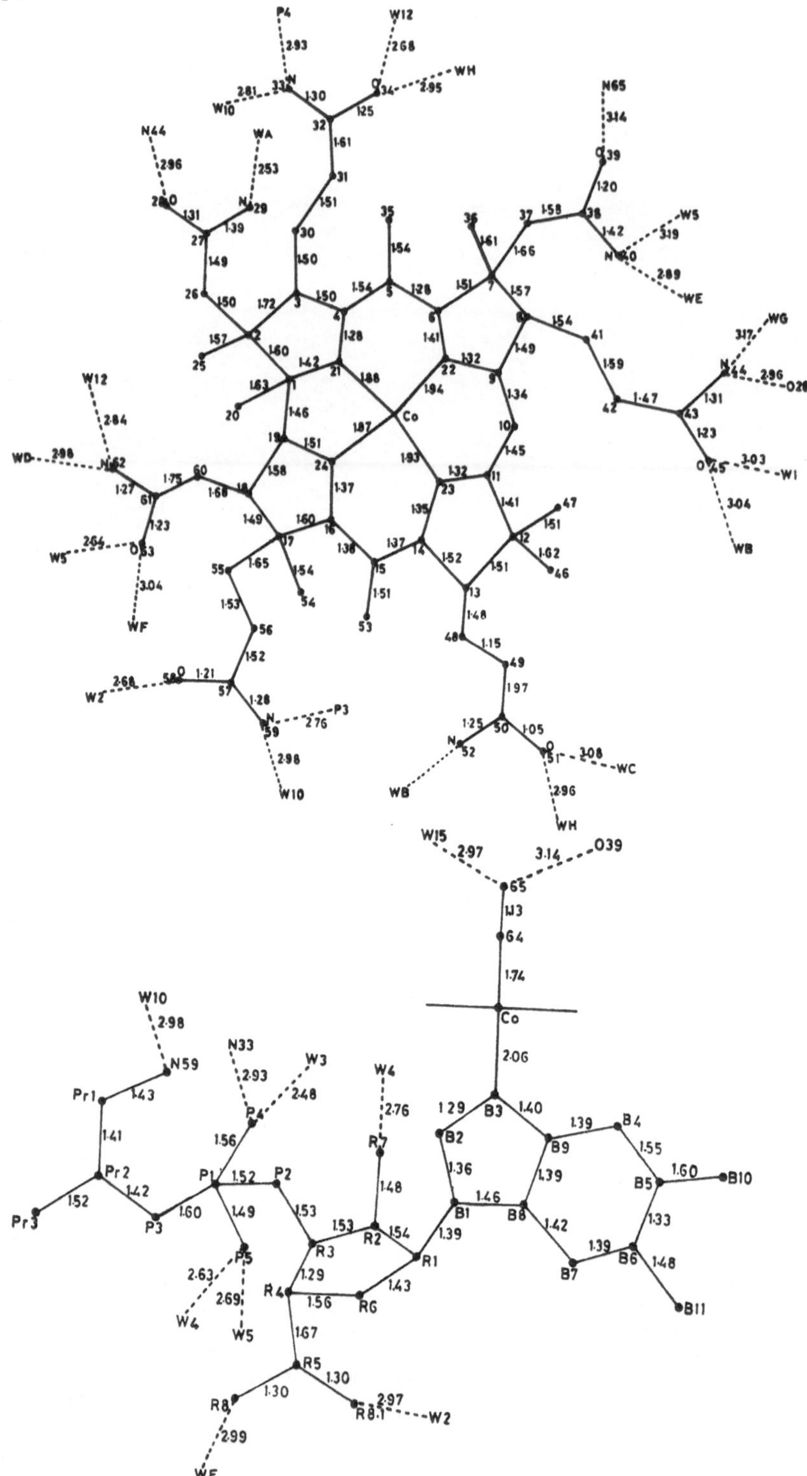

P4
W12
WH
W10
N65
N44
WA
Co
W5
WE
WG
O28
WD
WI
WB
WF
W2
P3
WC
W15
O39
W4
W3
N33
N59
Pr1
Pr2
Pr3
P1
P2
P3
P4
P5
R1
R2
R3
R4
R5
R6
R7
R8
R8.1
B1
B2
B3
B4
B5
B6
B7
B8
B9
B10
B11

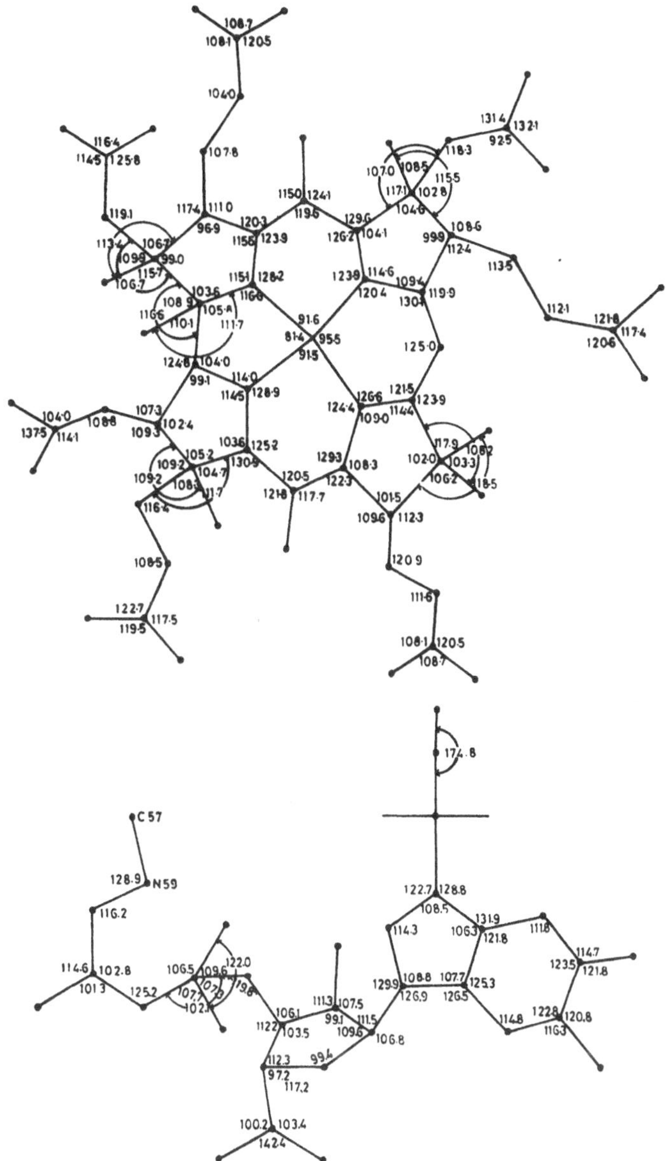

Fig. 3. Bond lengths and angles in neovitamin B_{12}. Mean bond length σ values range from 0.017 for Co-N at the centre to 0.09 Å for amide $C\text{-}NH_2$ at the periphery of the molecule.

1. Structure Reports, 27, 1042.
2. Ibid., 29, 754.

μ-OXO-BIS[α,β,γ,δ-TETRAPHENYLPORPHINATOIRON(III)]

$C_{88}H_{56}Fe_2N_8O$ $O(FeTPP)_2$

A.B. HOFFMAN, D.M. COLLINS, V.W. DAY, E.B. FLEISCHER, T.S. SRIVASTAVA and J.L. HOARD, 1972. J. Amer. Chem. Soc., 94, 3620-3626.

Orthorhombic, C2ca, a = 15.197, b = 25.077, c = 18.074 Å, D_m = 1.28, Z = 4. Mo radiation, R = 0.0709 for 3501 reflexions.

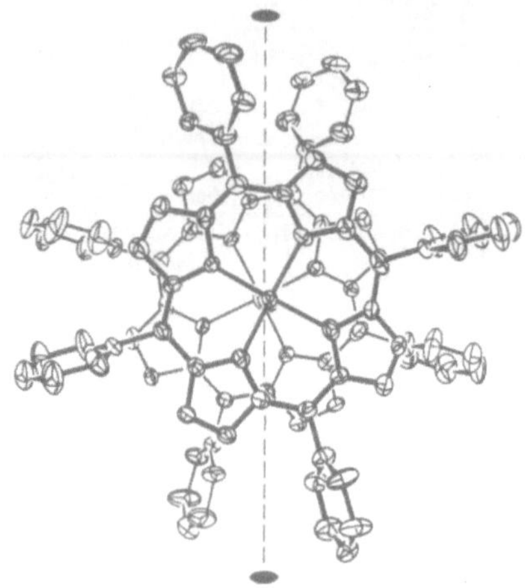

Fig. 1. Perspective drawing of the $O(FeTPP)_2$ molecule. The required twofold axis, which passes through the bridging oxo oxygen atom, requires structural equivalence of the upper and lower halves of the oligomer.

A perspective drawing of the structure is given in Fig. 1. The binuclear $N_4FeOFeN_4$ coordination group of the oligomer approximates D_{4d}-$\bar{8}$2m symmetry with an Fe-O bond length of 1.763(1) Å, an averaged Fe-N distance of 2.087(3) Å, and an FeOFe angle of 174.5(1)°. The square-pyramidal five coordination around each iron atom is entirely typical for a high-spin ferric porphyrin, the strong antiferromagnetic interaction of the ferric ions through the Fe-O-Fe linkage notwithstanding. Thus the ferric ion is displaced 0.50 Å from the mean plane of the porphinato nitrogen atoms toward the axial oxo ligand. As is usual in crystalline porphyrins, the porphine skeleton departs significantly from planarity. Bond lengths in the skeleton are in good agreement with those in other iron porphyrins.

TETRAHYDROPENTALENOLACTONE BROMOHYDRIN HEMIACETONATE

$C_{15}H_{21}BrO_5.\frac{1}{2}(C_3H_6O)$

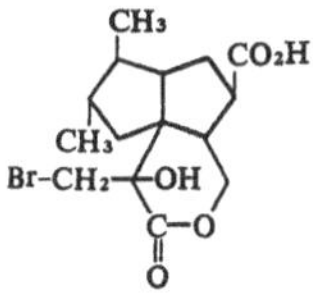

D.J. DUCHAMP and C.G. CHIDESTER, 1972. Acta Cryst., B28, 173-180.

Monoclinic, C2, a = 19.007, b = 7.886, c = 12.223 Å, β = 102.47°, D_m = 1.453, Z = 4. Cu radiation, R = 0.077 for 1818 reflexions.

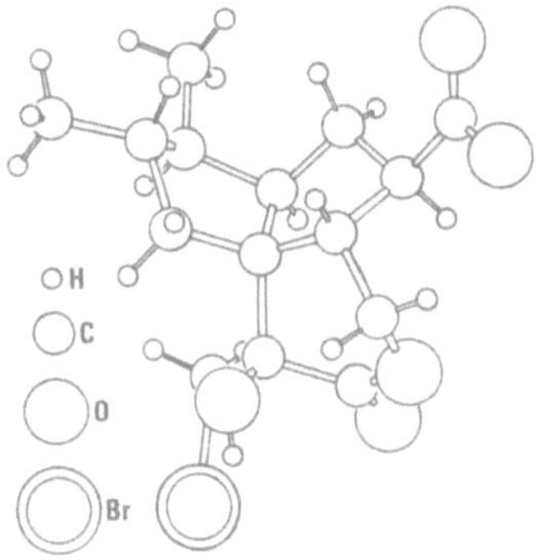

Fig. 1. Absolute configuration and conformation of tetrahydropentalenolactone bromohydrin.

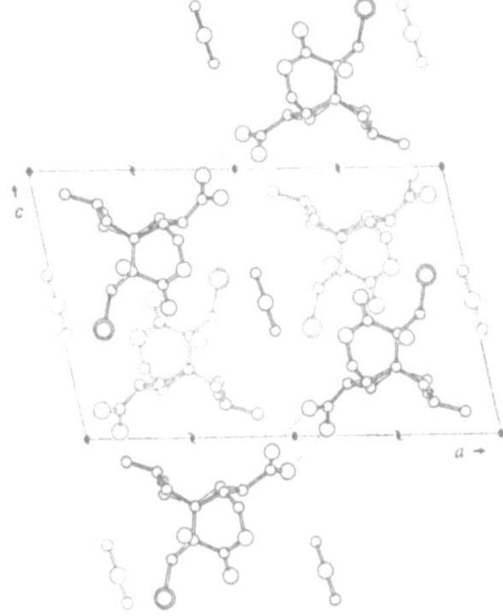

Fig. 2. The tetrahydropentalenolactone bromohydrin hemiacetonate structure viewed down b. Dotted lines indicate hydrogen bonds.

The absolute configuration and conformation of the molecule are shown in Fig. 1. Bond distances and angles are normal. The packing arrangement (Fig. 2) involves an acetone molecule on a twofold axis and a disordered carboxyl group hydrogen bonding across another twofold axis with O-H...O distances 2.60(2) and 2.66(2) Å. Another O-H...O hydrogen bond, 2.79(1) Å, occurs between the hydroxyl group and the lactone carbonyl oxygen of a neighbouring molecule.

PUROMYCIN DIHYDROCHLORIDE PENTAHYDRATE

$C_{22}H_{41}C\ell_2N_7O_{10}$

M. SUNDARALINGAM and S.K. ARORA, 1972. J. Mol. Biol., 71, 49-70.

Orthorhombic, $P2_12_12_1$, a = 28.642, b = 13.211, c = 7.836 Å, D_m = 1.418, Z = 4. Cu radiation, R = 0.055 for 1968 reflexions.

Puromycin as its dihydrochloride exists as a dication (see Fig. 1 for structure and geometry); N(1) of the adenine base and the amino group of the amino acid are protonated. Both the adenosyl- and p-methoxyphenylalanyl residues occur in the preferred conformations. The nucleoside is in the anti conformation, χ = 19.3°, with C(3')endo-C(2')exo ($^3T_2$) puckering for the ribose, and the gauche-gauche conformation about the C(4')-C(5') bond. The amino acid is in the syn conformation about the C(2)-C(3) bond. The peptide group is approximately planar and it, too, exhibits the preferred conformation relative to the ribose. The symmetry related puromycin molecules form a zig-zagging "wall" normal to the ab plane (Fig. 2). The water molecules and chloride ions fill the channels between the walls. There are about twelve hydrogen bonds per molecule of puromycin. The chloride ions show tetrahedral coordination, while the water molecules show both three and four coordination with variable hydrogen bonding geometries. The adenine bases and the p-methoxyphenyl rings form alternating stacks with inter-planar spacing of 3.4 Å.

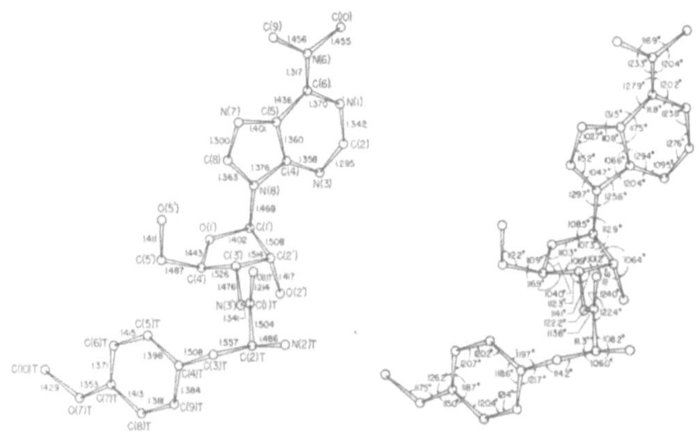

Fig. 1. The bond distances (σ = 0.008 Å) and angles (σ = 0.4°) in puromycin dihydrochloride.

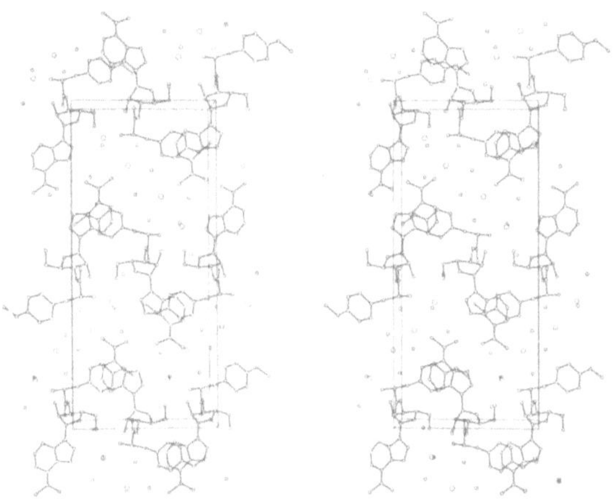

Fig. 2. Puromycin dihydrochloride pentahydrate: a drawing showing the molecular packing as viewed along c. The large circles indicate the halide ions while the small circles indicate the water molecules.

CAESIUM CHLOROTHRICOLIDE METHYL ESTER TRIHYDRATE

$C_{30}H_{43}O_{11}Cs$ $C_{30}H_{37}O_8Cs.3H_2O$

M. BRUFANI, S. CERRINI, W. FEDELI, F. MAZZA and R. MUNTWYLER, 1972. Helv. Chim. Acta, 55, 2094-2102.

Triclinic, P1, a = 11.698, b = 11.533, c = 14.931 Å, α = 107.3, β = 125.0, γ = 94.3°, Z = 2. Mo radiation, R = 0.064 for 3710 reflexions.

The structure and absolute stereochemistry of chlorothricolide methyl ester derived from this X-ray study are shown in Fig. 1. The two independent molecules in the unit cell have different conformations of the macrocyclic ring (Fig. 2). Main differences occur in the torsion angles at C(15)-C(16) [126.5 and -25.6°] and at C(17)-C(18) [101.2° and -101.2°]. The differences arise from the C(16)-C(17) double bond having different orientations in the two molecules. In the crystal structure there is an extensive system of hydrogen bonding involving the anions and water molecules. Each Cs ion is surrounded by six oxygen atoms (from both anions and water molecules) with Cs...O in the range 2.87 - 3.28 Å.

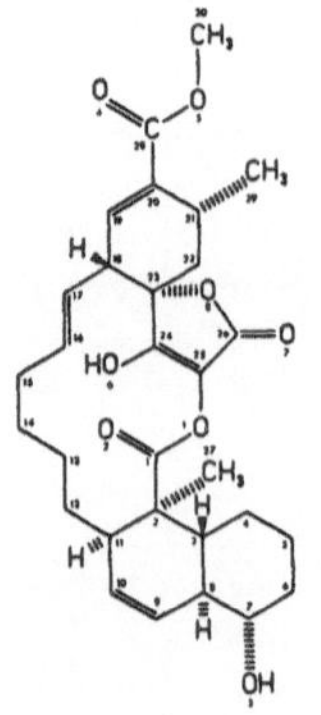

Fig. 1. The structure and absolute stereochemistry of chlorothricolide methyl ester.

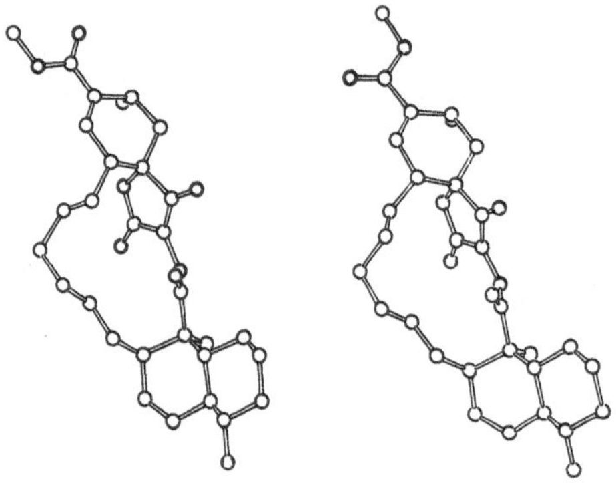

Fig. 2. The conformations of the two independent chlorothricolide methyl ester anions.

KINAMYCIN C p-BROMOBENZOATE BENZENE SOLVATE

$C_{37}H_{29}N_2O_{11}$ $C_{31}H_{23}N_2O_{11}.C_6H_6$

A. FURUSAKI, M. MATSUI, T. WATANABÉ, S. OMURA, A. NAKAGAWA and T. HATA, 1972. Israel J. Chem., 10, 173-187.

Monoclinic, $P2_1$, a = 18.404, b = 21.299, c = 9.049 Å, β = 90.07°, Z = 4. Cu radiation, R = 0.089 for 5387 reflexions.

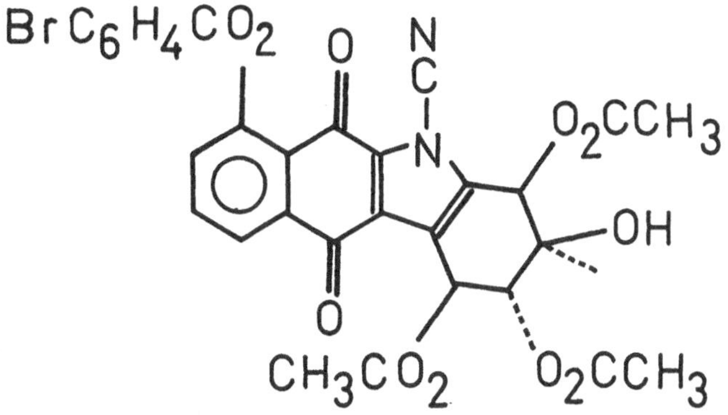

Fig. 1. The structure and absolute stereochemistry of kinamycin C p-bromobenzoate.

The conformations of the two independent molecules in the asymmetric unit are almost identical. The molecular skeleton and absolute stereochemistry are shown in Fig. 1. The cyclohexene ring adopts a half-chair conformation. Bond lengths and angles are essentially normal. In the crystal structure there is O-H...O hydrogen bonding (2.78 Å) along *c*. The benzene molecules occupy crevices between layers.

ISOZYGOSPORIN A p-BROMOBENZOATE ISOPROPYL ALCOHOL SOLVATE

$C_{40}H_{48}BrNO_8$ $C_{37}H_{40}BrNO_7.C_3H_8O$

Y. TSUKUDA and H. KOYAMA, 1972. J. Chem. Soc., Perkin II, 739-744.

Orthorhombic, $P2_12_12_1$, a = 22.535, b = 12.332, c = 13.917 Å, D_m = 1.281, Z = 4. Mo radiation, R = 0.098 for 2318 reflexions.

The analysis establishes the structure and absolute configuration as (I).

I

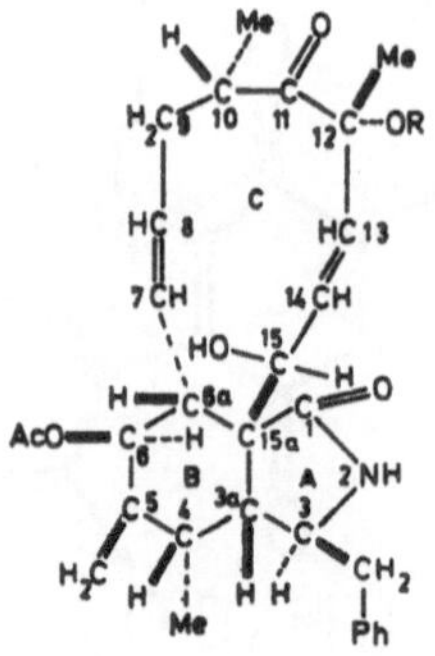

R = -OC·C_6H_4·Br-*p*

The skeleton of the molecule consists of a five-membered ring A, a six-membered ring B, and an eleven-membered ring C. For the most part the corresponding bond distances are not significantly different from expected values. Ring A is a slightly distorted envelope and ring B has a distorted boat conformation with atoms C(5), C(6), C(3a), C(15a) in the plane. The γ-lactam ring A is cis-fused to the six-membered ring B, and the junction between ring B and ring C is trans. The most significant intermolecular distances are N(2)-H...O(1') = 2.971 Å, and O(1)..H-O(7) = 2.655 Å, and these represent typical hydrogen bonds, the former involving the isopropyl alcohol solvate [which surprisingly is planar].

NONACTIN

$C_{40}H_{64}O_{12}$

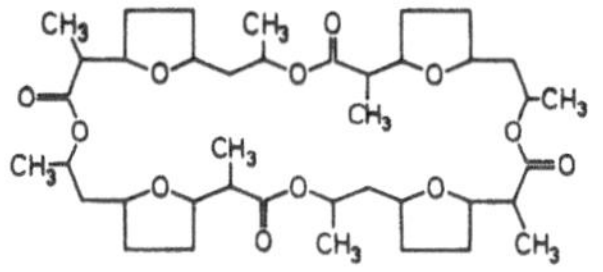

M. DOBLER, 1972. Helv. Chim. Acta, 55, 1371-1384.

Orthorhombic, Pbam, a = 47.57, b = 31.40, c = 5.69 Å, D_m = 1.11 - 1.15, Z = 8. Mo radiation, R = 0.103 for 1344 reflexions.

Crystals of nonactin itself give diffraction patterns with unusual symmetry and extinctions (1); the crystal structure is of the OD (order-disorder) type, which makes a detailed analysis difficult. The conformation of free nonactin (Fig. 1) is similar to that of the K^+-complex (2) in that both have approximate S_4 symmetry. The free molecule is, however, much flatter (17x17x8.5 Å) than the complexed molecule (15x15x12 Å). The conformational changes involved in passing from free to complexed nonactin affect only eight torsion angles of the 32-membered ring. The other 24 torsion angles remain virtually unchanged. The shape of the uncomplexed nonactin molecule can be described as a torus. The central hole is large enough to allow close approach of a hydrated K^+-ion. Stepwise removal of water molecules may be coupled to the conformational changes involved in complexation.

The packing of the molecules in the unit cell cannot be described in a straight-forward way. The crystal structure is built from perfectly ordered layers in the b, c directions, but the structure of these layers is such that successive pairs of layers may be stacked in two ways, equivalent with respect to neighbouring layers but distinct with respect to more distant ones. This characterizes the structure as an OD (order-disorder) structure.

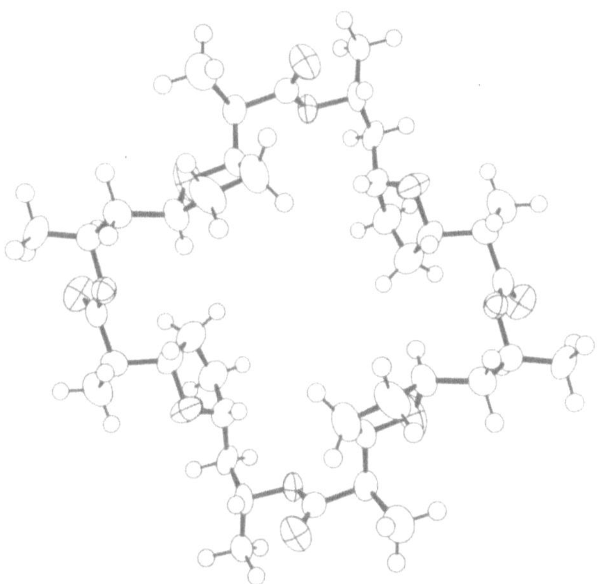

Fig. 1. The nonactin molecule viewed down the twofold axis.

1. J. DOMINGUEZ, J.D. DUNITZ, H. GERLACH and V. PRELOG, 1962. Helv. Chim. Acta, 45, 129.
2. M. DOBLER, J.D. DUNITZ and B.T. KILBOURN, 1969. Helv. Chim. Acta, 52, 2573.

8-AZAESTRADIOL

$C_{17}H_{23}NO_2$

J.N. BROWN and L.M. TREFONAS, 1972. J. Amer. Chem. Soc., 94, 4311-4314.

Monoclinic, $P2_1/c$, a = 9.782, b = 19.957, c = 10.947 Å, β = 137.61°, D_m not given, Z = 4. Cu radiation, R = 0.04 for 1373 reflexions.

The steroid skeleton is in the natural configuration (with hydrogen at 9 and 14 α, and the methyl at 13 β). Bond distances and angles (Fig. 1) are similar to those in estradiol derivatives. The hydroxyl hydrogen atoms link the molecules in the crystal via O-H...N and O-H...O hydrogen bonds (H...N = 1.92, H...O = 1.78 Å)

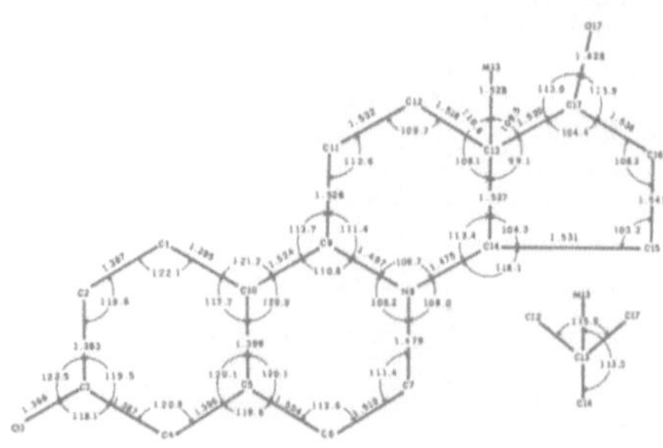

Fig. 1. 8-Azaestradiol with distances and angles (σ = 0.005 Å and 0.3°).

3-HYDROXYESTRA-1,3,5(10)-TRIEN-17-ONE

(ESTRONE)

$C_{18}H_{22}O_2$

T.D.J. DEBAERDEMAEKER, 1972. Cryst. Struct. Comm., 1, 39-42.

Orthorhombic, $P2_12_12_1$, a = 18.520, b = 7.815, c = 10.083 Å, Z = 4. Mo radiation, R = 0.09 for 842 reflexions.

The intramolecular angles and distances are in good agreement with their theoretical values. The cohesion of the crystal is due to a hydrogen bond between the carbonyl group of one molecule C(17)-O(19) and the hydroxyl group of another C(3')-O(20'). The phenyl ring and the atoms C(9) and C(6), all lie in the same plane. The cyclohexane ring is in the form of a chair and the cyclopentane ring in the form of an envelope (C(14) out of the plane).

ESTRADIOL HEMIHYDRATE

$C_{18}H_{24}O_2.1/2H_2O$

B. BUSETTA and M. HOSPITAL, 1972. Acta Cryst., B28, 560-567.

Orthorhombic, $P2_12_12_1$, a = 12.055, b = 19.280, c = 6.632 Å, D_m = 1.20, Z = 4. Cu radiation, R = 0.065 for 1024 reflexions.

Fig. 1. Bond lengths (σ = 0.008 Å) and angles (σ = 0.8°) in estradiol.

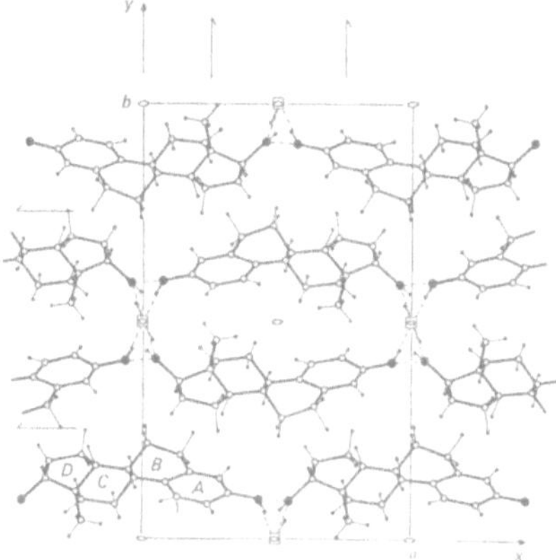

Fig. 2. The crystal structure of hemihydrated estradiol viewed along c.

Molecular dimensions are in Fig. 1. Ring A is essentially planar and ring B is in half-chair conformation (C(8) is 0.74 Å from the plane of C(5), C(6), C(7), C(9), C(10)). The molecular conformation corresponds approximately to that found for estriol (1). The water molecule lies on a crystallographic two-fold axis and forms hydrogen bonds with O(3) and O(17). The molecular packing is shown in Fig. 2. There are three independent hydrogen bonds, O(3)-H...O(water) = 2.792, O(17)-H...O(3) = 2.774, and O(water)-H...O(17) = 2.855 Å.

1. A. COOPER, D. NORTON and H. HAUPTMAN, 1969. Acta Cryst., B25, 814; Structure Reports, 34B.

17β-HYDROXY-8[9→10β]abeo-ESTR-4-EN-3,10-DIONE

$C_{18}H_{24}O_3$

M.O. CHANEY and N.D. JONES, 1972. Cryst. Struct. Comm., 1, 197-202.

Monoclinic, $P2_1$, a = 11.204, b = 9.124, c = 7.907 Å, β = 110.58°, D_m = 1.252, Z = 2. Cu radiation, R = 0.080 for 1300 reflexions.

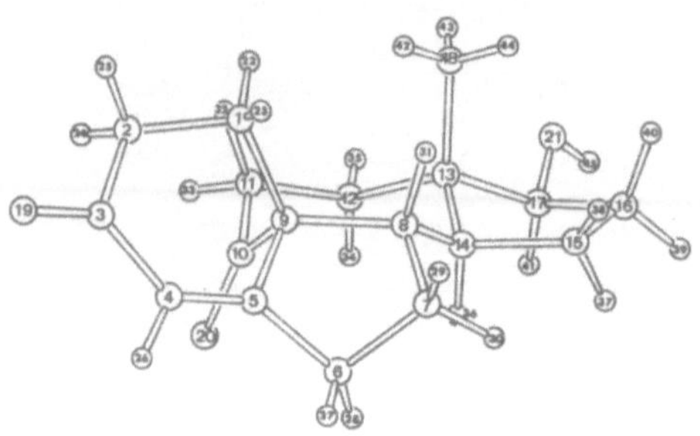

Fig. 1. The structure of 17β-hydroxy-8[9→10β]abeo-estr-4-en-3,10-dione.

The bond distances and angles are in good agreement with the expected values. Beacuse of the C(4)-C(5) double bond, the A ring assumes a skewed-chair conformation (Fig. 1). The 5-membered B and D rings are in an envelope conformation, with C(6) and C(13) out of their respective planes. The molecules are held together along the b direction in the crystal by one moderately strong hydrogen bond (2.809 Å) between the hydroxyl oxygen, O(21), on one molecule and the carbonyl oxygen, O(19), on another. The O(21)-H...O(19) intermolecular hydrogen bond angle is 151°.

17β-HYDROXYESTR-5(10)-EN-3-ONE

$C_{18}H_{26}O_2$

R.R. SOBTI, S.G. LEVINE and J. BORDNER, 1972. Acta Cryst., B28, 2292-2297.

Orthorhombic, $P2_12_12_1$, a = 9.575, b = 8.087, c = 19.495 Å, D_m = 1.18, Z = 4. Cu radiation, R = 0.040 for 775 reflexions.

The conformation of ring A is not the expected half-chair but the semiplanar conformation (Fig. 1) similar to that found in the 17β-iodoacetate derivative (1).

The bond distances and angles have normal steroid values. The only close-contact distance is 2.88 Å between the alcoholic oxygen, O(17A), and the carbonyl oxygen, O(3A). It has been assumed that crystal packing forces do not cause the unusual conformation of ring A.

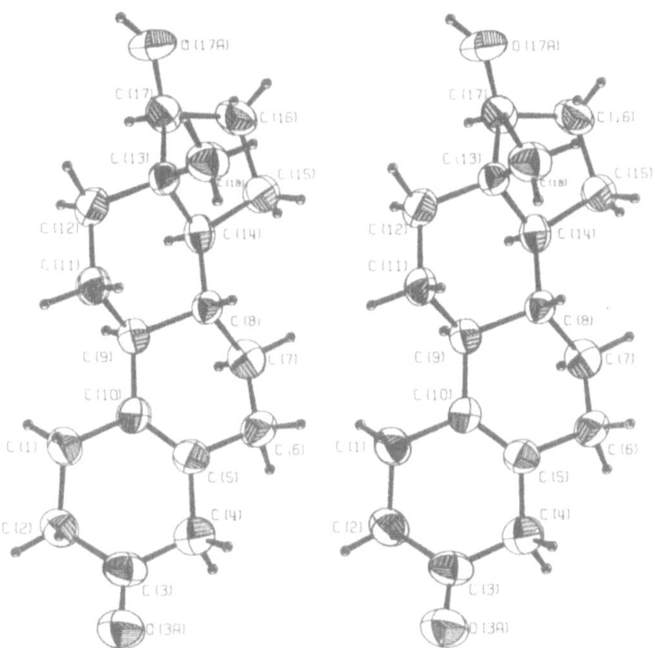

Fig. 1. A stereoscopic view of the 17β-hydroxyestr-5(10)-en-3-one molecule. Atom C(2) lies -0.67 Å from the plane through carbon atoms 1,4,5,6,9,10.

1. J. BORDNER, R.R. SOBTI and S.G. LEVINE, 1971. J. Amer. Chem. Soc., 93, 5588.

ANDROST-4-EN-3,17-DIONE

$C_{19}H_{26}O_2$

B. BUSETTA, G. COMBERTON, C. COURSEILLE and M. HOSPITAL, 1972. Cryst. Struct. Comm., 1, 129-133.

Orthorhombic, $P2_12_12_1$, a = 12.978, b = 16.949, c = 7.365 Å, Z = 4. Cu radiation, R = 0.044 for 1408 reflexions.

The molecular packing (Fig. 1) is governed by van der Waals forces. Ring D is close to a β-envelope conformation. Bond distances are in accord with those found in related steroids.

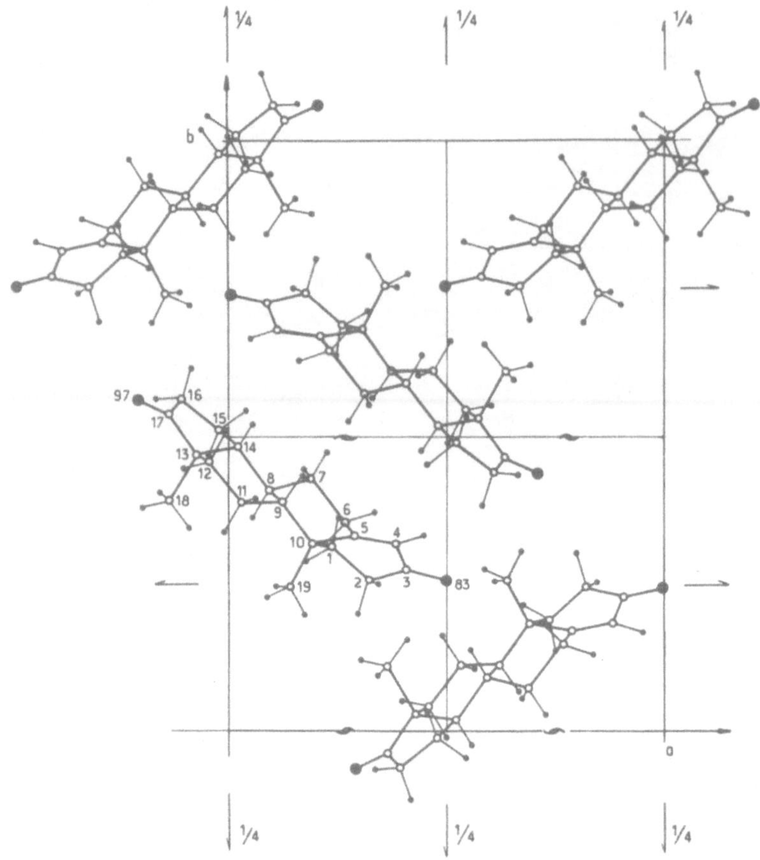

Fig. 1. Projection of the structure of androst-4-en-3,17-dione along c.

17α-HYDROXYANDROST-4-EN-3-ONE

(EPITESTOSTERONE)

$C_{19}H_{28}O_2$

N.W. ISAACS, W.D.S. MOTHERWELL, J.C. COPPOLA and O. KENNARD, 1972. J. Chem. Soc., Perkin II, 2335-2339.

Orthorhombic, $P2_12_12_1$, a = 11.423, b = 20.151, c = 7.003 Å, D_m = 1.14, Z = 4. Cu radiation, R = 0.058 for 1646 reflexions.

In the molecule (Fig. 1) ring A is in a half-chair conformation whereas rings B and C have chair conformations. Ring D is in a slightly distorted half-chair conformation. All ring junctions are trans-trans. The molecule is convex towards the methyl groups at C(10) and C(13). Bond lengths are close to expected values. The molecules are linked head-to-tail in the crystal structure by hydrogen bonds between the 17-hydroxy and 3-keto groups, with O-H...O = 2.899 Å.

Fig. 1. Perspective view of the epitestosterone molecule.

16β,17β-DIBROMOANDROSTANE

$C_{19}H_{30}Br_2$

N. MANDEL and J. DONOHUE, 1972. Acta Cryst., B28, 308-312.

Monoclinic, $P2_1$, a = 11.528, b = 11.128, c = 7.645 Å, β = 110.61°, D_m not given, Z = 2. Cu radiation, R = 0.085 for 1356 reflexions.

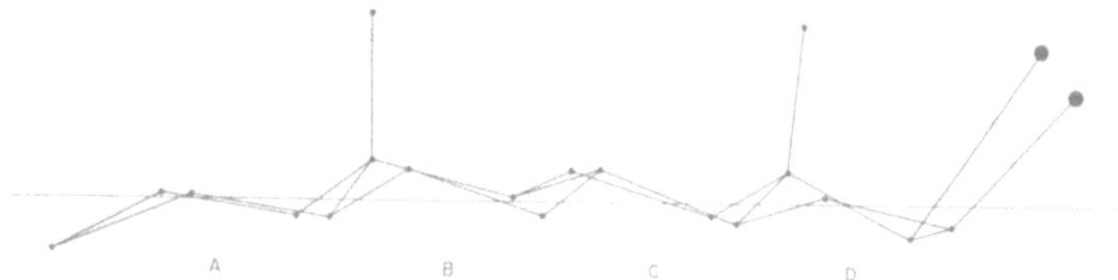

Fig. 1. Molecular skeleton and the best plane of the whole molecule (excluding CH_3 and Br groups) in 16β,17β-dibromoandrostane.

The structure analysis has shown that the addition product of hydrogen bromide to 17-bromo-5α-androst-16-ene has the bromine atoms cis and on the β-side of the molecule. Bond distances and angles are normal. Rings A, B, and

C are in chair conformation, while ring D is somewhere between a half-chair and a β-envelope. The molecular skeleton (Fig. 1) is concave to the α-side. Intermolecular distances are normal.

TESTOSTERONE HYDRATE

$C_{19}H_{30}O_3$

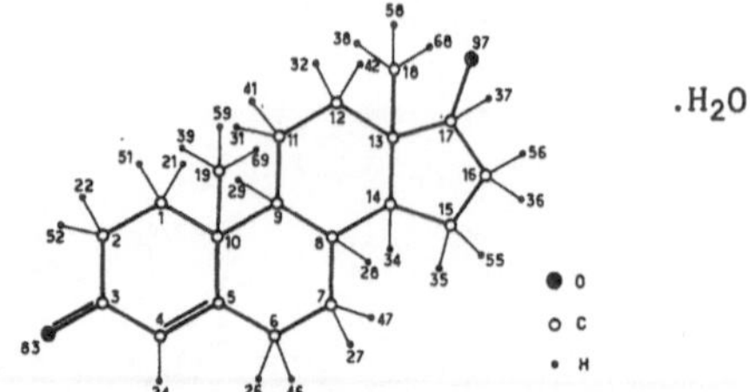

$.H_2O$

B. BUSETTA, C. COURSEILLE, F. LEROY and M. HOSPITAL, 1972. Acta Cryst., B28, 3293-3299.

Monoclinic, $P2_1$, a = 9.622, b = 8.099, c = 11.148 Å, β = 97°49', Z = 2. Cu radiation, R = 0.045 for 618 reflexions.

In the crystal structure of testosterone (Fig. 1), ring A adopts a conformation which minimizes the interactions between the hydrogens attached to C(1) and C(11) (C...C = 2.457 Å); unexpectedly, there is very little π-conjugation in the A ring as the atoms C(5), C(10), C(6), C(4), C(3), and O(83) do not form a plane and bonds C(3)-O(83) 1.234 and C(4)-C(5) 1.345 Å (Fig. 2) are very near pure double bonds. The pentagonal ring D adopts a conformation between that of a half-chair and an envelope. In the b-direction, the water molecule is hydrogen-bonded to the oxygen atom from the 17-β hydroxyl function while in the [10$\bar{1}$] direction, the hydroxyl hydrogen is hydrogen-bonded to the ketone group, O-H...O = 2.693, 2.829, and 2.981 Å.

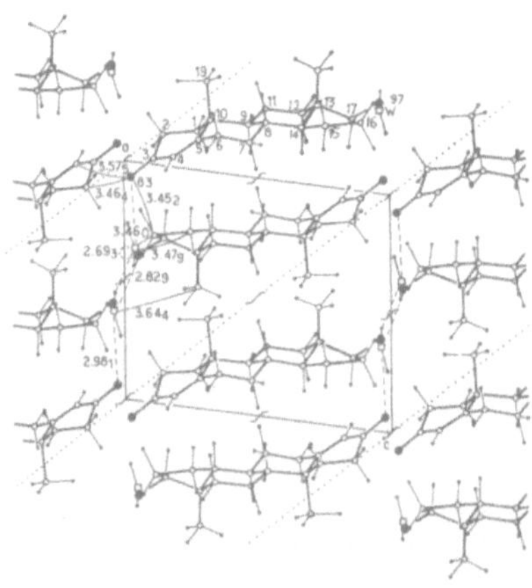

Fig. 1. Projection of the testosterone structure down b.

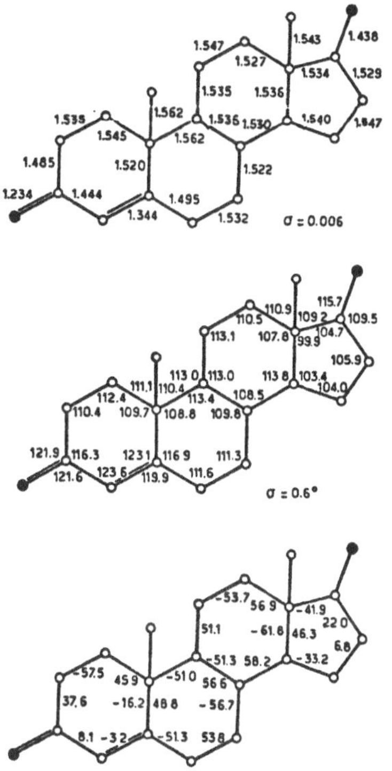

Fig. 2. Bond lengths, valence, and torsion angles for testosterone.

3α,17β-DIHYDROXY-5α-ANDROSTANE

$C_{19}H_{32}O_2$

G. PRECIGOUX, B. BUSETTA, C. COURSEILLE and M. HOSPITAL, 1972. Cryst Struct. Comm., 1, 265-268.

Monoclinic, $P2_1$, a = 12.327, b = 7.191, c = 10.347 Å, β = 114.14°, Z = 2. Cu radiation, R = 0.055 for 1837 reflexions.

In this molecule rings A, B, and C have chair conformations, while ring D is near half-chair. The structure occurs in (001) sheets in which the molecules are linked by hydrogen bonds around the twofold axis (Fig. 1). Each hydroxy-group is involved in two hydrogen bonds, one in the a-direction (2.849 Å) and the other in the c-direction (2.962 Å); the different sheets are linked by weak van der Waals interactions.

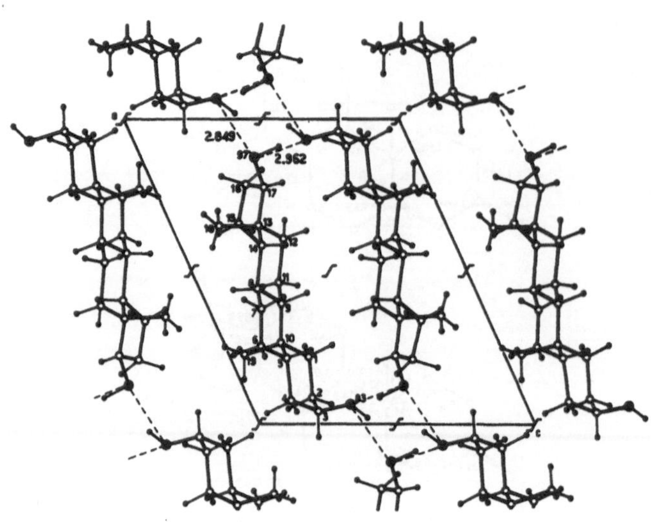

Fig. 1. Projection of the 3α,17β-dihydroxy-5α-androstane structure along b.

17β-HYDROXY-ANDROSTAN-3-ONE MONOHYDRATE

$C_{19}H_{32}O_3$ $C_{19}H_{30}O_2.H_2O$

B. BUSETTA, C. COURSEILLE, J.M. FORNIES-MARQUINA and M. HOSPITAL, 1972, Cryst. Struct. Comm., 1, 43-46.

Monoclinic, $P2_1$, a = 9.406, b = 8.074, c = 11.549 Å, β = 99.1°, Z = 2. Cu radiation, R = 0.053 for 1850 reflexions.

Bond distances and angles within the molecule are in good agreement with expected values. Rings A, B, and C have chair conformations while ring D has a conformation between a half-chair and an α-envelope. The structure (Fig. 1) is organized in (101) sheets in which steroid molecules are bound by two different hydrogen bond systems from the water molecule to (i) the OH function and (ii) the C(3) keto function (with OH...O = 3.02 Å).

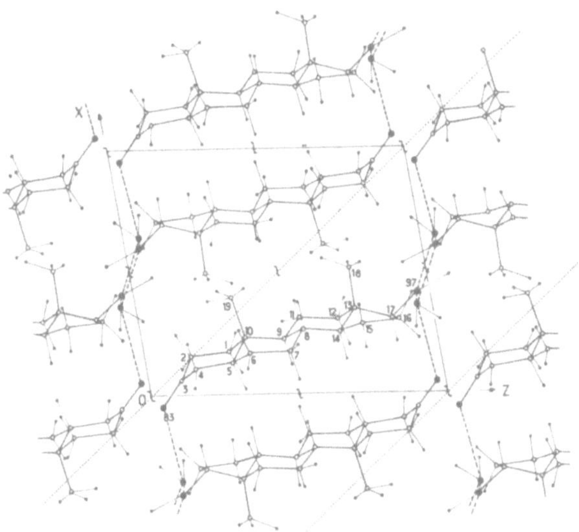

Fig. 1. Projection of the 17β-hydroxyandrostan-3-one monohydrate structure along b.

19-NOR-RETROTESTOSTERONE ACETATE

$C_{20}H_{28}O_3$

B. BUSETTA, C. COURSEILLE and M. HOSPITAL, 1972. Cryst. Struct. Comm., 1, 235-238.

Orthorhombic, $P2_12_12_1$, a = 13.180, b = 18.160, c = 7.191 Å, Z = 4. Cu radiation, R = 0.042 for 1654 reflexions.

Bond lengths and angles are in good agreement with expected values. The B/C ring junction is cis and the configuration at C(10) is α. The conformation of ring D is almost a half-chair.

2α-BROMO-17α-METHYL-5α,14β-ANDROSTAN-3α-OL

$C_{20}H_{33}BrO$

A. CHIARONI and C. PASCARD-BILLY, 1972. Acta Cryst., B28, 1085-1091.

Orthorhombic, $P2_12_12_1$, a = 6.427, b = 12.536, c = 23.323 Å, D_m = 1.265, Z = 4. Cu radiation, R = 0.101 for 1020 reflexions.

The bond lengths in the structure are normal, the average C-C distance being 1.543 Å. The C(2)-Br distance is 2.00 Å and C(3)-O = 1.40 Å. The A, B, and C rings all have the chair conformation and are trans-trans-fused (Fig. 1). The five-membered D ring adopts the envelope conformation, with C(13) 0.62 Å off the four atom plane, and is cis fused to the C ring. The two methyl groups, C(18) and C(20), are in a trans relative position. All intermolecular distances correspond to van der Waals interactions.

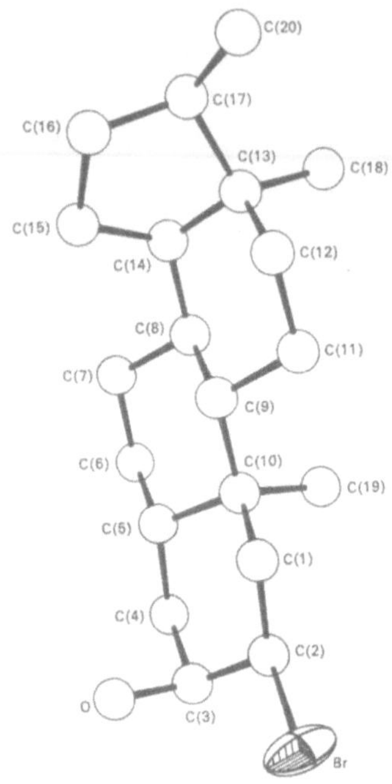

Fig. 1. Perspective view of 2α-bromo-17α-methyl-5α,14β-androstan-3α-ol.

17α,21β-DIHYDROXY-4-PREGNENE-3,11,20-THIONE

(CORTISONE)

$C_{21}H_{28}O_5$

J.P. DECLERCQ, G. GERMAIN and M. van MEERSSCHE, 1972. Cryst. Struct. Comm., 1, 13-15.

Orthorhombic, $P2_12_12_1$, a = 10.040, b = 23.649, c = 7.784 Å, Z = 4. Cu radiation, R = 0.058 for 1468 reflexions.

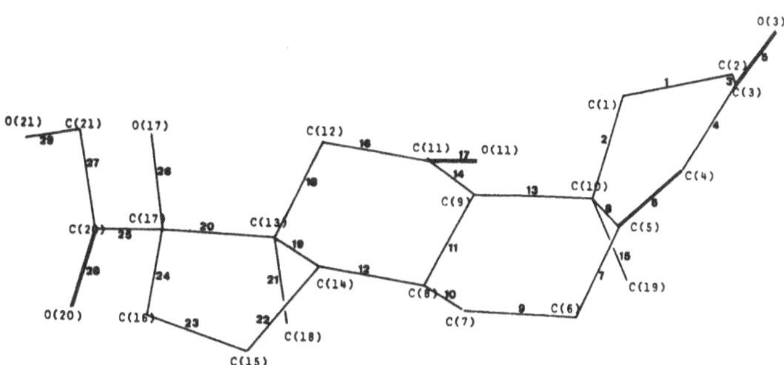

Fig. 1. Cortisone.

The structure of this steroid (Fig. 1) is very similar to that of the 17α-hydroxyprogesterone (1). The same orientation of the molecules is found in both compounds, but the packing is different. The torsional angles in the B ring are all greater than ±51°, while values as small as ±41° were found in 4-chlorocortisone (2). It is thus confirmed that this decrease is due to van der Waals interactions of the chlorine atom.

1. This volume, p. 537.
2. W.L. DUAX, A. COOPER and D.A. NORTON, 1971. Acta Cryst., B27, 1; Structure Reports, 37B, 296.

9α-FLUORO-Δ^4-PREGNENE-11β,17α,21-TRIOL-3,20-DIONE

(9α-FLUOROCORTISOL)

$C_{21}H_{29}FO_5$

I. L. DUPONT, O. DIDEBERG and H. CAMPSTEYN, 1972. Cryst. Struct. Comm., 1, 177-180.

II. Idem, 1972. Acta Cryst., B28, 3023-3032.

Orthorhombic, $P2_12_12_1$, a = 10.088, b = 23.723, c = 7.663 Å, Z = 4. Cu radiation, R = 0.046 for 1685 reflexions.

Bond distances and angles are in good agreement with expected values. The conformation of the side chain is very similar to that found (I) in cortisone (Fig. 1). The torsional angles O(20)-C(20)-C(17)-C(16) and O(21)-C(21)-C(20)-O(20) are 26.3 and 3.3°. Cohesion of the crystal is due to the hydrogen bonds O(21)...O(17) (2.776 Å), O(21)...O(3) (2.772 Å) and F...O(11) (3.003 Å), and to van der Waals interactions.

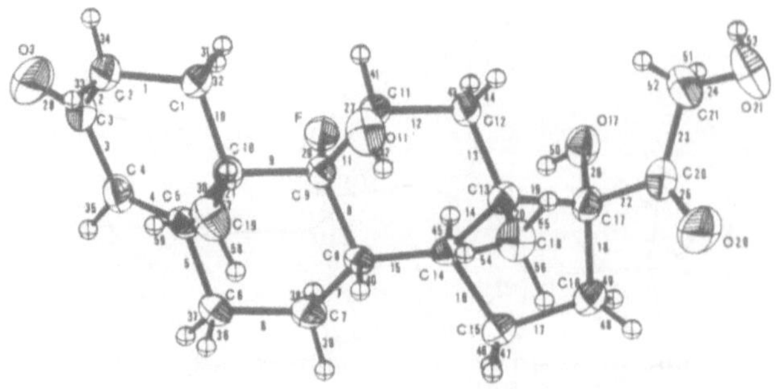

Fig. 1. 9α-Fluorocortisol.

1. Preceding report.

4-PREGNEN-3,20-DIONE

(PROGESTERONE)

$C_{21}H_{30}O_2$

I. H. CAMPSTEYN, L. DUPONT and O. DIDEBERG, 1972. Cryst. Struct. Comm., 1, 219-222.

II. Idem, 1972. Acta Cryst., B28, 3032-3042.

Orthorhombic, $P2_12_12_1$, a = 12.559, b = 13.798, c = 10.340 Å, Z = 4. Cu radiation, R = 0.047 for 1595 reflexions.

All bond distances and angles are in good agreement with the theoretical values. Torsional angles in rings A and B are different from their values in 6β-bromoprogesterone (1). The calculation of Altona parameters ($\Delta = 1°, \phi_m = 46°$) denotes half-chair conformation of ring D. The conformation of the side chain is similar to that found in the 6β-brominated compound: dihedral angle between C(17)-C(20)-C(21) and C(16)-C(17)-C(20) is 8.6°. As the molecule contains no hydroxyl function, cohesion of the crystal is due only to van der Waals interactions.

1. E.M. GOPALKRISHNA, A. COOPER and D.A. NORTON, 1969. Acta Cryst., B25, 639. Structure Reports, 34B.

17α-HYDROXY-4-PREGNENE-3,20-DIONE

(17α-HYDROXYPROGESTERONE)

$C_{21}H_{30}O_3$

J.P. DECLERCQ, G. GERMAIN and M. van MEERSSCHE, 1972. Cryst. Struct. Comm., 1, 9-11.

Orthorhombic, $P2_12_12_1$, a = 9.831, b = 23.468, c = 7.839 Å, Z = 4. Cu radiation, R = 0.049 for 1547 reflexions.

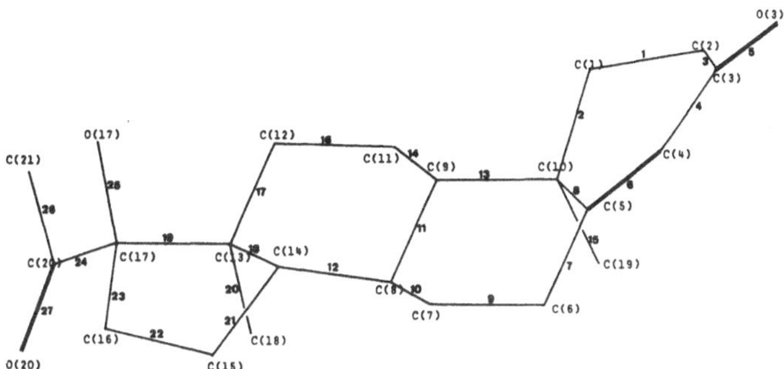

Fig. 1. Molecular conformation of 17α-hydroxyprogesterone.

Bond distances in the molecule are normal; the molecular configuration is shown in Fig. 1. Hydrogen bonding occurs between atoms O(3) and HO(17) of two molecules. The conformation of the side chain is very similar to that found in 6α-methylprednisolone (1). The present steroid having no hydroxyl group on C(21), the influence of hydrogen bonding on the side chain conformation is very weak. Torsional angles in the C ring are greater than those of 12α-bromo-11β-hydroxyprogesterone (2), where the 11β-hydroxyl group gives rise to van der Waals interactions with the methyl group C(18).

1. This volume p. 546.

2. Structure Reports, 33B, 193.

21-HYDROXY-4-PREGNEN-3,20-DIONE

(DESOXYCORTICOSTERONE)

$C_{21}H_{30}O_3$

O. DIDEBERG, H. CAMPSTEYN and L. DUPONT, 1972. Cryst. Struct. Comm., 1, 323-326.

Monoclinic, $P2_1$, a = 8.718, b = 12.509, c = 8.480 Å, β = 98.01°, D_m = 1.22, Z = 2. Cu radiation, R = 0.042 for 1563 reflexions.

The interatomic distances and bond angles are in good agreement with already published values. However, the torsional angles in rings A and D (Fig. 1) are different from their values in progesterone (1). Ring D has a conformation between a half-chair and β-envelope form. The dihedral angle between C(17)-C(20)-C(21) and C(16)-C(17)-C(20) is 10.7°, and between O(20)-C(20)-C(21) and O(21)-C(21)-C(20) is 3.6°. Cohesion of the crystal is due to van der Waals interactions.

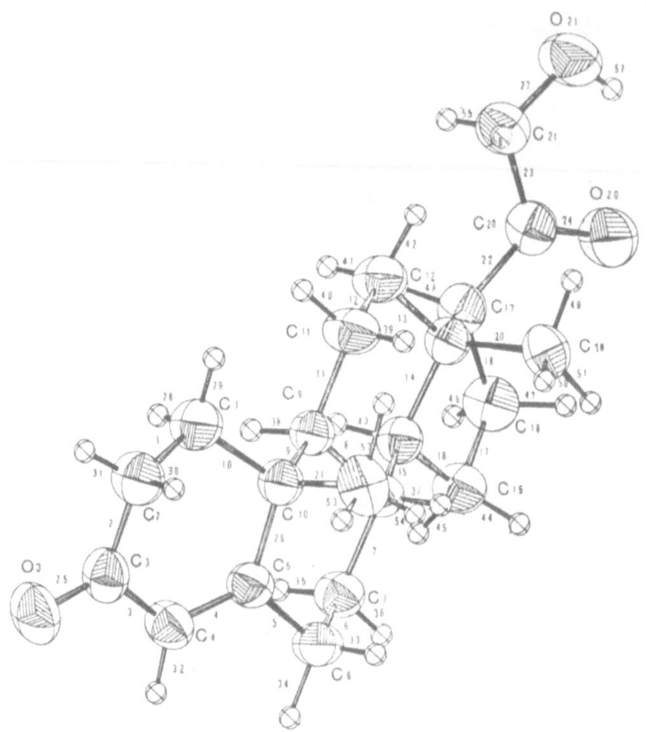

Fig. 1. Desoxycorticosterone.

1. This volume p. 536.

4-PREGNENE-17α,21-DIOL-3,20-DIONE

$C_{21}H_{30}O_4$

L. DUPONT, O. DIDEBERG and H. CAMPSTEYN, 1972. Cryst. Struct. Comm., 1, 357-361.

Monoclinic, $P2_1$, a = 11.472, b = 7.536, c = 10.698 Å, β = 96.07°, Z = 2. Cu radiation, R = 0.061 for 1454 reflexions.

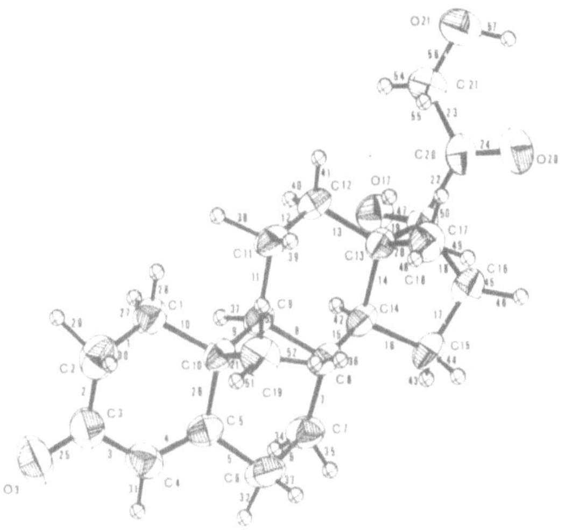

Fig. 1. Molecular conformation of 4-pregnene-17α,21-diol-3,20-dione.

The conformation of the molecule is shown in Fig. 1. All bond distances and angles are in good agreement with the expected values. Some torsional angles in ring A are rather different (3 to 6°) from their values in 17α-hydroxyprogesterone (1). Ring D has a β-envelope form (Δ = -37.7°; ϕ_m = 47.0°) which is quite different from the conformation in 17α-hydroxyprogesterone, progesterone (2) and desoxycorticosterone (3). Conformation of the side chain is described by the following torsional angles: C(16)-C(17)-C(20)-C(21) 149.0°; O(20)-C(20)-C(21)-O(21) -12.1°; O(17)-C(17)-C(20)-C(21) 24.8°; C(16)-C(17)-C(20)-O(20) -32.4°; O(17)-C(17)-C(20)-O(20) -156.6°; C(17)-C(20)-C(21)-O(21) 166.5°. Cohesion of the crystal is due to van der Waals interactions and to hydrogen bonds: O(21)-H(O21)...O(3) = 2.843 Å; O(17)-H(O17)...O(3) = 3.074 Å.

1. This volume, p. 537.
2. Ibid., p. 536.
3. Preceding report.

ALDOSTERONE MONOHYDRATE

$C_{21}H_{30}O_6$ $C_{21}H_{28}O_5.H_2O$

W.L. DUAX and H. HAUPTMAN, 1972. J. Amer. Chem. Soc., 94, 5467-5471.

Monoclinic, $P2_1$, a = 12.199, b = 6.0345, c = 13.181 Å, β = 107.31°, Z = 2. Cu radiation, R = 0.052 for 1749 reflexions.

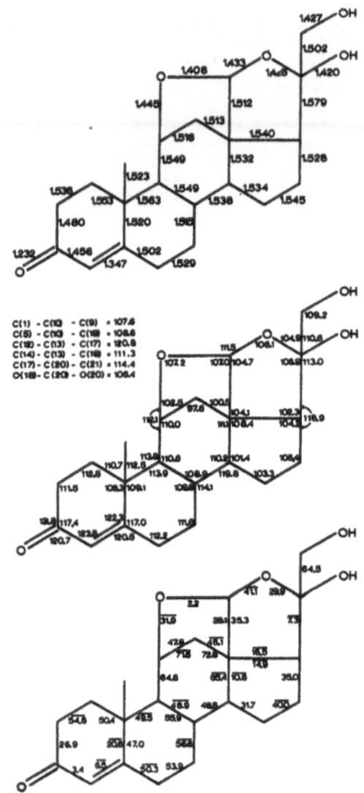

Fig. 1. Bond lengths (σ = 0.005 Å), valency angles (σ = 0.3°) and torsion angles in aldosterone.

Aldosterone occurs in the monohydrated form as the 18-acetal-20-hemiketal structural isomer. Details of molecular geometry are in Fig. 1. The addition of two five-membered rings to the four-ring steroid nucleus introduces considerable strain into the C ring as indicated by the C(11)-C(12)-C(13) valency angle of 97.6°. Despite this strain and the nearly total eclipsing of the bonds to C(17) and C(20) introduced by the hemiketal formation, the crystalline hydrate is very stable and the thermal motion is no greater than that generally observed in steroid structures. Both hydroxyl groups of the 17 side chain participate in strong hydrogen bonds in the solid (with O-H...O = 2.696 and 2.859 Å). Infinite chains of aldosterone molecules are formed by head-to-tail hydrogen bonding between molecules which are translationally equivalent in the c-direction. Each water molecule is hydrogen bonded to three crystallographically distinct steroid molecules.

3β-ACETOXY-17α-IODO-Δ^5-ANDROSTENE

$C_{21}H_{31}IO_2$

H
I
CH_3COO

H.-C. MEZ and G. RIHS, 1972. Helv. Chim. Acta, 55, 375-380.

Orthorhombic, $P2_12_12_1$, a = 6.430, b = 10.563, c = 29.369 Å, D_m = 1.47, Z = 4. Mo radiation, R = 0.051 for 2352 reflexions.

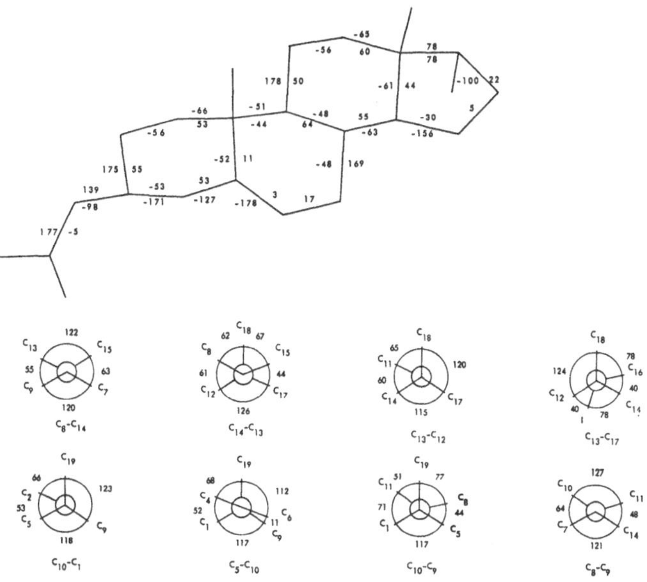

Fig. 1. Torsion angles in 3β-acetoxy-17α-iodo-Δ^5-androstene.

The bond lengths and angles have normal values; the torsion angles are shown in Fig. 1. Rings A and C are chairs, B a half-chair and D an envelope.

20(S)-HYDROXYPREGN-4-EN-3-ONE

(20(S)-HYDROXYPROGESTERONE)

$C_{21}H_{32}O_2$

N.W. ISAACS, W.D.S. MOTHERWELL, J.C. COPPOLA and O. KENNARD, 1972. J. Chem. Soc., Perkin II, 2331-2335.

Tetragonal, $P4_1$, a = 9.422, c = 20.789 Å, D_m = 1.13, Z = 4. Cu radiation, R = 0.047 for 1462 reflexions.

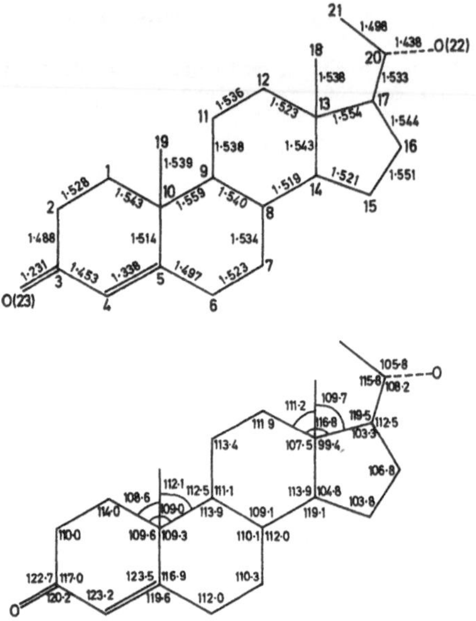

Fig. 1. Bond distances and angles in 20(S)-hydroxy-progesterone. Mean σ values are 0.006 Å for bond lengths and 0.3° for angles not involving H atoms.

Bond distances and angles in the molecule are shown in Fig. 1. All ring junctions are trans-trans. The A ring is in a half-chair conformation whereas rings B and C have chair conformations. Ring D is a distorted C(13) envelope. The molecules are linked in the crystal structure head-to-tail by a hydrogen bond between the C(20) hydroxy group and the C(3) keto group (O-H...O = 2.792 Å).

ESTRADIOL - PROPANOL SOLVATE

$C_{21}H_{32}O_3$ $C_{18}H_{24}O_2.C_3H_8O$

B. BUSETTA, C. COURSEILLE, S. GEOFFRE and M. HOSPITAL, 1972. Acta Cryst., B28, 1349-1351.

Orthorhombic, $P2_12_12_1$, a = 12.215, b = 24.251, c = 6.671 Å, D_m not given, Z = 4. Cu radiation, R = 0.061 for 2213 reflexions.

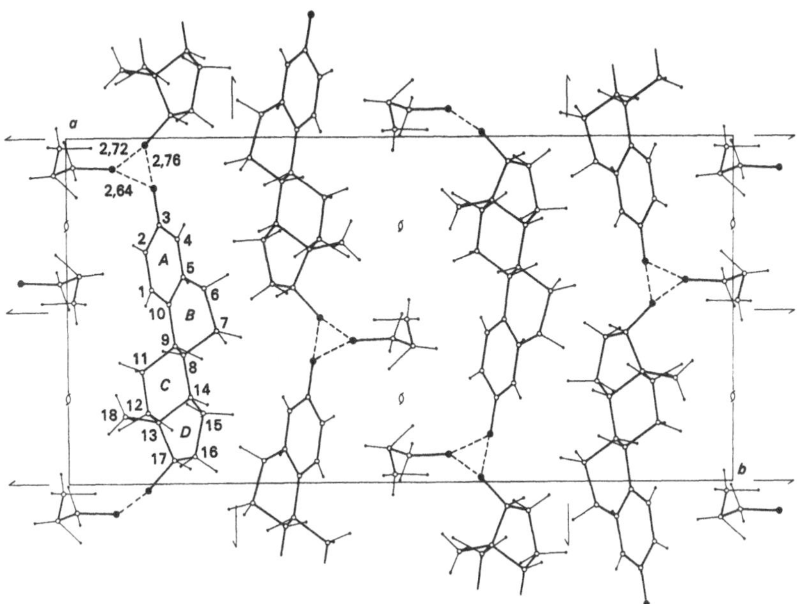

Fig. 1. Projection of the structure of the estradiol-propanol complex along c.

The structure of the complex is in Fig. 1. Ring B has a half-chair conformation, ring C has the normal chair form, and ring D is described as being in a half-chair form. There is very little difference between the molecular conformation found in the propanol solvate and that in the water solvate (1).

1. This volume, p. 524.

17-ETHYLENEDIOXY-3-METHOXY-6,7,8-METHYLIDYNE-1,3,5(10)-ESTRATRIENE

$C_{22}H_{26}O_3$

H.P. WEBER and E. GALANTAY, 1972. Helv. Chim. Acta, 55, 544-553.

Orthorhombic, $P2_12_12_1$, a = 29.58, b = 6.50, c = 9.16 Å, D_m = 1.26, Z = 4. Mo radiation, R = 0.074 for 1054 reflexions.

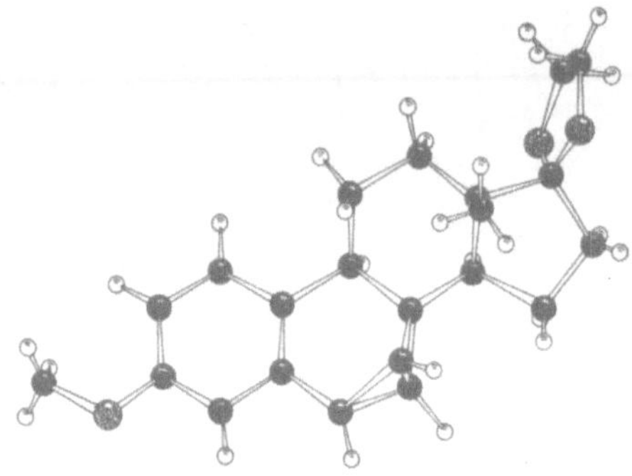

Fig. 1. Perspective view of 17-ethylenedioxy-3-methoxy-6,7,8-methylidyne-1,3,5(10)-estratriene.

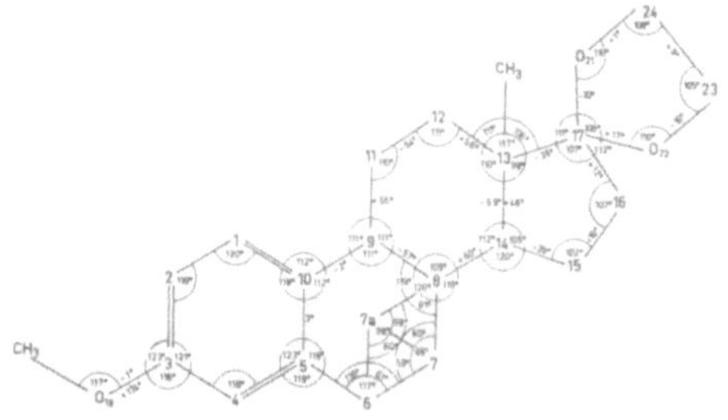

Fig. 2. Bond angles (σ = 0.8°) and torsion angles (σ = 1.4°) for 17-ethylenedioxy-3-methoxy-6,7,8-methylidyne-1,3,5(10)-estratriene.

A view of the molecule is given in Fig. 1; torsion angles are in Fig. 2. Rings A and B together with the methoxy group, form a planar system with the two apex atoms of the bicyclobutyl system C(7) and C(7a) symmetrically above and below the plane by 0.76(1) Å. The bicyclobutyl system has approximate C_{2v} symmetry with an angle of 122° between the planes of the two three-membered rings. The five C-C bonds in the system do not differ significantly from their mean value, 1.504(6) Å. The C-ring is in a nearly undistorted chair conformation (see torsion angles, Fig. 2). Ring D has almost the same conformation (approximate C_2 symmetry) as found in many other steroids. The five-membered acetal ring is slightly non-planar; as seen from the torsion angles (Fig. 2) it occurs as a rather flattened envelope conformation (approximate C_s symmetry) with O(2) as the flap. There is a close intramolecular contact of 3.2 Å between the 13β-methyl and C(7), the bicyclobutyl apex-atom on the β-side. This may explain why exclusively α-attack (i.e. attack on the unhindered C(7a) apex-atom) is observed in the hydrogenolysis of this system.

17β-BROMOACETOXY-3-METHOXY-8α-METHYL-1,3,5(10),6-ESTRATETRAENE

$C_{22}H_{27}BrO_3$

H.P. WEBER and E. GALANTAY, 1972. Helv. Chim. Acta, 55, 544-553.

Orthorhombic, $P2_12_12_1$, a = 22.83, b = 10.47, c = 8.35 Å, D_m = 1.37, Z = 4. Mo radiation, R = 0.096 for 767 reflexions.

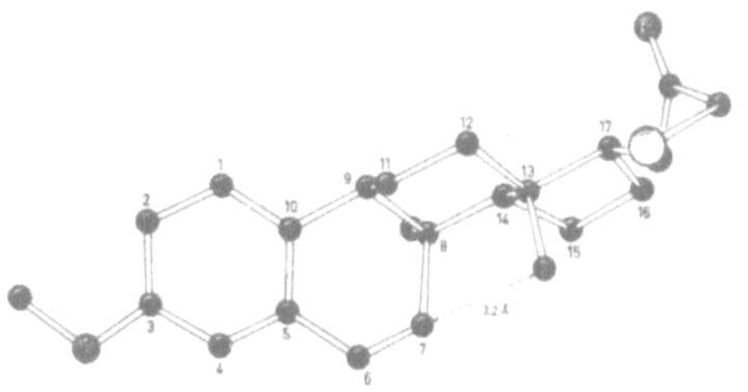

Fig. 1. Perspective view of 17β-bromoacetoxy-3-methoxy-8α-methyl-1,3,5(10),-6-estratetraene.

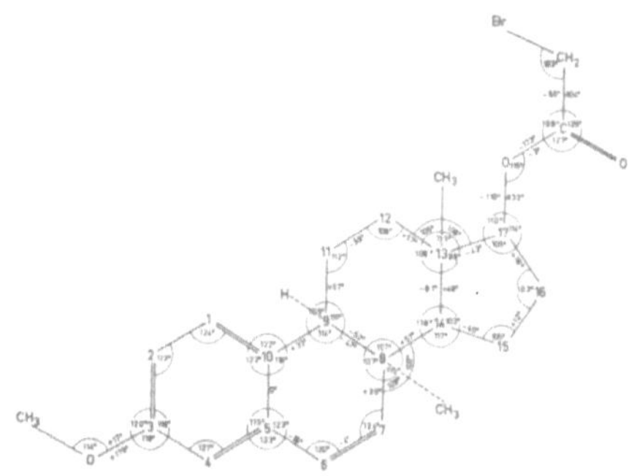

Fig. 2. Bond angles (σ = 2.8°) and torsion angles (σ = 3.2°) in 17β-bromo-acetoxy-3-methoxy-8α-methyl-1,3,5(10),6-estratetraene.

A feature of the 8-iso-steroid skeleton (Fig. 1) is the cis-fusion of rings B and C which brings the mean plane through rings A and B almost perpendicular to the mean plane through rings C and D. As a consequence the 13β-methyl group comes into close contact with C(7), the interatomic distance being 3.2 Å. The Δ^6-double bond should then be considerably hindered from both the α- and the β-sides by the 8α- and 13β-methyl groups respectively. The O-methyl group is in the plane of the aromatic A-ring. The conformation of ring B has approximate C_2-symmetry (2-fold axis through the middle of bonds C(5)-C(6) and C(8)-C(9), and the torsion angles (Fig. 2) are in reasonable agreement with those derived from a microwave investigation of 1,3-cyclohexadiene (1). The conformations of rings C and D are very similar to those observed in other steroids in spite of the B/C cis-fusion.

1. S.S. BUTCHER, 1965. J. Chem. Phys., 42, 1830.

6α-METHYL-11β,17α,21β-TRIHYDROXY-1,4-PREGNADIENE-3,20-DIONE

$C_{22}H_{30}O_5$

J.P. DECLERCQ, G. GERMAIN and M. van MEERSSCHE, 1972. Cryst. Struct. Comm., 1, 5-7.

Monoclinic, C2, a = 21.597, b = 9.202, c = 10.916 Å, β = 114.39°, Z = 4. Cu radiation, R = 0.036 for 1502 reflexions.

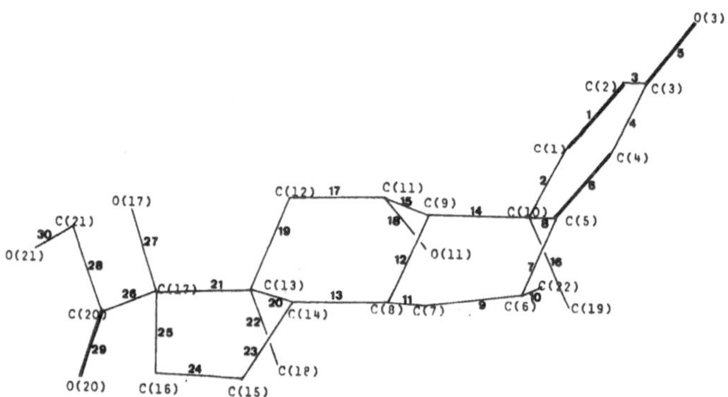

Fig. 1. Molecular configuration of $C_{22}H_{30}O_5$ (the compound is also called 6α-methylprednisolone).

The molecular configuration is in Fig. 1. The side chain conformation of the molecule is in good agreement with the results found in 4-chlorocortisone (1). In the C ring torsional angles involving bonds 12 and 15 are decreased, due to a van der Waals interaction between O(11) and the methyl group C(18). Molecules are joined together by formation of three hydrogen bonds: O(3)...O(17), O(3)...O(21), and O(11)...O(21). Bond lengths are normal.

1. W.L. DUAX, A. COOPER and D.A. NORTON, 1971. Acta Cryst., B27, 1; Structure Reports, 37B, 296.

17β-TRIMETHYLSILOXY-4-ANDROSTEN-3-ONE

(SILANDRONE)

$C_{22}H_{36}O_2Si$

C.M. WEEKS, H. HAUPTMAN and D.A. NORTON, 1972. Cryst. Struct. Comm., 1, 79-82.

Orthorhombic, $P2_12_12_1$, a = 13.222, b = 14.092, c = 11.523 Å, Z = 4. Cu radiation, R = 0.103 for 1456 reflexions.

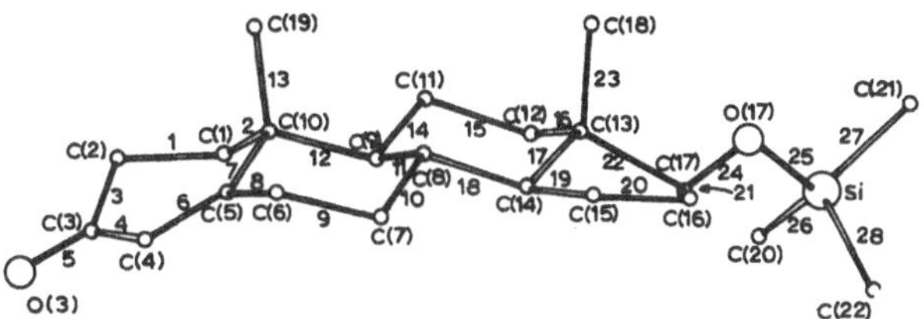

Fig. 1. Molecular conformation of silandrone.

The geometry of the silandrone molecule (Fig. 1) appears to be normal when compared to similar 4-androsten-3-one molecules. The angles about the silicon atom are tetrahedral, and the O-Si distance of 1.62 Å and an average C-Si distance of 1.87 Å closely resemble the average distances of 1.612 (O-Si) and 1.865 (C-Si) observed in silicates.

21β-ACETOXY-17α-HYDROXY-4-PREGNENE-3,11,20-TRIONE (FORM II)

(CORTISONE ACETATE)

$C_{23}H_{30}O_6$

J.P. DECLERCQ, G. GERMAIN and M. van MEERSSCHE, 1972. Cryst. Struct. Comm., 1, 59-62.

Orthorhombic, $P2_12_12_1$, a = 11.048, b = 7.102, c = 27.095 Å, Z = 4. Cu radiation, R = 0.081 for 1191 reflexions.

The molecular conformation is shown in Fig. 1. The addition of an acetoxy group on the side chain of cortisone (1) does not modify the conformation of this chain very much. A mean difference of 4° is found for torsional angles involving the C(17)-C(20) bond and of 9° for those involving the C(20)-C(21) bond. One hydrogen bond between atoms H(O17) and O(22) joins the molecules together. Bond distances correspond to normal values.

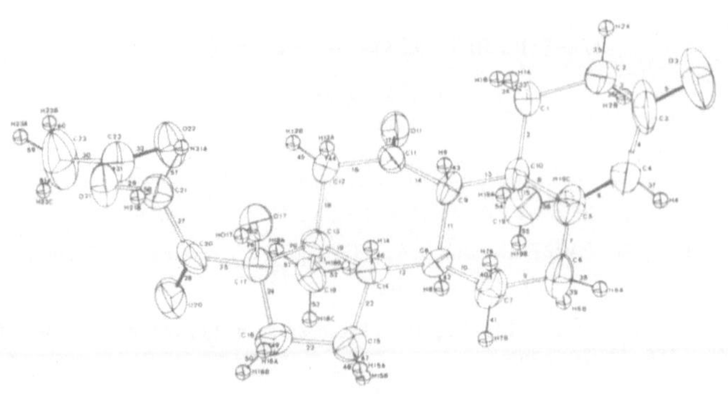

Fig. 1. Stereochemistry of cortisone acetate.

1. This volume, p. 534.

N-CYANO-N-METHYL-18-AMINO-20-BROMO-C-NOR-D-HOMO-PREGNANE

$C_{23}H_{37}BrN_2$

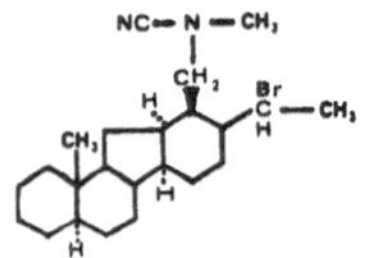

J. GUILHEM, 1972. Acta Cryst., B28, 291-298.

Monoclinic, $P2_1$, a = 10.72, b = 6.90, c = 15.06 Å, β = 101.5°, D_m = 1.27, Z = 2. Cu radiation, R = 0.090 for 1804 reflexions.

This structure determination confirms the α configuration of the hydrogen atoms at C(12) and C(13). Bond distances and angles are normal. The geometry at the trigonal nitrogen is intermediate between planar and pyramidal. The molecular packing is controlled by van der Waals forces.

7α-ACETYLTHIO-3-OXO-17β-4-PREGNENE-21,17β-CARBOLACTONE (SPIRONOLACTONE)

$C_{24}H_{32}O_4S$

I. O. DIDEBERG and L. DUPONT, 1972. Cryst. Struct. Comm., 1, 99-102.

II. Idem, 1972. Acta Cryst., B28, 3014-3022.

Orthorhombic, $P2_12_12_1$, a = 9.979, b = 35.573; c = 6.225 Å, Z = 4. Cu radiation, R = 0.066 for 1671 reflexions.

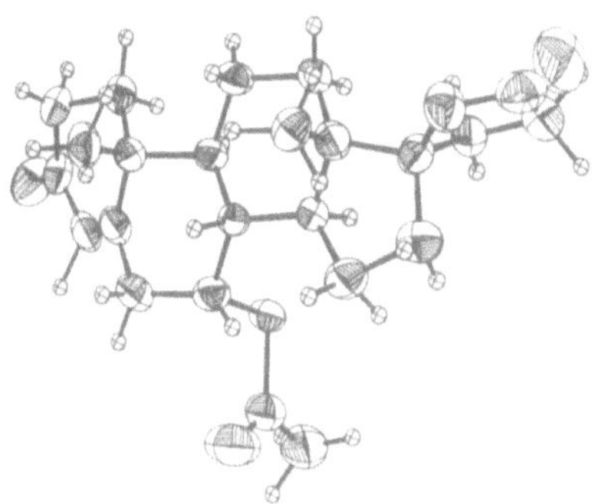

Fig. 1. Conformation of spironolactone.

This chemically modified steroid (Fig. 1) inhibits the mineralocorticoid actions of aldosterone. The A ring is highly distorted because of the Δ^4 double bond and the ketone group. Rings B and C are chair shaped. Ring D is a distorted half-chair (Δ = -18.14 or 18.44 [both values quoted], ϕ_m = 46.6°) and ring E is a half-chair. The molecules are held in the crystal by van der Waals forces.

2β-METHYL-19-NORTESTOSTERONE p-BROMOBENZENESULPHONATE

$C_{25}H_{31}BrO_4S$

V. CODY and W.L. DUAX, 1972. Cryst. Struct. Comm., 1, 439-442.

Orthorhombic, $P2_12_12_1$, a = 9.127, b = 36.15, c = 7.162 Å, Z = 4. Cu radiation, R = 0.06 for 2360 reflexions.

The A-ring of the steroid (Fig. 1) is found to be in the typical half-chair conformation in contrast to the twist conformation previously assigned for 2β-methyl-19-nortestosterone from the interpretation of ORD data. In light of the apparent stability of the observed structure and CD measurements subsequent to the X-ray determination, a reinterpretation of the ORD spectra giving greater attention to its fine structure is suggested in order to resolve the ambiguity between solid structure and solution structure assignment. There is considerable flattening of the steroid nucleus at the A/B-ring junction compared to the normally observed conformations of 4-en-3-one steroids as 19-methyl substitution causes steric crowding prohibiting such flattening. None of the values for bond lengths and valency angles is significantly different from those observed in similar steroids.

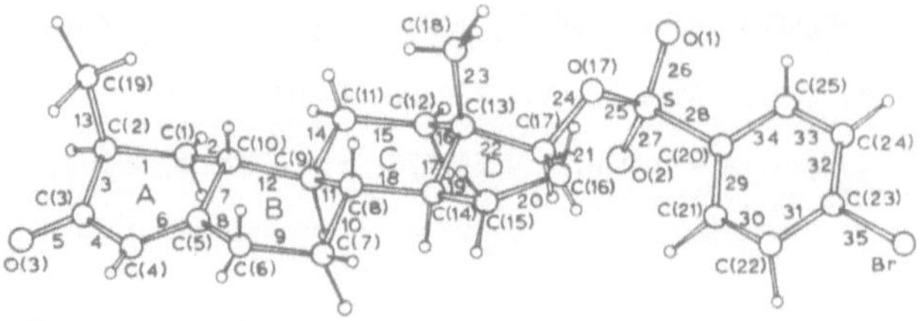

Fig. 1. The stereochemistry of the 2β-methyl-19-nortestosterone derivative, $C_{25}H_{31}BrO_4S$.

3β-p-BROMOBENZOYLOXYANDROST-5-EN-17-ONE

$C_{26}H_{31}BrO_3$

J.C. PROTHEINE, C. ROMERS and E.W.M. RUTTEN, 1972. Acta Cryst., B28, 849-857.

At 20°C.
Monoclinic, $P2_1$, a = 17.566, b = 8.360, c = 7.838 Å, β = 102.71°, D_m = 1.397, Z = 2. Mo radiation, R = 0.049 for 3340 reflexions.

At -180°C.
a = 17.433, b = 8.257, c = 7.769 Å, β = 102.08°, Cu radiation, R = 0.112 for 2262 reflexions.

No significant differences between the structures at the two temperatures were observed. Bond length data are in Fig. 1 and torsion angles in Fig. 2. Rings A and C have chair conformations whereas ring B is a distorted half-chair. Ring D is a nearly ideal half-chair. The molecular packing is shown in Fig. 3. The intermolecular contacts are normal; those at -180°C are smaller (by amounts up to 0.12 Å) than those at room temperature.

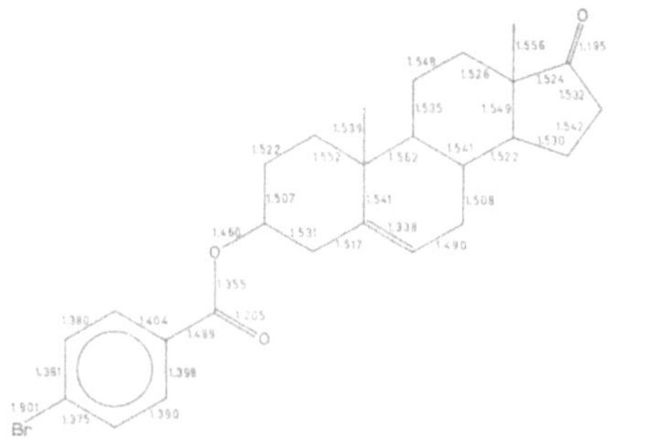

Fig. 1. Bond lengths in the androst-5-en-17-one derivative, $C_{26}H_{31}BrO_3$, at 20°C after correction for thermal motion.

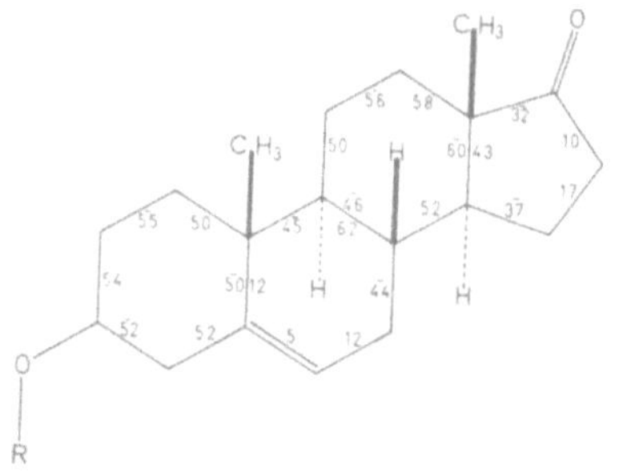

Fig. 2. Torsion angles in $C_{26}H_{31}BrO_3$ at 20°C.

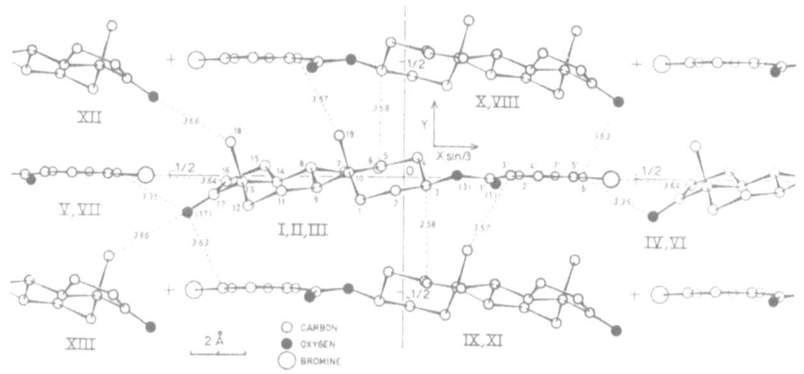

Fig. 3. Projection of 3β-p-bromobenzoyloxyandrost-5-en-17-one at 20°C along c.

3,5-CYCLOSTEROID DERIVATIVES

R.C. PETTERSEN, O. KENNARD and W.G. DAUBEN, 1972. J. Chem Soc., Perkin II, 1929-1935.

6-OXO-3α,5-CYCLOANDROSTAN-17-YL p-BROMOBENZOATE

$C_{26}H_{31}BrO_3$

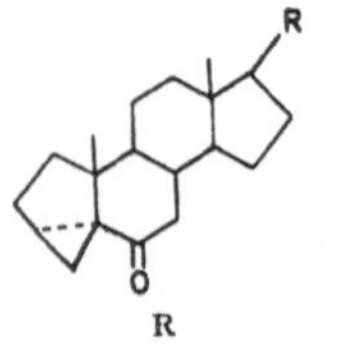

R

O

R

R = O·CO·C_6H_4·Br

Monoclinic, $P2_1$, a = 6.45, b = 28.69, c = 6.690 Å, β = 109.97°, D_m = 1.28, Z = 2. Cu radiation, R = 0.17 for 1643 reflexions.

6-OXO-3β,5-CYCLOANDROSTAN-17-YL ACETATE

$C_{21}H_{30}O_3$

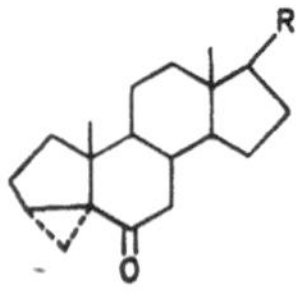

R

O

R = OCOMe

Orthorhombic, $P2_12_12_1$, a = 7.423, b = 11.184, c = 22.110 Å, D_m = 1.18, Z = 4. Cu radiation, R = 0.10 for 1888 reflexions.

Both compounds possess the normal steroidal nucleus for rings B,C, and D (Fig. 1). The introduction of the C(3)-C(5) bond into the steroidal nucleus changes greatly the general shape of ring A and the two different orientations of the hydrogen at C(3) considerably affect the geometry of the two molecules in the region of the A and B rings. The valency angles at C(5) are distorted in both molecules to accommodate the strain of the 3,5-cyclization (ranging from 59(2) and 60.0(4)° for C(3)-C(5)-C(4) to 114(2) and 129.9(6)° for C(3)-C(5)-C(6) in the α and β forms respectively). The large increase in the C(3)-C(5)-C(6) bond angle in the β-form may be due to a reduced p-character in the C(3)-C(5) bond in this form. The C(5)-C(6) bond length is 1.45(3) in the α-form and 1.48(1) in the β-form. The crystal packing of the β-form is unusual in that twofold screw axes parallel to <u>c</u> pass through rings A and D respectively (Fig. 2).

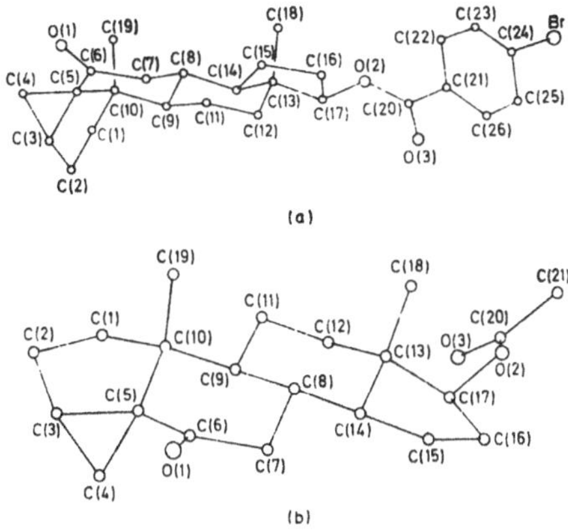

Fig. 1. Configurations of 3,5-cyclosteroids: (a) 6-oxo-3α,5-cycloandrostan-17-yl p-bromobenzoate; note that the 'new' ring A is bent away from the general plane of the steroid nucleus; and (b) 6-oxo-3β,5-cycloandrostan-i7-yl acetate.

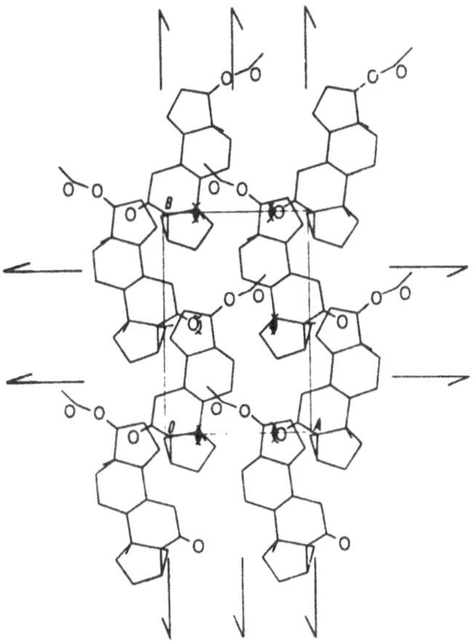

Fig. 2. Packing diagram of 6-oxo-3β,5-cyclo-androstan-17-yl acetate projected on (001). Note the unusual placement of the molecule with respect to the two-fold screw axes parallel to <u>c</u>.

CHOLIC ACID - ETHANOL ADDITION COMPOUND

$C_{26}H_{46}O_6$ $C_{24}H_{40}O_5.C_2H_6O$

P.L. JOHNSON and J.P. SCHAEFER, 1972. Acta Cryst., B28, 3083-3088.

Orthorhombic, $P2_12_12_1$, a = 14.661, b = 11.759, c = 15.066 Å, D_m = 1.16, Z = 4. Cu radiation, R = 0.113 for 2110 reflexions.

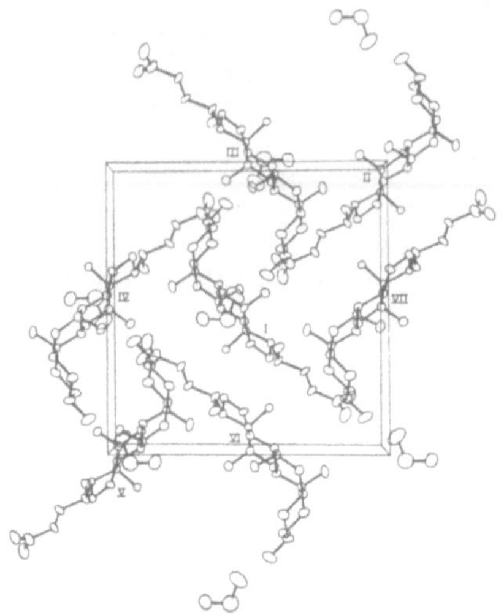

Fig. 1. Packing diagram of the cholic acid - ethanol complex.

Rings A,B, and C of the cholic acid molecule have chair conformations; the only unusual torsional angle is C(2)-C(3)-C(4)-C(5) (68.1°). The D ring is in a conformation between a half-chair and a β-envelope but close to the β-envelope. In the crystal structure (Fig. 1) there is an intricate hydrogen bonding scheme involving three cholic acid molecules and one ethanol molecule linked by five hydrogen bonds O-H...O (2.55 - 2.83 Å). Bond lengths and angles in the molecules are normal.

DESOXYCHOLIC ACID p-BROMOANILIDE

$C_{30}H_{44}O_3BrN$

J.P. SCHAEFER and L.L. REED, 1972. Acta Cryst., B28, 1743-1748.

Orthorhombic, $P2_12_12_1$, a = 11.942, b = 30.62, c = 7.585 Å, D_m = 1.30, Z = 4. Cu radiation, R = 0.132 for 2453 reflexions.

A cage-like dimeric structure, with a polar interior cavity and a non-polar exterior surface, characterizes the molecular association in the crystal structure (Fig. 1). Bond distances and angles have normal values. The absolute stereochemistry was determined and agrees with that deduced chemically.

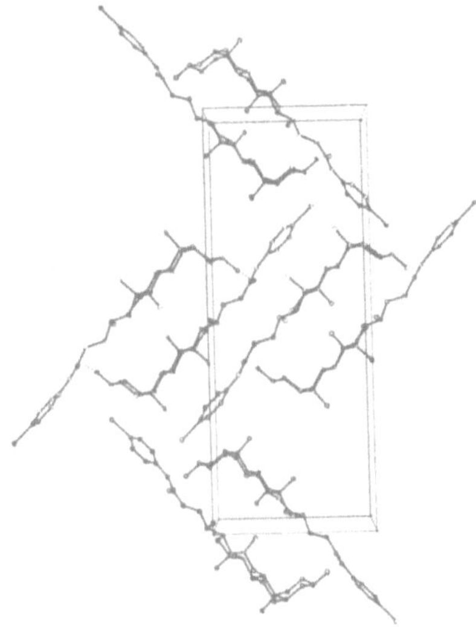

Fig. 1. A view of the contents of a unit-cell in the crystal of desoxycholic acid p-bromoanilide. The intermolecular OH...O hydrogen bond (2.96 Å) between the C(3) hydroxyl group and the carbonyl oxygen of the C(17) side chain is indicated by dotted lines.

TETRAHYDROVERALKAMINE HYDROIODIDE

$C_{27}H_{48}INO_2$

E. HÖHNE, I. SEIDEL, G. ADAM, K. SCHREIBER and J. TOMKO, 1972. Tetrahedron, 28, 4019-4026.

Orthorhombic, $P2_12_12_1$, a = 21.809, b = 7.187, c = 17.026 Å, Z = 4. Cu radiation, R = 0.109 for 1534 reflexions (Weissenberg films).

The analysis was undertaken to establish the conformation of ring C in this steroidal alkaloid (Fig. 1). In the event ring C was found to be disordered in the crystal structure with a chair : boat ratio of 2 : 1. No lists of bond lengths or angles are given. The molecules are linked in the crystal via O-H...O, O-H...I, and N-H...I hydrogen bonds of lengths 2.64, 3.49, and 3.59 Å respectively.

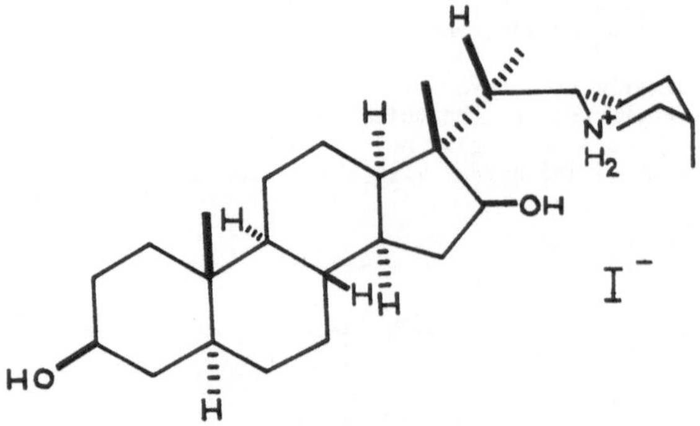

Fig. 1. Tetrahydroveralkamine hydroiodide.

ANHYDROBROMONITROCAMPHANE

$C_{10}H_{14}BrNO$

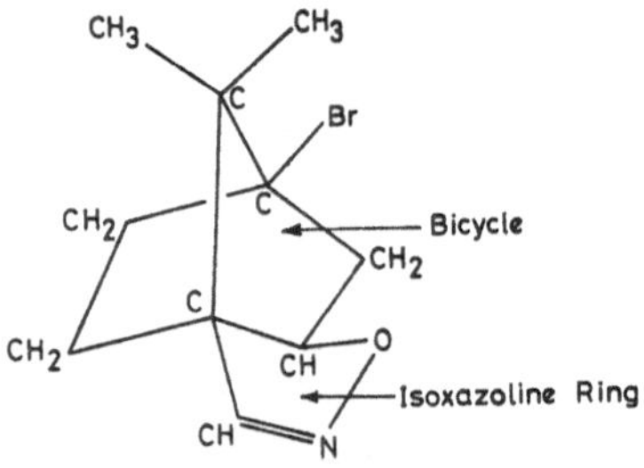

G.L. DWIVEDI and R.C. SRIVASTAVA, 1972. Acta Cryst., B28, 2567-2576.

Orthorhombic, $P2_12_12_1$, a = 10.364, b = 9.408, c = 10.499 Å, D_m = 1.51, Z = 4. Cu radiation, R = 0.097 for 496 reflexions.

The structure is highly disordered so that each type of site is statistically occupied by two optically isomeric forms of the molecule in two orientations (Fig. 1) and an approximate (100) pseudomirror is produced. Bond lengths and angles have high standard deviations (0.01 Å and 1°) because of the disorder but are comparable with those in other camphane structures.

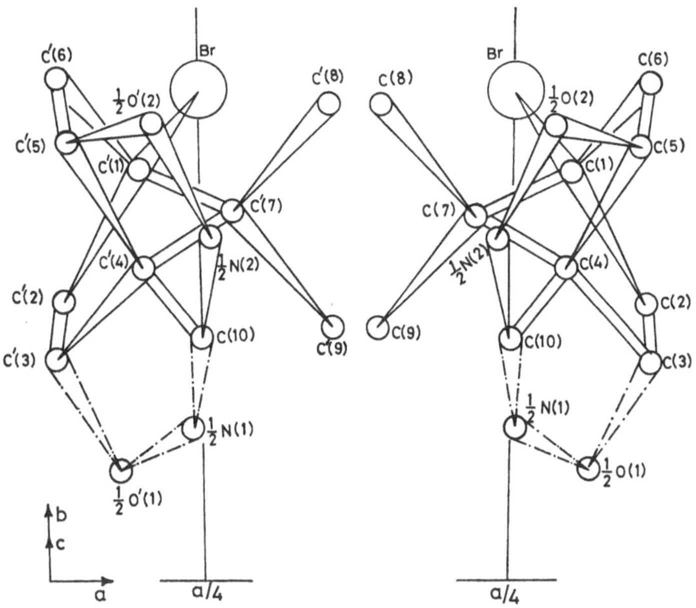

Fig. 1. A perspective drawing of each orientation of the anhydrobromonitrocamphane molecule when seen in a plane parallel to (100). One of the two probable isoxazoline rings of each orientation is shown by the broken lines.

DIHYDROFUKINOLIDOL SULPHITE

$C_{15}H_{22}O_5S$

A. FURUSAKI and T. WATANABÉ, 1972. Bull. Chem. Soc. Japan, 45, 2288-2299.

Orthorhombic, $P2_12_12_1$, Z = 4.

	Isomer 1	Isomer 2
a (Å)	12.544	12.919
b	14.457	14.375
c	8.545	8.496

Cu radiation, films, visual intensities, R = 0.108 and 0.112 for 1773 and 1754 reflexions for isomers 1 and 2 respectively.

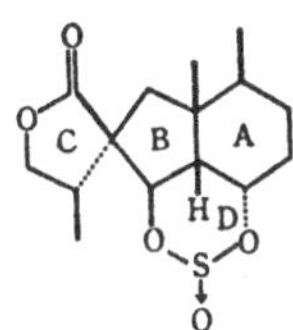

Fig. 1. Dihydrofukinolidol sulphite.

The two isomers (Fig. 1) differ only in the configuration at the sulphur atom. The six-membered ring containing the sulphite group has a chair conformation in isomer 1, and a twisted boat conformation in isomer 2, with the exocyclic S-O bond axially oriented in both isomers.

7-BROMOCYCLO[3:15]LONGIFOLANE

$C_{15}H_{23}Br$

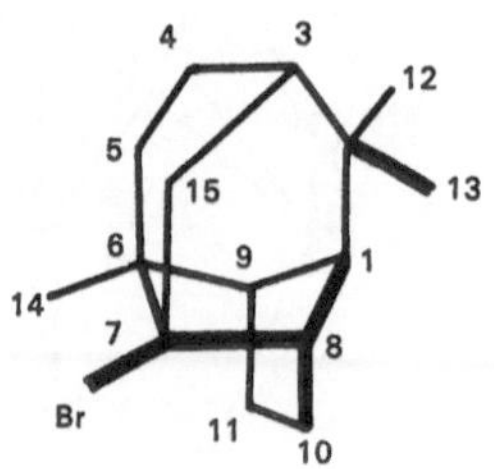

J.C. THIERRY and R. WEISS, 1972. Acta Cryst., B28, 3228-3234.

Orthorhombic, $P2_12_12_1$, a = 20.53, b = 6.901, c = 9.251 Å, D_m = 1.42, Z = 4. Cu radiation, R = 0.043 for 748 reflexions.

The absolute configuration of the molecule is shown in Fig. 1. The mean C-C distance is 1.550 Å and the C-Br distance is 1.969 Å. The seven-membered ring, C(1)-C(6), C(9), has a twist-chair conformation. The bridge C(8)-C(1)-C(9) angle of the [2,2,1] heptane system is 93.1° and this system can be defined as two planes consisting of C(9), C(6), C(7), C(8) and C(9), C(11), C(10), C(8). There are only van der Waals contacts between molecules in the crystals.

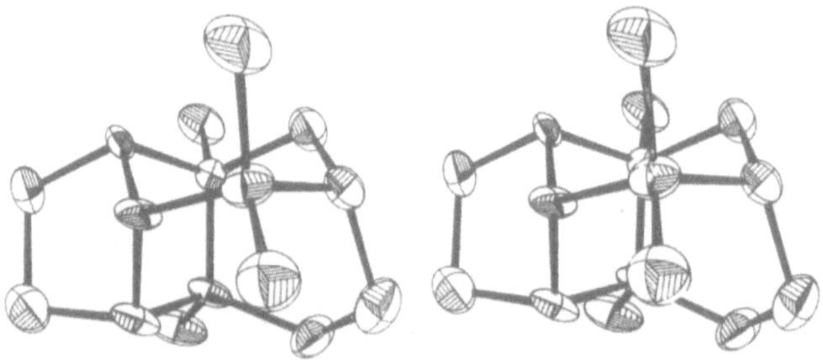

Fig. 1. The molecular structure of 7-bromocyclo[3:15]longifolane.

3α-BROMOLONGIFOLENE

$C_{15}H_{23}Br$

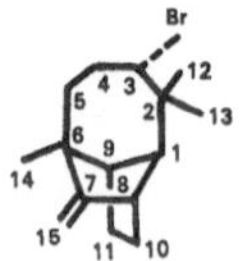

J.C. THIERRY and R. WEISS, 1972. Acta Cryst., B28, 3249-3257.

Orthorhombic, $P2_12_12_1$, a = 8.950, b = 10.21, c = 15.260 Å, D_m = 1.34, Z = 4.
Cu radiation, R = 0.044 for 664 reflexions.

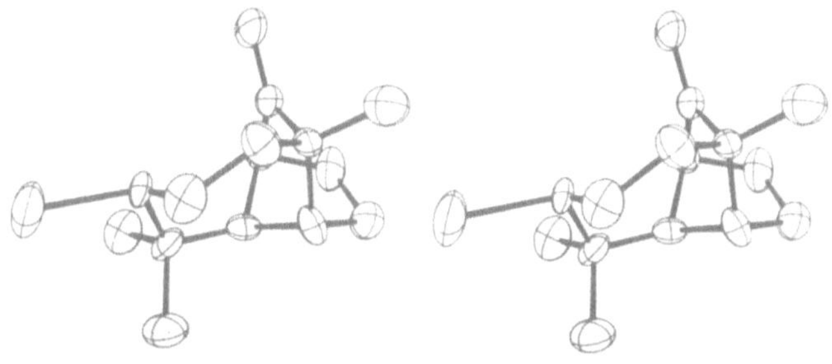

Fig. 1. View of 3α-bromolongifolene.

In this molecule (Fig. 1) the seven-membered ring is in twist-chair conformation with bonds C(3)-Br and C(1)-C(2) anti-parallel. Bond lengths are for the most part normal (e.g., mean C-C = 1.55(2) Å). The C-Br distance is 2.019(8) Å. Some of the angles have values considerably different from tetrahedral as a result of bridged ring formation e.g., C(8)-C(1)-C(9) = 93.6(6), C(1)-C(9)-C(11) = 98.7(6)°. The bicyclo[2,2,1]heptane system can be defined as consisting of two planes, C(9), C(6), C(7), C(8) and C(9), C(11), C(10), C(8), with dihedral angles of 127.5 and 120.0°, respectively, with the plane C(9), C(1), C(8). The molecules are held in the crystal by normal van der Waals contacts.

ω-BROMOLONGIFOLENE

$C_{15}H_{23}Br$

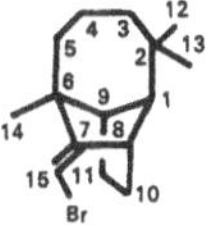

J.C. THIERRY and R. WEISS, 1972. Acta Cryst., B28, 3249-3257.

Orthorhombic, $P2_12_12_1$, a = 11.68, b = 12.87, c = 9.31 Å, D_m = 1.35, Z = 4.
Cu radiation, R = 0.049 for 659 reflexions.

In this molecule (Fig. 1) the seven-membered ring system C(1)-C(6), C(9) adopts a twist-chair conformation with a twofold axis through C(5) and C(1)-C(2). The mean C-C bond is 1.56 Å while the C-Br distance is 1.904(14) Å. The bicyclo-[2,2,1]heptane system has a bridgehead C(8)-C(1)-C(9) angle of 92.8(6)°; this system can be defined as consisting of two planes C(9), C(6), C(7), C(8) and C(9), C(11), C(10), C(8) with dihedral angles of 130.6 and 120.5°, respectively, with the plane C(9), C(1), C(8). The molecules are held in the crystal by normal van der Waals contacts.

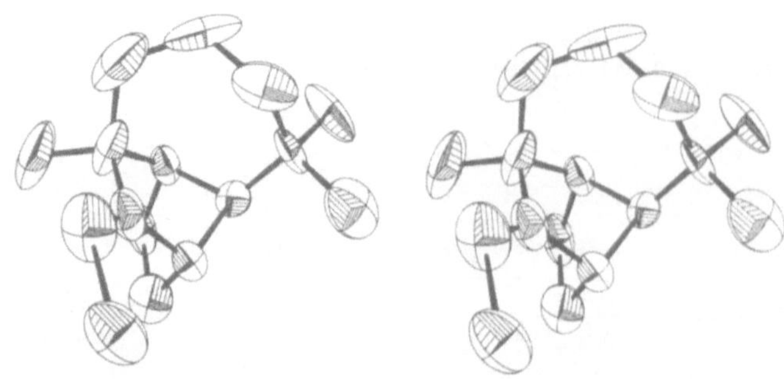

Fig. 1. The molecular structure of ω-bromolongifolene.

3α-BROMO-7βH-LONGIFOLANE

$C_{15}H_{25}Br$

J.C. THIERRY and R. WEISS, 1972. Acta Cryst., B28, 3234-3241.

Monoclinic, $P2_1$, a = 10.080, b = 7.787, c = 9.461 Å, β = 112.31°, D_m = 1.39, Z = 2. Cu radiation, R = 0.048 for 951 reflexions.

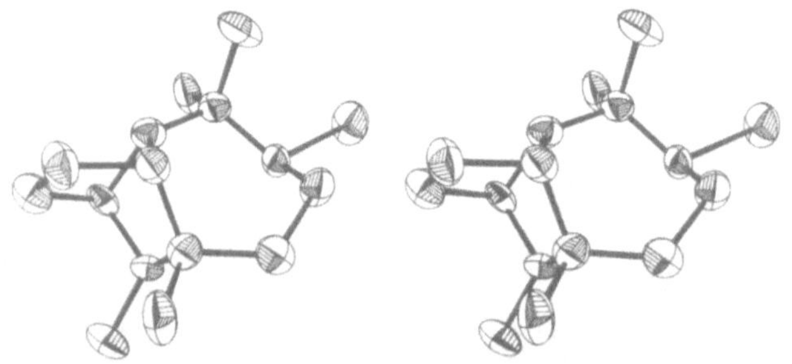

Fig. 1. The molecular structure of 3α-bromo-7βH-longifolane.

The absolute configuration of 3α-bromo-7βH-longifolane is shown in Fig. 1. The mean C-C single bond is 1.539 Å. The C-Br bond length, 2.013(7) Å, is significantly longer than the usual value, 1.96 Å, for this type of bond. The seven-membered ring, C(1)-C(6),C(9), has C_2 symmetry and a twist-chair conformation. The bicyclo[2,2,1]heptane system has a bridgehead angle of 91.75(41)° for C(8)-C(1)-C(9); this system can be defined as two planes consisting of C(9), C(6), C(7), C(8) and C(9), C(11), C(10), C(8) which make dihedral angles of 127.43° and 118.57°, respectively, with the plane C(9), C(1), C(8). The C(3)-Br and

C(1)-C(2) bonds are anti-parallel. There is a strong transannular interaction between the hydrogen atom H(3,1) attached to C(3) and H(7,1) attached to C(7). The molecules are held in the crystal structure by van der Waals contacts.

CENTAUREPENSIN

$C_{19}H_{24}Cl_2O_7$

A.T. HEWSON, R.C. PETTERSEN and O. KENNARD, 1972. Cryst. Struct. Comm., 1, 383-388.

Monoclinic, $P2_1$, a = 10.478, b = 9.254, c = 11.507 Å, β = 113.16°, D_m = 1.41, Z = 2. Cu radiation, R = 0.052 for 1548 reflexions.

The analysis was undertaken to establish the structure and absolute stereochemistry of the molecule. The results are summarized in the skeletal formula above. Bond lengths and angles are normal; the molecules are held in the crystal by O-H...O hydrogen bonds (2.90 and 2.94 Å).

ELEPHANTOL p-BROMOBENZOATE

$C_{22}H_{19}BrO_7$

A.T. McPHAIL and G.A. SIM, 1972. J. Chem. Soc., Perkin II, 1313-1316.

Orthorhombic, $P2_12_12_1$, a = 10.64, b = 30.34, c = 6.41 Å, Z = 4. Cu radiation, R = 0.119 for 1690 reflexions.

Fig. 1. Structure of elephantol p-bromobenzoate (R = BrC_6H_4CO).

The derivative has the structure and absolute stereochemistry shown in Fig. 1. The ten-membered carbocyclic ring adopts a conformation which is characterized by a short C(1)...C(5) transannular separation of 2.98 Å and by both C(14)- and C(15)-substituents being on the β-face of the molecule. The trans-ethylenic group in the ten-membered ring is distinctly distorted from planarity, the C(2)-C(1)-C(10)-C(9) torsion angle being 163°.

ISOCOLLYBOLIDE

$C_{22}H_{20}O_7$

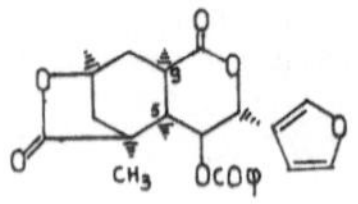

C. PASCARD-BILLY, 1972. Acta Cryst., B28, 331-337.

Orthorhombic, $P2_12_12_1$, a = 6.490, b = 12.408, c = 23.960 Å, D_m = 1.35, Z = 4. Cu radiation, R = 0.096 for 1883 reflexions.

The molecular structure is shown in Fig. 1. Bond lengths and angles are normal. The molecular packing and some intermolecular distances are shown in Fig. 2.

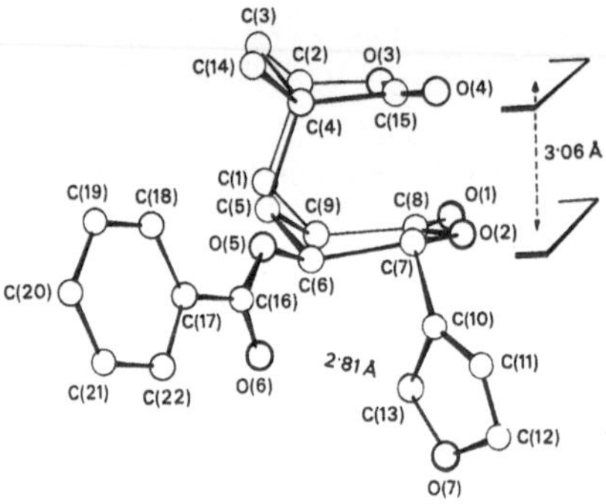

Fig. 1. Configuration of isocollybolide.

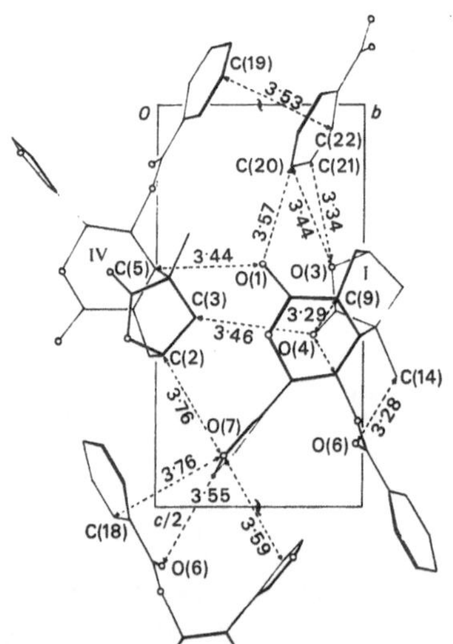

Fig. 2. Molecular packing viewed along a and the shorter intermolecular contacts in isocollybolide.